Air Pollution Control Engineering

VOLUME 1
HANDBOOK OF ENVIRONMENTAL ENGINEERING

Air Pollution Control Engineering

Edited by

Lawrence K. Wang, PhD, PE, DEE
Zorex Corporation, Newtonville, NY
Lenox Institute of Water Technology, Lenox, MA

Norman C. Pereira, PhD
Monsanto Company (Retired), St. Louis, MO

Yung-Tse Hung, PhD, PE, DEE
Department of Civil and Environmental Engineering
Cleveland State University, Cleveland, OH

Consulting Editor

Kathleen Hung Li, MS

HUMANA PRESS TOTOWA, NEW JERSEY

999 Riverview Drive, Suite 208
Totowa, New Jersey 07512

humanapress.com

For additional copies, pricing for bulk purchases, and/or information about other Humana titles, contact Humana at the above address or at any of the following numbers: Tel.: 973-256-1699; Fax: 973-256-8341; E-mail: humana@humanapr.com

This publication is printed on acid-free paper. ∞
ANSI Z39.48-1984 (American Standards Institute)
Permanence of Paper for Printed Library Materials.

Cover design by Patricia F. Cleary.

eISBN 1-59259-778-5

Printed in the United States of America. 10 9 8 7 6 5 4 3 2 1

Library of Congress Cataloging-in-Publication Data

Air pollution control engineering / edited by Lawrence K. Wang, Norman C. Pereira, Yung-Tse Hung; consulting editor, Kathleen Hung Li.
p. cm.—(Handbook of environmental engineering ; v. 1)
Includes bibliographical references and index.
ISBN 1-58829-161-8 (alk. paper)
1. Air–pollution. 2. Air quality management. I. Wang, Lawrence K. II. Pereira, Norman C. III. Hung, Yung-Tse. IV. Series: Handbook of environmental engineering (2004) ; v. 1.
TD170 .H37 2004 vol. 1
[TD883]
628 s–dc22
[628.5] 2003023596

Preface

The past 30 years have seen the emergence of a growing desire worldwide to take positive actions to restore and protect the environment from the degrading effects of all forms of pollution: air, noise, solid waste, and water. Because pollution is a direct or indirect consequence of waste, the seemingly idealistic goal for "zero discharge" can be construed as an unrealistic demand for zero waste. However, as long as waste exists, we can only attempt to abate the subsequent pollution by converting it to a less noxious form. Three major questions usually arise when a particular type of pollution has been identified: (1) How serious is the pollution? (2) Is the technology to abate it available? and (3) Do the costs of abatement justify the degree of abatement achieved? The principal intention of the *Handbook of Environmental Engineering* series is to help readers formulate answers to the last two questions.

The traditional approach of applying tried-and-true solutions to specific pollution problems has been a major contributing factor to the success of environmental engineering, and has accounted in large measure for the establishment of a "methodology of pollution control." However, realization of the ever-increasing complexity and interrelated nature of current environmental problems renders it imperative that intelligent planning of pollution abatement systems be undertaken. Prerequisite to such planning is an understanding of the performance, potential, and limitations of the various methods of pollution abatement available for environmental engineering. In this series of handbooks, we will review at a tutorial level a broad spectrum of engineering systems (processes, operations, and methods) currently being utilized, or of potential utility, for pollution abatement. We believe that the unified interdisciplinary approach in these handbooks is a logical step in the evolution of environmental engineering.

The treatment of the various engineering systems presented in *Air Pollution Control Engineering* will show how an engineering formulation of the subject flows naturally from the fundamental principles and theory of chemistry, physics, and mathematics. This emphasis on fundamental science recognizes that engineering practice has in recent years become more firmly based on scientific principles rather than its earlier dependency on empirical accumulation of facts. It is not intended, though, to neglect empiricism when such data lead quickly to the most economic design; certain engineering systems are not readily amenable to fundamental scientific analysis, and in these instances we have resorted to less science in favor of more art and empiricism.

Because an environmental engineer must understand science within the context of application, we first present the development of the scientific basis of a particular subject, followed by exposition of the pertinent design concepts and operations, and detailed explanations of their applications to environmental quality control or improvement. Throughout the series, methods of practical design calculation are illustrated by numerical examples. These examples clearly demonstrate how organized, analytical reasoning leads to the most direct and clear solutions. Wherever possible, pertinent cost data have been provided.

Our treatment of pollution-abatement engineering is offered in the belief that the trained engineer should more firmly understand fundamental principles, be more aware of the similarities and/or differences among many of the engineering systems, and exhibit greater flexibility and originality in the definition and innovative solution of environmental pollution problems. In short, the environmental engineers should by conviction and practice be more readily adaptable to change and progress.

Coverage of the unusually broad field of environmental engineering has demanded an expertise that could only be provided through multiple authorships. Each author (or group of authors) was permitted to employ, within reasonable limits, the customary personal style in organizing and presenting a particular subject area, and, consequently, it has been difficult to treat all subject material in a homogeneous manner. Moreover, owing to limitations of space, some of the authors' favored topics could not be treated in great detail, and many less important topics had to be merely mentioned or commented on briefly. All of the authors have provided an excellent list of references at the end of each chapter for the benefit of the interested reader. Because each of the chapters is meant to be self-contained, some mild repetition among the various texts is unavoidable. In each case, all errors of omission or repetition are the responsibility of the editors and not the individual authors. With the current trend toward metrication, the question of using a consistent system of units has been a problem. Wherever possible the authors have used the British system (fps) along with the metric equivalent (mks, cgs, or SIU) or vice versa. The authors sincerely hope that this doubled system of unit notation will prove helpful rather than disruptive to the readers.

The goals of the *Handbook of Environmental Engineering* series are (1) to cover the entire range of environmental fields, including air and noise pollution control, solid waste processing and resource recovery, biological treatment processes, water resources, natural control processes, radioactive waste disposal, thermal pollution control, and physicochemical treatment processes; and (2) to employ a multithematic approach to environmental pollution control since air, water, land, and energy are all interrelated. No consideration is given to pollution by type of industry or to the abatement of specific pollutants. Rather, the organization of the series is based on the three basic forms in which pollutants and waste are manifested: gas, solid, and liquid. In addition, noise pollution control is included in one of the handbooks in the series.

This volume of *Air Pollution Control Engineering*, a companion to the volume, *Advanced Air and Noise Pollution Control*, has been designed to serve as a basic air pollution control design textbook as well as a comprehensive reference book. We hope and expect it will prove of equally high value to advanced undergraduate or graduate students, to designers of air pollution abatement systems, and to scientists and researchers. The editors welcome comments from readers in the field. It is our hope that this book will not only provide information on the air pollution abatement technologies, but will also serve as a basis for advanced study or specialized investigation of the theory and practice of the unit operations and unit processes covered.

The editors are pleased to acknowledge the encouragement and support received from their colleagues and the publisher during the conceptual stages of this endeavor. We wish to thank the contributing authors for their time and effort, and for having patiently borne our reviews and numerous queries and comments. We are very grateful to our respective families for their patience and understanding during some rather trying times.

The editors are especially indebted to Dr. Howard E. Hesketh at Southern Illinois University, Carbondale, Illinois, and Ms. Kathleen Hung Li at NEC Business Network Solutions, Irving, Texas, for their services as Consulting Editors of the first and second editions, respectively.

Lawrence K. Wang
Norman C. Pereira
Yung-Tse Hung

Contents

Contributors

WEI-YIN CHEN, PhD • *Department of Chemical Engineering, University of Mississippi, University, MS*

JOSÉ RENATO COURY, PhD, MENG • *Department of Chemical Engineering, Universidade Federal de Sao Carlos, Sao Carlos, Brazil*

JAMES E. ELDRIDGE, MA, MS, MENG • *Lantec Product, Agoura Hills, CA*

YUNG-TSE HUNG, PhD, PE, DEE • *Department of Civil and Environmental Engineering, Cleveland State University, Cleveland, OH*

GREGORY T. KLEINHEINZ, PhD • *Department of Biology and Microbiology, University of Wisconsin-Oshkosh, Oshkosh, WI*

KATHLEEN HUNG LI, MS • *Senior Technical Writer, NEC Business Network Solutions, Inc., Irving, TX*

NORMAN C. PEREIRA, PhD (RETIRED) • *Monsanto Company, St. Louis, MO*

REINALDO PISANI JR., DEng, MEng • *Centro de Ciencias Exatas, Universidade de Ribeirao Preto, Ribeirao Preto, Brazil*

THOMAS T. SHEN, PhD • *Independent Environmental Advisor, Delmar, NY*

JERRY R. TARICSKA, PhD, PE • *Environmental Engineering Department, Hole Montes, Inc., Naples, FL*

LAWRENCE K. WANG, PhD, PE, DEE • *Zorex Corporation, Newtonville, NY and Lenox Institute of Water Technology, Lenox, MA*

CLINT WILLIFORD, PhD • *Department of Chemical Engineering, University of Mississippi, University, MS*

PHILLIP C. WRIGHT, PhD • *Department of Chemical and Process Engineering, University of Sheffield, Sheffield, UK*

CHUNG-SHIN J. YUAN, PhD • *Institute of Environmental Engineering, National Sun Yat-Sen University, Kaohsiung, Taiwan*

1
Air Quality and Pollution Control

Lawrence K. Wang, Jerry R. Taricska, Yung-Tse Hung, and Kathleen Hung Li

CONTENTS

1. INTRODUCTION

The Engineer's Joint Council on Air Pollution and Its Control defines air pollution as "the presence in the outdoor atmosphere of one or more contaminants, such as dust, fumes, gas, mist, odor, smoke or vapor in quantities, of characteristics, and of duration, such as to be injurious to human, plant, or property, or which unreasonably interferes with the comfortable enjoyment of life and property."

Air pollution, as defined above, is not a recent phenomenon. Natural events always have been the direct cause of enormous amounts of air pollution. Volcanoes, for instance, spew lava onto land and emit particulates and poisonous gases containing ash, hydrogen sulfide (H_2S), and sulfur dioxide (SO_2) into the atmosphere. It has been estimated that all air pollution resulting from human activity does not equal the quantities released during three volcanic eruptions: Krakatoa in Indonesia in 1883, Katmai in Alaska in 1912, and Hekla in Iceland in 1947.

From: *Handbook of Environmental Engineering, Volume 1: Air Pollution Control Engineering*
Edited by: L. K. Wang, N. C. Pereira, and Y.-T. Hung © Humana Press Inc., Totowa, NJ

Lightning, another large contributor to atmospheric pollution, activates atmospheric oxygen (O_2) to produce ozone (O_3), a poisonous gas [ozone in the upper atmosphere, however, acts as a shield against excessive amounts of ultraviolet (UV) radiation, which can cause human skin cancer]. In addition to the production of ozone, lightning is the indirect cause of large amounts of combustion-related air pollution as a result of forest fires. The Forest Service of the United States Department of Agriculture reported that lightning causes more than half of the over 10,000 forest fires that occur each year.

For centuries, human beings have been exposed to an atmosphere permeated by other natural pollutants such as dust, methane from decomposing matter in bogs and swamps, and various noxious compounds emitted by forests. Some scientists claim that such natural processes release twice the amount of sulfur-containing compounds and 10 times the quantity of carbon monoxide (CO) compared to all human activity.

Why, then, is society so perturbed by air pollution? The concern stems from a combination of several factors:

1. Urbanization and industrialization have brought together large numbers of people in small areas.
2. The pollution generated by people is most often released at locations close to where they live and work, which results in their continuous exposure to relatively high levels of the pollutants.
3. The human population is still increasing at an exponential rate.

Thus, with rapidly expanding industry, ever more urbanized lifestyles, and an increasing population, concern over the control of man-made air pollutants is now clearly a necessity. Effective ways must be found both to reduce pollution and to cope with existing levels of pollution.

As noted earlicr, natural air pollution predates us all. With the advent of *Homo sapiens*, the first human-generated air pollution must have been smoke from wood burning, followed later by coal.

From the beginning of the 14th century, air pollution from coal smoke and gases had been noted and was of great concern in England, Germany, and elsewhere. By the beginning of the 19th century, the smoke nuisance in English cities prompted the appointment of a Select Committee of the British Parliament in 1819 to study and report on smoke abatement.

Many cities in the United States, including Chicago, St. Louis, and Pittsburgh, have been plagued with smoke pollution. The period from 1880 to 1930 has often been called the "Smoke Abatement Era." During this time, much of the basic atmospheric cleanup work started. The Smoke Prevention Association was formed in the United States near the turn of the 20th century, and by 1906, it was holding annual conventions to discuss the smoke pollution problem and possible solutions. The name of the association was later changed to the Air Pollution Control Association (APCA).

The period from 1930 to the present has been dubbed the "Disaster Era" or "Air Pollution Control Era." In the most infamous pollution "disaster" in the United States, 20 were killed and several hundred made ill in the industrial town of Donora, Pennsylvania in 1948. Comparable events occurred in the Meuse Valley, Belgium in 1930 and in London in 1952. In the 1960s, smog became a serious problem in California, especially in Los Angeles. During a 14-day period from November 27 to December 10, 1962, air pollution concentrations were extremely high worldwide, resulting in "episodes" of high respiratory incidents in London, Rotterdam, Hamburg, Osaka, and New York. During this period, people in many other cities in the United States experienced serious pollution-related illnesses, and as a result, efforts to clean up the air were started in the cities of Chicago, New York, Washington, DC, and Pittsburgh. The substitution of less smoky fuels, such as natural gas and oil, for coal, for power production and for space heating accounted for much of the subsequent improvement in air quality.

Air quality in the United States depends on the nature and amount of pollutants emitted as well as the prevalent meteorological conditions. Air pollution problems in the highly populated, industrialized cities of the eastern United States result mainly from the release of sulfur oxides and particulates. In the western United States, air pollution is related more to photochemical pollution (smog). The latter form of pollution is an end product of the reaction of nitrogen oxides and hydrocarbons from automobiles and other combustion sources with oxygen and each other, in the presence of sunlight, to form secondary pollutants such as ozone and PAN (peroxy acetyl or acyl nitrates).

Temperature inversions effectively "put a lid over" the atmosphere so that emissions are trapped in relatively small volumes and in correspondingly high concentrations. Los Angeles, for example, often suffers a very stable temperature inversion and strong solar input, both ideal conditions for the formation of highly localized smog. Rain and snow wash out the air and deposit the pollutants on the soil and in water. "Acid rain" is the result of gaseous sulfur oxides combining with rain water to form dilute sulfuric acid and it occurs in many cities of eastern United States.

2. CHARACTERISTICS OF AIR POLLUTANTS

Air pollutants are divided into two main groups: particulates and gases. Because particulates consist of solids and/or liquid material, air pollutants therefore encompass all three basic forms of matter. Gaseous pollutants include gaseous forms of sulfur and nitrogen. Gaseous SO_2 is colorless, yet one can point to the bluish smoke leaving combustion operation stacks as SO_2 or, more correctly, SO_3 or sulfuric acid mist. Nitric oxide (NO) is another colorless gas generated in combustion processes; the brown color observed a few miles downwind is nitrogen dioxide (NO_2), the product of photochemical oxidation of NO. Although the properties of gases are adequately covered in basic chemistry, physics, and thermodynamics courses, the physical behavior of particulates is less likely to be understood. The remainder of this section is thus devoted to the physical properties of particulate matter, not gaseous pollutants.

Particulates may be subdivided into several groups. Atmospheric particulates consist of solid or liquid material with diameters smaller than about 50 μm (10^{-6} m). Fine particulates are those with diameters smaller than 3 μm. The term "aerosol" is defined specifically as particulates with diameters smaller than about 30–50 μm (this does not refer to the large particulates from aerosol spray cans). Particulates with diameters larger than 50 μm settle relatively quickly and do not remain in the ambient air.

The movement of small particles in gases can be accounted for by expressions derived for specific size groups: (1) The smallest group is the molecular kinetic group and includes particles with diameters much less than the mean free path of the gas molecules (l); (2) next is the Cunningham group, which consists of particles with diameters about equal to l; (3) the largest is the Stokes group, which consists of particles with diameters much larger than l. The reported values of l are quite varied, however, for air at standard conditions (SC) of 1 atm and 20°C and range from 0.653×10^{-5} to 0.942×10^{-5} cm. One can also estimate l for air at a constant pressure of 1000 mbar using

$$l = 2.23 \times 10^{-8} T \tag{1}$$

where l is the mean free path of air (cm) and T is the absolute temperature (K).

One also can estimate the terminal settling velocity of the various size spherical particles in still air. The Stokes equation applies for that group and gives accuracy to 1% when the particles have diameters from 16 to 30 μm and 10% accuracy for 30–70 μm:

$$v_s = d^2 g\, \rho_p / 18\mu_g \tag{2}$$

where v_s is the terminal settling velocity (cm/s), d is the diameter of the particle (cm), g is the gravitational acceleration constant (980 cm/s^2), ρ_p is the density of the particle (g/cm^3), and μ_g is the viscosity of the gas (g/cm s, where μ_g for air is 1.83×10^{-4}).

Particles in the Cunningham group are smaller and tend to "slip" through the gas molecules so that a correction factor is required. This is called the Cunningham correction factor (C), which is dimensionless and can be found for air at standard conditions (SC):

$$C = 1 + \left\{\left(T\left(2 \times 10^{-4}\right)\right)/d_1\right\}\left\{2.79 + 0.894 \exp - \left[d_1\left(2.47 \times 10^{-3}\right)/T\right]\right\} \tag{3}$$

where T is the absolute temperature (K) and d_1 is the particle diameter (μm). When Eq. (2) is multiplied by this factor, accuracy is within 1% for particles for 0.36–0.80 μm and 10% for 1.0–1.6 μm. Particles of the molecular kinetic size are not amenable to settling because of their high Brownian motion.

Liquid particulate and solids formed by condensation are usually spherical in shape and can be described by Eqs. (1)–(3). Many other particulates are irregularly shaped, so corrections must be used for these. One procedure is to multiply the given equations by a dimensionless shape factor (K):

$$K = 0.843 \log\left(K'/0.065\right) \tag{4}$$

where K' is the sphericity factor and

$K = 1$ for spheres
$K = 0.906$ for octahedrons
$K = 0.846$ for rod-type cylinders
$K = 0.806$ for cubes and rectangles
$K = 0.670$ for flat splinters

Concentrations of air pollutants are usually stated as mass per unit volume of gas (e.g., μg/m^3, or micrograms of pollutant per total volume of gases) for particulates and as a volume ratio for gases (e.g., ppm, or volume of pollutant gas per million volumes of total gases). Note that at low concentrations and temperatures (room conditions) frequently present in air pollution situations, the gaseous pollutants (and air) may be considered as ideal gases. This means that the volume fraction equals the mole fraction equals the pressure fraction. This relationship is frequently useful and should be remembered.

Special methods must be used to evaluate the movement of particulates under conditions in which larger or smaller particles are present, of nonsteady state, of nonrectangular

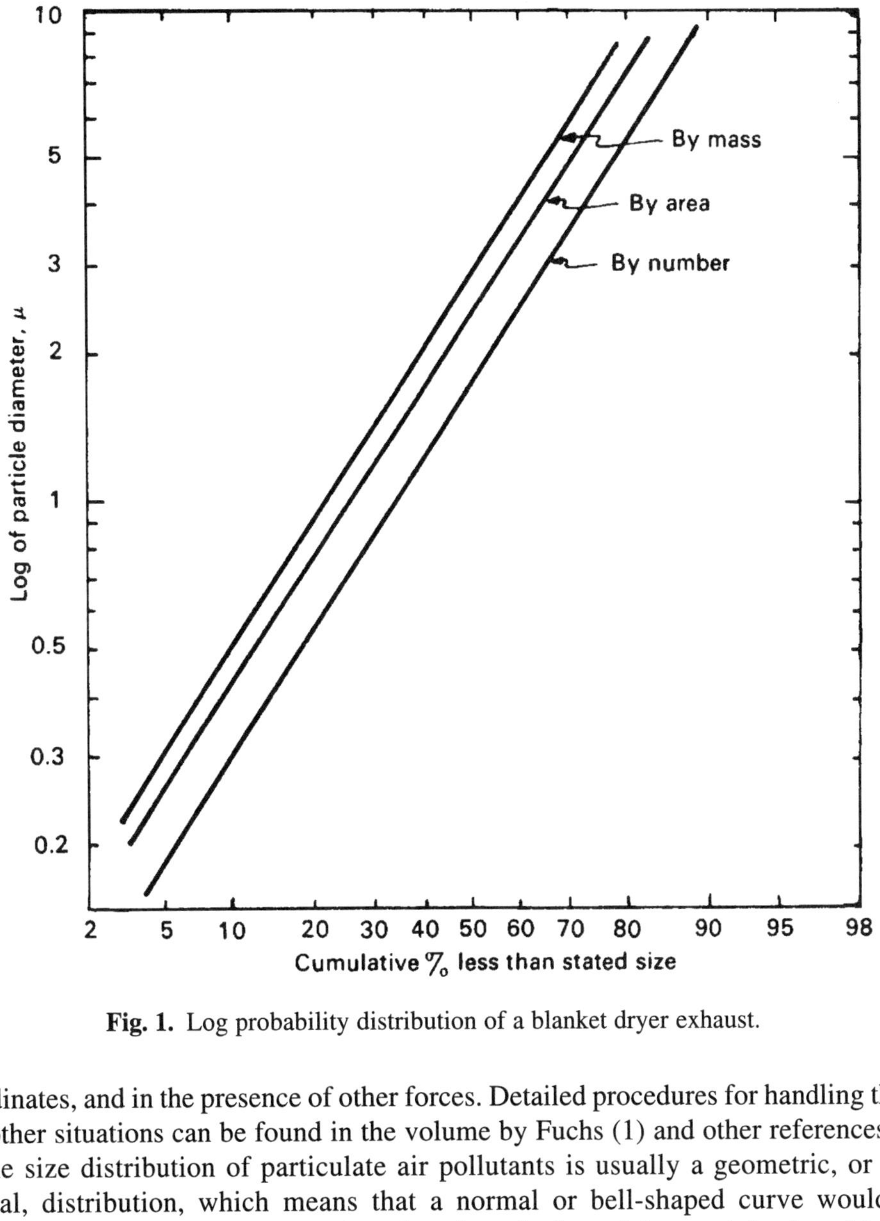

Fig. 1. Log probability distribution of a blanket dryer exhaust.

coordinates, and in the presence of other forces. Detailed procedures for handling these and other situations can be found in the volume by Fuchs (1) and other references.

The size distribution of particulate air pollutants is usually a geometric, or log-normal, distribution, which means that a normal or bell-shaped curve would be obtained if size frequency were plotted against the log of the particle size. Also, if the log of the particle size were plotted versus a cumulative percentage value, such as mass, area, or number, straight lines would be obtained on a log probability graph, as shown in Fig. 1. The values by mass in Fig. 1 were the original samples, and the surface area and number curves can be estimated mathematically, as was done to obtain the other lines shown. Of course, these data could be measured directly (e.g., by optical techniques).

The mean diameter of such a sample is obtained by noting the 50% value and must be reported as a mean (d_{50}) by either mass, area, or number. In Fig. 1, the mass mean is 3.0 μm. The standard deviation can be obtained from the ratio of diameter for 84.13%

($d_{84.13}$) and 50%, or the ratios for 50% and 15.87% ($d_{15.87}$). This geometric standard deviation (σ_g) becomes:

$$\sigma_g = d_{84.13}/d_{50} = d_{50}/d_{15.87} \quad (5)$$

In Fig. 1, σ_g is 3.76. Note that the slopes of the curves (σ_g) should be similar for all three methods of expressing the same material.

If the particulate matter is composed of more than one material or if it is a single substance in different physical structures, it will most likely be bimodal in size distribution. This can be true for material in the stack effluent and mixtures in the free (ambient) atmosphere. For example, combustion-flue gases contain particulates composed of a large fraction mainly entrained as partially unburned fuel, plus a smaller fraction consisting of ash. Particulates sampled from a stoker-fired, chain grate boiler (2) are shown in Fig. 2. Note how this material must be plotted as two intercepting lines on log probability coordinates.

As shown in Fig. 2, atmospheric particulates are also bimodal in size distribution (3). These data are plotted as Δmass/Δlog diameter versus the log of diameter to amplify the bimodal distribution character. In general, atmospheric particulates consist of a submicron group (<1 μm) and a larger group. Although the data shown in Fig. 3 are typical for the United States, similar results are obtained throughout the world, as reported, for example, in Japan (4) and Australia (5). These authors note that atmospheric sulfates and nitrates dominate the smaller group, which by mass accounts for 40% and include particles with diameters from 0.5 to 1.5 μm, which account for another 40%.

3. STANDARDS

3.1. *Ambient Air Quality Standards*

Ambient air is defined as the outside air of a community, in contrast to air confined to a room or working area. As such, many people are exposed to the local ambient air 24 h a day, 7 d a week. It is on this basis that ambient air quality standards are formulated. The current standards were developed relatively quickly after the numerous episodes of the 1960s. The feeling of many people were summed up by President Johnson's statement in 1967: "If we don't clean up this mess, we'll all have to start wearing gas masks," and "This country is so rich that we can achieve anything if we make up our mind what we want to do." There are those who believe that some requirements in the standards disregard costs of control compared with costs of benefits, but all benefits and costs cannot be accurately assessed. Even so, we would be reluctant to put dollar values on our own health and life.

The Clean Air Act Amendments of 1970 (Public Law 91-601, signed December 31, 1970) include ambient air standards that consist of two parts: primary standards, which are intended for general health protection, and the more restrictive secondary standards, which are for protection against specific adverse effects on health and welfare. "Welfare" here includes plants, other animals, and materials. The primary standards are effective as of 1975, and the secondary standards are effective as of June 1, 1977. An abbreviated list of these 1997 standards for particulate matter and categories of gaseous pollutants is

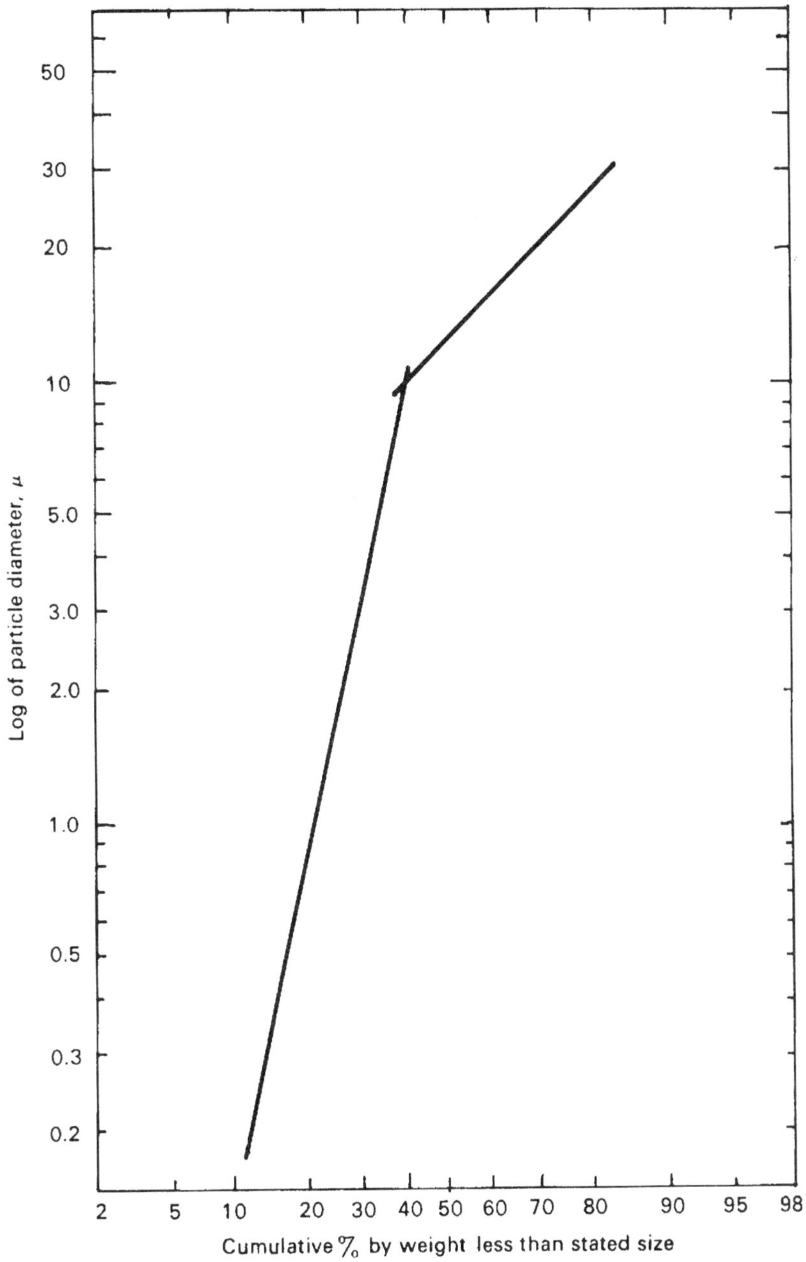

Fig. 2. Bimodal mass distribution of flue gas dust from a chain grate boiler.

given in Table 1. The parts per million values are by volume and are calculated from the mass per unit volume for the specific chemical substance noted.

The Federal Clean Air Act (CAA) has been amended periodically in an attempt to adjust the law to current needs when economic and technological feasibility factors are considered. Table 1 includes some of the modifications from the original standards. Other standards on ambient air have been set to limit the amount of ambient air

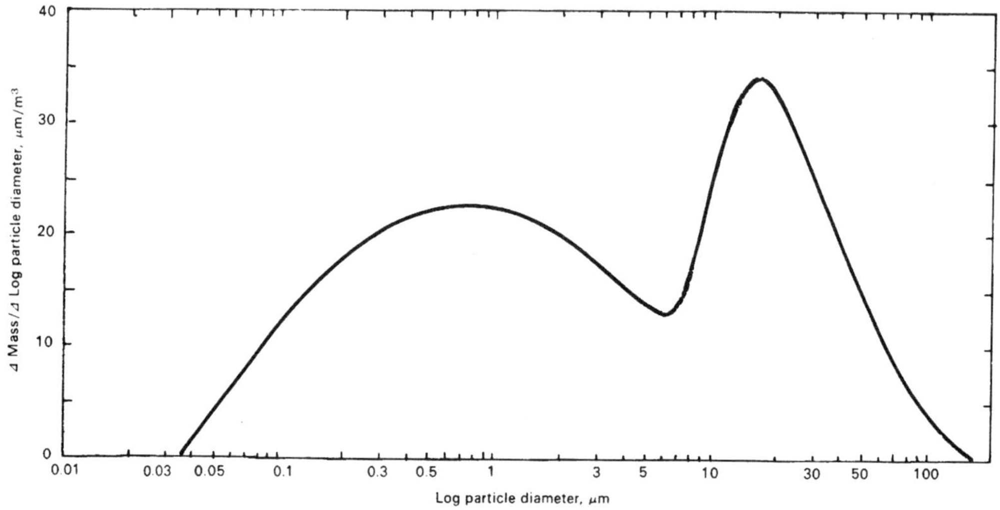

Fig. 3. Typical bimodal mass distribution of atmospheric aerosols.

degradation permissible for various locations. For example, areas where ambient air quality levels are below the standard maximums are protected so that pollution levels cannot increase to the maximum values. The most restrictive requirements apply to national forest and park recreational areas.

The natural background level of SO_2 in the United States is considered to be about 0.002 ppm. Measured high levels of SO_2 have been as follows: Donora disaster, 5720 μg/m^3 (2.2 ppm); London 1962, 3830 μg/m^3 (1.45 ppm), and Chicago 1939, about 1000 μg/m^3 (0.4 ppm). Annual averages of SO_2 in several United States cities in 1968 were approximately as follows: New York City, 0.13 ppm; Chicago, 0.08 ppm; Washington DC, 0.04 ppm; and St. Louis, 0.03 ppm. Since these data were reported, the air quality in these cities has been improving, as noted in the next subsection.

3.2. *Emission Standards*

Emission standards relate to amounts of pollutants that are released from a source. In general, emission standards for existing sources of air pollution are set by each state in an attempt to reduce ambient air pollution levels to the ambient standards. Various diffusion modeling techniques are used to develop emission standards. The final plans developed by each state showing how the ambient standards levels will be obtained are submitted to the United States Environmental Protection Agency (US EPA) for approval as State Implementation Plans (SIPs). If a state delays in preparing and obtaining US EPA approval of the SIP, the federal government will prepare the SIP for that particular state.

Air pollution source emission limits delineated by the US EPA for new installations are called "Standards of Performance for New Stationary Sources." These standards are intended to cover the major pollution emitters and include 19 types of new stationary source. The Federal Register (6) outlines standards for steam generators, incinerators, Portland cement plants, nitric acid plants, and sulfur acid plants. The federal government also establishes transportation source emission limits.

Table 1
National Ambient Air Quality Standards, 1997

Pollutant	Standard value[a]	Standard type
Carbon Monoxide (CO)		
8-h Average	9 ppm (10 mg/m^3)	Primary
1-h Average	35 ppm (40 mg/m^3)	Primary
Nitrogen Dioxide (NO_2)		
Annual arithmetic mean	0.053 ppm (100 $\mu g/m^3$)	Primary and secondary
Ozone (O^3)		
8-h Average	0.12 ppm (235 $\mu g/m^3$)	Primary and secondary
1-h Average	0.08 ppm (157 $\mu g/m^3$)	Primary and secondary
Lead (Pb)		
Quarterly average	1.5 $\mu g/m^3$	Primary and secondary
Particulate (PM 10) *Particles with diameters of 10* μm *or less*		
Annual arithmetic mean	50 $\mu g/m^3$	Primary and secondary
24-h Average	150 $\mu g/m^3$	Primary and secondary
Particulate (PM 2.5) *Particles with diameters of 2.5* μm *or less*		
Annual arithmetic mean[b]	15 $\mu g/m^3$	Primary and secondary
24-h Average[b]	65 $\mu g/m^3$	Primary and secondary
Sulfur Dioxide (SO_2)		
Annual arithmetic mean	0.03 ppm (80 $\mu g/m^3$)	Primary
24-h Average	0.14 ppm (365 $\mu g/m^3$)	Primary
3-h Average	0.50 ppm (1300 $\mu g/m^3$)	Secondary

[a]Parenthetical value is an approximately equivalent concentration.
[b]The ozone 8-h standard and the PM 2.5 standards are included for information only. A 1999 federal court ruling blocked implementation of these standards, which the US Environmental Protection Agency (US EPA) proposed in 1997. US EPA has asked the US Supreme Court to reconsider that decision. The Updated Air Quality Standards website has additional information.

States may adopt air quality and/or emission regulations that are more stringent than those specified by the federal government, and some have done this. Often, these regulations are open ended in that each situation is evaluated independently in view of the particular situation and the currently best available demonstrated control technology.

The emission limit for hazardous substances is also being established by the federal government. This includes regulations on emissions of cadmium, beryllium, mercury, asbestos, chlorine, hydrogen chloride, copper, manganese, nickel, vanadium, zinc, barium, boron, chromium, selenium, pesticides, and radioactive substances. Other substances may be added to the list at the discretion of the US EPA administrator.

The US EPA was formed on December 2, 1970 by order of President Richard M. Nixon with consent of Congress in an attempt to consolidate federal pollution control activities. In addition to setting standards and timetables for compliance with the standards, this agency conducts research, allocates funds for research and for construction of facilities, and provides technical, managerial, and financial assistance to state, regional, and municipal pollution control agencies.

Since the passage of the Clean Air Act and formation of the US EPA, significant reductions in the level of air pollution have been made in the United States. Since 1977, the US EPA has submitted annual reports to Congress. It has been noted in these reports that sulfur dioxide emissions have been cut significantly, but the reduction in automotive emissions has been offset by increasing motor vehicle use and fuel consumption, so that total nationwide reductions are not as high as they are per vehicle mile. The 2003 US EPA air emission standards can be found on the agency's website (www.epa.gov).

4. SOURCES

As previously noted, there is often much difficulty and little agreement in how to accurately classify the various emissions. The US EPA, in an extensive attempt, classified the estimated emissions in the United States in a 433-page document (7). In response to public demand, the US EPA summarized air pollutant emissions in the United States in 1998. These emissions are listed in seven categories in Table 2, which also includes data on natural and miscellaneous sources: forest fires, agricultural burning, structural fires, and coal refuse fires. The values in parentheses represent the percentage of total pollutants emitted.

Much of the data in Table 2 comes from such sources as State Emission Inventories. However, it is sometimes necessary to estimate emissions by using "emission factors," which are published values of expected emissions from a particular source and are usually expressed as quantity of pollutant per unit weight of raw material consumed or product produced. The most complete listing of emission factors is found in the US EPA publication (8), which is periodically updated. The 2003 update can be found on the US EPA website.

As shown in Table 2, highway and off-highway transportation account for most of the total pollutants emitted. Fuel combustion emissions from electrical utilities, industries, and other categories are other major sources of air pollution emissions.

Fossil fuels, especially coal, contain sulfur. When burned, most of the sulfur is converted to SO_2. Most of the SO_2 pollution (77%) comes from fuel combustion sources. Eastern coal has a high sulfur content, compared to coal from the West, with values as high as 6%. The weighted average is in the 2.5–3.5% sulfur range. The content of western coal is lower in sulfur, with a weighted average of about 0.5–1.0%. However, the heating value of this coal is lower, and so a direct comparison should not be made between the two types of coal based only on sulfur content. It is estimated that 87% of the coal is used from the eastern reserves. To reduce sulfur emission, a greater percentage should come from the western coal reserve. As recycling and conservation increase, pollution from the waste disposal and recycling category should also decrease. Much of this material could be used to produce energy and thus reduce the use of high-pollution fuels.

There are over 20,000 major stationary sources of air pollution in the United States. They include mainly power plants, industries, and incinerators. Over 80% of these stationary sources have been either in compliance with US EPA standards or are meeting an abatement schedule.

5. EFFECTS

One of the requirements of Document PL 91-601 was that the US EPA publish criteria documents related to the effects of various air pollutants. A number of these docu-

Table 2
Estimated Summary of Air Pollutant Emission in the United States in 1998

	Pollutants (short tons/yr) (%)						
Source category	CO	NO_x	VOC	SO_2	PM_{10}	$PM_{2.5}$	NH_3
Fuel combustion, elec. util.	417 (0)	6,103 (25)	54 (0)	13,217 (67)	302 (1)	165 (2)	8 (0)
Fuel combustion, industrial	1,114 (1)	2,969 (12)	161 (1)	2,895 (15)	245 (1)	160 (2)	47 (1)
Fuel combustion, other	3,843 (4)	1,117 (5)	678 (4)	609 (3)	544 (2)	466 (6)	6 (0)
Chemical and allied product	1,129 (1)	152 (1)	396 (2)	299 (2)	65 (0)	39 (0)	165 (3)
Manufacturing							
Metals processing	1,495 (2)	88 (0)	76 (2)	444 (2)	171 (0)	112 (1)	5 (0)
Petroleum and related industries	368 (0)	138 (1)	496 (3)	345 (2)	32 (0)	18 (0)	35 (1)
Other industrial processes	632 (1)	408 (2)	450 (2)	370 (2)	339 (1)	187 (2)	44 (1)
Solvent utilization	2 (0)	2 (0)	5,278 (29)	1 (0)	6 (0)	5 (0)	0 (0)
Storage and transport	80 (0)	7 (0)	1,324 (7)	3 (0)	94 (0)	32 (0)	1 (0)
Waste disposal and recycling	1,154 (1)	97 (0)	433 (2)	42 (0)	310 (1)	238 (3)	86 (2)
Highway vehicles	50,386 (56)	7,765 (32)	5,325 (29)	326 (2)	257 (1)	197 (2)	250 (5)
Off-highway vehicles	19,914 (22)	5,280 (22)	2,461 (13)	1,084 (6)	461 (1)	413 (5)	10 (0)
Natural sources	0 (0)	0 (0)	14 (0)	0 (0)	5,307 (15)	796 (10)	34 (1)
Miscellaneous	8,920 (10)	328 (1)	772 (4)	12 (0)	26,609 (77)	5,549 (66)	4,244 (86)
Total	**89,454 (100)**	**24,454 (100)**	**17,917 (100)**	**19,647 (100)**	**34,741 (100)**	**8,379 (100)**	**4,936 (100)**

Source: US EPA.

ments were printed and used as the basis for establishing ambient air quality standards. Now, the validity of these criteria, as well as other related data and reports, has become open to question by industry-appointed lawyers, doctors, and others. They defend that pollution does not differ from any other substance contacted by living matter: Small concentrations and dosages may be beneficial, whereas excessive amounts are usually harmful. The problem lies in deciding what "excessive" means. (In an extreme sense, this term relates not only to living plants and animals but also to material objects, as there are those who claim that all matter, including rocks and so forth can be shown to be living.)

It goes without saying that air pollution is harmful to all living things and their environment. Air pollution can be a contributing factor to chronic bronchitis, emphysema, and lung cancer. It can increase the discomfort of those suffering from allergies, colds, pneumonia, and bronchial asthma. It also can cause dizziness, headaches, eye, nose and throat irritations, increased nasal discharges, nausea and vomiting, coughing, shortness of breath, constricted airway passages, chest pains, cardiac problems, and poison in the stomach, bloodstream, and organs.

Many of the air pollution effects observed on people and animals come from disaster occurrences. In these situations, in which SO_2, particulates, and other pollutants were present in high concentrations, illness and death rates rose. In the Meuse Valley, the Belgian disaster victims were mainly older persons with heart and lung problems. In Donora, Pennsylvania, nearly half of the town's population became ill, severity increasing with age. Those who died were older persons with cardiac or respiratory problems. In London, a similar situation occurred, and in addition a number of prize animals being exhibited in London at that time died or were adversely affected. In one London episode, 52 of 351 animals were severely affected with acute bronchitis, emphysema, or heart failure, or combinations of these.

Plants vary widely in their resistance to pollution damage. Certain species are very resistant to one pollutant and highly sensitive to another, whereas in other species, the reverse could be true. Other contributing factors include plant age, soil, moisture, nutrient levels, sunlight, temperature, and humidity. In general, plants are more sensitive to air pollution than humans. Using SO_2 as an example, plants that are particularly affected by this pollutant include alfalfa, barley, cotton, wheat, apple, and many soft woods. Resistant crops are potatoes, corn, and the maple tree. Chronic injury occurs at concentrations of 0.1–0.3 ppm SO_2; acute injury occurs above 0.3 ppm. Damage can range from retarded growth to complete destruction of the vegetation. Aesthetic as well as true economic cost can have definite associations with this problem. Laboratory studies have shown that nearly all pollutants can have adverse effects on plants. It is important to note that in a noncontrolled situation, it is difficult to determine whether damage is caused by air pollution, crop disease, bacteria, insects, soil nutrient deficiencies, lack of moisture, or mechanical damage because the effects of many of these can appear similar.

Material damage resulting from air pollution can be extensive because nearly everything is bathed continuously in air. Corrosion and erosion of metals is a common example. To list a few problems, pollution deteriorates painted surfaces, oxidizes rubber (causing it to stress crack), paper, clothes, and other material, reacts with stone and masonry, and just plain "dirties" surfaces.

One indirect effect of air pollution on the environment is the "greenhouse effect" phenomenon. Here, the presence of pollution in the atmosphere helps produce a stable atmospheric layer. Incoming solar radiation passes through the layer and warms the earth. The layer retards convection and radiation processes, resulting in heat buildup. Conversely, the pollution layer could prevent incoming radiation from reaching the surface and produce cooling.

Acid rain pollution has not been adequately investigated, but the acidity of rain downwind from fossil fuel power stations has been measured at values of pH 3 and less, which is 300 times the acidity of normal rain. Normal rain in the United States is acid, with an average pH of about 5.5. This could result from sulfur, nitrogen, and/or carbon oxides. Particulates in the atmosphere can react to form secondary pollutants such as sulfites/sulfates and nitrites/nitrates. It has already been pointed out that these materials dominate the submicron group of bimodally distributed atmospheric aerosols, and it is these small particulates (about 0.2 μm) that are most detrimental when inhaled by humans. Atmospheric particulates act as nucleation sites that cause abnormalities in rainfall. They also cause haze and reduced visibility.

A final example of a possible adverse effect of atmospheric pollutants on the environment has already been mentioned: the fluorocarbon–ozone problem, which may result in ozone destruction and consequent increased radiation levels that could cause an increase in skin cancer. As is true with many of the other effects discussed, more study is needed to fully evaluate this potential hazard.

6. MEASUREMENTS

Measurements of air pollution generally fall into two broad categories: ambient and source. Well-designed procedural, setup, and analytical techniques are minimum requirements to obtain meaningful data for both types. Unfortunately, too many insignificant data are reported, and the problem often becomes one of sorting out the good from the bad.

Several points apply to measurements made in both categories. As previously noted, gaseous air pollutants and air are treated as ideal gases, and the ideal gas law can be used:

$$PV = n\,RT \tag{6}$$

where P is absolute pressure, V is volume, n is number of moles, R is the gas constant, and T is absolute temperature.

Dalton's law of partial pressure is also used:

$$P_A = y_A\,P_T \tag{7}$$

where P_A is the partial pressure of component A, y_A is the mole fraction of component A, and P_T is the total pressure. The sum of all the individual partial pressures equals the total pressure:

$$P_T = P_A + P_B + P_C + \ldots. \tag{8}$$

It is important that consistent units be used in these equations. A convenient constant to remember is the volume of ideal gas at standard temperature and pressure (STP): 22.4 L/g mol (359 ft^3/lb mol). Conditions of STP are 1 atm pressure and 0°C (273 K). Using this constant and Eq. (6) enables one to derive values of the gas constant (R) in any convenient units. For example,

$$R = 82.05 \text{ atm cm}^3/\text{g mol °K}$$

or

$$R = 4.968 \times 10^4 \text{ lb}_m \text{ ft}^2/\text{lb-mol °R}$$

(where lb_m means pounds of mass). R also equals 1.987 cal/g mol K.

Both ambient and source particulates occur in a distribution of size; that is, they are polydisperse. These size distributions are usually log-normal and can be plotted on log probability coordinates, as shown in Fig. 1. A probability plot of any sample containing several types of material or material that has been treated by different techniques will most likely be two or more straight intersecting lines. For example, a probability plot of a pure crystalline substance should be a single line; if some of the crystals were thermally shocked by rapid cooling at the walls of the crystallizer or if some were mechanically ground by the agitator, the plot may show as two or more intersecting lines.

6.1. Ambient Sampling

The US EPA announced ambient air-monitoring methods in 1971 (9), in 1973 (10), and in 2004 (www.epa.gov). These announcements provide information on sampling procedures, rates, times, quantities, operating instructions, and calibration methods. The basic reference methods for gases are often wet chemistry analytical procedures, which include the use of 24-h bubbler systems and very precise laboratory analyses. Accepted equivalent methods include instrumental techniques, which are to be used under specific conditions and must be calibrated. Briefly, the reference methods for gases are as follows:

1. SO_2, pararosaniline method
2. CO, nondispersive infrared methods
3. Photochemical oxidants, neutral-buffered potassium iodide photochemical method
4. Hydrocarbon, flame ionization methods
5. NO_2, Saltzman method

The reference method for suspended particulates in the atmosphere is the "high values sampler," which is discussed in a government publication (9).

Numerous instrumental methods are available for measuring atmosphere gaseous pollutants, several of which are noted in Table 3. The systems are grouped into categories according to the detection principle used, and some suppliers of the systems are noted.

The proper placement of ambient-air-monitoring systems can be as important as the analytical method selected for obtaining good samples. The site location will influence what is sampled. To obtain "typical" ambient air data, locations not directly adjacent to roadways and other concentrated sources should be used and there should be nonrestricted airflow around the site.

Many devices are located outside in the ambient air and, as such, minimize losses resulting from sample lines. High-volume samplers are always taken outdoors, and many bubblers are enclosed in protective, heated cases and kept outside. Some devices

Table 3
Partial Listing of Available Ambient Air-Monitoring Systems for Gaseous Pollutants

	SO_2	NO_x	CO	HC	H_2S
Chemical electrode					
Beckman Instruments, Inc.	×	×			
Geomet, Inc.	×	×			
Orion Research, Inc.	×				
Chemiluminescence					
AeroChem Research Laboratories, Inc.		×			
Beckman Instruments, Inc.		×			
Bendix Corp.		×			
Geomet, Inc.		×			
LECO Corp.		×			
Scott Research Laboratories, Inc.		×			
Thermo Electron Corp.		×			
Colorimetric					
Bendix Corp.					
Drager Werk, AG	×	×	×	×	×
Unico Environmental Instruments, Inc.	×	×	×	×	×
Conductometric					
Bristol Co.		×			
Calibrated Instruments, Inc.		×			
Leeds & Northrup Co.		×			
Research Appliance Co.		×			
Tracor, Inc.		×			
Correlation spectrometry					
Barringer Research Ltd.	×				
Bausch & Lomb	×				
CEA Instruments	×				
Environmental Measurements, Inc.	×				
Coulometric					
Beckman Instruments, Inc.	×				
Curtis Instruments, Inc.	×				
Geomet, Inc.	×				
Electrochemical cell					
Dynasciences Corp.	×	×	×		
International Biophysics Corp.	×	×	×	×	×
Theta Sensors, Inc.	×	×	×		
Flame ionization detection					
Bendix Corp.				×	
Delphi Industries				×	
GOW–MAC Instrument Co.				×	
Mine Safety Appliances Co.				×	

(Continued)

Table 3 *(Continued)*

	SO_2	NO_x	CO	HC	H_2S
Gas chromatographic (FID, FPD, and TC)					
Applied Science Laboratories, Inc.		×	×		
Beckman Instruments, Inc.		×	×		
Bendix Corporation		×	×		
Byron Instruments, Inc.		×	×		
GOW–MAC Instrument Co.		×	×		
Hewlett-Packard, Avondale Div.		×	×		
Perkin-Elmer Corp.		×	×		
Tracor, Inc.	×	×	×		×
NDIR					
Beckman Instruments, Inc	×	×	×	×	
Bendix Corp.	×	×	×	×	
Calibrated Instruments, Inc.	×	×	×		
Leeds & Northrup Co.	×		×		
Mine Safety Appliance Co.	×	×	×	×	
Scott Research Laboratories, Inc.		×			
NDUV					
Beckman Instruments, Inc.		×			
E.I. du Pont de Nemours & Co., Inc.	×	×			
Teledyne Analytical Instruments	×	×			
UV fluorescence					
REM Scientific, Inc.	×				

Abbreviations: FPD, flame photometric detector; FID, flame ionization detector: TC, thermal conductivity detector; GC, gas chromatograph; NDIR, nondispersive infrared; UV, ultraviolet; NDUV, nondispersive UV.

must be placed inside to safeguard the systems. This could require the use of relatively long sampling tubes, which can result in a potential error by absorption, adsorption, or fallout of the pollutants. In order to minimize these problems, a molecular diffusion system should be used to bring samples close to the instruments (11). This requires the installation of a large vertical duct or probe through which inlet air can be passed in laminar flow, as shown in Fig. 4.

This system shows a 15-cm inlet duct with a 150-L/min airflow rate. The top of the duct is covered to keep debris from falling into the system and should be located about 2 m above the surface of the roof to prevent pickup of dust raised from the roof by localized turbulent eddies. Sample ports using approx 1.5-cm-diameter tubing and taking flows of about 5 L/min can then be located in the duct close to the sampling instruments. Note that if many small samples are needed, the duct size and flow should be increased to provide adequate air for truly representative samples. Unused air from the duct blower is exhausted outside. All sample lines require periodic checking and cleaning. Note that the ends of the small lines shown are gas sample probes with the tips pointing away from the moving airstream to reduce the chance of picking up entrained particulate matter.

The relevant data include initial and final airflows, instrument readouts, time, dates, type of analytical system used, solution preparation dates, and dry weights (e.g., filter

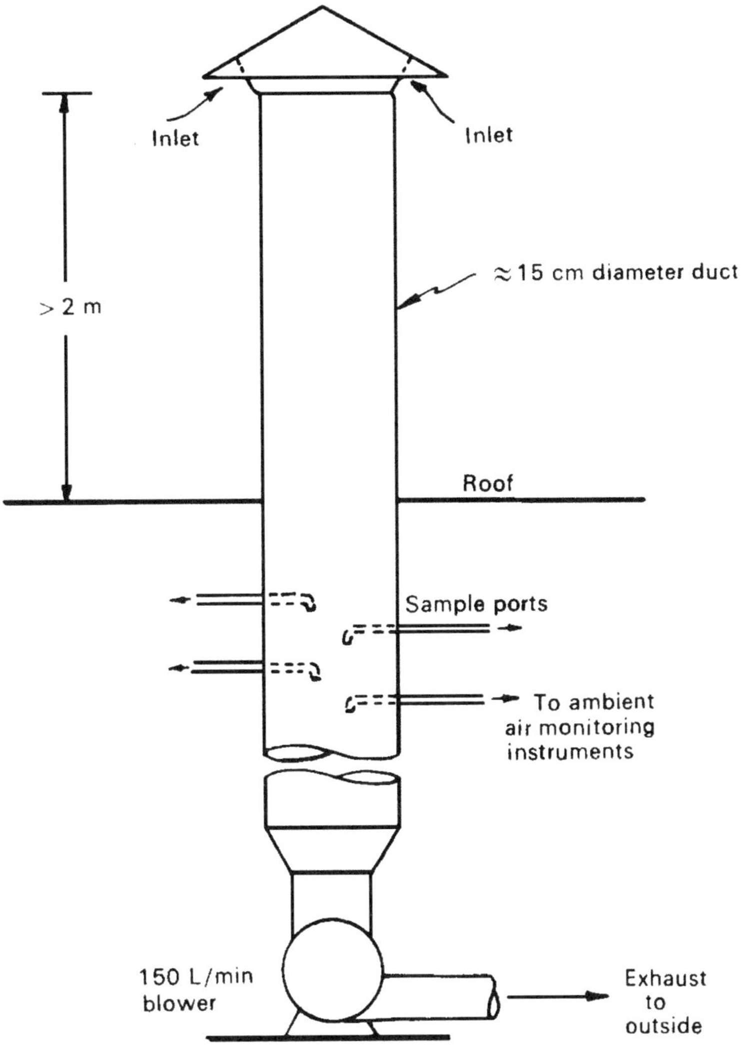

Fig. 4. Example of a molecular diffusion sampling manifold.

papers). In addition, ambient air data that could be noted simultaneously include weather conditions: wind speed and direction, precipitation, temperature, barometric pressure, relative humidity, and solar radiation (12).

6.2. Source Sampling

Pollutants released from an emission source are measured by proper sampling of the exhaust gases, which is often complicated by the difficulties and dangers involved. The sample locations may be hundreds of feet above the ground; the gases may be extremely hot; residual electrical charges might be present, requiring equipment grounding to prevent the buildup of dangerous potentials; and the gases could contain poisonous or toxic substances or active bacteria. Furthermore, the analytical techniques may be extremely complex, even inadequate for the specific requirements. These, plus the atmospheric problems of wind, precipitation, temperature, and humidity, often make stack testing an unenviable occupation.

Stack or source testing usually requires obtaining the following minimum data:

1. Gas velocity
2. Gas temperature (dry and wet bulb)
3. Static pressure in the duct
4. Barometric pressure
5. Inside diameter or area of the duct
6. Concentration of desired pollutants, which may include size and size distribution of particulate
7. Emission source, name, and location
8. Date and time
9. Wind speed and direction
10. Control system operating conditions (pressure drop, temperature, liquid flow rate, and type)
11. Process operating conditions, including charge rate

Two procedures should be evaluated before an actual source test is undertaken. If the system is a typical classical operation, it may be possible to obtain an estimate of the amount of emissions from a listing of emission factors (8). To supplement these data, it may even be possible to obtain data on size and size distribution from other sources such as the *Scrubber Handbook* (13) or the McIlvaine Company manuals (14). The second procedure consists of making an opacity method using the Ringelmann Smoke Chart. This old but valuable approximation procedure developed by Professor Maximilian Ringelmann in 1897 uses five charts ranging from white to black to indicate the degree of opacity. For example, a white chart with a 20% apparent grayness of a plume blends with the apparent grayness of the chart. Charts and instructions for using this method are given in a Bureau of Mines circular (15).

The source sampling problems noted suggest that sampling costs could be high. However, there is no substitute for good emission data, especially if control equipment must be specified and installed. The expenditure of several thousands of dollars at his stage could save many times that amount in control equipment capital and operating costs. In addition, the control system designed for a specific facility has a high chance of working compared with "guesstimation" procedures.

6.3. Sample Locations

The sample ports in a typical full-sized installation can be simply constructed by installing "close" 4-in.-diameter pipe nipples in the stack or duct at the point where the samples are to be taken. The nipples should not protrude inside the stack or duct systems where they could disturb the gas flow patterns. The 4-in.-diameter nipples are required to permit the installation of standard-size test devices. When not in use, they can be sealed with an installing cap. Heavy-wall nipples should not be used because some devices will not pass through them. The typical installation will require a minimum of four nipples at equal distance around the stack.

Gas flow patterns inside a pipe are influenced by bends, openings, location of the blower, and location of obstructions. It is important that the sample location be chosen in such a manner as to minimize flow irregularities. An engineering rule of thumb is to choose the longest straight section in the area where the sample is to be taken. Ideally, the sample location should be at least 15-pipe diameters downstream from the last bend or obstruction and 10-pipe diameters upstream from any opening, bend, or obstruction. The US EPA has suggested guidelines (6) that can be followed for increasing the number of traverse points at any sample location, depending on how near obstructions are to sample locations. These instructions are essential for good particulate sampling.

The traverse locations at the sample point are chosen so that all samples are taken from a single plane perpendicular to the flow of gas. For a circular duct, traverses are made on two lines that intersect at right angles in the plane. Each point of the traverse is chosen to represent the center of an equal-area segment. A minimum of 12 equal areas with the traverse points located at the centroid of each area is suggested (6), as shown in Fig. 5. These points are located at the following distances from the inside wall: 4.4,

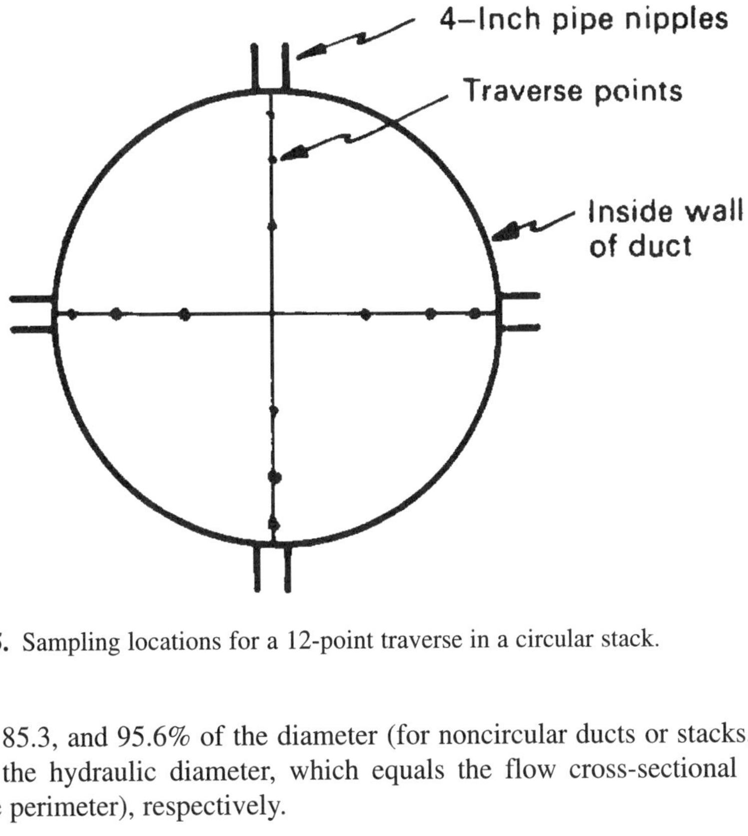

Fig. 5. Sampling locations for a 12-point traverse in a circular stack.

14.7, 29.5, 70.5, 85.3, and 95.6% of the diameter (for noncircular ducts or stacks, the diameter equals the hydraulic diameter, which equals the flow cross-sectional area divided by inside perimeter), respectively.

6.4. *Gas Flow Rates*

The gas volumetric flow rate and pollutant concentrations are needed to determine emission rates and to size control equipment. The volumetric flow rate expressed in terms such as cubic meters per second (m^3/s) or cubic feet per minute (ft^3/min) can be obtained by measuring the weighted-average gas velocity multiplied by the inside diameter of the duct. The average of velocities measured at the traverse points, as discussed in the previous subsection, provides an acceptable weighted-average velocity

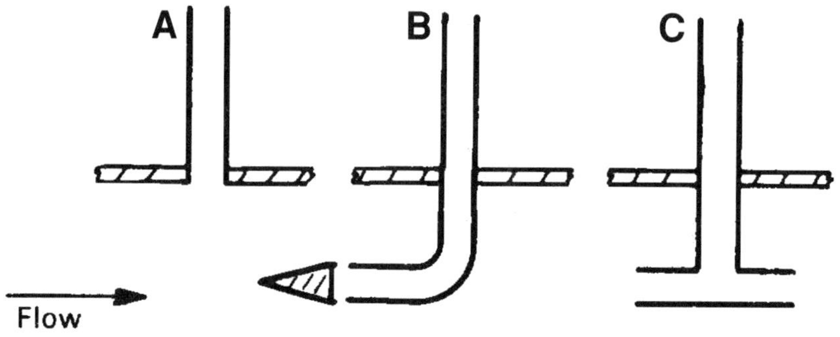

Fig. 6. Static-pressure-sensing devices: (**A**) wall type; (**B**) static tip; (**C**) low pressure.

for the system. Gas velocities may be obtained by measuring either the gas kinetic or the velocity pressure.

Total pressure (P_T), which includes static pressure (P_s) plus velocity pressure (P_v), is measured by placing an impact tube so that it faces directly into a gas stream. Static pressure must be measured separately and subtracted from this total pressure to obtain

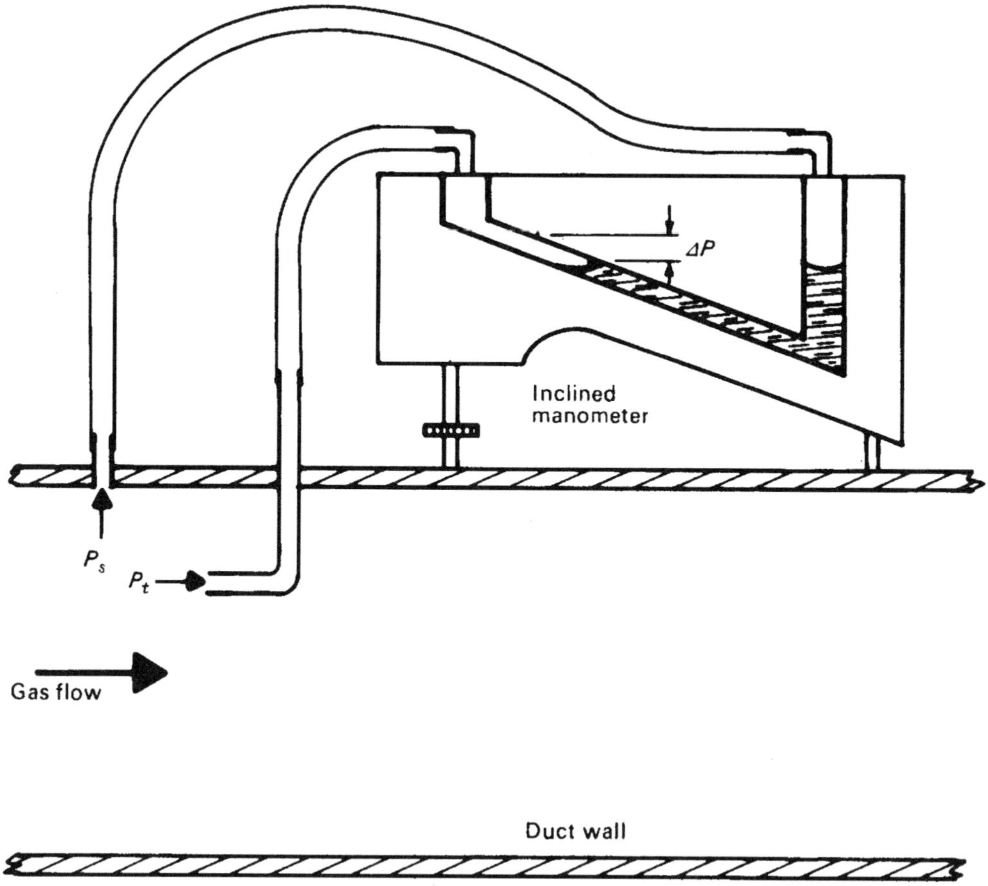

Fig. 7. Pressure-drop measurement.

the velocity pressure. Methods of obtaining static pressure are shown in Fig. 6. The simple through-the-wall tap (Fig. 6A) is a sharp, burr-free tubing located perpendicular to and flush with the inside wall. This is good for nonturbulent systems, and like all sampling devices, it must be kept free of liquid (condensate, entrained liquid, etc.). Method B utilizes a pipe with radially drilled holes, and because it is located away from the wall disturbances, it is good for velocities up to 12,000 ft/min. Systems with high dust loads may require a device, shown by Method C, which gives a rapid response and also responds best to low pressure.

A smooth, sharp-edged impact tube facing directly into the gas stream, as shown in Fig. 7, can be attached to the "high" side of a manometer and a static pressure connection attached to the "low" side. This shows velocity pressure (P_v) directly on the manometer. An inclined manometer as shown in Fig. 7 can be used for improving accuracy in reading a low-pressure drop (ΔP). Any other type of pressure gage or manometer can be used. The connections between the tubes and the gage must be kept tight and free of liquid.

Two general types of combination static–total pressure tubes are used. These units, called Pitot tubes, are shown in Fig. 8. A good standard Pitot tube has no correction factor (C); that is, $C = 1$. The S-type (Stauscheibe or reverse tube) has a correction factor of about 0.8. Note that static pressures can be obtained using the Pitot tube by properly

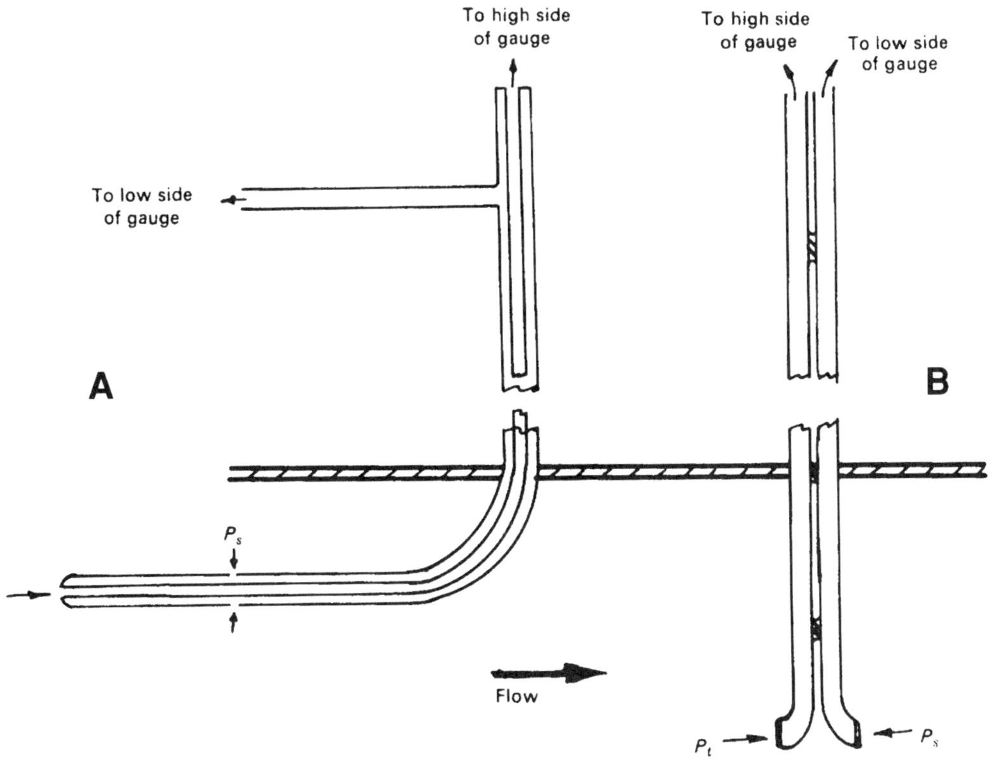

Fig. 8. Pitot tubes: (**A**) standard; (**B**) type S.

connecting only the static connection to the pressure gage. Using a standard or previously calibrated Pitot tube one can obtain a test Pitot tube correction factor:

$$C_{\text{test}} = C_{\text{std}}\left(\Delta P_{\text{std}} / \Delta P_{\text{test}}\right)^{0.5} \tag{9}$$

Pitot tubes cannot be used for gas velocities below about 120 m/min (400 ft/min). One can use Pitot tubes to measure the gas velocity by applying one of the following equations:

For compressible fluids,

$$v = C\left\{\left[2\, g_c\gamma/(\gamma-1)\right]\left(P_D/\rho_D\right)\left[\left(P/P_D\right)^{(\gamma-1)/\gamma} - 1\right]\right\}^{0.5} \tag{10}$$

where C is the Pitot tube correction factor, v is the velocity, g_c is a dimensional constant, (32.174 lb_m ft / lb_f s^2, or 1 kg m / N s^2), P_D is the absolute pressure in the duct or stack, static plus barometric pressures, P is the impact tube pressure plus the barometric pressure, γ is the ratio of gas-specific heats (at constant pressure and constant volume), and ρ_D is the density of gas in the duct or stack.

The generalized Pitot equation is

$$v = C\left[2g_c\left(\Delta P/\rho_1\right)\right]^{0.5} \tag{11}$$

where $\Delta P = P_t - P_s$.

For air, Eq. (10) simplifies to (in metric units)

$$v_1 = 147.4\ C\left(T_1\Delta P_1\right)^{0.5} \tag{12a}$$

where v_1 is the air velocity (m/min), T_1 is the absolute temperature (K), and ΔP_1 is the pressure drop (centimeters of water), or (in English units)

$$v_2 = 174.6\ C\left(T_2\Delta P_2\right)^{0.5} \tag{12b}$$

where v_2 is the air velocity (ft/min), T_2 is the absolute temperature (°R), and ΔP_2 is the pressure drop (inches of H_2O).

The average velocities can be obtained by averaging the velocities obtained for each ΔP as calculated by Eqs. (10)–(12b) or, as implied by the equations, the $(\Delta P)^{0.5}$ can be averaged and a single average velocity calculated.

Multiplying the duct inside cross-sectional area by this average velocity results in an actual volumetric flow rate (e.g., actual cubic feet per minutes). To correct this to SC of 1 atm and 20°C, the ideal gas law as shown in Eq. (6) can be applied to obtain normal cubic meters per second [(normal)m^3/s] or standard cubic feet per minute (std ft^3/min) as follows:

$$Q_a\left(293/T_D\right)\left(P_D/76\right) = (\text{normal})\ \text{m}^3/\text{s}\ \text{ or std ft}^3/\text{min}\ \text{ or scfm} \tag{13}$$

where Q_a is the actual volumetric flow rate, T_D is the absolute temperature in the duct (K), and P_D is the absolute pressure in the duct (cm Hg).

6.5. *Relative Humidity*

The amount of water vapor in a gas stream can be determined by several methods. One standard method for gases is to use wet–dry-bulb temperature measurements. This method consists mainly of air or nitrogen and does not have high temperature or

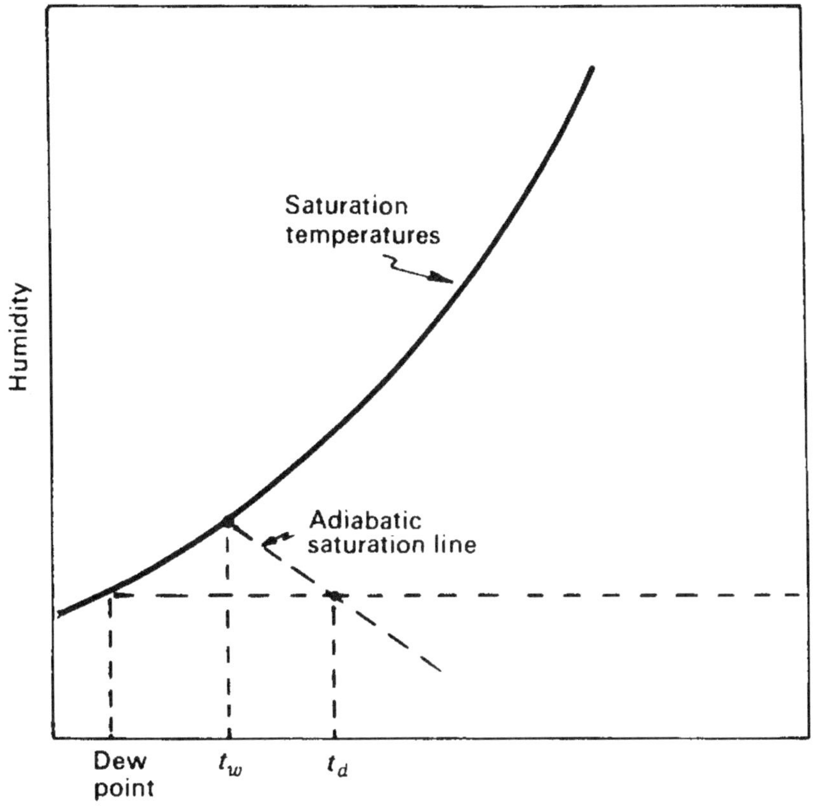

Fig. 9. Flue gas moisture determined by the psychrometric chart method.

high velocity and is near atmospheric pressure. Moisture on the wick on the wet bulb must be present during the readings. From a psychrometric chart, the humidity (H) in mass of water per mass of dry air can be obtained as shown in Fig. 9. The wet-bulb temperature (t_w) is read at the saturation line, and the adiabatic saturation line is followed until the dry-bulb temperature (t_d) is reached. This gives the humidity. Water content is expressed as percentage of moisture by volume (W). For air, this is

$$W = 161\ H/(1 + 1.61H) \tag{14}$$

An alternate procedure for determining water content is the condensation method. An ice bath is used in most sample trains to condense moisture. The gas leaving the ice bath is then saturated with water vapor but contains no liquid droplets (i.e., it is dry). Pumps and meters can run dry, so this eliminates moisture problems during sampling. The total volume of water vapor in sampled gas (Q_w) is equal to the volume of condensed water (Q_c) plus the volume of saturated water vapor at the meter (Q_s). Q_s can be using

$$Q_s = \left(P_w / P_m\right)Q_m \tag{15}$$

where P_w is the saturation vapor pressure of water at the meter temperature from steam tables, P_m is the absolute pressure at the meter, meter static pressure plus barometric pressure, and Q_m is the metered gas volume. The value of Q_c in cubic centimeters is found by converting the grams of water condensate to vapor:

$$Q_c = (M)(22,400\ \text{cm}^3/18\ \text{g})(T_m/273)(76/P_m) \tag{16}$$

where M is the mass of condensate (g), T_m is the meter absolute temperature (K), and P_m is the absolute pressure at meter (cm Hg). Finally, the percentage of water content is

$$W = 100(Q_w/Q_T) \tag{17}$$

where Q_T is the total volume of metered gas at meter conditions.

Use of the Carrier equation is another method for determining water content in flue gases when the wet-bulb temperature is >180°F at nonstandard pressure. This equation includes factors for transfer to heat by conduction and radiation and accounts for diffusion and vaporization. It is accurate at temperatures up to about 400°F and can be used with flue gases containing up to about 15% CO_2. The actual water vapor pressure (P_a) is given in inches of Hg:

$$P_a = P_w - [(P_D - P_w)(t_D - t_w)/(2,830 - 1.44\ t_w)] \tag{18}$$

where P_w is the saturation vapor pressure of water at duct wet-bulb temperature from steam tables (inches of Hg), P_D is the absolute pressure in duct (inches of Hg), t_D is the dry-bulb temperature in the duct (°F), and t_w is the wet-bulb temperature in the duct (°F).

The percentage of water content becomes

$$W = 100(P_a/P_D) \tag{19}$$

6.6. *Sample Train*

Sampling methods and systems have been discussed thoroughly in various publications. However, the two major variations are the US EPA Method 5 test train (6) and the American Society of Mechanical Engineers (ASME) Performance Test Code (PTC) 21 method as described in ref. 16 and as supplemental by PTC 27 in 1957 and PTC 28 in 1965. These systems are shown in Figs. 10 and 11, respectively. The basic differences are as follows. In the US EPA Method 5, an ice bath is used in the train, which can result in the condensation of insoluble material after the filtering stage. These condensables are considered as particulates. In the US EPA Method 5, the meter is placed after an airtight pump, whereas in the ASME method the meter is run under vacuum by placing it before the pump. In the ASME method, the meter is considered to be airtight (or at least as leak-free as an airtight pump). It is important that all trains be inspected for leaks and operated leak-free.

Gas samples are withdrawn proportionally, which means that as the gas flow in the duct changes, the sample rate is changed proportionally to provide properly weighted results. Isokinetic sampling is used in collecting particulates and consists of drawing the sample into the sample probe at a velocity equal to the velocity in the duct where the sample's tip is located. This means that duct velocities must be taken simultaneously with the samples and at the same locations. It is important to be able to relate sample

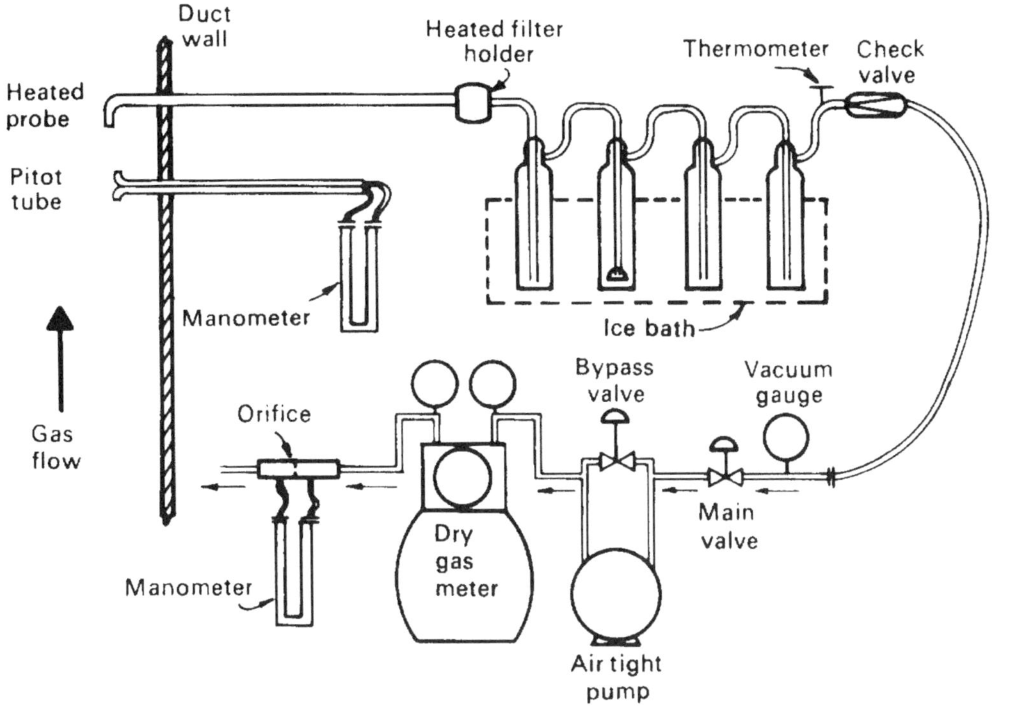

Fig. 10. EPA Method 5 source test train.

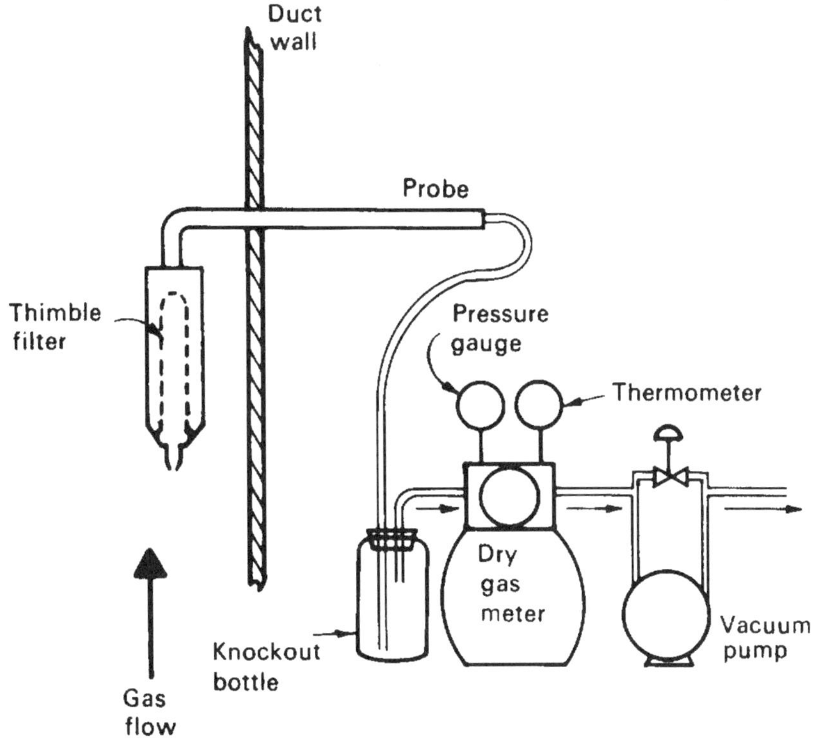

Fig. 11. ASME PTC 21 source test train.

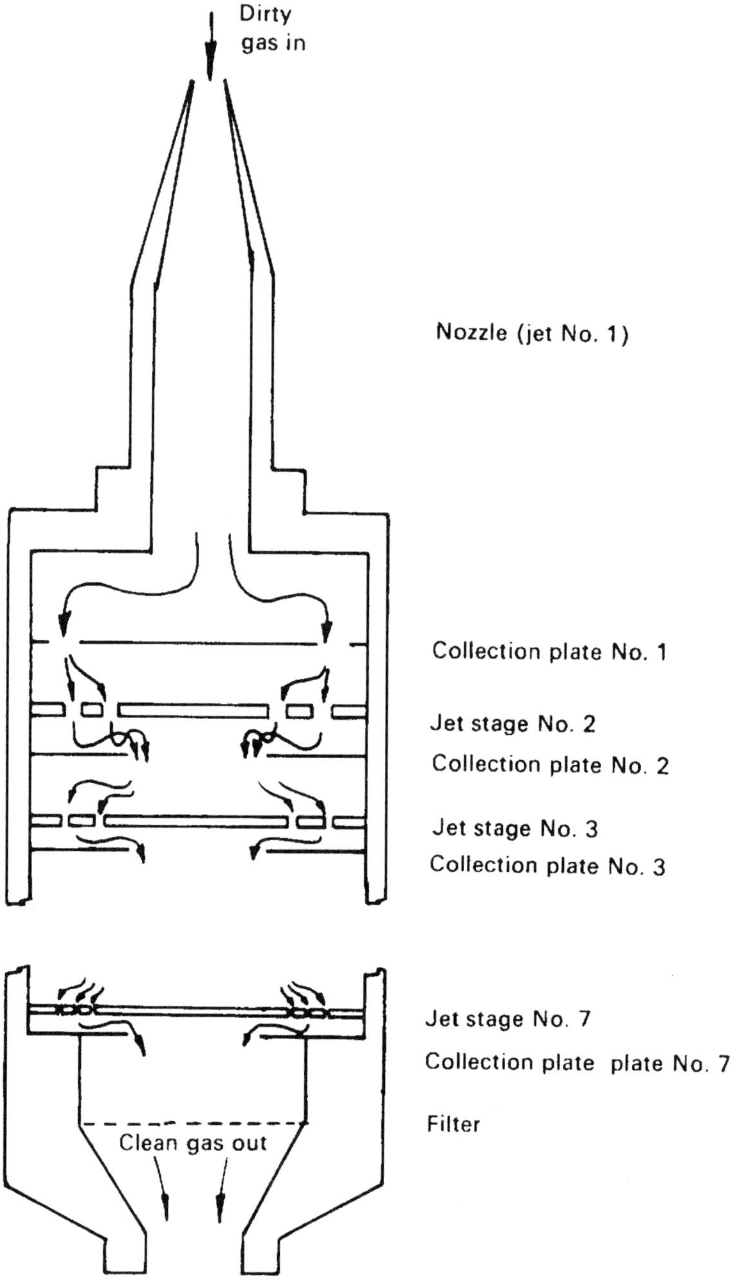

Fig. 12. Schematic diagram of a Mark III source test cascade impactor.

meter flow rates to probe-tip flow rates by accounting for pressure, temperature, and moisture changes, as discussed in the previous subsections.

Sample tip sizes can be changed to allow samples to be taken at a reasonable rate (about 0.5–3.0 ft^3/min). Gas probes are not subject to this problem because these samples are taken proportionally, as noted. Not only must the sampling rate be maintained isokinetic, but the particulate probes must be pointed directly into the gas flow and traverse the entire duct. Usually, 5–10 min per traverse point are needed and at least 50 ft^3/min of gas should be sampled. Two or more duplicate complete runs are desirable and may be required for each system sampled.

6.7. Determination of Size Distribution

Normal size distribution has been discussed in Section 2, and a sample log probability distribution plot is given in Fig. 1. It has been further pointed out that accurate size and size distribution data are required in order to properly specify air pollution control systems. It is difficult, however, to obtain accurate size data, especially when a large portion of the particulates are fine (less than 3 μm).

Various size distribution techniques may be used, but the most accurate procedure for fine particles is aerodynamic sizing, which consists of sizing the material in flight in the duct. Methods for doing this include mechanical, optical, and condensation techniques and a number of commercial sizing devices. The mechanical devices are rugged, highly portable, and suited for field work. The University of Washington Mark III impactor (17) is one type of mechanical size-classification device. A schematic diagram of this cascade impactor is shown in Fig. 12, and Table 4 lists the design parameters of the unit. The overall size of this unit is about 7.5 cm in diameter and 24 cm long.

Table 4
University of Washington Mark III Source Test Cascade Impactor Jet Quantities and Dimensions

Stage	No. of jets	Jet diameter (in.)	Jet depth[a] (in.)	Jet-to-plate clearance (in.)	Ratio of jet depth to jet diameter	Ratio of jet-to-plate distance to jet diameter
1	1	0.7180	1.50	0.56	2.09	0.78
2	6	0.2280	0.125	0.255	1.60	1.80
3	12	0.0960	0.125	0.125	1.97	1.97
4	90	0.0310	0.125	0.125	4.03	4.03
5	110	0.0200	0.063	0.125	3.15	6.25
6	110	0.0135	0.030	0.125	2.22	9.26
7	90	0.0100	0.030	0.125	3.00	12.50

[a]Jet depth is the thickness of metal that was drilled to make jet.

Particulates are separated aerodynamically in cascade impactors by causing the particles to strike collection plates. The plates in the first stages are constructed with large holes, and the gas flows through them at low velocities. Therefore, large particles are captured on these plates. The gas flows through progressively smaller holes, resulting in the collection of finer and finer material at higher velocities in successive stages. The impacted material usually stays on the collection plates after impaction. A very light grease film can be applied to improve the adhesion of particles to the collection plates.

The impactors must be thoroughly cleaned before and after each use. The collection plates and backup filters are desiccated and weighed before and after use to provide data on the size of the particles. The impactor must be brought to operating temperature

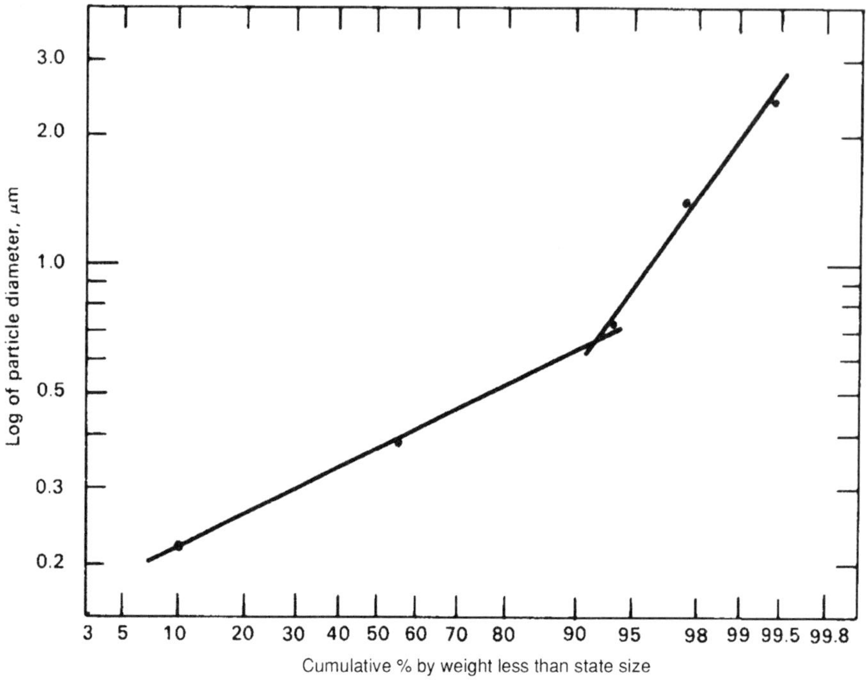

Fig. 13. Typical size distribution data for particulates in scrubbed flue gas.

before use to prevent condensation, and it is secured to the end of a probe, which is, in turn, connected to a sample train. The impactor becomes a filter, and the rest of the train may be arranged as shown in either Fig. 10 or 11.

Size distribution data that can be obtained with cascade impactors is shown in Fig. 13. These data were taken with a Mark III cascade impactor (2). The particulates are fly ash in flue gas after being passed through a Venturi scrubber operating at a low (6.25 in. of H_2O) pressure drop. The inlet dust concentration for this scrubber is shown in Fig. 1. Note that there is still a bimodal dust distribution, but the scrubber removed more large particles than small particles. The slopes of the lines representing each size group have changed, with the larger group changing more.

7. GAS STREAM CALCULATIONS

7.1. General

When examining potential air emissions, one should consider the following types: volatile organic compounds (VOCs), semivolatile organic compounds (SVOCs), particulate matter (PM), and metals. Additionally, it is also important to examine potential emission sources; for example, any *ex situ* treatment process that will likely result in VOCs (if present in the waste/soil) and PM emissions from material handling activities (e.g., excavation, transport, waste feeding) and soil/waste storage, as well as emissions from the treatment process itself (18, 19). At many manufacturing facilities, it is common for one pollution control system to serve several air emission sources. In these cases, the combined emission parameters must be determined from the mass and heat balances. This section provides calculation procedures for combining emission stream and single-emission-stream parameters.

7.2. Emission Stream Flow Rate and Temperature Calculations

Gas flow rates must be converted to standard conditions (77°F, 1 atm) before flow rates can be added together. The calculation procedures for converting gas flow rates to SC and then combining gas flow rates are presented below. (Note: Because the emissions are approximately at atmospheric conditions, pressure corrections are not necessary in the following examples.)

$$Q_{e1} = 537\ Q_{e1,a}/(460 + T_{e1}) \tag{20}$$

where Q_{e1} is the flow rate of gas stream 1 in cubic feet per minute at standard conditions (scfm), $Q_{e1,a}$ is the flow rate of gas stream 1 in at cubic feet per minute actual conditions (acfm), and T_{e1} is the temperature of gas stream 1 (°F).

The flow rate for each emission stream is converted to SC by using Eq. (20). After all emission streams are converted to SC, the total volumetric flow rate for the combine emission stream at standard condition (Q_e) is determined by summing these flow rates, as follows:

$$Q_e = Q_{e1} + Q_{e2} + Q_{e3} + \ldots + Q_{en} \tag{21}$$

where Q_e is the flow rate of the combined gas stream at standard conditions (scfm).

To convert combined volumetric flow rate at standard conditions (Q_e) to flow rate at actual conditions ($Q_{e,a}$), the temperature of the combined gas stream (T_e) must be determined. This is accomplished by first determining the enthalpy (sensible heat content) of each individual stream. The calculation procedures are presented as follows:

$$H_{s1} = Q_{e1}(T_{e1} - 77°\text{F})(0.018\ \text{Btu/ft}^3\ °\text{F}) \tag{22}$$

where H_{s1} is the sensible heat content of gas stream 1 (Btu/min) and T_{e1} is the temperature of gas stream 1 (°F). Repeat this calculation for each emission stream. The total sensible heat content for the combined emission stream is determined by summation of sensible heat contents, which is shown in the following:

$$H_s + H_{s1} + H_{s2} + H_{s3} + \ldots + H_{sn} \tag{23}$$

where H_s is the sensible heat content of the combined gas stream (Btu/min). The temperature of the combined stream can now be determined from

$$T_e = (H_s/Q_e)/(0.018\ \text{Btu/ft}^3\,°\text{F}) \tag{24}$$

where T_e is the temperature of the combined gas stream (°F).

The factor 0.018 Btu/ft^3 °F is obtained by multiplying specific heat at constant pressure ($c_p = 0.240$ Btu/lb °F) by the density ($\gamma = 0.0739$ lb/ft^3) of air at 77°F. With the temperature of the combined gas stream (T_e) and the flow rate of the combined gas stream at SC, the actual combined gas stream flow rate is determined as follows:

$$Q_{e,a} = Q_e(460 + T_e)/537 \tag{25}$$

where $Q_{e,a}$ is the flow rate of the combined gas stream at actual conditions (acfm).

7.3. *Moisture Content, Dew Point Content, and Sulfur Trioxide Calculations*

When examining the potential gas stream, the moisture content of the combined stream must be calculated as a volume percent. The procedures for calculating the volume percent of moisture for the combined gas stream require that the volume moisture of each stream be converted to the lb-mol basis, added together, and then divided by the total combined gas stream volumetric flow rate (Q_e). Steps for calculating moisture content of gas stream are provided below as on both a volume percent and a mass percent basis. To determine the dew point, the mass percent of moisture in the gas stream must be calculated.

The procedures for converting the volume percent of moisture to lb-mol is

$$M_{e1,\text{l-m}} = Q\left(M_{e1}/100\%\right)\left(\text{lb-mol}/392\text{ scf}\right) \tag{26}$$

where M_{e1} is the percent of moisture content of gas stream 1 (% volume) and $M_{e1,\text{l-m}}$ is the moisture content of gas stream 1 (lb-mol/min). For each gas stream in the emission stream, this calculation is repeated and the lb-mol/min of each gas stream is summed to determine the combined gas stream, as shown in Eq. (27). The volume percent of the combined stream is determined using Eq. (28).

$$M_{e,\text{l-m}} = M_{e1,\text{l-m}} + M_{e2,\text{l-m}} + \ldots + M_{en,\text{l-m}} \tag{27}$$

$$M_e = (M_{e,\text{l-m}}/Q_e)(392\text{ scf/lb-mol})(100\%) \tag{28}$$

where $M_{e,\text{l-m}}$ is the moisture content of the combined gas stream (lb-mol/min) and M_e is the moisture content of the combined gas stream (% volume).

To determine the dew point, the moisture content of the combined gas stream must be presented on mass basis ($M_{e,\text{m}}$), which is calculated as

$$M_{e,\text{m}} = (M_{e,\text{l-m}})(18\text{ lb/lb-mol}) \tag{29}$$

where $M_{e,\text{m}}$ is the moisture content of the combined gas stream (lb/min). Equation (30) is used to calculate the amount of dry air in the combined gas stream:

$$\text{DA}_e = Q_e(\text{lb-mol/392 scf})(29\text{ lb/lb-mol}) \tag{30}$$

where DA_e is the dry air content of combined gas stream (lb/min). The psychrometric ratio is calculated as

$$\text{Psychrometric ratio }\left(\text{lb of water/1b dry air}\right) = M_{e,\text{m}}/\left(DA_e - M_{e,\text{m}}\right) \tag{31}$$

Using Table 5 with the psychrometric ratio and the gas stream temperature, the dew point temperature of the combined gas stream is determined.

When the gas stream contains sulfur trioxide (SO_3), the dew point temperature of the gas stream increases. If the concentration of trioxide is not considered when determining the dew point temperature of the gas stream, condensation may occur, resulting in a gas stream containing liquid droplets. This, in turn, will combine with SO_3 to form sulfuric acid and cause severe corrosion of metal and deterioration of fabric components in air pollution control equipment (bag house), fans, and filter. Therefore, to protect air

Table 5
Dew Point Temperature

	Gas stream temperature (°F)										
	70	80	90	100	120	140	160	180	200	220	240
Psychrometric ratio	Dew Point Temperature (°F)										
0.000	0	0	0	0	0	0	0	0	0	0	0
0.005	54	58	61	65	70	76	81	86	89	93	96
0.010	62	65	68	71	77	82	86	90	94	97	100
0.015	68	72	75	77	82	86	90	94	97	100	103
0.020		77	80	82	87	91	94	97	100	103	106
0.025			85	87	91	94	98	101	103	106	109
0.030			89	91	95	98	100	104	107	109	111
0.035				95	98	101	104	107	109	110	114
0.040				98	101	104	107	109	111	114	116
0.045					104	107	109	112	114	116	118
0.050					107	109	112	114	116	118	120
0.055					109	112	114	116	118	120	122
0.060					111	114	116	118	120	122	124
0.065					114	116	118	120	122	124	125
0.070					116	118	120	122	123	125	130
0.075					118	120	122	124	125	130	150
0.080					119	122	123	125	130	140	170
0.085						123	125	130	143	168	182
0.090						124	130	140	162	180	205
0.095						128	140	165	180	205	225

Source: US EPA.

pollution equipment, the amount of SO_3 in the gas stream must be considered when determining the dew point temperature for the gas stream. The acid dew point temperature can be determined when the parts per million volume bases (ppmv) of SO_3 and the volume percentage of moisture (vol%) in the gas stream are known. Using Fig. 14, the acid dew point can be determined. The figure provides dew points for three volume percentages of moisture levels: 2%, 5%, and 10% H_2O. Other dew point temperatures for moisture percentages less than 10% and greater than 2% can be estimated using the provided moisture level curves.

The SO_3 content of each individual stream is converted to lb-mol by using Eq. (32):

$$S_{e1,\text{l-m}} = Q_{e1}(S_{e1}/10^6)(\text{lb-mol}/392\ \text{scf}) \tag{32}$$

where S_{e1} is the SO_3 content of gas stream 1 (ppmv) and $S_{e1,\text{l-m}}$ is the SO_3 content of gas stream 1 (lb-mol/min). Repeat this converting procedure for each gas stream. After all streams are converted, then these values are summed to obtain the combined gas stream content as follows:

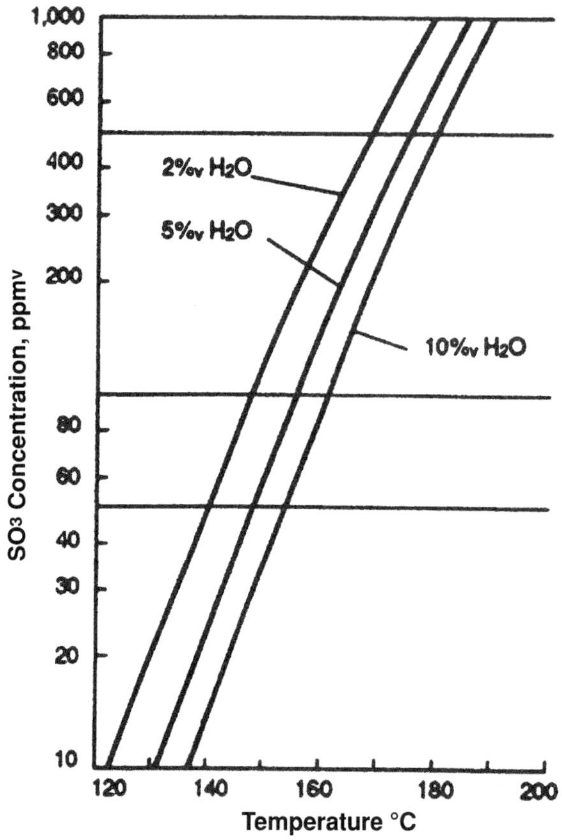

Fig. 14. "Acid" drew points in stack gases. (From US EPA.)

$$S_{e,\text{l-m}} = S_{e1,\text{l-m}} + S_{e2,\text{l-m}} + \ldots + S_{en,\text{l-m}} \tag{33}$$

$$S_e = S_{e,\text{l-m}}(10^6/Q_e)(392 \text{ scf/lb-mol}) \tag{34}$$

where $S_{e,\text{l-m}}$ is the SO_3 content of the combined gas stream (lb-mol/min) and S_e is the SO_3 content of the combined gas stream (ppmv). Using the SO_3 content (S_e) and moisture content (M_e) for the combined gas stream, the acid dew point can be obtained by entering these values in Fig. 14.

7.4. *Particulate Matter Loading*

Usually, the particulate matter is reported in grams per actual cubic feet (g/acf). Outlined below are the procedures to convert the reported particulate matter in a gas stream from g/acf to lb/h.

$$W_{e1,1} = W_{e1,g}\, Q_{e1,a}(60 \text{ min/h})(\text{lb}/7{,}000 \text{ g}) \tag{35}$$

where $W_{e1,g}$ is the particulate loading for gas stream 1 (g/acf) and $W_{e1,1}$ is the particulate loading for gas stream 1 (lb/h). This procedure is repeated for each gas stream and the results in lb/h for each gas stream are summed as shown in Eq. (36):

$$W_{e,1} = W_{e1,1} + W_{e2,1} + \ldots + W_{en,1} \tag{36}$$

where $W_{e,1}$ is the particulate loading for the combined gas stream (lb/h).

Equation (37) shows how particulate loading in the combined gas stream is converted from lb/h to g/acf:

$$W_{e,g} = \left(W_{e,1}/Q_{e,a}\right)\left(7{,}000 \text{ g/1b}\right)\left(\text{h/60 min}\right) \tag{37}$$

where $W_{e,g}$ is the particulate loading for the combined gas stream (g/acf).

7.5. Heat Content Calculations

Utilizing the heat of combustion for each component in gas stream 1, the heat content of the gas stream can be calculated using

$$h_{e1} = (0.01)\sum_{i=1}^{n} y_{e1,i}\, h_{e1,i} \tag{38}$$

where h_{e1} is the heat content in gas stream 1 (Btu/scf), $y_{e1,i}$ is the volume percent of component i in gas stream 1 (vol%), $h_{el,i}$ is the heat of combustion of component i in gas stream 1 (Btu/scf) (see Table 6), and n is the number of components in gas stream 1.

The combined gas stream heat content is determined from the heat content of the individual gas streams, with emission streams as follows

$$H_e = (0.01)\sum_{j=1}^{m} y_{ej}\, h_{ej} \tag{39}$$

where H_e is the combined emission stream heat content (Btu/scf), y_{ej} is the volume percent of stream j in the combined gas stream (vol%), h_{ej} is the heat content of gas stream j in the combined gas stream (Btu/scf) (see previous discussion), and m is the number of individual gas streams in the combined gas stream. To convert the heat content of a stream to Btu/lb, multiply the heat content of a stream Btu/scf by the density of the emission stream at SC (typically 0.0739 lb/ft^3).

7.6. Dilution Air Calculations

Dilution air can be used to decrease the heat content of the emission stream. Equation (40) is used to determine the quantity of dilution air (Q_d) required to lower the heat content of an emission stream (18):

$$Q_d = Q_e\left[\left(h_e/h_d\right) - 1\right] \tag{40}$$

where Q_d is the dilution airflow rate (scfm), h_e is the emission stream heat content before dilution (Btu/scf), h_d is the emission stream heat content after dilution (Btu/scf), and Q_e is the emission stream flow rate before dilution (scfm). After the dilution quantity is determined, the concentrations of the various components and flow rate of the emission stream must be adjusted as follows:

$$O_{2,d} = O_2\left(h_d/h_e\right) + 21\left[1 - \left(h_d/h_e\right)\right] \tag{41}$$

Table 6
Heats of Combustion and Lower Explosive Limit (LEL) Data of Selected Compounds

Compound	LEL (ppmv)	Net heat of combustion[a,b] (Btu/scf)
Methane	50,000	882
Ethane	30,000	1,588
Propane	21,000	2,274
n-Butane	16,000	2,956
Isobutane	18,000	2,947
n-Pentane	15,000	3,640
Isopentane	14,000	3,631
Neopentane	14,000	3,616
n-Hexane	11,000	4,324
Ethylene	27,000	1,472
Propylene	20,000	2,114
n-Butene	16,000	2,825
I-Pentene	15,000	3,511
Benzene	13,000	3,527
Toluene	12,000	4,196
Xylene	11,000	1,877
Acetylene	25,000	1,397
Naphthalene	9,000	5,537
Methyl alcohol	67,000	751
Ethyl alcohol	33,000	1,419
Ammonia	160,000	356
Hydrogen sulfide	40,000	583

Source: Data from *Steam/Its Generation and Use,* The Babcock & Wilcox Company, New York, 1995 and *Fire Hazard Properties of Flamable Liquids, Gases, Volatile Solids*—1977. National Fire Protection Association. Boston, MA, 1977.

[a]Lower heat of combustion.

[b]Based on 70°F and 1 atm.

$$M_{e,d} = M_e(h_d/h_e) + 2[1 - (h_d/h_e)] \tag{42}$$

$$Q_{e,d} = Q_e(h_e/h_d) \tag{43}$$

where $O_{2,d}$ is the oxygen content of the diluted emission stream (vol%), $M_{e,d}$ is the moisture content of the diluted emission (vol%), and $Q_{e,d}$ is the flow rate of the diluted emission stream (scfm).

In Eq. (41), the factor 21 represents the volumetric percentage of oxygen in air and the factor 2 in Eq. (42) is the volumetric percentage of moisture in air at 70ºF and 80% humidity. Note that the calculations for moisture content are presented as a point of reference, but are not usually required for the equations used in this handbook. After the dilution is completed, the hazardous air pollutant (HAP) emission stream characteristics are reassigned as follows:

$O_2 = O_{2,d} =$ ______________ %

$M_e = M_{e,d} =$ ______________ %

$$h_e = h_d = \_\_\_\_\_\_\_\_\_\_ \text{ Btu/scf}$$

$$Q_e = Q_{e,d} = \_\_\_\_\_\_\_\_\_\_ \text{ scfm}$$

8. GAS STREAM CONDITIONING

8.1. General

Gas conditioning equipment is installed upstream of the control device, to ensure that the control device operates efficiently and economically. This equipment is used to temper or treat the gas stream prior to its entering the control device. A mechanical dust collector, wet or dry gas cooler, and gas preheater can be used as preconditioning equipment. Typically, when the control device is a fabric filter or electrostatic precipitator and the gas stream contains significant amounts of large particles, a mechanical dust collector is installed upstream of the control device to remove large particles (20). To protect the fabric used in the fabric-filter control device from damaging high gas stream temperatures, a gas cooling device could be installed upstream of this device to lower the gas stream temperature to temperatures within the operating temperature of the fabric. The gas cooler device can also be used to reduce the volume of gas stream or maximize the collection of HAPs by electrostatic precipitators or fabric filters. The elimination of moisture condensation problems can be accomplished by using a gas preheater, which will increase the temperature of the gas stream entering the control device. This manual presents discussions on gas conditioning equipment, but not equipment designs. The latter are readily available from vendors and common literature sources.

8.2. Mechanical Collectors

The removal of heavy dust particles from a gas stream can be accomplished with a mechanical dust collector, such as a cyclone. This device utilizes centrifugal force to separate the dust particles from the gas stream. The efficiency of the cyclone is dependent on the gas velocity entering the cyclone and the diameter of the inlet of the cyclone. In theory, as the inlet velocity increases or as the diameter of the inlet decreases, the greater the collection efficiency is and the greater the pressure drop through the cyclone is. Particles above 20–30 μm in size can be effectively removed for the gas stream by the cyclone. This removal will reduce the loading and wear on downstream control equipment.

In evaluating the use of a cyclone as a control for an emission stream, one must first determine the size distribution for the particulates in the gas stream. If the gas stream has significant amounts of particles above 20–30 μm, the installation will require that filters or electrostatic precipitators be installed upstream of the cyclone to reduce the loading to the cyclone. When utilizing a wetted Venturi scrubber as a control for an emission stream, a preconditioning device is not generally required even when the gas stream contains large particles (20–30 μm in size). When using a nonwetted Venturi scrubber as the control device for an emission stream, the emission stream must be free of particles that could clog the nozzles. Therefore, the emission stream either has to be free of particulates or a pretreatment device such as mechanical dust equipment must used to treat the emission prior to the nonwetted Venturi scrubber (20,21).

8.3. Gas Coolers

To maximize the collection of HAPs by electrostatic precipitators and fabric filters, the gas stream volume can be reduced by utilizing a gas cooler. Control devices, such as the Venturi scrubber, are less sensitive to high gas-emission-stream temperature, because these types of device cool the gas emission stream prior to particle collection. As a result of the temperature decrease of the gas emission stream, HAPs in vapor form will also decrease. When cooling the gas emission stream, care must be exercised to ensure that the temperature of the gas emission stream is maintained above the dew point temperature. A good standard of practice is to maintain a 50–100°F cushion above the dew point to account for process fluctuations. Procedures for calculating the dew point temperature for a gas emission stream were presented in previous sections.

Coolers are available as dry or wet types. The dry-type cooler cools the emission gas stream by radiating heat to the atmosphere. Spray chambers are utilized by wet-type coolers to add humidity and cool the emission gas stream with evaporating water. Another way to cool the emission gas stream is to add cooler dilution air. The cost, dew point temperature, and downstream control device must be considered when selecting gas-cooling equipment. If the downstream control device is a fabric filter, then a wet-type cooler would not be appropriate for cooling, as this type of cooler would increase the possibility of condensation with the fabric-filter system.

When a gas cooler is used, the gas stream parameters will have to be recalculated using standard industrial equations. For instance, when a wet-type gas cooler is used, then a new actual gas flow rate and moisture content for the emission gas stream will have to be calculated.

8.4. Gas Preheaters

The temperature of the emission gas stream can be increased by using a gas preheater. Increasing the temperature of the emission gas stream reduces the likelihood of condensation. In fabric filters, condensation can plug or blind fabric pores. Additionally, condensation can increase the corrosion of metal surfaces in a control device. To overcome these problems, a gas preheater can be used to increase the temperature of the emission gas stream above the dew point temperature. Three methods are commonly utilized to raise the gas emission temperature: direct-fired afterburners, heat exchangers, and stream tracking. Direct-fire afterburners preheat the gas stream by using a flame produced from burning an auxiliary fuel. Additionally, the flame also combusts organic constituents in the emission gas stream that might otherwise blind the filter bags in a downstream control device. An shell-and-tube arrangement is used by a heat exchanger to preheat the gas emission stream. The stream-tracking method runs an emission gas stream line inside of a steam line to preheat the emission stream. This method is typically only employed when a steam line is available at the site.

When the emission streams contain HAPs, the preheating of the stream should be raised to only 50–100°F above the dew point temperature to minimize the vapor components of the HAP. This allows the downstream control devices, such as a bag house or an electrostatic precipitator, to control the HAP as effectively as possible. Emission stream parameters must be recalculated using a standard industrial equation when the temperature of the gas stream is preheated, because when the gas stream temperature increases, it increases the actual gas flow rate of the emission stream.

9. AIR QUALITY MANAGEMENT

9.1. *Recent Focus*

9.1.1. *Emission Sources*

A recent focus of air quality management (12,18–61) has been on reducing natural and man-made airborne contaminants from various sources: (1) point source hazardous air emissions, (2) non-point-source fugitive hazardous emissions, (3) greenhouse or global warming gases, (4) ozone-depleting gases, (5) indoor emissions that release asbestos, microorganisms, radon gases, VOCs, lead, and so forth, (6) odor emissions, (7) vehicle emissions, (8) wildfire emissions, and (9) terrorists' emissions of airborne infectious and/or toxic contaminants.

In this handbook, the chapters entitled "Fabric Filtration," "Cyclones," "Electrostatic Precipitation," "Web and Dry Scrubbing," "Condensation," "Flare Process," "Thermal Oxidation," "Catalytic Oxidation," "Gas-Phase Carbon Adsorption," and "Gas-Phase Biofiltration" introduce the new technologies for removal of the point source hazardous air emissions in detail. Another chapter, "Emerging Air Pollution Control Technologies," introduces various new technologies for the treatment of non-point-sources fugitive hazardous emissions and vehicle emissions.

Indoor and odor pollution problems are addressed in detail in the chapters entitled "Ventilation and Air Conditioning," "Indoor Air Pollution Control," "Noise Pollution," and "Noise Control." Additional literature of indoor and odor pollution control can be found elsewhere (39–43,59).

Discussions of the greenhouse or global warming gases, and the ozone-depleting gases are covered in Sections 9.2 and 9.4. The readers are also referred to a chapter entitled "Carbon Sequestration" in ref. 59.

Terrorist-launched emissions of airborne biocontaminants and toxic gases cannot be prevented nor controlled easily and cost-effectively (49–53). Rademakers (49) introduced the biological warfare detection technologies and new decontamination methods. Ziegler (50) introduced the procedures to deal with a terrorist incident emitting airborne pathogenic microorganisms or toxic gases. Gudia (51) presented an overview of issues related to environmental regulations as we attempt to deal with possible future terrorist events. Abkowitz (52) raised communication issues related to dealing with possible future emergencies.

As stated previously, lightning is the main cause of producing bad ozone in the tropospheric zone and is also the indirect cause of large amounts of combustion-related air pollution as a result of forest fires. In the United States alone, there are over 10,000 forest fires annually, which are mainly caused by lightning. Natural air pollution sources, such as volcanic eruptions and forest fires, produce much more airborne pollutants than all man-made airborne pollutants combined. Although we have no control over volcanic eruptions, perhaps attention should be paid to management of forest fires. In January 2003, the United States government proposed steps to prevent wildfires (62).

9.1.2. Airborne Contaminants

Of various airborne contaminants, organic gaseous emissions are the most important recent focus. Air emission standards have been developed by the US Office of Air Quality Planning and Standards (OAQPS) to address organic emissions from several waste-management sources. The unit operations and processes for removing organic airborne contaminants include wet and dry scrubbing, condensation, flare, thermal oxidation, catalytic oxidation, gas-phase carbon adsorption, gas-phase biofiltration, and so forth presented in Chapters 5–12.

Waste-management sources also contribute other types of air emission such as inorganic gaseous (metals) and particulate matter (PM), which are subject to federal regulation under other programs. This can be illustrated by the program developed by the US Office of Solid Waste (US OSW), which has standards for metal emissions from industrial boilers and furnaces. At landfills and hazardous waste-treatment, storage and disposal facilities (TSDFs), US OSW has general requirements that limit blowing dust (particulate matter). Additionally, the US EPA has developed the following document that deals with the control of emissions from TSDFs: Hazardous Waste TSDF—Fugitive Particulate Matter Air Emission Guidance Document EPA-450/3-89-019 (22). Fabric filtration (Chapter 2), cyclones (Chapter 3), electrostatic precipitation (Chapter 4), and wet scrubbing (Chapter 5) are the processes for removal of PM and inorganic contaminants (metals). The readers are referred to ref. 59 dealing with the following important subjects for removing inorganic and PM contaminants:

1. Atmospheric modeling and dispersion
2. Desulfurization and SO_x/H_2S emission control
3. Carbon sequestration
4. Control of nitrogen oxides during stationary combustion
5. Control of heavy metals in emission streams
6. Ventilation and air conditioning.

Infectious airborne pollutants are various pathogenic microorganisms, including bacteria, virus, and fungus, and can be present indoors (39–41), or both indoor and outdoor when there is a bioterrorist's attack (49–52). Finally, radon gases are radioactive airborne pollutants, and noise is transmitted through air. The solutions to the problems of airborne infectious bacteria, virus, fungus, radon, and noise can be found elsewhere (59).

9.2. Ozone

9.2.1. Ozone Layer Depletion and Protection

Depending on what part of the atmosphere contains ozone, it can either benefit or harm human health and the environment. Figure 15 illustrates this relationship. Ozone occurs naturally in the upper (stratosphere) and the lower atmospheres (troposphere). Ozone in the stratosphere protects us from the sun's radiation; ozone in the troposphere, however, can have adverse health effects and other negative environmental impact. The greenhouse gases are CO_2, H_2O, CH_4, NO_2, and chloroflurocarbons (CFCs), the concentrations of which are increasing and causing global warming. CFC gases destroy the ozone in the stratosphere, thus reducing the ozone layer's radiation protection effect (20).

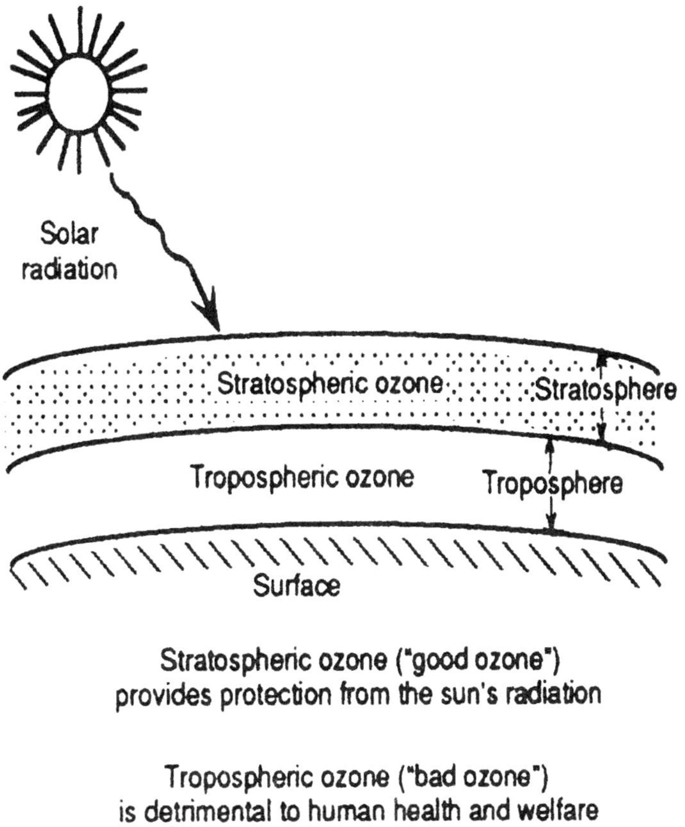

Solar radiation
Stratospheric ozone
Stratosphere
Tropospheric ozone
Troposphere
Surface
Stratospheric ozone ("good ozone")
provides protection from the sun's radiation
Tropospheric ozone ("bad ozone")
is detrimental to human health and welfare

Fig. 15. Ozone in the atmosphere. (From US EPA.)

As illustrated in Fig. 15, the stratospheric ozone, which provides protection from the sun's radiation, is the "good ozone," whereas the tropospheric ozone, which is detrimental to human health and welfare, is the "bad ozone." For our health or long-term survival, we must protect the stratospheric ozone in the ozone layer. CFCs are the major ozone-depleting substances. Other ozone-depleting substances that also reach the stratospheric ozone layer include carbon tetrachloride, methyl chloroform, and halons. Recent major scientific findings and observations (37,38) include the following: (1) Record ozone depletion was observed in the mid-latitudes of both hemispheres in 1992–1993, and ozone values were 1–2 % lower than would be expected from an extrapolation of the trend prior to 1991; (2) the Antarctic ozone "holes" of 1992 and 1993 were the most severe on record, and a substantial Antarctic ozone "hole" is expected to occur each austral spring for many more decades; (3) ozone losses have been detected in the Arctic winter stratosphere, and their links to halogen chemistry have been established; (4) the link between a decrease in stratospheric ozone and an increase in surface ultraviolet (UV) radiation has been further strengthened; (5) the ozone depletion potential (ODP) for CFC-11 is designated to be 1, and the ODP for methyl bromide is calculated to be about 0.6; (6) methyl bromide continues to be viewed as another significant ozone-depleting compound; (7) stratospheric ozone losses cause a global-mean negative radiative forcing; the ozone-depleting gases (CFCs, carbon tetrachloride, methyl chloroform, methyl bromide, etc.) have been used extensively in industrial applications, including refrigeration, air conditioning, foam blowing, cleaning of electronic components, and as solvents; (8) many countries have decided to discontinue the production of CFCs, halons, carbon tetrachloride, and methyl chloroform, and industry has developed many "ozone-friendly" substitutes for protection of the stratospheric ozone layer; and (9) in the domestic refrigeration industry, HFC134A and HFC152A have been used as the substitutes for CFC; in commercial refrigeration industry, HFC134A, HCFC22, HCFC123, and ammonia have been used as the substitute for CFC; and in mobile air conditioning systems, only HFC134A is recommended as the substitute for CFC.

9.2.2. Photochemical Oxidants

The formation of ozone in the troposphere (lower level) is simplistically illustrated in Fig. 16. The primary constituents are nitrogen oxides (NO_x), organic compounds, and solar radiation. Nitrogen oxide emissions are primarily from combustion sources, including both stationary and nonstationary types. Coal-burning power plants are the major stationary source for NO_x, whereas transportation modes, such as automobiles, trucks, and buses, are the major nonstationary source for NO_x. Another source for organic compounds is waste-management operations.

When nitrogen oxides and organic compounds are exposed to sunlight, a series of complex chemical reactions occur to form two principal byproducts: ozone (O_3) and an aerosol that, among other things, limits visibility. This mixture of ozone and aerosol is described as photochemical smog. The respiratory system can be negatively affected when humans are exposed to ozone. Possible effects include inflammation of the lungs, impaired breathing, reduced breathing capacity, coughing, chest pain, nausea, and general irritation of the respiratory passages. The long-term exposure to ozone could result in increased susceptibility to respiratory infections, permanent damage to lung tissue, and severe loss of breathing capacity. The effects of ozone are more severe on the very young, elderly, and those with pre-existing respiratory conditions than on the normal, healthy, adult population. It has been shown, however, that young, healthy individuals who exercise outdoors can also exhibit negative health effects when exposed to ozone.

Urban areas can be subjected to an oxidizing type of pollution, which is the result of a chemical reaction of NO_x and HC in sunlight and produces O_3, PAN, and other complex compounds. This pollution is described as ozone and referred to as photochemical oxidants. Because this pollution is considered a secondary pollutant, transport is a concern. Ozone is a regional concern because it can impact an area 250 km from the source.

Researchers showed the existence and the extent of impact of ozone on human health. In 1998, a study examined several hundred deceased persons in Los Angeles who were victims of automobile accidents but were otherwise healthy. It was found that about half had lesions on their lungs, which is a characteristic of lung disease in the early stages. One of the causes for the observed lesions was attributed to the victims' exposure to the levels of ozone in the Los Angles area.

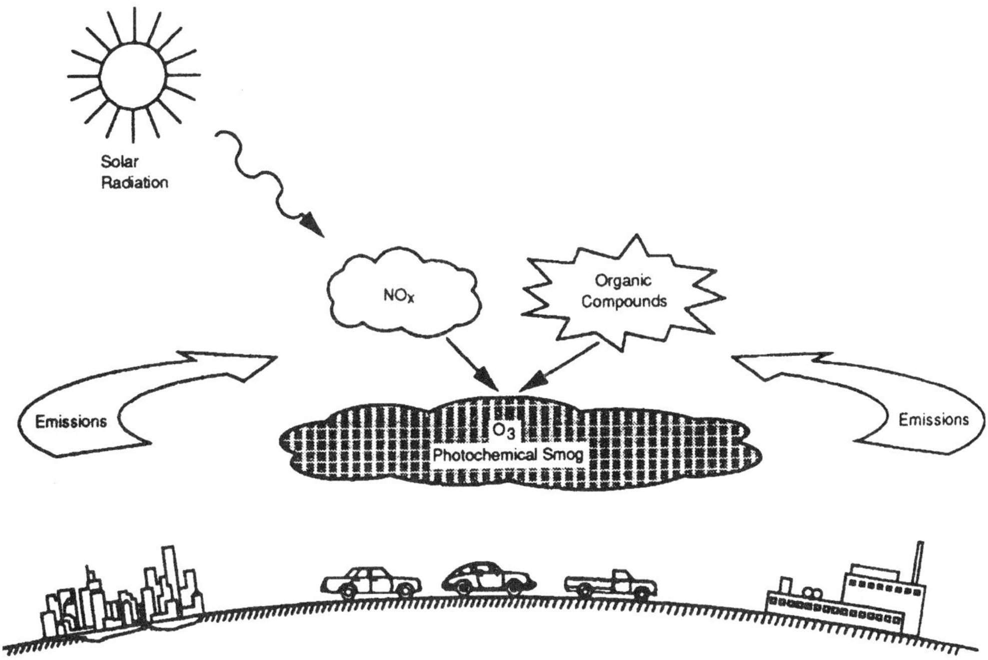

Fig. 16. Tropospheric ozone formation. (From US EPA.)

Other negative impacts are associated with ozone exposure, such as materials sustaining damage from ozone exposure. The useful life of synthetic and rubber compounds become significantly shorter when they are exposed to ozone-laden environment. Additionally, reduction in crop fields, lower forest growth rate, and premature leaf droppage may occur from ozone exposure. The US EPA has estimated that the damage to commercial crops and forests resulting from ozone exposure ranged between 2 and 3 billion dollars. Reduction in visibility by photochemical smog can also be considered as a negative impact on society.

As established under the Clean Air Act (CAA), the air quality standards for air pollutants including ozone are the responsibility of the US EPA. To illustrate the extent of the ozone problem in the United States, one can compare the health-based ambient air quality standard with the air-quality-monitoring data reported for areas throughout the United States. Based on an hourly average not to be exceeded more than once annually, the national ambient air quality (NAAQ) standard for ozone is 0.12 ppm. When this ambient air standard is compared to historical monitoring data, over 60 areas nationwide have routinely exceeded this standard. It is estimated that over 100 million people live in these areas. However, recent data have indicated that ozone levels have shown some improvement in these areas, but the ambient air quality standard for ozone is still being exceeded in many areas that contain a significant portion of the total population of the United States. It has been determined that some these areas may not attain the ambient air quality standard for many years. To address these "nonattainment" areas, Congress amended the CAA in 1990. These nonattainment areas do not attain the ambient air quality standards for several criteria pollutants, include ozone. The amendment

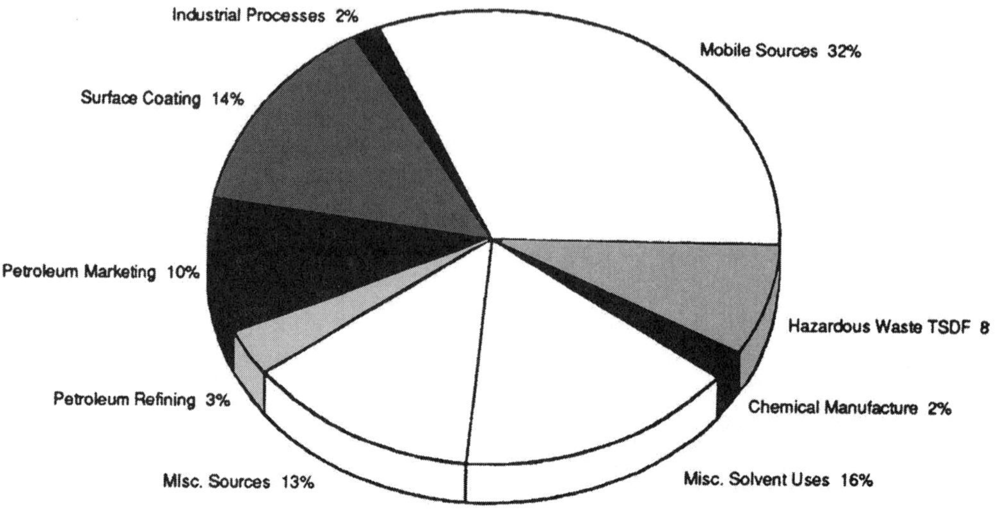

Fig. 17. Sources of nationwide VOC emissions. (From US EPA.)

requires that areas with extreme ozone nonattainment problems have 20 yr to achieve the ozone ambient air quality standard. For the more sensitive populations, a more stringent standard for ozone has been requested.

These nonattainment areas include the largest urban areas, such as the Los Angeles area, Chicago, Houston, and the Northeast corridor. These areas are classified as "hot spots" for ozone, but there are many areas across the United States that have ozone problems. It has been observed in some of the national parks, for example, that the ozone levels occasionally exceed the ambient air quality standard.

The US EPA considers VOCs one of primary ingredients for the formation of ozone. Figure 17 shows the relative contribution of various source categories to total nationwide emissions of VOCs. As shown in Fig. 17, one significant source for VOCs is hazardous waste TDSFs, which contribute about 8% of the total VOC emission in the United States.

9.3. *Air Toxics*

Air toxics are described as air pollutants that cause cancer or other human health effects. The CAA amendments of 1990 specifically identify 190 compounds as air toxics. As required by the CAA, the US EPA must investigate and potentially regulate these air toxics. Air toxic compounds include radon, asbestos, and organic compounds.

Radon is a naturally occurring, colorless, odorless gas formed from the normal radioactive decay of uranium in the earth's rocks and soils. Exposure to radon through inhalation has been demonstrated to increase risk for lung cancer. Radon gas typically enters buildings via soil or groundwater migration. The best technology for radon gas is activated carbon adsorption (27,32,36).

Generally, the inhalation of asbestos is from occupational exposure to asbestos when asbestos material is being applied, as well as its manufacturing and the demolition of buildings. Exposure to asbestos through inhalation has shown a higher than expected incidence of bronchial cancer. Various technologies for the control of airborne asbestos have been reported in US EPA Report No. TS-799 (28), a United Nations report (27), and elsewhere (29–32).

Air toxics emit from existing point and area sources. Large point sources include chemical plants, petroleum refineries, and power plants. The small point sources of air toxic emission, such as dry cleaners, are more widespread than large point sources. Air toxics emissions are also attributed to waste-management sources; the US EPA OSW has shown that there are 2600 to 3000 potential TSDFs.

Acute (short-term) or chronic (long-term) exposure to an air pollutant has characteristic health effects. The neurological, respiratory, and reproduction systems can be affected by exposure to air toxics. Exposure to benzene, for instance, can result in cancer. The US EPA has developed two methods to identify or quantify the impact of carcinogenic air toxics: individual risk and population risk. Individual risk is expressed as a statistical probability to show an individual's increased risk of contracting cancer when exposed to a specific concentration of a pollutant over a 70-yr lifetime. Population risk, which is expressed as number of cancer incidences per year expected nationwide, shows the risk as result of exposure to a pollutant.

9.4. Greenhouse Gases Reduction and Industrial Ecology Approach

9.4.1. Industrial Ecology

Industrial ecology seeks to balance industrial production and economic performance with the emerging understanding of both local and global ecological constraints. As a result, industrial ecology is now a branch of systems science for sustainability, or a framework for designing and operating industrial systems as sustainable and interdependent with natural systems (33).

9.4.2. Global Warming

Over the past 50 yr, global warming has been attributed to greenhouse gases, such as carbon, water vapor, methane, nitrogen dioxide, CFCs, and so forth. It has been projected that average temperatures across the world could climb between 1.4°C and 5.8°C over the next century. A major cause for this projected global warming is the increased carbon dioxide emission by industries and automobiles. At the source, carbon dioxide emission can be easily removed from industrial stacks by a scrubbing process that utilizes alkaline substances. The long-term effect of global warming, projected in the UN Environmental Report released in February 2001, may cost the world about $304 billion (US) a year down the road. This projected cost is based on the following anticipated losses: (1) human life loss and property damages as a result of more frequent tropical cyclones; (2) land loss as a result of rising sea levels; (3) damages to fishing stocks, agriculture, and water supplies; and (4) disappearance of many endangered species (33).

According to a 2001 Gallup poll, 57% of Americans surveyed stated that where economic growth conflicts with environmental interests, the interest of the environment should prevail. On the other hand, the same survey discovered that only 31% of those polled think global warming would pose a serious threat to themselves or their way of life. The results of this poll indicate that both environmental and economical interests are important to

Americans. Some of the existing removal processes, although very simple in theories and principles, are considered to be economically unfeasible by industry and government leaders. For example, carbon dioxide could be easily removed by a wet scrubbing process, but the technology is not considered cost-effective, because the only reuse is the solution in the process. In response, President Bush decided not to regulate carbon dioxide emission at industrial plants. He also rejected the Kyoto international global warming treaty, but US EPA Administrator Christine Todd Whitman stated: "We can develop technologies, market-based incentives and other innovative approaches to global climate changes."

9.4.3. Carbon Dioxide Reuse

An industrial ecology approach to carbon dioxide has been extensively studied (decarbonization) by Wang and his associates (25,26,33) at the Lenox Institute of Water Technology in Massachusetts. Their studies showed that decarbonization is technically and economically feasible when the carbon dioxide gases from industrial stacks are collected for in-plant reuse as chemicals for tanneries, dairies, water-treatment plants, and municipal wastewater plants. It is estimated that tannery wastewater contains about 20% of organic pollutants. Using the tannery's own stack gas (containing mainly carbon dioxide), dissolved proteins can be recovered from the tannery wastewater. Recovery of protein can also be accomplished at a dairy factory. By bubbling dairy factory stack gas containing mainly carbon dioxide through dairy factory wastewater stream, about 78% of the protein in the stream can be recovered. Stack gas containing mainly carbon dioxide can be used at a water-treatment softening plant as a precipitation agent for hardness removal. Neutralization and warming agent can be accomplished at a municipal wastewater-treatment plant by using stack gas containing carbon dioxides. At plants that produce carbon dioxide gas, a large volume of carbon dioxide gases can be immediately reused as chemicals in various in-plant applications, which may save chemical costs, produce valuable byproducts, and reduce the global warming problem.

9.4.4. Vehicle Emission Reduction

A second industrial ecology approach is to develop a new generation of vehicles capable of traveling up to 80 mpg while reducing nitrogen oxides, carbon dioxide, and hydrocarbon levels. Specifically, a "supercar" is to be developed to meet the US EPA's Tier 2 emission limits (33). There are growing health concerns about persistent bioaccumulative toxics that are produced from the combustion of coal, wood, oil, and current vehicle fuels (46).

The issues of energy versus environment have been continuously discussed by many scientists and policy-makers (44–46). In the United States, automakers are racing to build hybrid vehicles and fuel-cell vehicles (53). On January 29, 2003, President George W. Bush announced a $1.2 billion Freedom Fuel Program to speed the development of hydrogen-powered vehicles in 17 yr using fuel-cell technology (58,61,65). Fuel cells create energy out of hydrogen and oxygen, leaving only harmless water vapor as a byproduct of the chemical process. For automobiles, this would end their damaging air pollution and eliminate American dependence on foreign oil. Menkedick discusses the energy and the emerging technology focus (46).

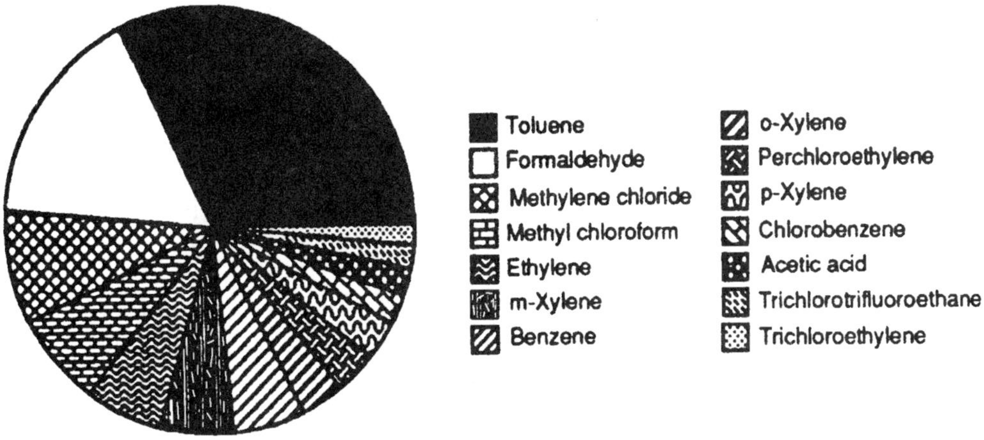

Fig. 18. Top 14 VOC/HAP chemicals. (From US EPA.)

9.4.5. Planting Fast-Growing Trees

A third industrial ecology approach is to plant Loblolly pines (*Pinus taeda*) in areas with high CO_2 concentrations. These faster-growing trees will respond more to elevated CO_2 levels than will slower-growing hardwoods, because of faster photosynthesis and plant growth. The Loblolly pine is the species most frequently grown for timber production in the United States. Its wood has a wide range of uses, such as building material pulpwood and fuel (33).

9.5. Environmental Laws

As required by the Superfund Amendments and Reauthorization Act (SARA), Section 313, major US industries began reporting the amounts of toxic chemicals they released into the air, land, and water. In 1987, industries reported that about 2.4×10^9 lb of toxic pollutants were released into the air. The US EPA has estimated that air toxics accounted for between 1600 and 3000 cancer deaths per year, and the average urban individual lifetime risk of contracting cancer as a result of exposure to air toxics is estimated to be as high as 1 in 1000 persons.

Air toxics emissions from TSDFs have been preliminarily estimated to have a national population risk of about 140 cancer cases per year. Exposure to these air toxics from TSDFs has also been estimated to have a maximum individual risk of 2 in 100 persons contracting cancer.

Figure 18 shows the top 14 VOCs and air toxics, which are referred to as hazardous air pollutants (HAPs) on a mass emission basis. During the examination of air emission, the toxicity of each compound and degree of exposure that occur must be considered (e.g., time and concentration). The most emitted VOC is toluene, but this is less toxic than benzene, a carcinogen linked to leukemia.

Several major environmental laws have been established to address organic air emissions. The CAA was created to address major air pollution problems in the United States. Additional environmental laws include RCRA, amended by the Hazardous and Solid

Table 7
Standards Development Under Section 3004(n)

Phase I	Total organics Process vents and equipment leaks Promulgated 6/21/90 (55 FR 25454)
Phase II	Total organics Tanks, surface impoundments, containers, and miscellaneous units Proposal package in OMB
Phase III	Individual constituent standards, as needed to supplement Phase I and Phase II standards Early Work Group stage

Source: US EPA.

Waste Amendments and the Comprehensive Environmental Response, Compensation, and Liability Act (CERCLA) as amended by SARA.

Most of the new air emission standards, discussed in this chapter, are being developed under RCRA. As required under Section 3004(n), the US EPA Administrator is directed to protect public health and welfare by establishing the standards for monitoring and controlling the air emission from TSDFs. The implementation of these standards is conducted under RCRA's permitting systems for hazardous waste-management units.

The US EPA is developing the RCRA 3004(n) air standards under a three-phase program, as shown in Table 7. Phase I develops the organic emissions standards from process vents associated with specific noncombustion waste-treatment processes (e.g., stream stripping and thin-film evaporation units). Additionally under this phase, organic emissions standards are developed for equipment leaks from pumps, valves, and pipe fittings. On June 21, 1990, the final standards for these sources were promulgated. See Subparts AA and BB in the Code of Federal Regulations (CFR), Title 40, Parts 264 and 265 (40 CFR 264 and 265). Under Phase II, the organic emission from tanks, surface impoundments, containers, and miscellaneous units are established. The proposed standards for these sources were proposed in July 1991 and published in (new) Subpart CC in 40 CFR, Parts 264 and 265. Even after the implementation of Phase I and II organic standards, current analyses indicate a potential residual risk problem. As a result, Phase III will develop individual constituent standards as necessary to bring the residual maximum individual risk to within acceptable range (10^{-6} to 10^{-4}). Proposed Phase III standards are planned to be concurrent with promulgation of Phase II standards.

Also established under RCRA is the Corrective Action Program, which requires solid-waste-management units to go through a site-specific facility evaluation. Air emissions must also be included in the site-specific evaluation and risk assessment. Additionally, air emissions are affected by the land disposal restrictions (LDR), which were promulgated under RCRA. Unless certain treatment requirements are met, LDR prohibits the depositing of hazardous waste on or into land disposal sources such as landfills, surface impoundments, and waste piles. When the hazardous waste is treated to meet LDR requirements, air emission may result if the treatment process is not prop-

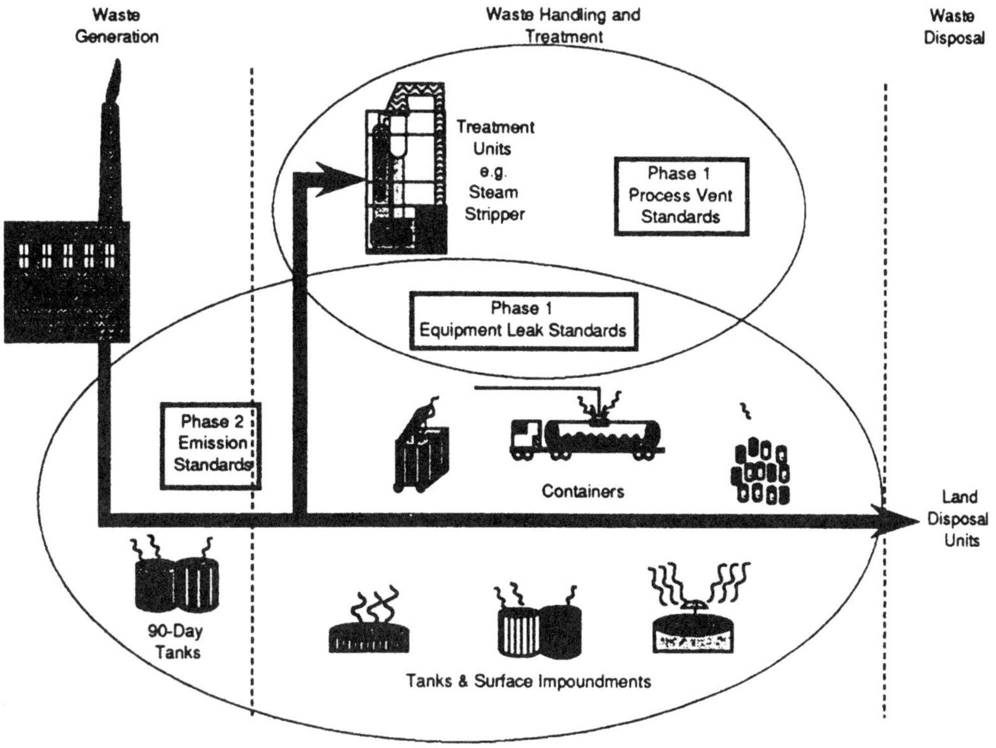

Fig. 19. Phases I and II RCRA air standards overlaid onto hazardous waste management. (From US EPA.)

erly controlled. To prevent cross-media pollution, the RCRA 3004(n) air standards work in concert with the LDR.

Another RCRA program establishes location standards for the siting of new facilities. These standards also require consideration of air emissions.

Figure 19 shows an overview of hazardous waste management. After a hazard is generated, it may go through a series of different processes and waste-management units before disposal. For example, hazardous waste may be stored or treated in tanks and containers. Various types of containers can be used for storage, including 55-gal drums, dumpsters, tank trucks, and railcars. Treatment of the hazardous waste to meet LDR can occur early in the waste-management process, or just prior to disposal. Additionally, hazardous waste management can occur at the generator site (on-site) or at commercial TSDF (off-site). A storage and transfer station may be used to handle waste off-site prior to its being transported to another location for final treatment and disposal.

As shown in Figure 19, the coverage of the Phase I and II RCRA air standards is overlaid onto the hazardous-waste-management units. As required in Phase I standards, organic air emissions from process vents from treatment units specifically identified in the standards are limited. Additionally, Phase I limits the air emission from equipment leaks at waste-management units.

Table 8
Clean Air Act (66)

National Ambient Air Quality Standards (NAAQS)
Criteria pollutants
PM, SO_2, CO, NO_x, O_3, Pb
New Source Performance Standards (NSPS)
Criteria pollutants
Designated pollutants (e.g., total reduced sulfur)
National Emission Standards for Hazardous Air Pollutants (NESHAP)

Source: US EPA.

To address organic air emissions from tanks, surface impoundments, and containers, coverage of the RCRA air standards would be expanded by Phase II standards. These standards are designed to contain (or suppress) potential organic emissions from escaping prior treatment. As dictated by the standards, operators would be required, for example, to cover open tanks containing organic waste unless it can be demonstrated that the concentration of organic material in the waste is below a specific level. Because the control requirements are initiated by the organic concentration of the waste in the container, these standards are described as "waste-based" rules. According to the Phase II RCRA air standards, waste treatment is not required; however, treatment is required under LDR. For benzene waste, the national emissions standards for hazardous air pollutants (NESHAP) require containment-type control prior to treatment. This requirement is similar to requirements under Phase II RCRA air standards. Additionally, NESHAP requires waste containing benzene to meet treatment requirements.

Table 8 lists the major regulatory programs established under the CAA. These programs address organic air emissions, ozone precursors, and air toxics. The US EPA, as discussed previously, establishes the NAAQS for "criteria" pollutants. States are then required to set standards to attain and maintain NAAQS. Because ozone is a criteria pollutant, states regulate VOCs (ozone precursors) on a source-by-source basis. Additionally, Section 111 of CAA sets new source performance standards (NSPS) for emissions of criteria pollutants. These sources include new, modified, or reconstructed stationary sources. Other pollutants may also be addressed by NSPS. These pollutants are classified as "designated pollutants." Designated pollutants are noncriteria pollutants that are identified by the US EPA for regulation under CAA Section 111(d) based on impact on health and welfare. Total reduced sulfur (TRS) and sulfuric acid mist are examples of designated pollutants (66).

Section 112 of the CAA establishes the NESHAP standards, which identify and limit hazardous pollutant emissions from both existing and new stationary sources. The 1990 CAA amendments substantially change Section 112. Prior to 1990, Section 112 required the US EPA to first list the pollutant as hazardous and then to establish standards to protect public health "with an ample margin of safety." The amended Section 112 requires the US EPA to establish technology-based standards for the sources of 190 hazardous pollutants listed in the new law. Additionally, the law requires further action by US EPA

Table 9
CERCLA/SARA (Superfund)

• Site-specific risk analysis required for removal and remediation actions
• Removal and remediation actions must comply with federal and state laws that applicable or relevant and appropriate (ARARS)
• Toxic release inventory required by SARA Title 313

Source: US EPA.

to establish a more stringent standard, if a risk assessment at later time indicates that technology-based standards are not adequately protective.

Recently completed was the NESHAP for benzene waste operations. This was the last NESHAP set under the "old" Section 112. In March 1990, it was promulgated and codified in 40 CFR 61, Subpart FF. It applies to the following emission sources: chemical plants, petroleum refineries, coke byproduct recovery facilities, and TSDFs. The rule establishes a compliance deadline of March 7, 1992 for which existing facilities must install the required control.

The Comprehensive Environment Response, Compensation, and Liability Act mandates the cleanup of inactive contaminated sites. Table 9 indicates that CERCLA has several aspects that provide control of organic emissions. The process required for a Superfund site cleanup is a site-specific risk analysis conducted prior to a removal and remediation action. Under this analysis, consideration of air emissions resulting from the cleanup must be incorporated in the cleanup. This is illustrated by a cleanup of groundwater contaminated with organics. This cleanup could use a groundwater strip-

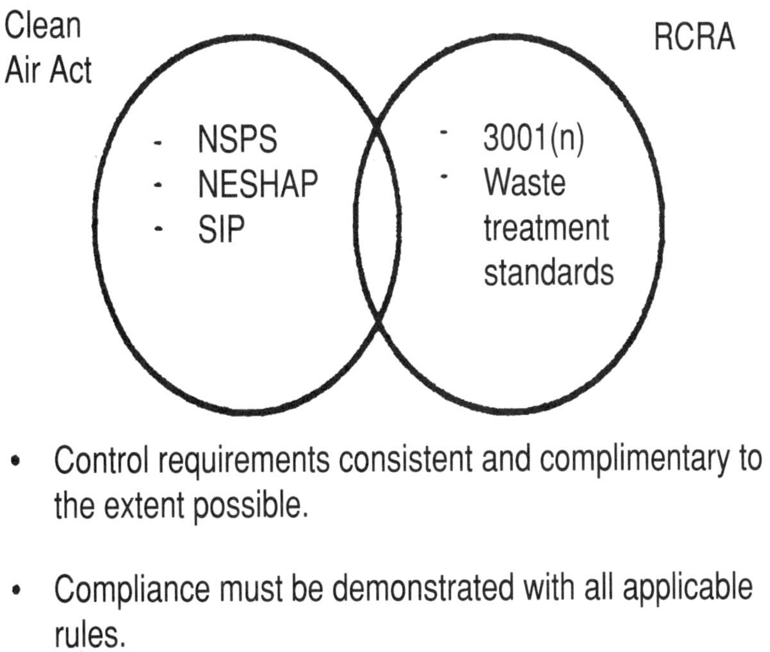
Clean
Air Act
RCRA
- NSPS
- NESHAP
- SIP
- 3001(n)
- Waste treatment standards
• Control requirements consistent and complimentary to the extent possible.
• Compliance must be demonstrated with all applicable rules.

Fig. 20. Overlap of statutory coverage for air emission sources. (From US EPA)

ping process to remove the organic contaminants, which would result in a potential cross-media problem if air emissions created by this process were not controlled. The cleanup process (removal and remediation) is also required to comply with applicable rules and requirements (ARARs). Additionally, for some cleanup operations under the Superfund, Phase I RCRA air standards may be ARARs.

Another important tool used to address air toxics is toxic release inventory, required by SARA Title 313. Generally, this inventory assists in improving US EPA's knowledge of the sources of air toxics. Recently, the US EPA reviewed this inventory to identify sources of the 190 toxic pollutants listed under the CAA of 1990.

Figure 20 shows the overlap of statutory coverage of air emission sources with various laws. This overlap may in some instances result in the same source being subject to regulations with different control requirements. This conflict is the result of regulations being developed under laws with different mandates. For example, CAA requires technology-based standards for NSPS, whereas RCRA 3004(n) air standards are risk based. Compliance under this circumstance must be demonstrated with all applicable rules. The US EPA will try to make consistent and complementary control requirements of rules that apply to the same sources. This overlap is illustrated in Fig. 20 for the coverage of air emissions from storage tanks in waste management. For example, three separate rules may cover storage tanks containing benzene waste located at chemical plants, petroleum refineries, coke byproduct plants, and certain TSDFs. First, NESHAP regulations, as described in 40 CFR 61 Part FF, would apply. Second, NSPS regulations, as described in 40 CFR Part 60, Subpart Kb, would apply to new, modified, or reconstructed tanks containing volatile organic liquids (VOLs) and tanks above a certain size limits. Third, Phase II air standards, as described in RCRA, would apply to tanks in which organic hazardous waste is managed. Therefore, depending on the particular physical characteristics of the tank, these standards could be covered by the benzene waste NESHAP, the VOL storage NSPS, and RCRA Phase II air standards. This overlap would have minimal ramifications on owner/operators because control requirements would be the same for all three. Readers are referred to the literature for the additional discussions on the US Clean Air Act compliance (47,48,56,66).

10. CONTROL

In the subsequent chapters of this volume, the ways in which emissions diffuse and become diluted in the atmosphere and methods for controlling air pollution emissions are discussed. Once the basic diffusion mechanisms and their theory of control are mastered, it is important to understand how to implement this information effectively in real situations (23,24). When determining the control option to meet a specific regulation, there will always be several available alternatives. One must consider factors such as adverse environmental impact, economics, and effect on the process (27,33,34,63,64). The following discussion illustrates how these factors may influence the choice of control devices.

One must remember that the improper control of air pollutants can result in other environmental problems. For example, the byproduct discharge from a control system can create odors and other varieties of air pollution, water pollution, or solid-waste disposal problems. As a result, it may be necessary to provide auxiliary facilities to dewater or even completely dry the slurries, deodorize wastes, and cover dry discharges to prevent the escape of fugitive dust. Incorporation of air pollution control systems is often a convenient method of "closing the loop" in a process, and recycling the by-product of the control system should be examined.

Economic considerations are linked closely to the end use of the byproduct. When considering recycling of the byproduct, one needs to examine the potential market. It has been shown that fly ash, for example, can be used in construction material (lightweight strong building blocks, concrete, and asphalt) and the demand in some markets has exceeded the supply.

The economical analysis of control system must include both capital and operational costs. Significant factors in this economical analysis include the cost of money (interest rate), the age of the existing process facilities, and/or the expected life of the processing system with pollution controls. Additionally, both the capital and operational costs must be examined. The operating cost for a control system is often related to the cost of purchase and installations of the control equipment. Often, control equipment with a high-cost capital has a low operational cost. Conversely, it is also true that control equipment with a lower capital cost has a high operational cost. The gas pumping system (blowers, etc.) is the single largest energy-related operating expense for a control system. Therefore, a control system that requires a high-pressure gas pumping system will have high energy demands that result in high operating expenses.

Another economical consideration is the actual operation and maintenance procedures, which can influence the operating costs significantly. Control equipment requires frequent and periodical preventive maintenance care and inspection to ensure the useful life of control equipment. This work will be performed by either the operator or the maintenance personnel. For example, blowers, pumps, and other parts in a control unit require routine lubrication, adjustment of belts and seals, and inspection. Periodically, a control unit requires complete inspection of the entire unit. Additionally, some control units may also require a complete shutdown in order to purge the system so that it can be entered for inspection.

To maintain performance and assist in the maintenance of the equipment, provisions should be made for obtaining sample measurements, including the building in of sample and velocity ports and pressure taps into equipment. Routine measurement of pressure drop across the control unit, pressure in the system, and gas and liquid flow rates are minimal requirements for ensuring proper operation of control equipment. It is important during the startup of the control that baseline information on the control equipment be taken and recorded. This information should include measuring and recording the outlet and inlet static pressures at the blower and current draw from the blower motor. These startup measurements are compared to the regularly made pressure drops and current draws to help in troubleshooting.

To handle the maximum process emission rate without inducing adverse pressure (negative or positive) on the process, the control system must be sized properly. Important considerations in sizing control equipment are temperature fluctuations and humidity changes. Changes in the temperature and humidity of the emission can significantly affect the volume of gas required for treatment by a control unit. Section 6 presented the calculation procedures for estimating the effects of these changes. Quantitative gas stream calculations were presented in Section 7.

11. CONCLUSIONS

The US President's Council on Environmental Quality (CEQ) annual reports since 1973 have shown a positive investment return on control equipment. These reports show that an investment return of nearly 15% from the purchase and installation of control equipment can be obtained in the United States. Even though there are numerous ways that data can be interpreted to estimate investment return, these annual reports provide positive reinforcement for providing control equipment. We are moving in the right direction.

In order to make sound decisions on the selection of air pollution control equipment, it is necessary to have solid data on air pollutants. Such data can be obtained by diligent, careful work, knowledge of how the system behaves, and following proper test procedures. The cost for collecting and reporting data sometimes seems excessive, but it is necessary to obtain the starting point for selecting and designing adequate control equipment. Some control systems have demonstrated reliable control for essentially every pollution control problem. On the other hand, problems with control equipment can be associated with improper design, installation, operation, and maintenance, which result in excessive costs and poor performance. Optimization of pollution control technology, by expanding research and development and providing adequate training, can help to reduce some of the problems with control equipment. The latest information on the air quality, environmental laws, control equipment, process systems, monitoring technologies, and so forth can be obtained periodically from the Internet (57,65–66).

12. EXAMPLES

12.1. Example 1

Determine the heat content of an emission stream (gas stream 1) from a paper-coating operation. Gas stream 1 has the following components: 100 ppmv of methane and 960 ppmv of toluene. Let subscripts 1 and 2 denote the methane and toluene components of gas stream 1, respectively.

Solution

Equation (38) becomes

$$h_{e1} = (0.01)[(y_{e1,1})(h_{e1,1}) + (y_{e1,2})(h_{e1,2})] \qquad (38)$$

The ppmv for each gas is converted to % volume as follows:

Methane

$$y_{e1,1} = (100 \text{ ppmv})\ (\%/10{,}000 \text{ ppmv}) = 0.0100\%$$

Toluene

$$y_{e1,2} = (960 \text{ ppmv})\ (\%/10{,}000 \text{ ppmv}) = 0.0960\%$$

The heat content of each component is obtained from Table 6:

Methane: $h_{e1,1} = 882$ Btu/scf
Toluene: $h_{e1,2} = 4196$ Btu/scf

These values are substituted into Eq. (38) for heat content and yields

$$h_{e1} = 4.11 \text{ Btu/scf or } (4.11 \text{ Btu/scf})/(0.0739 \text{ lb/ft}^3) = 55.6 \text{ Btu/lb}$$

12.2. *Example 2*

Outline the step-by-step procedures for calculation of air dilution requirements.

Solution

Step 1: Determine the dilution airflow rate from Eq. (40) to decrease the heat content of the emission stream from h_e to h_d as follows:

$$Q_d = Q_e \, [(h_e/h_d) - 1] \tag{40}$$

$$Q_d = \underline{\hspace{3cm}} \text{ scfm}$$

Step 2: Determine the concentrations of the various components in the diluted emission stream as follows:

$$O_{2,d} = O_2\left(h_d/h_e\right) + 21\left[1 - \left(h_d/h_e\right)\right] \tag{41}$$

$$O_{2,d} = \underline{\hspace{3cm}} \ \%$$

$$M_{e,d} = M_e \, (h_d/h_e) + 2[1 - (h_d/h_e)\,] \tag{42}$$

$$M_{e,d} = \underline{\hspace{3cm}} \ \%$$

$$Q_{e,d} = Q_e \, (h_e/h_d) \tag{43}$$

$$Q_{e,d} = \underline{\hspace{3cm}} \text{ scfm}$$

Step 3: Redesignate emission stream characteristics as follows:

$$O_2 = O_{2,d} = \underline{\hspace{3cm}} \ \%$$

$$M_e = M_{e,d} = \underline{\hspace{3cm}} \%$$

$$h_e = h_d = \underline{\hspace{3cm}} \text{Btu/scf}$$

$$Q_e = Q_{e,d} = \underline{\hspace{3cm}} \text{ scfm}$$

NOMENCLATURE

atm	Atmosphere
Btu	British thermal units
C	Cunningham correction factor
CFC	Chlorofluorocarbon
CAA	Clean Air Act
CERLA	Comprehensive Environmental Response, Compensation, and Liability Act
CEQ	Council on Environmental Quality
DA_e	Dry air content of combined gas stream (lb/min)
g	Gravitational acceleration (980 cm/s^2)
H	Humidity, mass water/mass of air
H_e	Combined emission stream heat content (Btu/scf)

H_s	Total sensible heat for the combined emission stream (Btu/min)
H_{s1}	Sensible heat content of gas stream 1 (Btu/min)
H_{s2}	Sensible heat content of gas stream 2 (Btu/min)
h_d	Emission stream heat content after dilution (Btu/scf)
h_e	Emission stream heat content before dilution (Btu/scf)
h_{e1}	Heat content in gas stream 1 (Btu/scf)
$h_{e,j}$	Heat content of gas stream *j* in the combined gas stream (Btu/scf)
$h_{e1,i}$	Heat of combustion of component *i* in gas stream 1 (Btu/scf)
HAP	Hazardous air pollutant
K	Shape factor (dimensionless)
l	Mean free path of air (cm)
LDR	Land disposal restrictions
M_e	Percent of moisture content of the combined gas stream (vol%)
$M_{e,d}$	Percent of moisture content of diluted emission (vol%)
$M_{e,\mathrm{m}}$	Moisture content of the combined gas stream (lb/min)
M_{e1}	Percent of moisture content of gas stream 1 (vol%)
$M_{e1,\mathrm{l\text{-}m}}$	Moisture content of gas stream 1 (lb-mol/min)
n	Number
NAAQ	National ambient air quality
NESHAP	National emission standard for hazardeous air pollutant
OAQPS	Office of Air Quality Planning and Standards
$O_{2,d}$	Percent volume of oxygen content of diluted emission stream
P	Absolute pressure
PAN	Peroxy acetyl nitrate
ppm	Part per million
ppmv	Part per million volume
PM	Particulate matter
R	Gas constant
SARA	Superfund Amendments and Reauthorization Act
S_e	SO_3 content of combined gas stream (ppmv)
S_{e1}	SO_3 content of gas stream 1 (ppmv)
$S_{e1,\mathrm{l\text{-}m}}$	SO_3 content of gas stream 1 (lb-mol/min)
SVOC	Semivolatile organic compound
T	Absolute temperature (K)
T_e	Temperature of the combined gas stream (°F)
T_{e1}	Temperature of gas stream 1 (°F)
TRS	Total reduced sulfur
TSDF	Treatment, storage, and disposal facility
Q_d	Dilution air flow rate (scfm)
Q_e	Flow rate of the combined stream before dilution (scfm)
$Q_{e,a}$	Flow rate of combined stream at actual conditions (acfm)
$Q_{e,d}$	Flow rate of diluted emission stream (scfm)
Q_{e1}	Flow rate of gas stream 1 at standard conditions (scfm)
Q_{e2}	Flow rate of gas stream 2 at standard conditions (scfm)
Q_{e3}	Flow rate of gas stream 3 at standard conditions (scfm)

$Q_{e1,a}$	Flow rate of gas stream 1 at actual conditions (acfm)
V	Volume (ft^3)
VOC	Volatile organic compound
v_s	Terminal settling velocity (cm/s)
W	Water content (vol % of moisture)
$W_{e1,g}$	Particulate loading for gas stream 1 (g/acf)
$W_{e1,1}$	Particulate loading for gas stream 1 (lb/h)
$W_{e,g}$	Particulate loading for the combined gas stream (g/acf)
$W_{e,1}$	Particulate loading for the combined gas stream (lb/h)
$y_{e,j}$	Percent volume of component *j* in the combined gas stream (vol %)
$y_{e1,i}$	Percent volume of component *i* in gas stream 1 (vol %)
μm	Micrometer

REFERENCES

1. N. A. Fuchs, *The Mechanics of Aerosols,* Pergamon, New York, 1964.
2. H. E. Hesketh, *Aerosol capture efficiency in scrubbers,* 68th Annual APCA Meeting, 1975, paper 75–50.6.
3. D. A. Lundgren, and H. J. Paulus, *The mass distribution of large atmospheric particles,* 66th Annual APCA Meeting, 1973, paper 73–163.
4. S. Kadowaki, *Atmos. Environ.* **10(1)**, 39 (1976).
5. S. J. Mainwaring, and S. Harsha, *Atmos. Environ.* **10(1)**, 57 (1976).
6. Code of Federal Regulations. *EPA Regulations on Standards of Performance for New Stationary Sources. Federal. Register,* 40 CFR Parts 60, p. 24,876, December 23, 1971.
7. US EPA, *1972 National Emission Report,* EPA-450/2-74-012, US Environmental Protection Agency, Washington, DC, 1974.
8. US EPA, *Compilation of Air Pollutant Emission Factors,* 2nd ed. with Supplement No. 5, US EPA Publication No. AP-42, US Environmental Protection Agency, Washington, DC, 1975.
9. US EPA, *EPA Regulations on National Primary and Secondary Ambient Air Quality Standards, Fed. Register*, 40 CFR Parts 50, p.22,384, US Environmental Protection Agency, Washington, DC, November, 1971.
10. US EPA, *Ambient Air Monitoring Equivalent and Reference Methods,* Fed. Register, Vol. 28, No. 187, 40 CFR Parts 50–53, pp. 28,438–28,448, US Environmental Protection Agency, Washington, DC, October 12, 1973.
11. V. M. Yamada, and R. J. Carlson, *Environ. Sci. Technol.* **3,** 483–484 (1969).
12. H. E. Hesketh, in *Handbook of Environmental Engineering* (L. K. Wang, and N. C. Pereira, eds.), Humana Press, Totowa, NJ, 1979, Vol. 1, pp. 3–39.
13. S. Calvert, J. Goldshmid, and D. Leith, *Scrubber Handbook,* APT, Inc., Riverside, CA, 1972.
14. McIlvaine Co., *The Fabric Filter, The Scrubber, and the Electrostatic Precipitator Manuals* McIlvaine Co., Northbrook, IL, 1976.
15. Bureau of Mines, *Ringelmann Smoke Chart,* Bureau of Mines Information Circular 8333, 1967.
16. ASME, *Determining dust concentrations in a gas stream,* ASME PTC 21, 1941.
17. M. J. Pilat, D. S. Ensor, and J. C. Bosch, *Atmos. Environ.* **4**, 671–679 (1970).
18. US EPA, *Control Technologies for Hazardous Air Pollutants,* EPA/625/6-91/014, US Environmental Protection Agency, Washington, DC, 1991.
19. US EPA, *Control of Air Emissions from Superfund Sites,* EPA/625/R-92/012, US Environmental Protection Agency, Washington, DC, 1992.
20. US EPA, *Organic Air Emissions from Waste Management Facilities,* EPA/625/R-92/003, US Environmental Protection Agency,Washington, DC 1992

21. US EPA, *Control Techniques for Fugitive VOC Emissions from Chemical Process Facilities*, EPA/625/R-93/005, US Environmental Protection Agency, Cincinnati, OH, 1994.
22. US EPA, *Hazardous Waste TSDF—Fugitive Particulate Matter Air Emissions Guidance*, Document. EPA/450/3-89-019, US Environmental Protection Agency,Washington DC, 1989.
23. M. H. S. Wang, L. K. Wang, and T. Simmons, *J. Environ. Manage.* **9**, 61–87 (1979).
24. L. K. Wang, M. H. S. Wang, and J. Bergenthal, *J. Environ. Manage.* **12**, 247–270 (1981).
25. L. Nagghappan, *Leather tanning effluent treatment,* Master thesis, Lenox Institute of Water Technology, Lenox, MA (2000), p. 167.
26. J.A. Ohrt, *Physicochemical pretreatment of a synthetic industrial dairy waste,* Master thesis, Lenox Institute of Water Technology, Lenox, MA, 2001, p. 62 (L. K. Wang, advisor).
27. L. K. Wang, M. H. S. Wang, and P. Wang, *Management of Hazardous Substances at Industrial Site,* UNIDO Registry No. DTT/4/4/95, United Nations Industrial Development Organization, Vienna, Austria (1995), p. 105.
28. US EPA, *Managing asbestos in place—a building owner's guide to operations and maintenance programs for asbestos-containing materials,* Report No. TS-799, US Environmental Protection Agency, Washington, DC, 1990.
29. US GPO, *Occupational Safety and Health Guidance Manual for Hazardous Waste Site Activities,* Publication No. DHHS-NIOSH, US Government Printing Office, Washington, DC, 1985.
30. US GPO, *Asbestos in the Home,* US Government Printing Office, Washington, DC, 1989.
31. L. K. Wang, and J. Zepka, *An investigation of asbestos content in air for Eagleton School,* Report No. PB86-194172/AS US Department Commerce, National Technical Information Service, Springfield, VA, 1984, p. 17.
32. L. K. Wang, J. V. Krouzek, and U. Kounitson, *Case Studies of Cleaner Production and Site Remediation.* Training Manual No. DTT-5-4-95, United Nations Industrial Development Organization, Vienna, Austria, 1995.
33. L. K. Wang, and S. L. Lee, *Utilization and reduction of carbon dioxide emissions: an industrial ecology approach,* 2001 Annual Conference of Chinese American Academic and Professional Society (CAAPS), 2001.
34. J. H. Ausubel, *Resources*, **130(14)** (1998).
35. P. Barnes, *Environ. Protect.* **12(6)**, 22–28 (2001).
36. J. Wilson, *Water Condit. Purif.* 102–104 (2001).
37. World Meteorological Organization, *Scientific Assessment of Ozone Depletion,* United Nations Environment Programme, Nairobi, Kenya, 1994.
38. UNEP, *Action on Ozone,* United Nations Environment Programme, Nairobi, Kenya, 1993.
39. A. Neville, *Environ. Protect.* **13(5)**, 6 (2002).
40. C. Cook, and D. McDaniel, *Environ. Protect.* **13(5)**, 24 (2002).
41. K. Abbott, and H. Alper, *Environ. Protect.* **13(6)**, 48 (2002).
42. H. Goraidi, *Environ. Protect.* **13(9)**, 18 (2002).
43. L. G Garner, *Environ. Protect.* **13(9)**, 37 (2002).
44. M. R. Harris, *Environ. Protect.* **13(1)**, 32 (2002).
45. A. Neville, *Environ. Protect.* **13(6)**, 6 (2002).
46. J. R. Menkedick, *Environ. Protect.* **13(6)**, 30 (2002).
47. B. S. Forcade, *Environ. Protect.* **13(1)**, 16 (2002).
48. P. Zaborowsky, *Environ. Protect.* **13(4)**, 26 (2002).
49. L. Rademakers, *Environ. Protect.* **13(3),** 20 (2002).
50. J. P. Zeigler, *Environ. Protect.* **13(5),** 19 (2002).
51. J. F. Guida, *Environ. Protect.* **13(9),** 25 (2002).
52. M. D. Abkowitz, *Environ. Protect.* **13(11),** 44 (2002).
53. G. Gray, *Environ. Protect.* **13(11),** 40 (2002).
54. USDA, US Dept of Agriculture, Washington, DC website, www.usda.gov, 2003.
55. NRDC, Natural Resources Defense Council website www.nrdc.gov, 2003.

56. B. S. Forcade, *Environ. Protect.* **14(1),** 22–28 (2003).
57. J. C. Bolstridge, *Chem. Eng. Prog.* **99(1)** 50–56 (2003).
58. J. Mazurek, *USA Today,* pp.B1 and B2, Jan. 30 (2003).
59. L. K. Wang, N. C. Pereira, and Y. T. Hung (eds.), *Advanced Air and Noise Pollution Control,* Humana Press, Totowa, NJ (2004).
60. H. Roenfeld, *Times Union,* p.B5, Feb. 5 (2003).
61. Anon., *Chem. Eng. Prog.* **98(2),** 21–22 (2002).
62. A. Ellis, *Environ. Protect.* **14(1),** 10 (2003).
63. K. Wark, and C. F. Warner, *Air Pollution, Its Origin and Control,* Donnelley, New York, 1976.
64. A. Weiser, *Polluti. Eng.* **9(2),** 27–30 (1977).
65. US EPA. EPA/Daimler Chrysler/UPS Fuel Cell Delivery Vehicle Initiative. www.epa.gov. US Environmental Protection Agency, Washington, DC, 2004.
66. US EPA. Air Quality Planning and Standards, www.epa.gov. US Environmental Protection Agency, Washington, DC, 2004.

2
Fabric Filtration

Lawrence K. Wang, Clint Williford, and Wei-Yin Chen

Contents

1. INTRODUCTION

Fabric filtration is a physical separation process in which a gas or liquid containing solids passes through a porous fabric medium, which retains the solids. This process may operate in a batch or semicontinuous mode, with periodic removal of the retained solids from the filter medium. Filtration systems may also be designed to operate in a continuous manner. As with other filtration techniques, an accumulating solid cake performs the bulk of the filtration. Importantly, an initial layer of filter cake must form at the beginning of the filtration operation (1,2).

Fabric filtration effectively controls environmental pollutants in gaseous or liquid streams. In air pollution control systems, it removes dry particles from gaseous emissions; in water pollution control, filtration removes suspended solids; in solid-waste disposal, filtration concentrates solids, reducing the landfill area required. Often, filtration processes simultaneously reduce air, water, and solid-waste disposal problems. An air pollution control system might, for example, remove particles and/or gases from an emission source and might consist of a scrubbing device that removes particulates by impaction and the gases by chemical absorption. The reaction products of gases and chemicals can produce a crystalline sludge. A fabric filter may also be used to remove solids from water so that the water can be recycled. As a result, effluent slurry does not present a water pollution problem. Effective use (optimization) of a fabric-filter system would minimize problems with waste disposal.

From: *Handbook of Environmental Engineering, Volume 1: Air Pollution Control Engineering*
Edited by: L. K. Wang, N. C. Pereira, and Y.-T. Hung © The Humana Press, Inc., Totowa, NJ

Although fabric filtration is suitable for removing solids from both gases and liquids, it is often important that the filter remain dry when gases are filtered, and likewise, it may be desirable to prevent the filter from drying out when liquids are filtered. In the gas system, many solids are deliquescent, and if moisture is present, these materials will have a tendency to pick up moisture and dissolve slightly, causing a bridging or blinding of the filter cloth. The result is a "mudded" filter fabric. In such cases, it is often impossible to remove this material from the cloth without washing or scraping the filter. If the cake on the cloth is allowed to dry during liquid filtration, a reduction in the porosity of the cake as well as a partial blinding of the filter could result, which could then reduce the rate of subsequent filtration.

2. PRINCIPLE AND THEORY

In section 1, it was stated that the fabric itself provides the support, and true filtering usually occurs through the retained solid cake that builds up on the fabric. This is especially true for woven fabrics; however, felts themselves actually can be considered as the filtering media. It has also been stated that the cake must be removed periodically for continued operation. The resistance to fluid flow through the fabric therefore consists of cloth resistance and cake resistance and is measured as a pressure drop across the filter. Cleaned cloth resistance is often reported, although this in itself is not the new or completely clean cloth resistance. Once the filter has been used and cleaned a few times, a constant minimum resistance is achieved, which consists of the clean cloth resistance and the residual resistance resulting from deposited material that remains trapped in the cloth pores. This resistance may remain constant for the life of the fabric. Changes in this resistance usually indicate either plugging of the pores or breaking of the filter. Clean cloth resistances may be obtained from suppliers. However, it is best to obtain the steady-state values by empirical measurements. An example of clean cloth resistance, expressed according to the American Standards of Testing and Materials (ASTM) permeability tests for air, ranges from 10 to 110 $ft^3/min\text{-}ft^2$ (3–33.5 $m^3/min\text{-}m^2$) with a pressure differential of 0.5 in. (1.27 cm) H_2O. In general, at low velocities, the gas flow through the fabric filter is viscous, and the pressure drop across the filter is directly proportional to flow:

$$\Delta P_1 = K_1 v \tag{1}$$

where ΔP_1 is the pressure drop across fabric (inches of water [cm H_2O]), K_1 is the resistance of the fabric [in. H_2O/ft/min (cm H_2O/m/min)], and v is gas flow velocity [ft/min (m/min)].

In practice, the fabric resistance K_1 is usually determined empirically. It is possible to estimate a theoretical value of this resistance coefficient from the properties of cloth media. Darcy's law states that

$$\Delta P_1 = -(vK/\mu) + \rho g \tag{2}$$

where K is the Kozeny permeability coefficient, μ is viscosity, ρ is density, and g is gravitational acceleration. Note that necessary constants need to be applied to make the equation dimensionally consistent. Values of the permeability coefficient K found in literature range between 10^{-14} and 10^{-6} ft^2 (10^{-15} and 10^{-8} m^2). Values of K may also be estimated using the relation

$$K = \varepsilon^3/cS^2 \tag{3}$$

where ε is porosity or fraction void volume (dimensionless), c is a flow constant, K is the Kozeny coefficients, and S is the specific surface area per unit volume of porous media [ft^{-1} (m^{-1})]. Values of the Kozeny constant can be estimated using the free-surface model (2). Assuming a random orientation averaging two cross-flow fibers and one parallel fiber and assuming that a cloth medium behaves like a bed of randomly oriented cylinders, the constant for flow parallel to the cylinder is obtained by

$$c = 2\varepsilon^3 \Big/ \left\{(1-\varepsilon)\left[2\ln\frac{1}{1-\varepsilon} - 3 + 4(1-\varepsilon) - (1-\varepsilon)^2\right]\right\} \tag{4}$$

and when flow is at right angles to the cylinder,

$$c = 2\varepsilon^3 \Big/ \left\{(1-\varepsilon)\left[\ln\left(\frac{1}{1-\varepsilon}\right) - \frac{1-(1-\varepsilon)^2}{1+(1-\varepsilon)^2}\right]\right\} \tag{5}$$

As the system is operated, cake deposits on the fabric, producing an additional flow resistance proportional to the properties of the granular cake layer. The resistance to fluid flow owing to cake build-up usually amounts to a significant portion of the total flow resistance. This resistance increases with time as the cake thickness increases. This additional resistance (ΔP_2) is typically of the same order of magnitude as the residual resistance (ΔP_1) and can be expressed as

$$\Delta P_2 = K_2 v^2 L t \tag{6}$$

where ΔP_2 is the change in pressure drop over time interval t [in. H_2O (cm H_2O)], K_2 is the cake-fabric filter resistance coefficient,

$$\left[\frac{\text{in. of water}}{(\text{lb}_\text{m}\text{dust}/\text{ft}^2)(\text{ft}/\text{min})}\right] \text{or} \left[\frac{\text{cm of water}}{(\text{kg dust}/\text{m}^2)(\text{m}/\text{min})}\right]$$

v is fluid velocity [ft/min (m/min)], L is inlet solids concentration [lb/ft^3 (kg/m^3)], and t is time (min). An expression for the cake–fabric filter resistance coefficient using the Kozeny–Carman procedure has been derived for determining flow through granular media (2):

$$K_2 = \left(3.2\times10^{-3}\right)\left(\frac{k}{g}\right)\left(\frac{\mu_f S^2}{\rho_p}\right)\left(\frac{1-\varepsilon}{\varepsilon^3}\right) \tag{7}$$

where k is the Kozeny–Carman coefficient, which equals approx 5 for a wide variety of fibrous and granular materials up to a porosity equal to about 0.8, ε is the porosity or fraction void volume in cake layer (dimensionless), μ_f is fluid viscosity [lb_m/(s ft)], ρ_P is the true density of solid material (lb_m/ft^3), and the S is the specific surface area/unit volume of solids in the cake layer (ft^{-1}). This equation shows that as the particles being filtered become smaller in diameter, the porosity of the cake decreases and consequently, K_2 increases. The net result of the larger cake–fabric filter resistance coefficient (K_2) is that the pressure drop increases as porosity decreases.

Table 1
Dust-Fabric Resistance Coefficients for Certain Industrial Dusts on Cloth-Type Air Filters

	K_2 (in. water per lb of dust per ft^2 per ft per m in of filtering velocity)[a] for particle size less than the following						
	Coarse		Medium[b]			Fine[b]	
Dust	~800 μm	~100 μm	~44 μm	<90 μm	<45 μm	<20 μm	<2 μm
Granite	1.58	2.20				19.80	
Foundry	0.62	1.58	3.78				
Gypsum			6.30			18.90	
Feldspar			6.30			27.30	
Stone	0.96			6.30			
Lamp black							47.20
Zinc oxide							15.70[c]
Wood				6.30		25.20	
Resin (cold)	0.62		11.00				
Oats	1.58		9.60	8.80			
Corn	0.62		1.58	3.78			

[a] $\frac{\text{in. water}}{(\text{lb/ft}^2)(\text{ft/min})} = \frac{1.75\ \text{cm water}}{(\text{kg/m}^2)(\text{m/min})}$

[b]Theoretical size of silica, no correction made for materials having other densities.

[c]Flocculated material, not dispersed; size actually larger.

Source: ref. 2.

The value of the dust–fabric filter resistance coefficient is necessary to predict the operating pressure drop in new fabric-filter installations. This information, with filter velocity and time between cleaning cycles, then may be used to estimate optimum operational procedures, which affect both installation and operating expenses. Some typical dust–fabric resistance coefficients for air–dust filter systems are given (2) in Table 1. The resistance coefficients calculated by Eq. (7) do not always agree with the values obtained from operating systems using Eq. (6). Some engineering data (2–4) are summarized in Table 2 for several particle sizes ranging from 0.1 to 100 μm for solids with a density of 2 g/cm^3. The specific area is estimated assuming spherical particles and standard conditions (SC) of 70°F (21.1°C) and 1 atm pressure. These data are taken from industrial cloth-type air filters.

The above equations and tables show that the various parameters of pressure drop, velocity inlet loading, and time are closely coupled with the physical properties of both the fluid and the solids being filtered. The value of K_2 also depends on the size distribution of the particles, which is often neglected when estimating porosity. Particles usually exhibit a log-normal (geometric) probability distribution. Two materials with the same mass mean size could be quite different in size distribution (geometric deviation), which would affect the porosity of the cake. The shape of the particles, which is not accounted for in the theoretical equations, is also significant and influences both cake porosity and fluid flow drag.

Table 2
Comparison of Calculated and Observed Dust-Fabric-Filter Resistance Coefficients (K_2) (4)

				Resistance coefficient (K_2), in. $H_2O/(lb/ft^2)(ft/min)$	
Particle size (μm)	S (ft^{-1})	Porosity ε	$\frac{1-\varepsilon}{\varepsilon^3}$	Calculated using Eqs. (6) and (7)	Observed
0.1	1.83×10^7	0.25	48.0	41,200	715
1	1.83×10^6	0.40	9.38	705	180
10	1.83×10^5	0.55	2.70	2.32	12
100	1.83×10^4	0.70	0.878	7.56×10^{-3}	0.2

When no data are available, it has been shown that it is possible to estimate values of the resistance coefficient; however, it is more desirable to obtain the coefficient by actual measurements [operating data and Eq. (6)] when this is possible. Once the coefficient is known, any one of the parameters in Eq. (6) can be determined by specifying the remaining variables.

Empirically derived values for the resistance coefficient also may differ for similar systems under different operating conditions. For example, if the cake is composed of hard, granular particles, it will be rigid and essentially incompressible. As the filtration process continues, there is no deformation of the particles and the porosity remains constant. On the other hand, if the cake is extremely soft, it can be deformed, resulting in a different effective porosity as filtration continues. The amount of cake buildup, which is a function of gas velocity, inlet solids concentration, and time, must be considered when attempting to obtain a meaningful value of K_2 for similar systems.

An equally perplexing problem is the fact that there is no standardized filtration rating test procedure. Ratings such as "nominal," "absolute," and "mean flow pore" serve largely to describe filter systems, but they do not provide a rational basis for filtration engineering and analysis.

Fabric filters consist of a porous filtration medium, in which the pores are not all uniform in size. Therefore, attempts are made in the rating procedures to take this into consideration; for example, the mean flow pore system exerts air pressure to one side of a porous filter, and the pressure is noted at which the first bubble appears on the wetted medium. This is called the bubble point and corresponds to the largest pore in the filter. The distribution of pores in the medium would be expected to be log-normal and obtaining the pressure corresponding to the smallest pore is quite a different story. Recently, Cole (5) suggested a "summation of flow" rating, in which an attempt is made to define the pore size at which about 16% of the flow goes through larger pores.

A common laboratory technique for obtaining empirical data for liquid fabric filters is to use a device called a filter leaf. In the test procedure, the filter fabric is secured over a backup screen and inserted in the test system. Unfortunately, this procedure is not standardized, although Purchas (6) has proposed a standardized test procedure for liquid filtration tests. This procedure consists of obtaining a 1-cm-thick cake when utilizing a pressure differential of 1 atm. The result for a given fabric–solid combination would be a "standard cake formation time expressed in minutes." In gas filtration tests, the most common method for expressing new fabric resistance is to measure the gas volumetric flow rate at a 0.5-in. (1.27 cm) H_2O pressure drop.

3. APPLICATION

3.1. General

The use of fabrics as a porous filter medium in both liquid and gas cleaning systems has been stated, and the separation of solids from liquids will be discussed in detail in other chapters of this handbook series. The major emphasis of this section is on gas cleaning, and, in most applications, the gas considered is air.

3.2. Gas Cleaning

Filters used to clean gases are categorized in this section in five different ways according to the energy required, the fabric employed, the type of cycle, the service, and the application. The first category includes either *high-energy* or *low-energy* filters, depending on whether the filters are operated at high or low filter pressure drops. For any given application involving filters, a high-energy system is usually more efficient, but, ultimately, this depends on the size, size distribution, and type of material being filtered. Energy and efficiency are not always directly related and will be discussed below.

High-energy systems generally consist of pulse-jet devices, whereas low-energy cleaning systems utilize shaking and reverse flow. Note that this classification also describes the cleaning method used to remove dust from the bags. In the pulse-jet systems, blasts of air are blown through jet nozzles in pulses to free the dust from the fabric, as shown in Fig. 1. Note that the cleaning jet is introduced into the Venturi nozzle to expand and clean the bag.

The low-energy systems are split approx 50–50 between continuous and intermittent-type collectors. Shaking, as the word states, simply implies mechanically flexing the bag to clean it. Reverse-flow applications consist of introducing air into *sections* of the filter system in the opposite direction from normal gas flow to blow the dust off the bags. There is a third category, in which no cleaning energy is utilized. This applies to units designed for situations in which the media are disposable.

Fabric filters can be divided generally into two basic types, depending upon the fabric: felt (unwoven) and woven. Felt media are normally used in high-energy cleaning systems; woven media are used in low-energy devices. Felt fabrics are tighter in construction (i.e., less porous), and for this reason, they can be considered to be more of a true filter medium and should be kept as clean as possible to perform satisfactorily as a filter. In contrast, the woven fabric is, in general, only a site upon which the true filtering occurs as the dust layer builds up, through which the actual filtering takes place. In addition, a third type of fabric filter is nonwoven disposable configuration material, which is used as a vacuum cleaner with disposable bags.

Filter systems can also be categorized as either *continuous* or *intermittent collectors*. In a continuous collector, the cleaning is accomplished by sectionalizing the filter so that, while one part is being cleaned, the rest of the filter is still in operation. Under these conditions, the gas flow through the device and the overall pressure drop across the device are essentially constant with time. In contrast, there must be an interruption

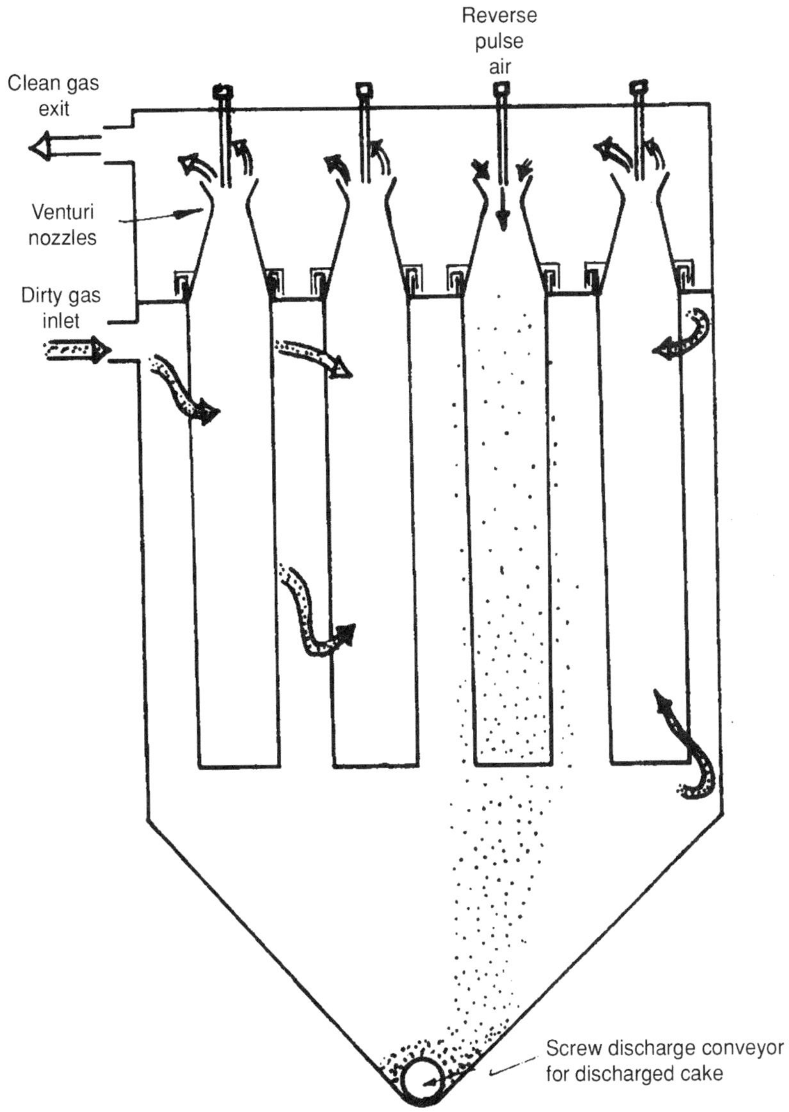

Fig. 1. Schematic diagram of a reverse-pulse baghouse.

in the gas flow while the cleaning process takes place in intermittent collectors. In these systems, gas flow is greatest immediately after the filter medium has been cleaned and decreases as the cake builds up. A typical cycle for an intermittent system is operating for 0.25–4 h and cleaning for 5 min.

A fourth major way in which fabric filters can be classified is by *service*. Particulate removal is the major service performed by fabric filters. However, they also can be used for gaseous control by adsorption and chemical adsorption (chemisorption), which are well-proven industrial techniques. For example, solid alumina can be used to adsorb chlorine; gaseous ammonia can be injected to react with sulfur oxides to form a solid particulate, which can be filtered; sodium and/or calcium compounds can be added as precoats to react with and adsorb sulfur oxides; and activated carbon can be introduced to remove odors.

There is another basic service distinction between process and nonprocess work. Process functions may include the removal of material from air-conveying systems in which product collection is the primary function. A nonprocess application would be the removal of nuisance dust, where only a small amount of the product would actually encounter the filter. However, because of pollution control considerations, the same care and attention should be paid to nonprocess applications that have been given process collectors in the past.

The fifth and final classification of fabric filters is by *application*. These classes include temperature solids concentration type of pollution in the inlet gas moisture content suction, pressure applications size of filter and filter efficiency. The use of glass fiber media makes it possible to operate filters at temperatures up to about 550°F (288°C). A number of different fabric filter media and their characteristics are given in Table 3. Work is currently in progress to develop higher-temperature media, as indicated in the table.

Dust loading is defined as the concentration of solids in the inlet gas stream. Obviously, as dust loading increases, the amount of cake will increase for a given volumetric flow rate of gas. In order to maintain the necessary gas approach velocity and be able to operate an intermittent filter for a reasonable filter cycle time, it may be desirable to reduce the inlet dust loading. One method of doing this is to install mechanical collection devices in front of the fabric filter to remove large-diameter solid material. Gas conditioning, which can consist of introducing air as a diluent, could, in effect, reduce dust loading. However, this process is used more often to reduce inlet temperature and/or humidity.

It is a wise precaution to operate gas cleaning filter systems above the dew point temperature. It has been pointed out that if some dusts become wet, they will bridge and mud (plug) the filter. Methods of keeping the system above the dew point include insulating the filter, heating either the filter and/or the gas, and using warm, dry dilution gas.

Fabric filters can be used in systems that operate at either positive or negative pressures. Some systems are operated at pressures over 200 psi (1.38×10^6 N/m^2), and vacuum systems commonly operate at up to 15 in. (0.38 m) Hg. The most common operating range is ±20 in. (0.508 m) H_2O.

3.3. *Efficiency*

Fabric filters are extremely efficient solids removal devices and operate at nearly 100% efficiency. Efficiency depends on several factors (10,11):

1. Dust properties
   a. Size: particles between 0.1 and 1.0 μm in diameter may be more difficult to capture.
   b. Seepage characteristics: Small, spherical solid particles tend to escape.
   c. Inlet dust concentration: The deposit is likely to seal over sooner at high concentrations.
2. Fabric properties
   a. Surface depth: Shallow surfaces form a sealant dust cake sooner than napped surfaces.
   b. Weave thickness: Fabrics with high permeabilities, when clean, show lower efficiencies. Also, monofilament yarns, without fibrils protruding into the yarn interstices, show lower efficiencies than "fuzzier" staple yarns having similar interstitial spacing.

Table 3
Characteristics of Several Fibers Used in Fabric Filtration

Fiber type[a]	Max operating temp. (°F)	Resistance[b]				
		Abrasion	Mineral acids	Organic acids	Alkalis	Solvent
Cotton[c]	180	VG	P	G	P	E
Wool[d]	200	F/G	VG	VG	P/F	G
Modacrylic[d] (Dynel™)	160	F/G	E	E	E	E
Polypropylene[d]	200	E	E	E	E	G
Nylon polyamide[d] (Nylon 6 and 66)	200	E[e]	F	F	E	E
Acrylic[d] (Orlon™)	260	G	VG	G	F/G	E
Polyester[d] (Dacron[f])	275	VG	G	G	G	E
(Creslan™)	250	VG	G	G	G	E
Nylon aromatic[d] (Nomex™)	375	E	F	G	E	E
Fluorocarbon[d] (Teflon™, TFE)	450	F/G	E[g]	E[g]	E[g]	E[g]
Fiberglass[c]	500	F/G[h]	G	G	G	E
Ceramics[i] (Nextel 312™)	900+	—	—	—	—	—

[a]Fabric limited.
[b]P = poor resistance, F = fair resistance, G = good resistance, VG = very good resistance, E = excellent resistance.
[c]Woven fabrics only.
[d]Woven or felted fabrics.
[e]Considered to surpass all other fibers in abrasion resistance.
[f]Dacron dissolves partially in concentrated H_2SO_4.
[g]The most chemically resistant of all these fibers.
[h]After treatment with a lubricant coating.
[i]The ceramic fiber market is a very recent development. As a result, little information on long-term resistance and acid and alkali performance has been documented.
Source: Data from refs. 7–9.

c. *Electrostatics*: Known to affect efficiency. (Particles, fabrics, and gas can all be influenced electrostatically and proper combination can significantly improve efficiency in both gas and liquid filtering systems.)

3. *Dust cake properties*
   a. Residual weight: The heavier the residual loading, the sooner the filter is apt to seal over.
   b. Residual particle size: The smaller the base particles, the smaller (and fewer) are the particles likely to escape.
4. *Air properties*. Humidity: with some dusts and fabrics, 60% relative humidity is much more effective than 20% relative humidity.
5. *Operational variables*
   a. Velocity: Increased velocity usually gives lower efficiency, but this can be reversed depending on the collection mechanisms, for example, impaction and infusion.

b. Pressure: Probably not a factor except that increase of pressure after part of the dust cake has formed can fracture it and greatly reduce efficiency until the cake reseals.
c. Cleaning: Relatively unstudied but discussed in the following sections.

It is important to stress that all of the considerations discussed thus far can be optimized only when the system is properly operated and maintained. Several of the factors mentioned earlier under operational variables are significant enough to merit further discussion in the following section.

4. ENGINEERING DESIGN

4.1. *Pretreatment of an Emission Stream*

The temperature of the emission stream should remain 50–100°F above the stream dew point. An emission stream too close to its dew point can experience moisture condensation, causing corrosion and bag rupture. Acid gases (e.g., SO_3) exacerbate this problem. Procedures for determining the dew point of an emission stream are provided in Chapter 1. If the emission stream temperature does not fall within the stated range, pretreatment (i.e., emission stream preheating or cooling) is necessary, as discussed in Chapter 1. Pretreatment alters emission stream characteristics, including those essential for baghouse design: emission stream temperature and flow rate. Therefore, after selecting an emission stream temperature, the new stream flow rate must be calculated. The calculation method depends on the type of pretreatment performed and should use appropriate standard industrial equations. Also, emission streams containing appreciable amounts of large particles (20–30 μm) typically undergo pretreatment with a mechanical dust collector. Chapter 1 also describes the use of mechanical dust collectors.

All fabric-filter systems share the same basic features and operate using the principle of aerodynamic capture of particles by fibers. Systems vary, however, in certain key details of construction and in the operating parameters. Successful design of a fabric filter depends on key design variables (7–26).

- Filter bag material
- Fabric cleaning method
- Air-to-cloth ratio
- Baghouse configuration (i.e., forced or induced draft)
- Materials of construction

4.2. *Air-to-Cloth Ratio*

The filtration velocity, or air-to-cloth (A/C) ratio, is defined as the ratio of actual volumetric air flow rate to the net cloth area. This superficial velocity can be expressed in units of feet per minute or as a ratio. A/C ratios of 1:1 to 10:1 are available in standard fabric-filter systems. Low-energy shaker and reverse-flow filters usually operate at A/C ratios of 1:1–3:1, whereas the high-energy reverse-pulse units operate at higher ratios.

Particulate collection on a filter fabric occurs by any or all mechanisms of inertial impaction, interception, and diffusion, as shown in Fig. 2. Inertial impaction occurs for particles above about 1 μm in diameter when the gas stream passes around the filter fiber, but the solid, with its high mass and inertia, collides with and is captured by the filter. Interception occurs when the particle moves with the gas stream around the filter fiber,

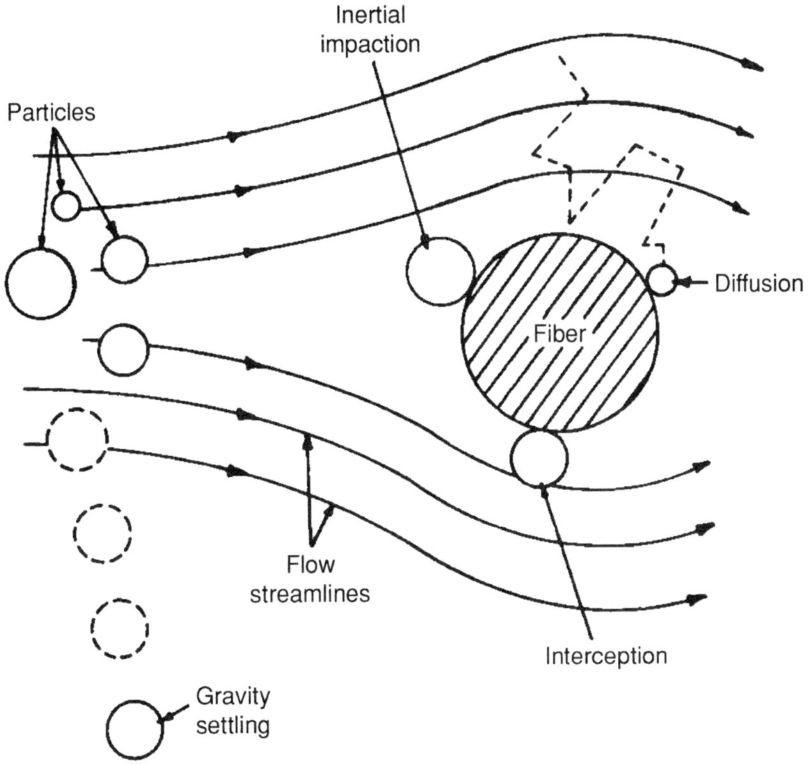

Fig. 2. Mechanisms for particle removal by a filter.

but touches and is captured by the filter. Diffusion consists of random particle motion in which the particles contact with and adhere to the fiber filters. Diffusion increases as particle size decreases and is only significant for submicron-diameter particles.

A high A/C ratio (filtering velocity) promotes particle capture by impaction. On the other hand, an excessive velocity will blow captured material off or through the fabric, in many cases the only support for the cake. This would reduce collection efficiency. As for filtering by diffusion, a higher air-to-cloth ratio reduces the residence time available for particle collection. "Normal" air-to-cloth ratios are about 3:1; "high" air-to-cloth ratios are 6:1 and above.

New filter fabrics having no buildup of solid material will often exhibit a pressure drop of 0.5 in. (1.27 cm) H_2O at normal air-to-cloth ratio ranges. This is called the fabric permeability and is often the same for woven and felted fabrics, although woven bags usually have a weight of 5–10 oz/yd^2 (170–340 g/m^2), and the much heavier and fuzzier felted bags have a weight of 10–20 oz/yd^2 (340–680 g/m^2). A/C ratios are not based on theoretical or empirical relationships, but on installation experience of industry and fabric-filter vendors. Recommended A/C ratios usually depend on a specific dust and a specific cleaning method.

Hand calculations using basic equations give only a general indication of the needed A/C ratio. In practice, tabulated values are frequently provided and are an approximation.

Table 4
Air-to-Cloth Ratios

Dust	Shaker/woven Reverse-air/woven	Pulse jet/felt
Alumina	2.5	8
Asbestos	3.0	10
Bauxite	2.5	8
Carbon black	1.5	5
Coal	2.5	8
Cocoa, chocolate	2.5	12
Clay	2.5	9
Cement	2.0	8
Cosmetics	1.5	10
Enamel frit	2.5	9
Feeds, grain	3.5	14
Feldspar	2.2	9
Fertilizer	3.0	8
Flour	3.0	12
Fly ash	2.5	5
Graphite	2.0	5
Gypsum	2.0	10
Iron ore	3.0	11
Iron oxide	2.5	7
Iron sulfate	2.0	6
Lead oxide	2.0	6
Leather dust	3.5	12
Lime	2.5	10
Limestone	2.7	8
Mica	2.7	9
Paint pigments	2.5	7
Paper	3.5	10
Plastics	2.5	7
Quartz	2.8	9
Rock dust	3.0	9
Sand	2.5	10
Sawdust (wood)	3.5	12
Silica	2.5	7
Slate	3.5	12
Soap detergents	2.0	5
Spices	2.7	10
Starch	3.0	8
Sugar	2.0	7
Talc	2.5	10
Tobacco	3.5	13
Zinc oxide	2.0	5

Note: Generally safe design values — application requires consideration of particle size and grain loading. A/C ratio units are (ft^3/min)/(ft^2 of cloth area).

Source: ref. 8.

Table 5
Factors to Obtain Gross Cloth Area from Net Cloth Area

Net cloth area, A_{nc} (ft^2)	Factor to obtain gross cloth area, A_{tc} (ft^2)
1–4,000	Multiply by 2
4,001–12,000	Multiply by 1.5
12,001–24,000	Multiply by 1.25
24,001–36,000	Multiply by 1.17
36,001–48,000	Multiply by 1.125
48,001–60,000	Multiply by 1.11
60,001–72,000	Multiply by 1.10
72,001–84,000	Multiply by 1.09
84,001–96,000	Multiply by 1.08
96,001–108,000	Multiply by 1.07
108,001–132,000	Multiply by 1.06
132,001–180,000	Multiply by 1.05

Source: ref. 8.

Computer software provides rigorous design. However, the purpose of this section is to provide the reader with some qualitative insight concerning the design and operation of fabric filters. Therefore, these programs are not discussed.

In addition to evaluating a particular fabric filter application, the A/C ratio and the emission stream flow rate ($Q_{e,a}$) are used to calculate net cloth area (A_{nc}):

$$\frac{Q_{e,a}}{\text{A/C ratio}} = A_{nc} \tag{8}$$

where $Q_{e,a}$ is the emission stream flow rate at actual conditions (acfm), A/C ratio is the air-to-cloth ratio, (acfm/ft^2 or ft/min) (from Table 4), and A_{nc} is the net cloth area (ft^2).

The net cloth area is the cloth area in active use at any point in time. Gross or total cloth area (A_{tc}), by comparison, is the total cloth area contained in a fabric filter, including that which is out of service at any point in time for cleaning or maintenance. In this text, costing of the fabric-filter structure and fabric filter bags uses gross cloth area. Table 5 presents factors to obtain gross cloth area from net cloth area:

$$A_{nc} \times \text{Factor} = A_{tc} \tag{9}$$

where Factor is the value from Table 5 (dimensionless) and, A_{tc} is the gross cloth area (ft^2). Fabric filters with higher A/C ratios require fewer bags and less space, and may be less expensive. However, the costs of more expensive (felted) bags, bag framework structure, increased power requirements, etc., may reduce the savings of high-A/C-ratio systems.

4.3. Fabric Cleaning Design

One removes the cake from the fabric by mechanically disturbing the system. This can be done by physically scraping the fabric, mechanically shaking it, or pneumatically or hydraulically reversing the flow of fluid through the fabric to clean the pores. For gas cleaning systems, the common cleaning methods include mechanical shaking, pulse cleaning, and reverse flow.

Fabric shaking combines stress in a normal direction to the dust–fabric interface (tension), stress directed parallel to the interface (shear), and stress developed during the warping, binding, or flexing of the fabric surfaces. Mechanical cleaning studies (10) indicate that dust removal efficiency is a function of the number of shakes, shaking frequency, shaking amplitude, and bag movement acceleration. In general, more dust is removed each time the bag is shaken. However, after about 100 shakes, very little extra dust can be removed, and 200 shakes are recommended as being optimum. At this point, often a maximum of only about 50% of the dust is removed. The shaking frequency is significant in that a resonance frequency can be set up when the fabric is mounted as a bag in a baghouse. More dust is removed at the resonance frequency, but, otherwise, it appears that the higher the frequency, the greater the amount of dust that is removed. In the shaker amplitude range 0–2 in. (0–5.08 cm), dust removal is increased with increased amplitude.

Filter capacity increases with bag shaking acceleration, up to 10 g. Beyond the acceleration range of 1.5–10 g, residual dust holding varies approximately with the inverse square root of the average bag acceleration. Other factors also affect fabric cleaning and filter capacity. These include initial bag tension, amount of cake deposited on the fabric, and cohesive forces binding dust to the fabric. The initial bag tension values should range between 0.5 and 5 lb_f (2–20 N).

Overcleaning requires additional energy and causes undue wear on the bag fabric. However, undercleaning a filter (e.g., by shaking less than the recommended 200 times), decreases system filtration capacity and adversely affects operating costs.

The amount of cleaning by pulsed-jet air varies directly with the rate of rise of the pressure differential across the bag. This should range from 1000 to 4000 in. (2500–10,000 cm) H_2O pressure drop per second. Residual resistance values after cleaning also depend on the dust–fabric combination. Mechanical shaking often augments the reversed-airflow cleaning of bags. This is especially applicable to woven fabric bags. Dust removal in woven bags during reverse flow is usually attributed to bag flexure. Reverse-flow cleaning is, in general, not a satisfactory cleaning technique. In fact, data indicate that in combined shaking–reverse-flow systems, mechanical shaking is responsible for essentially all of the cleaning. The main role played by the reverse air appears to be prevention of projection of dust into the clean air side of the system. Reverse-air cleaning velocities typically range from 4 to 11 ft/min with 0.3–3 ft^3 of gas required per square foot of bag area.

Selection of a cleaning method depends on the type of fabric used, the pollutant collected, and the experiences of manufacturers, vendors, and industry. A poor combination of filter-fabric and cleaning methods can cause premature failure of the fabric, incomplete cleaning, or blinding of the fabric. Blinding of a filter fabric occurs when the fabric pores are blocked and effective cleaning cannot occur. Blinding can result from moisture blocking the pores, increased dust adhesion, or high-velocity gas stream embedding of particles too deeply in the fabric. The selection of cleaning method may be based on cost, especially when more than one method is applicable. Cleaning methods are discussed individually below (13,14), with Table 6 containing a comparison of methods.

A summary of recommended A/C ratios by typical bag cleaning method for many dusts and fumes is found in Table 4. These ranges serve as a guide, but A/C ratios may

Table 6
Comparison of Fabric-Filter-Bag Cleaning Methods

Parameter	Mechanical shake	Reverse airflow	Pulse-jet individual bags	Pulse-jet compartmented bags
Cleaning on-line or off-line	Off-line	Off-line	On-line	Off-line
Cleaning time	High	High	Low	Low
Cleaning uniformity	Average	Good	Average	Good
Bag attrition	Average	Low	Average	Low
Equipment ruggedness	Average	Good	Good	Good
Fabric type[a]	Woven	Woven	Felt/woven[a]	Felt/woven[a]
Filter velocity	Average	Average	High	High
Power cost	Low	Low to medium	High to medium	Medium
Dust loading	Average	Average	Very high	High
Maximum temperature[b]	High	High	Medium	Medium
Collection efficiency	Good	Good	Good[c]	Good[c]

[a]With suitable backing, woven fabrics can perform similarly to felted.
[b]Fabric limited.
[c]For a properly operated system with moderate to low pressures, the collection efficiency may rival other methods.
Source: US EPA.

vary greatly from those reported. Fabric-filter size and cost will vary with A/C ratio. Lower A/C ratios, for example, require a larger and thus more expensive fabric filter.

4.4. Baghouse Configuration

Baghouses have two basic configurations, with gases either pushed through the system by a fan located on the upstream side (forced draft fan) or pulled through by a fan on the downstream side (induced draft fan). The former is called a positive-pressure baghouse; the latter, is called a negative-pressure or suction baghouse. Positive-pressure baghouses may be either open to the atmosphere or closed (sealed and pressure-isolated from the atmosphere). Negative-pressure baghouses can only be of the closed type. Only the closed suction design should be selected for a hazardous air pollutant application to prevent accidental release of captured pollutants. At temperatures near the gas stream dew point, greater care must be taken to prevent condensation, which can moisten the filter cake, plug the cloth, and promote corrosion of the housing and hoppers. In a suction-type fabric filter, infiltration of ambient air can occur, lowering the temperature below design levels (8).

4.5. Construction Materials

The most common material used in fabric-filter construction is carbon steel. In cases where the gas stream contains high concentrations of SO_3 or where liquid–gas contact areas are involved, stainless steel may be required. Stainless steel will increase the cost of the fabric filter significantly when compared to carbon steel. However, keeping the emission stream temperature above the dew point and insulating the baghouse should eliminate the need for stainless steel.

4.6. Design Range of Effectiveness

A well-designed fabric filter can achieve collection efficiencies in excess of 99%, although optimal performance of the system may not occur for a number of cleaning cycles as the new filter material is "broken in." The fabric filter collection efficiency depends on the pressure drop across the system, component life, filter fabric, cleaning method and frequency, and the A/C ratio (13,14,26).

Performance can be improved by changing the A/C ratio, using a different fabric, or replacing worn or leaking filter bags. Collection efficiency can also be improved by decreasing the frequency of cleaning or allowing the system to operate over a greater pressure drop before cleaning is initiated. Section 5.2 will discuss the above filtration performance parameters in detail.

5. OPERATION

5.1. General Considerations

Many times, optimization of the fabric filter's collection efficiency occurs in the field after construction. The following discussion does not pertain to the preliminary design of the fabric filtration control system; however, the information presented should be helpful in achieving and maintaining the desired collection efficiency for the installed control system.

5.2. Collection Efficiency

To discuss fabric-filter "collection efficiency" is somewhat of a misnomer because a properly operated system yields very constant outlet concentrations over a broad range of inlet loadings. As such, the system really does not operate as an efficiency device—meaning that the performance of a fabric filter is not judged by the percent particulate matter (PM) reduction from initial PM concentration. Outlet concentrations are not a strong function of inlet loading. Typical outlet concentrations range between 0.001 and 0.01 g/dscf, averaging around 0.003–0.005 g/dscf. However, the term "collection efficiency" applies to a fabric-filter system when describing performance for a given application. The above given outlet concentration usually corresponds to very high collection efficiencies (17).

A well-designed fabric filter can achieve collection efficiencies in excess of 99%, although optimal performance of a fabric-filter system may not occur for a number of cleaning cycles, as the new filter material achieves a cake buildup. The fabric-filter collection efficiency is related to the pressure drop across the system, component life, filter fabric, cleaning method and frequency, and A/C ratio. These operating parameters should be modified to meet the required fabric filter performance. Modifications to improve performance include changing the A/C ratio, using a different fabric, replacing worn or leaking filter bags, and/or modifying the inlet plenum to ensure that the gas stream is evenly distributed within the baghouse. Collection efficiency can also be improved by decreasing the frequency of cleaning or allowing the system to operate over a greater pressure drop before cleaning.

5.3. *System Pressure Drop*

The pressure drop across the fabric-filter system depends on the resistance to the gas stream flow through the filter bags and accumulating dust cake, amount of dust deposit prior to bag cleaning, efficiency of cleaning, and plugging or blinding of the filter bags. Normally, the design pressure drop is set between 5 and 20 in. of water. In practice, variations in pressure drop outside the design range may indicate problems within the fabric-filter system. Excessive pressure differentials may indicate (1) an increase in gas stream volume, (2) blinding of the filter fabric, (3) hoppers full of dust, thus blocking the bags, and/or (4) inoperative cleaning mechanism. Subpar pressure differentials may indicate (1) fan or motor problems, (2) broken or unclamped bags, (3) plugged inlet ducting or closed damper; and/or (d) leakage between sections of the baghouse. For these reasons, continuous pressure-drop monitoring is recommended.

As the dust cake builds up during filtration, both the collection efficiency and system pressure drop increase. As the pressure drop increases toward a maximum, the filter bags (or at least a group of the bags contained in one isolated compartment) must be cleaned to reduce the dust cake resistance. This cleaning must be timed and performed to (1) maintain the pressure drop and thus operating costs within reasonable limits, (2) clean bags as gently and/or infrequently as possible to minimize bag wear and to maximize efficiency, and (3) leave a sufficient dust layer on the bags to maintain filter efficiency and to keep the instantaneous A/C ratio immediately after cleaning from reaching excessive levels, if woven fabric with no backing is used. In practice, these various considerations are balanced using engineering judgment and field trial experience to optimize the total system operation. Changes in the process or in fabric condition through fabric aging will shift in the cleaning requirements of the system. This shift may require more frequent manual adjustments to the automatic control to achieve the minimum cleaning requirements.

5.4. *Power Requirements*

The cost of electricity depends largely on the fan power requirement. Equation (10) can estimate this requirement, assuming a 65 % fan motor efficiency and a fluid specific gravity of 1.00:

$$F_p = 1.81 \times 10^{-4}\left(Q_{e,a}\right)(P)(\text{HRS}) \tag{10}$$

where F_p is the fan power requirement (kWh/yr), $Q_{e,a}$ is the emission stream flow rate (acfm), P is the system pressure drop (in. H_2O), and HRS is the operating hours (h/yr). For mechanical shaking, Eq. (11) provides an estimate of the additional power:

$$P_{ms} = 6.05 \times 10^{-6}(\text{HRS})\left(A_{tc}\right) \tag{11}$$

where P_{ms} is the mechanical shaking power requirement (kWh/yr) and A_{tc} is the gross cloth area (ft^2). The annual electricity cost is calculated as the sum of F_p and P_{ms}, multiplied by the cost of electricity given in Table 10.

A pulse-jet system uses about 2 scfm of compressed air per 1000 scfm of emission stream. Thus, a 100,000 scfm stream will consume about 200 scfm. Multiplying by both 60 and HRS gives the total yearly consumption. Multiplying this value by the cost of compressed air given in Table 10 gives annual costs. For other cleaning mechanisms, this consumption is assumed to be zero.

Table 7
Comparison of Calculated Values and Values Supplied by the Permit Applicant for Fabric Filters

Process variables	Calculated value (example case)[a]	Reported value
Continuous monitoring of system pressure drop and stack opacity	Yes	—
Emission stream temp. range[b]	365–415°F	—
Selected fabric material	Fiberglass or Teflon™	—
Baghouse cleaning method	Mechanical shaking, reverse-airflow, pulse jet	—
A/C ratio	2.5 ft/min for mechanical shaking or reverse air; 5 ft/min for pulse jet	—
Baghouse configuration	Negative pressure	—

[a]Based on the municipal incinerator emission stream.

5.5. *Filter Bag Replacement*

The cost of replacement bags is obtained from Eq. (12):

$$C_{RB} = \left[C_B + C_L\right]CRF_B \tag{12}$$

where C_{RB} is the bag replacement cost ($/yr), C_B is the initial bag cost ($), C_L is the bag replacement labor [$ (C_L = 0.14A_{nc}$)], and, CRF_B is the capital recovery factor, 0.5762 (indicates a 2-yr life, 10 % interest). Because the bag replacement labor cost is highly variable, a conservative high cost of $0.14/ft^2 of net bag area has been assumed (8).

6. MANAGEMENT

6.1. *Evaluation of Permit Application*

One can use Table 7 to compare the results from this section and the data supplied by the permit applicant (13). The calculated values are based on the typical case. As pointed out in the discussion on fabric filter design considerations, the basic design parameters are generally selected without the involved, analytical approach that characterizes many other control systems. Therefore, in evaluating the reasonableness of any system specifications on a permit application, the reviewer's main task will be to examine each parameter in terms of its compatibility with the gas stream and particulate conditions and with the other selected parameters. The following questions should be asked:

1. Is the temperature of the emission stream entering the baghouse within 50–100°F above the stream dew point?
2. Is the selected fabric material compatible with the conditions of the emission stream (i.e., temperature and composition) (*see* Table 3)?
3. Is the baghouse cleaning method compatible with the selected fabric material and its construction (i.e., material type and woven or felted construction) (*see* Section 4.3 and Table 6)?
4. Will the selected cleaning mechanism provide the desired control?
5. Is the A/C ratio appropriate for the application (i.e., type of dust and cleaning method used) (*see* Table 4)?

6. Are the values provided for the gas flow rate, A/C ratio, and net flow area consistent?

   The values can be checked with the following equation:

$$\text{A/C ratio} = \frac{Q_{e,a}}{A_{nc}} \tag{8}$$

   where the variables are as described earlier.
7. Is the baghouse configuration appropriate; that is, is it a negative-pressure baghouse?

6.2. Economics

Fabric filtration systems are attractive in that they are highly efficient collection devices that can be operated at low-energy requirements. In addition, they usually have no water requirements so that the solid-waste-disposal problem may be significantly less than that for wet systems. On the other hand, fabric filtration systems are expensive in that they require a large amount of space for installation [about 1 ft^2 (0.1 m^2) of floor space per each 5 ft^3/min (0.14 m^3/min)] and have a large capital investment.

The highest maintenance component of fabric-filter systems is the fabric itself. In baghouses, the bags have an average life of 18–36 mo and account for 20–40% of the equipment cost. If the system is expected to have a 10-yr life, this means that the bags must be replaced anywhere from three to seven times during this lifetime. Causes of bag failure include blinding (mudding), caking, burning, abrasion, chemical attack, and aging. Prior discussion in this chapter indicated how these problems can be reduced by proper operating and maintenance procedures.

The Industrial Gas Cleaning Institute (IGCI), representing about 90% of all fabric-filter gas cleaning device manufacturers, estimated that about half of the filter systems in the United States are low energy and half are high energy.

This chapter mentions factors affecting the economics of filter systems. These factors include the composition of both the solids and the gas, the type of filter system desired, requirements for gas conditioning, and proper operating and maintenance procedures. Other factors that also influence the cost of fabric filtration systems are, for example, *special* properties of the gas stream (toxic, explosive, corrosive, and/or abrasive), space restrictions in the installing facility, and the nature of ancillary equipment, such as hoods, ducts, fans, motors, material-handling conveyors, airlocks, stacks, controls, and valves.

These costs (Tables 8–10) are averages of all industries, and actual operating and relative costs would depend on the specific application. Abrasive, corrosive, hot applications may have greater total costs plus proportionally greater replacement and labor costs. Equipment costs for a fabric-filter system can be estimated by either obtaining quotations from vendors, or using generalized cost correlations from the literature. Total capital costs (*see* Table 9) include costs for the baghouse structure, the initial complement of the bags, auxiliary equipment, and the usual direct and indirect costs associated with installing or erecting new structures. The price per square foot of bags by type of fabric and cleaning system appears in Table 8 (3rd quarter 1986 dollars). The prices represent a 10 % range and should be escalated using the index provided in *Chemical Engineering* (27). The annual costs (*see* Table 11) for a fabric-filter system consist of the direct and indirect operating costs. Direct costs include utilities (electricity, replacement

Table 8
Bag Prices (3rd quarter 1986 $/ft²)

Type of cleaning	Bag diameter (in.)	Type of material[a]						
		PE	PP	NO	HA	FG	CO	TF
Pulse jet, TR[b]	4½–5⅛	0.59	0.61	1.88	0.92	1.29	NA[c]	9.05
	6–8	0.43	0.44	1.56	0.71	1.08	NA	6.80
Pulse jet, BBR	4½–5⅛	0.37	0.40	1.37	0.66	1.24	NA	8.78
	6–8	0.32	0.33	1.18	0.58	0.95	NA	6.71
Shaker								
Strap top	5	0.45	0.48	1.28	0.75	NA	0.44	NA
Loop top	5	0.43	0.45	1.17	0.66	NA	0.39	NA
Reverse air with rings	8	0.46	NA	1.72	NA	0.99	NA	NA
Reverse air	8	0.32	NA	1.20	NA	0.69	NA	NA
w/o rings[d]	11½	0.32	NA	1.16	NA	0.53	NA	NA

Note: For pulse-jet baghouses, all bags are felts except for the fiberglass, which is woven. For bottom access pulse jets, the cage price for one cage can be calculated from the single-bag fabric area using the following:

In 50 cage lots:	$\$ = 4.941 + 0.163\ ft^2$	$\$ = 23.335 + 0.280\ ft^2$
In 100 cage lots:	$\$ = 4.441 + 0.163\ ft^2$	$\$ = 21.791 + 0.263\ ft^2$
In 500 cage lots:	$\$ = 3.941 + 0.163\ ft^2$	$\$ = 20.564 + 0.248\ ft^2$

[a]PE = 16-oz polyester; PP = 16-oz polypropylene; NO = 14-oz nomex; HA = 16-oz homopolymer acrylic; FG = 16-oz fiberglass with 10% TeflonTM; CO = 9-oz cotton; TF = 22-oz TeflonTM felt.

[b]Bag removal methods: TR = top bag removal (snap in); BBR = bottom bag removal

[c]NA = Not applicable

[d]Identified as reverse-air bags, but used in low-pressure pulse applications.

These costs apply to 4½-in.- or 5⅝-in.-diameter, 8-ft and 10-ft cages made of 11 gage mild steel and having 10 vertical wires and "Roll Band" tops. For flanged tops, add $1 per cage. If flow-control Venturis are used (as they are in about half of the pulse-jet manufacturers' designs), add $5 per cage.

For shakers and reverse air baghouses, all bags are woven. All prices are for finished bags and prices can vary from one supplier to another. For Gore-Tex™ bag prices, multiply base fabric price by factors of 3–4.5.

Source: ref. 8.

bags, and compressed air), operating labor, and maintenance costs. Indirect costs consist of overhead, administrative costs, property taxes, insurance, and capital recovery. Table 10 provides the appropriate factors to estimate these costs.

The bag replacement labor cost depends on such factors as the number, size, and type of bags, the accessibility of the bags, how much they are connected to the tube sheet, and so forth. As such, these costs are highly variable. For simplicity, assume a conservatively high cost of $0.14/ft² net bag area, per EPA guidance (8). Dust disposal typically comprises a large cost component and varies widely with site. The reader should obtain accurate, localized costs. These fall between $20/ton and $30/ton for nonhazardous waste, and 10 times this amount for hazardous material (8).

The cost of operating labor assumes a requirement of 3 h per 8 h shift and the wage rate is provided in Table 10. Supervisory costs are taken as 15 % of operator labor costs. The cost of maintenance assumes a labor requirement of 1 h per 8 h shift, and the wage

Table 9
Capital Cost Factors for Fabric Filters

Direct costs	Factor
Purchased Equipment Costs (PEC)	
Fabric filter	As estimated
Bags	As estimated
Auxiliary equipment	As estimated
(EC = sum of as estimated)	
Instruments and controls	0.10 EC
Taxes	0.03 EC
Freight	0.05 EC
PEC = 1.18 EC	
Installation Direct Costs (IDC)	
Foundation and supports	0.04 PEC
Erection and handling	0.50 PEC
Electrical	0.08 PEC
Piping	0.01 PEC
Insulation for ductwork[a]	0.07 PEC
Painting[b]	0.02 PEC
Site preparation (SP)	As required
Buildings (Bldg.)	As required
IDC = 0.72 PEC + SP + Bldg.	
Total direct cost (DC) = PEC + IDC = 1.72 PEC + SP + Bldg.	
Indirect Costs	
Engineering and supervision	0.10 PEC
Construction and field expense	0.20 PEC
Construction fee	0.10 PEC
Start-up fee	0.01 PEC
Performance test	0.01 PEC
Contingencies	0.03 PEC
Total Indirect Cost, IC = 0.45 PEC	
Total capital cost (TCC) = DC + IC = 2.17 PEC + SP + Bldg.	

[a]If ductwork dimensions have been established, cost may be established based on \$10–\$12/ft^2 of surface for field application. Fan housings and stacks may also be insulated.
[b]The increased use of special coatings may increase this factor to 0.06 PEC or higher.
Source: ref. 8.

rate is provided in Table 10. The cost of maintenance materials is assumed to equal the maintenance labor costs.

6.3. New Technology Awareness

A sanitary bag filter has been developed to enhance clean-in-place (CIP) capability (28). The entire system can be cleaned between product changes without changing the filter bags. The system eliminates crosscontamination of products while still efficiently collecting powdered pollutants from an air emission stream. Another gas filter has been developed using the ceramic-element technology. The controlled filtration layers trap larger particles in the outer layer and catch smaller ones in the inner layer, resulting in

Table 10
Annual costs for Fabric Filters

Cost item	Factor
Direct Costs (DAC)	
Utilities	
Electricity	\$0.059/kWh
Compressed air	\$0.16/10^3 scfm
Replacement parts, bags	
Operating labor	
Operator	\$12.96/h
Supervisor	15% of operator labor
Maintenance	
Labor	\$14.26/h
Material	100% of maintenance labor
Waste disposal	Variable
Indirect costs (IAC)	
Overhead	0.60(operating labor + maintenance)
Administrative	2% of TCC (total capital cost)
Property tax	1% of TCC
Insurance	1% of TCC
Capital recovery[a]	0.1175(TCC – 0.05C_L – 1.08 C_B)

[a]Capital recovery factor is estimated as $i(1+i)^n/[(1+i)^n - 1]$,
where i is the interest rate, (10%) and n is the equipment life, (20 yr).
Source: ref. 8.

a 99.999% removal rating for 0.003 μm at maximum flow rate. The ceramic medium is processed at temperatures above 2000°C, eliminating organic contaminants. It is capable of producing flow rates up to 2700 L/min (29).

The most recently developed filtration processes use membrane filtration media. Because substances that permeate nonporous membranes are reasonably volatile, application of a vacuum always causes the permeate to be desorbed from the membrane in the vapor state. Hence, the term "pervaporation" applies if the feed to the membrane filter is liquid, because the contaminant appears to evaporate through the membrane (30,31). If the feed is vapor, or a gas–vapor mixture, the process is called "vapor permeation" (30). More new technologies are reported elsewhere (32–36).

7. DESIGN EXAMPLES AND QUESTIONS

Example 1

The process flow diagram for a typical shaker fabric filter appears in Fig. 3. Give a general process description for the fabric filtration process.

Solution

Fabric filters are air pollution control devices designed for controlling particulate matter emissions from point sources. A typical fabric filter consists of one or more isolated compartments containing rows of fabric bags or tubes. Particle-laden gas passes up along the surface of the bags, then radially through the fabric. The upstream face of the bags retains particles while the clean gas stream vents to the atmosphere. The filter operates cyclically to alternate between long filtering periods and short cleaning periods. During cleaning, accumulated dust on the bags is removed from the fabric surface and deposited in a hopper for subsequent disposal.

Fabric filters collect particles ranging from submicron to several hundred microns in diameter, at efficiencies generally in excess of 99% Routinely, gas temperatures can be accommodated up to about 500°F, with surges to approx 550°F. Most of the energy use in a fabric-filter system derives from the pressure drop across the bags and associated hardware and ducting. Typical values of pressure drop range from about 5 to 20 in. of water column.

Example 2

Fabric filters are often categorized by the cleaning method for removing the dust cake. Three common types include (1) shaker filters, (2) reverse-air filters, and (3) pulse-jet filters. Describe and discuss (1) general cleaning methods and (2) the three types of fabric filter.

Solution:

1. *General cleaning methods*: As dust accumulates on the filtering elements, the pressure drop across the bag compartment increases until cleaning of the bags occurs. Cleaning is usually controlled by a timer or a pressure switch set at the specified maximum pressure drop. At this point, the bags in the compartment are cleaned to remove the collected dust, and the cycle is then repeated. The two basic mechanisms for bag cleaning involve flexing the fabric to break up and dislodge the dust cake, and reverse airflow through the fabric to remove the dust. These may be used separately or together. The three principal methods used for fabric cleaning are mechanical shaking (manual or automatic), reverse airflow, and pulse-jet cleaning. The first method uses only the fabric flexing mechanism; the latter two methods use a combination of the reverse-airflow and fabric flexing mechanisms.
2. *Three types of fabric filters*:
   a. In a shaker filter (*see* Fig. 3), the bags are hung in a framework that is oscillated by a motor-controlled timer. In this type of system, the baghouse is usually divided into several compartments. The flow of gas to each compartment periodically is interrupted, and the bags are shaken to remove the collected dust. The shaking action produces more wear on the bags than other cleaning methods. For this reason, the bags used in this type of filter are usually heavier and made from durable fabrics (13,26).
   b. In a reverse-airflow filter, gas flow to the bag is stopped in the compartment being cleaned and a reverse flow of air is directed through the bags. This approach has the advantage of being "gentler" than shaking allowing the use of more fragile or lightweight bags (13).
   c. The third type of baghouse, pulse-jet fabric filter, is by far the most common type for Superfund applications. In this type of system, a blast of compressed air expands the bag and dislodges collected particles. One advantage of pulse-jet fabric filters is that bags can be cleaned on line, meaning fewer bags (less capacity) are required for a given application (26).

Example 3

Discuss (1) mechanical shaking cleaning methods, (2) reverse-airflow cleaning methods, and (3) pulse-jet cleaning methods in detail.

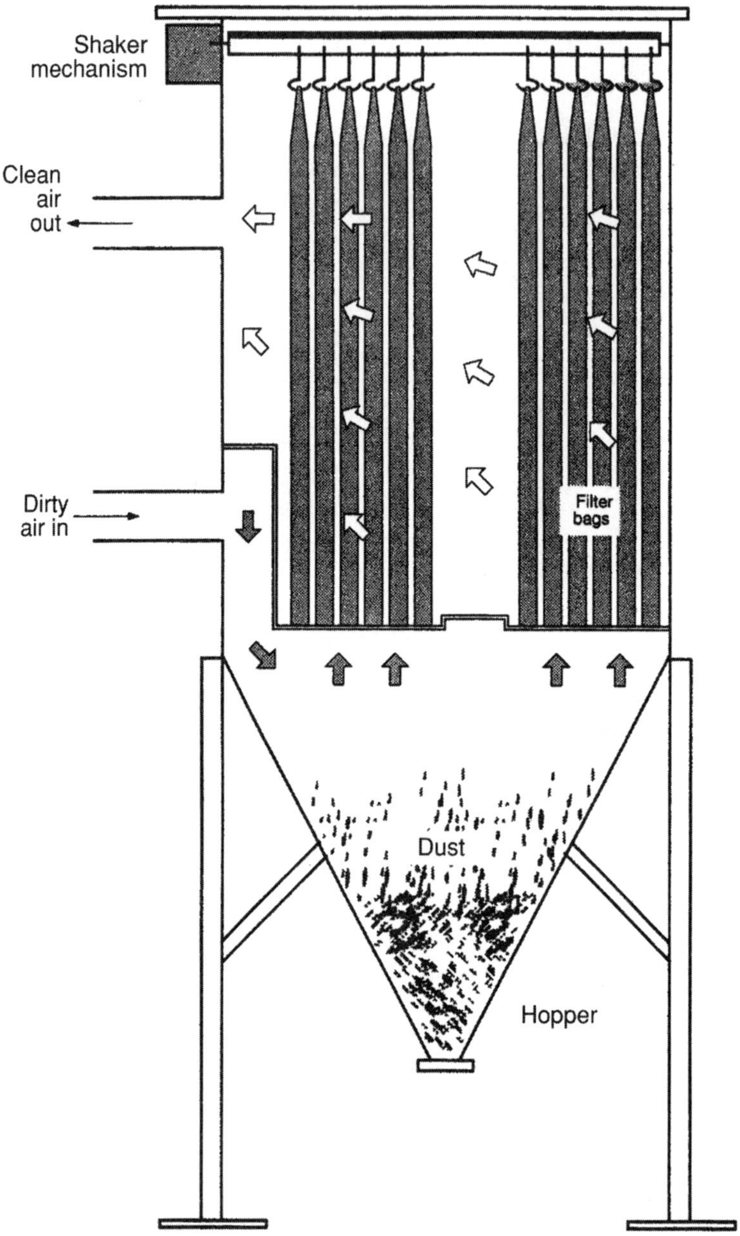

Fig. 3. Fabric-filter process flow diagram.

Solution

1. *Mechanical shaking cleaning method*: With mechanical shaking, bags hang on an oscillating framework that periodically shakes the bags at timed intervals or at a predefined pressure drop level (14,15,18). The shaker mechanisms produce violent action on the fabric-filter bags and, in general, produce more fabric wear than the other types of cleaning mechanism (16). For this reason, mechanical shaking is used in con-

junction with heavier more durable fabrics, such as most woven fabrics. Bags with fair to poor abrasion ratings in Table 3 (e.g., fiberglass) should not be chosen for fabric filters cleaned by mechanical shaking unless they are treated with a special coating (i.e., a backing) before use. Although shaking is abrasive to the fabric, it does allow a dust cake to remain on the fabric, thus maintaining high collection efficiency (15,22).

2. *Reverse-airflow cleaning method*: Reverse-airflow cleaning is used to flex or collapse the filter bags by allowing a large volume of low pressure air to pass countercurrent to the direction of normal gas stream flow during filtration (16,18). Reverse air is provided either by a separate fan or by a vent in the fan damper, which allows a backwash of air to clean the fabric filters. Reverse-airflow cleaning is usually performed off-line. It allows the use of fragile bags, such as fiberglass, or lightweight bags, and usually results in longer life for bags (16). As with mechanical shaking, woven fabrics are used. Because cleaning is less violent than with pulse-jet cleaning and is performed off-line, outlet concentrations are almost constant with varying inlet dust loading throughout the cleaning cycle. Reverse-airflow cleaning is, therefore, a good choice for fabric cleaning in hazardous air pollutant (HAP) control situations.
3. *Pulse-jet cleaning method*: In pulse-jet cleaning, a high-pressure air pulse enters the top of the bag through a compressed air jet. This rapidly expands the bag, vibrating it, dislodging particles, and thoroughly cleaning the fabric. The pulse of air cleans so effectively that no dust cake remains on the fabric to contribute to particulate collection. Because this cake is essential for effective collection on woven fabrics, felted fabrics are generally used in pulse-jet-cleaned fabric filters. Alternatively, woven fabrics with a suitable backing may be used. All fabric materials may be used with pulse-jet-cleaning, except cotton or fiberglass. Previously, mechanical shaking was considered superior to pulse-jet cleaning in terms of collection efficiency. Recent advances in pulse-jet cleaning have produced efficiencies rivaling those of mechanical shaking.

   Because the air pulse has such a high pressure (up to 100 psi) and short duration ($\leq$0.1 s), cleaning may also be accomplished on-line, but off-line cleaning is also employed. Extra bags may not be necessary to compensate for bags off-line during cleaning. Cleaning occurs more frequently than with mechanical shaking or reverse-airflow cleaning, which permits higher air velocities (higher A/C ratios) than the other cleaning methods. Furthermore, because the bags move less during cleaning, they may be packed more closely together. In combination, these features allow pulse-jet-cleaned fabric filters to be installed in a smaller space, at a lower cost, than fabric filters cleaned by other methods. This cost savings may be somewhat counterbalanced by the greater expense and more frequent replacement required of bags, the higher power use that may occur, and the installation of fabric-filter framework that pulse-jet cleaning requires (14,16,18).

Example 4

A new 8000-ft^3/min shaker-type filter installation is being designed to remove iron oxide from an electric furnace emission. Consider the gas to be air at 110°F with an inlet dust concentration of 0.8 gr/ft^3 (grains per cubic foot). The A/C ratio is 3 ft/min and the mass mean particle size is approx 1 µm. Other design parameters include the following.

From Table 2 for a 1-µm spherical particle:

S = specific surface area per unit volume of solids = 1.83×10^6 ft^{-1}

e = porosity = 0.40

Assume that the Kozeny–Carman coefficient k = 5 and

ρ_p = particle density = (5.18)(62.4) = 323 lb/ft^3

μ_f = air viscosity = 1.21×10^{-5} $lb_m/(s\ ft)$
g = 32.174 ft/s^2

Determine the following design variables:

1. Cake fabric-filter resistance coefficient, K_2
2. Filtration cycle time, t
3. Blower horsepower
4. Fabric-filter area
5. Solids removal rate

Solution

1. Using Eq. (7),

$$K_2 = \left(3.2\times10^{-3}\right)\left(\frac{k}{g}\right)\left(\frac{\mu_f S^2}{\rho_p}\right)\left(\frac{1-\varepsilon}{\varepsilon^3}\right)$$
$$= \left(3.2\times10^{-3}\right)\left(\frac{5}{32.174}\right)\frac{\left(1.21\times10^{-5}\right)\left(1.83\times10^{6}\right)^2}{323}(9.38)$$
$$= 585 \text{ in. } H_2O/\left(lb_m/ft^2\right)(ft/min)$$

Operating data in the literature (3) show that for an installation of this type, using Orlon fabric filters, $K_2 = 45$. This is obtained via Eq. (6) for an inlet dust loading of 0.8 gr/ft^3.

2. Assume that the filtration should operate so that the pressure drop increases by up to about 3 in. H_2O. The filtration cycle time can then be estimated by rearranging Eq. (6) (use $K_2 = 45$):

$$t = \frac{\Delta P_2}{K_2 v^2 L}$$
$$= \frac{(3 \text{ in. } H_2O)(7000 \text{ grains/lb})}{\left[45 \text{ in. } H_2O/\left(lb/ft^2\right)(ft/min)\right](3 \text{ ft/min})^2\left(0.8 \text{ grains/ft}^3\right)}$$
$$= 65 \text{ min}$$

Therefore, it would be necessary to shake the system about once an hour.

3. Considering that the residual fabric-filter resistance is also about 3 in. H_2O and there are other gas flow pressure losses, assume an overall ΔP of 7 in. H_2O. The size of the blower can be estimated (7) using 60% blower efficiency:

   Blower horserpower(HP) = $(3 \times 10^{-4})(\Delta P)(Q)$
   ΔP = 7 in. H_2O
   Q = 8000 ft^3/min
   HP = $(3 \times 10^{-4})(7)(8000) = 17$

4. The size of the filter area required is

$$\text{air-to-cloth ratio} = 3 \text{ ft/min} = \frac{3 \text{ ft}^3/\text{min}}{\text{ft}^2}$$

$$\text{filter area} = \frac{8000 \text{ ft}^3/\text{min}}{3 \text{ ft/min}} = 2670 \text{ ft}^2$$

5. The material handling system to remove the solids must be able to handle a maximum of

$$\left(\frac{8000\ \text{ft}^3}{\text{min}}\right)\left(\frac{0.8\ \text{grains}}{\text{ft}^3}\right)\left(\frac{\text{lb}}{7000\ \text{grains}}\right)\left(\frac{60\ \text{min}}{\text{h}}\right) = 55\frac{\text{lb}}{\text{h}}$$

This assumes 100% filter efficiency (1320 lb/d for 24-h operation).

Example 5

What is the "HAP Emission Stream Data Form" recommended by the US Environmental Protection Agency (US EPA)?

Solution

The "HAP Emission Stream Data Form" recommended by the US EPA is presented in Appendix 1.

Example 6

Prepare a step-by-step calculation procedure for design of a fabric filtration system.

Solution

1. Engineering data gathering for the HAP emission stream characteristics:
   1) Flow rate: $Q_{e,a}$ = _______________ acfm
   2) Moisture content: M_e = _______________ % (vol)
   3) Temperature: T_e = _______________ °F
   4) Particle mean diameter: D_p = _______________ μm
   5) SO_3 content = _______________ ppm (vol)
   6) Particulate content = _______________ grains/scf
   7) HAP content = _______________ % (mass)

2. Determine or decide the following engineering data for permit review and application:
   1) Filter fabric material _______________
   2) Cleaning method (mechanical shaking, reverse air, pulse jet) _______________
   3) Air-to-cloth ratio _______________ ft/min
   4) Baghouse construction configuration (open pressure, closed pressure, closed suction)
   5) System pressure drop range _______________ in. H_2O

3. Pretreatment Considerations:
   If the emission stream temperature is not from 50°F to 100°F above the dew point, pretreatment is necessary (see Chapter 1). Pretreatment will cause two of the pertinent emission stream characteristics to change; list the new values below.
   1) Maximum flow rate at actual conditions: $Q_{e,a}$ = _______________ acfm
   2) Temperature: T_e = _______________ °F

4. Fabric Filter System Design
   1) Fabric type(s) (use Table 3)
      a. _______________
      b. _______________
      c. _______________

   2) Cleaning method(s)
      a. _______________

b. ______________

3) Air-to-cloth ratio (Table 4) ______________ ft/min
4) Net cloth area, A_{nc}:

$$A_{nc} = Q_{e,a}/(\text{A/C ratio})$$

where A_{nc} is the net cloth area (ft^2), $Q_{e,a}$ = maximum flow rate at actual conditions (acfm) = $Q_e(T_e + 460)/537$ (which is to be used if given Q_e instead of $Q_{e,a}$), and A/C ratio = air-to-cloth ratio (ft/min)

A_{nc} = ______________/______________

A_{nc} = ______________ft^2

5) Gross cloth area, A_{tc}:

$$A_{tc} = A_{nc} \times \text{Factor}$$

where A_{tc} is the gross cloth area (ft^2) and Factor is the value from Table 5 (dimensionless).

A_{tc} = ______________×______________
A_{tc} = ______________ ft^2

6) Baghouse configuration ___________
7) Materials of construction ___________

5. Determination of baghouse operating parameters
   1) Collection efficiency (CE) = __________
   2) System pressure drop range __________ in. H_2O

Example 7

Fabric filtration is one of the selected control techniques for a municipal incinerator. Conduct a preliminary design for a fabric filtration system (select filter fabrics, decide cleaning method, and determine A/C ratio). The pertinent engineering data appear on the "HAP Emission Stream Data Form" (*see* Table 11).

Solution

1. Gather engineering data on HAP emission stream characteristics from Table 11:
   1) Flow rate, $Q_{e,a}$ = 110,000 acfm
   2) Moisture content, M_e = 5% vol
   3) Temperature, T_e = 400°F
   4) Particle mean diameter, D_p=1.0 µm
   5) SO_3 content = 200 ppm (vol)
   6) Particulate content = 3.2 gr/scf – flyash
   7) HAP content = 10% (mass) cadmium

2. Fabric-filter Preliminary Design. In this case, fabric selection depends on the emission stream temperature of 400°F, the SO_3 content of 200 ppmv, and the flyash particulate type. Table 3 indicates that filter fabrics capable of withstanding 400°F emission stream temperature are ceramics (Nextel 312™), nylon aromatic (Nomex), fluorocarbon (Teflon), and fiberglass. Because there is a high potential for acid damage (i.e., a high SO_3 content), however, Nomex bags should not be considered. To obtain an indication of the A/C ratio, use Table 4. This table shows that an A/C ratio of around 2.5 is

Table 11
Effluent Characteristics for a Municipal Incinerator Emission Stream

HAP EMISSION STREAM DATA FORM*

Company <u>Incineration Inc</u>.	Plant contact Mr. Phil Brothers
Location <u>(Street) 124 Main Stree</u>	Telephone No. (999) 555-5024
(City) <u>Somewhere</u>	Agency contact Mr. Ben. Hold
(State, Zip)	No. of Emission Streams Under Review 1

A.	Emission Stream Number/Plant Identification		#1/ Incineration	
B.	HAP Emission Source	(a) municipal incinerator	(b)	(c)
C.	Source Classification	(a) process point	(b)	(c)
D.	Emission Stream HAPs	(a) cadmium	(b)	(c)
E.	HAP Class and Form	(a) inorganic particulate	(b)	(c)
F.	HAP Content (1,2,3)**	(a) 10%	(b)	(c)
G.	HAP Vapor Pressure (1,2)	(a)	(b)	(c)
H.	HAP Solubility (1,2)	(a)	(b)	(c)
I.	HAP Adsorptive Prop. (1,2)	(a	(b)	(c)
J.	HAP Molecular Weight (1,2)	(a)	(b)	(c)
K.	Moisture Content (1,2,3)	5% vol	P.	Organic Content (1)***
L.	Temperature (1,2,3)	400°F	Q.	Heat/O_2 Content (1)
M.	Flow Rate (1,2,3)	110,000 acfm	R.	Particulate Content (3) 3.2 gr/acf, flyash
N.	Pressure (1,2)	atmospheric	S.	Particle Mean Diam. (3) 1.0 μm
O.	Halogen/Metals (1,2)	none/none	T.	Drift Velocity/SO_3 (3) 0.31 ft/sec/ 200 ppmv
U.	Applicable Regulation(s)			
V.	Required Control Level	assume 99.9% removal		
W.	Selected Control Methods	fabric filter, ESP, Venturi scrubber		

*The data presented are for an emission stream (single or combined streams) prior to entry into the selected control method(s). Use extra forms if additional space is necessary (e.g., more than three HAPs) and note this need.

**The numbers in parentheses denote what data should be supplied depending on the data on lines C and E:
1 = organic vapor process emission
2 = inorganic vapor process emission
3 = particulate process emission

***Organic emission stream combustibles less HAP combustibles shown on lines D and F.

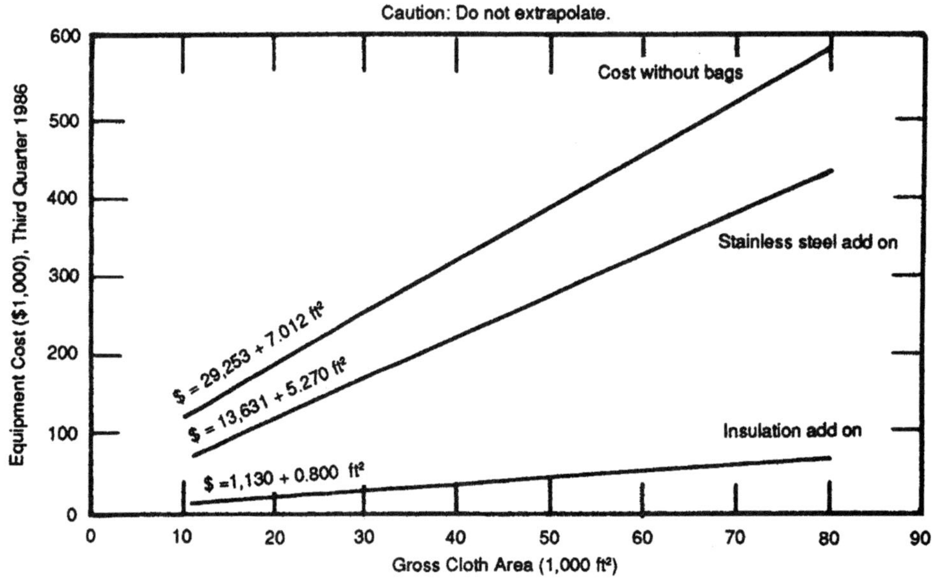

Fig. 4. Structure costs for reverse-air filters.

expected for mechanical shaking or reverse-air cleaning, and an A/C ratio of about 5.0 is expected for pulse-jet cleaning.

A fiberglass bag would provide the most protection during temperature surges (unless ceramics are used), and because fiberglass bags may be less expensive, it may be the fabric of choice for an installation with these emission characteristics. Fiberglass bags would require that reverse-air cleaning be used, unless a suitable backing allows pulse-jet cleaning. Teflon bags with mechanical shaking could also be a possibility (7,17). Limited information on the long-term effectiveness of ceramics has been documented. It is expected that ceramic fibers will have performance characteristics similar to the best synthetic fibers, but will cost significantly more.

Example 8

The HAP emission stream shown in Example 7 and Table 11 is to be treated by a reverse-air baghouse. Figure 4 is provided by the vendor for the cost of the baghouse structure. Determine the A/C ratio, net cloth area (A_{nc}), gross cloth area (A_{tc}), and the baghouse total capital cost (requiring stainless steel add-on and insulation).

Solution

1. From Table 4, flyash, the A/C ratio = 2.5.
2. Thus, A_{nc} = (110,000 acfm)/2.5 = 44,000 ft^2.
3. Obtain the total cloth area using Table 5. This table indicates that A_{nc} should be multiplied by 1.125 to obtain A_{tc}. Thus, A_{tc} = 44,000(1.125) = 49,500 ft^2. This value is used to obtain the structure cost.
4. Using Fig. 4, the structure cost equals $380,000 plus $270,000 for stainless-steel add-on, plus $40,000 for insulation. The total cost is then $380,000 + $270,000 + $40,000 = $690,000.

Table 12
Example Case Capital Costs

Direct costs	Cost ($)
Purchased Equipment Costs	
Fabric filter	$ 690,000
Bags	49,000
Auxiliary equipment	10,000
	$ 749,000
Instruments and controls	$ 74,900
Taxes	22,500
Freight	37,500
Purchased equipment cost (PEC)	$ 884,000
Installation Direct Costs	
Foundation and supports	$ 35,400
Erection and handling	442,000
Electrical	70,700
Piping	8,840
Insulation for ductwork	61,900
Painting	17,700
Site preparation (SP)	—
Buildings (Bldg.)	—
	$ 636,000
Total direct costs	$ 1,520,000
Indirect Costs	
Engineering and supervision	$ 88,400
Construction and field expense	177,000
Construction fee	88,400
Start-up fee	8,840
Performance test	8,840
Contingencies	26,500
Total indirect cost	$ 398,000
Total direct and indirect cost = Total capital cost (TCC)	$ 1,920,000

Table 8 is used to obtain the bag cost, C_B. From the previous example case, choose fiberglass bags with Teflon backing. Assume the bag diameter is 8 in. with rings. The bag cost is given as ($0.99/ft^2) × (49,500 ft^2) = $49,000.

Assume that auxiliary equipment costs obtained (see another chapter on cost estimation of air pollution control technologies) are $10,000. The equipment cost (EC) is then $690,000 + $49,000 + $10,000 = $749,000. Table 9 lists the purchased equipment cost (PEC):

Instrumentation = 0.10(EC) = $74,900
Taxes = 0.03(EC) = $22,500
Freight = 0.05(EC) = $37,500

The total PEC is then $749,000 + $74,900 + $22,500 + $37,500 = $884,000. Table 9 is then used to obtain the total capital cost (TCC) of the baghouse system. These costs are given in Table 12. Another Humana Press book (37) gives additional cost data.

Example 9

Assume that the waste generation ratio is 3.2 gr/ft^3 of HAP emission stream processed, the HAP emission stream flow is 110,000 ft^3/min, and the waste disposal cost is $200/ton-yr, determine the total annual waste disposal cost for a fabric filtration system.

Solution

1.

$$\text{Waste quantity generated} = 3.2\frac{\text{gr}}{\text{ft}^3} \times \frac{1\,\text{lb}}{7000\ \text{gr}} \times \frac{110,000\ \text{ft}^3}{\text{min}} \times \frac{60\ \text{min}}{\text{h}} \times \frac{6000\ \text{h}}{\text{yr}}$$

$$= 1.81 \times 10^7 \times \frac{\text{lb}}{\text{yr}}$$

2.

$$\text{Annual waste disposal cost} = 1.81 \times 10^7 \frac{\text{lb}}{\text{yr}} \times \frac{\text{ton}}{2000\ \text{lb}} \times \frac{200}{\text{ton/yr}}$$

$$= \$1,810,000$$

Example 10

The HAP emission stream shown in Example 7 and Table 11 is to be treated by a reverse-air baghouse. Assume the following are the given data:

1. Emission stream flow, Q_{acfm} = 110,000 acfm
2. System pressure drop = 10 in. H_2O
3. Annual operating hours, HRS = 6000 h/yr (assuming 8 h/shift)
4. Electricity cost = $0.059/kWh
5. Initial bag cost, C_B = $49,000 from Example 8
6. Net cloth area, A_{nc} = 44,000 ft^2 from Example 8
7. Operating labor and labor cost of baghouse at 3 h/shift and $12.96/h, respectively
8. Supervisory cost = 15% of total operating labor costs
9. Maintenance labor and cost of baghouse at 1 h/shift and $14.26/h, respectively
10. Maintenance cost = 100% maintenance labor cost
11. Waste generation rate = 3.2 gr/ft^3 of HAP emission stream processed. Waste generation cost = $1,810,600/yr from Example 9
12. Indirect annual cost = Table 10

Determine the following:

1. Total direct cost
2. Total indirect cost
3. Total annual cost

Solution

1. Total direct annual costs: Electricity usage is estimated using Eq. (10). Assume that the system pressure drop equals 10 in. H_2O.

$$F_p = 1.81 \times 10^{-4}(110,000)(10)(6,000)$$
$$= 1.19 \times 10^6$$

$$\text{Electricity cost} = \$0.059(1.19 \times 10^6) = \$70,200/\text{yr}$$

Because reverse air is used, $P_{ms} = 0$.
Bag replacement costs are obtained using Eq. (12):

$$C_{RB} = [49,000 + 0.14(44,000)]0.5762$$
$$= \$31,800/\text{yr}$$

Operating labor costs are estimated as

$$[(3\ \text{h/shift})/(8\ \text{h/shift})]6,000\ \text{h/yr} = 2,250\ \text{h/yr}$$
$$2,250\ \text{h/yr}\ (\$12.96/\text{h}) = \$29,200/\text{yr}$$

Supervisory costs are taken as 15% of this total, or $4370.
Maintenance labor costs are estimated as

$$[(1\ \text{h/shift})/(8\ \text{h/shift})]6,000\ \text{h/yr} = 750\ \text{h/yr}$$
$$750\ \text{h/yr}\ (\$14.26/\text{h}) = \$10,700/\text{yr}$$

Maintenance materials are taken as 100% of this total, or $10,700.
Waste disposal cost = $1,810,600/yr from Example 9
Total direct annual costs = $70,200 + $31,800 + $29,200 + $4370 + $10,700 + $10,700 + $1,810,000 = $1,970,000

2. Total indirect annual costs: These costs are obtained from the factors presented in Table 10 and the example case presented above.
   Overhead = 0.60($29,200 + $4370 + $10,700 + $10,700)
   = $33,000
   Administrative = 0.02($1,920,000)
   = $38,400
   Insurance = 0.01($1,920,000)
   = $19,200
   Property taxes = 0.01($1,920,000)
   = $19,200
   Capital recovery = 0.1175($1,920,000) – 1.08($49,000) – 0.05($0.14)(44,000)
   = $219,000
   Total indirect costs = $33,000 + $38,400 + $19,200 + $19,200 + $219,000
   = $329,000
   Total annual costs = $1,970,000 + $329,000
   = $2,200,000/yr

Example 11

The bag prices shown in Table 8 are for the third quarter 1986. Discuss how one can update the third quarter 1986 cost to the March 2002 cost, or any month in the future.

Solution

Using the following equation for equipment cost comparison:

$$\text{Cost}_b = \text{Cost}_a\ (\text{Index}_b)/(\text{Index}_a)$$

where Cost_b is the future cost ($), Cost_a is the old cost ($), Index_b is the future CE equipment cost index, and Index_a is the old CE equipment cost index. For instance, the CE (Chemical Engineering) equipment cost index for the third quarter 1986 can be obtained from the literature (32) to be 336.6. The March 2002 CE equipment cost index can also be obtained from a different issue of the same source (27). In turn, the March 2002 equipment costs can be calculated using the known values of $\text{Cost}_{9\text{-}1986}$, $\text{Index}_{3\text{-}2002}$, and $\text{Index}_{9\text{-}1986}$:

$$\text{Cost}_{3\text{-}2002} = \text{Cost}_{9\text{-}1986}\ (\text{Index}_{3\text{-}2002})/\text{Index}_{9\text{-}1986},$$

Readers are referred to ref. 37 for more detailed information on cost estimation.

NOMENCLATURE

A_{nc}	Net cloth area (ft^2)
A_{tc}	Gross cloth area (ft^2)
c	Flow constants
C_B	Initial bag cost (\$)
C_{RB}	Bag replacement cost (\$)
Cost	Equipment cost (\$)
CRF_B	Capital recovery factor
D_P	Particle mean diameter (μm)
DAC	Direct annual costs (\$)
ε	Porosity or fraction void volume (dimensionless)
F_P	Fan power requirement (kWh/yr)
g	Gravitational constant
HP	Horsepower
HRS	Operating hours (h/yr)
i	Interest rate
IAC	Indirect annual costs (\$)
Index	*Chemical Engineering* equipment cost index (dimensionless)
k	Kozeny–Carman coefficient (approx 5 for $0.8 \geq \varepsilon$)
K	Kozeny permeability coefficient
K_1	Resistance of the fabric (in. H_2O/ft/min)
K_2	Cake–fabric–filter resistance coefficient
L	Inlet solids concentration (lb_m/ft^3)
M_e	Moisture content (vol %)
μ	Viscosity
μ_f	Fluid viscosity
n	Equipment life (yr)
P_{ms}	Mechanical shaking power requirement (kWh/yr)
ΔP	Pressure drop (in. H_2O)
ΔP_1	Pressure drop across fabric (in. H_2O)
ΔP_2	Change in pressure drop due to cake build–up over time interval t (in. H_2O)
ρ	Density
ρ_p	True density of solid material (lb_m/ft^3)
Q	Volumetric flow rate (ft^3/min)
S	Specific surface area per unit volume of either porous filter media or solids in cake layer (ft^2/ft^3)
t	Time (min)

T_e	Emission stream temperature (°F)
TCC	Total capital costs ($)
v	Gas flow velocity (ft/min)

REFERENCES

1. J. Happel, and H. Brenner, *Low Reynolds Number Hydrodynamics with Special Applications to Particulate Media*, Prentice-Hall, Englewood Cliffs, NJ, 1965.
2. C. E. Williams, T. Hatch, and L. Greenberg, *Heating Piping Air Condit.* 12, 259 (1965).
3. C. E. Billings and J. E. Wilder, *Proc. EPA Symp. Control Fine-Particulate Emissions from Industrial Sources,* 1974.
4. C. E. Billings, and J. E. Wilder, *Handbook of Fabric Filter Technology*, Vol. 1., NTIS No. PB 200 648, (1970).
5. F. W. Cole, *Filtrat. Sep.* 17–25 (1975).
6. D. B. Purchas, *Proc. Filtration in Process Plant Design and Development: Liquid–Solids Separation* (1971).
7. *The Fabric Filter Manual*, Chap. III, The McIlvaine Company, Northbrook, IL, 1975.
8. US EPA, *OAQPS Control Cost Manual*, 4th ed., EPA 450/3–90–006 (NTIS PB90–16954), US Environmental Protection Agency, Washington, DC, 1990.
9. G. Parkins, *Chem. Eng.* **96(4)** (1989).
10. R. Dennis, and J. E. Wilder, *Fabric Filter Cleaning Studies*, EPA/650/2–75–009, US Environmental Protection Agency, Washington, DC, 1975.
11. H. E. Hesketh, *Understanding and Controlling Air Pollution*, 2nd ed., Ann Arbor Science, publ., Ann Arbor, MI, 1974.
12. S. A. Reigel, R. P. Bundy, and C. D. Doyle, *Pollut. Eng.* **5(5)** (1973).
13. US EPA, *Control Technologies for Hazardous Air Pollutants*, EPA/625–91/014, US Environmental Protection Agency, Washington, DC, 1994.
14. P. C. Siebert, *Handbook on Fabric Filtration,* ITT Research Institute, Chicago, IL, 1977.
15. US EPA, *Handbook of Fabric Filter Technology, Volume 1: Fabric Filter Systems Study*, APTD 0690 (NTIS PB 200648), US Environmental Protection Agency, Washington, DC, 1970.
16. US EPA, *Capital and Operating Costs of Selected Air Pollution Control Systems*, EPA/450/5–80–002 (NTIS PB80–157282), US Environmental Protection Agency, Washington, DC, 1978.
17. US EPA *Control Techniques for Particulate Emissions from Stationary Sources—Volume 2,* EPA/450/3–81–005b (NTIS PB83–127480), US Environmental Protection Agency, Washington, DC, 1982.
18. US EPA, *Control Techniques for Particulate Emissions from Stationary Sources—Volume 1*, EPA/450/3–81–005a (NTIS PB83–127498), US Environmental Protection Agency, Washington, DC, 1982.
19. W. Strauss, *Industrial Gas Cleaning*, 2nd ed., Pergamon, Oxford (1975).
20. US EPA, *Procedures Manual for Fabric Filter Evaluation*, EPA/600/7–78–113 (NTIS PB 283289), U.S. Environmental Protection Agency, Washington, DC, 1978.
21. US EPA, *Air Pollution Engineering Manual,* AP-40 (NTIS PB 225132), US Environmental Protection Agency, Washington, DC, 1973.
22. US EPA, *Particulate Control Highlights: Research on Fabric Filtration Technology*, EPA/600/8–78/005d (NTIS PB 285393), US Environmental Protection Agency, Washington, DC, 1978.
23. US EPA, *Baghouse Efficiency on a Multiple Hearth Incinerator Burning Sewage Sludge*, EPA 600/2/89–016 (NTIS PB89–190318), US Environmental Protection Agency, Washington, DC, 1990.

24. US EPA, *Handbook: Guidance on Setting Permit Conditions and Reporting Trial Burn Results*, EPA625/6–89–019, US Environmental Protection Agency, Washington, DC, 1989.
25. N. C. Durham, Company data for the municipal waste combustion industry. PES, Inc., 1990.
26. US EPA, *Control of Air Emission from Superfund Sites*, EPA/625/R-92/012, US Environmental Protection Agency, Washington, DC, 1992.
27. Anon. Equipment indices, *Chemical Engineering,* McGraw-Hill, New York, March (2002).
28. N. Niro, *Chem. Eng. Prog.*, **97(10)**, 24 (2001).
29. N. Swagelok, SCF Series gas filter, *Chem. Eng. Prog.*, **97(10)**, 53 (2001).
30. N. Wynn, *Chem. Eng. Prog.*, **97(10)**, 66–72 (2001).
31. L. K. Wang, J. V. Krouzek, and U. Kounitson, *Case Studies of Cleaner Production and Site Remediation,* Training Manual No. DTT–5–4–95, United Nations Industrial Development Organization, Vienna, 1995.
32. Anon. Equipment indices. *Chemical Engineering*, McGraw-Hill, New York, Sept. (1986).
33. US EPA, *Control Techniques for Fugitive VOC Emissions from Chemical Process Facilities*, EPA/625/R–93/005. US Environmental Protection Agency, Cincinnati, OH (1994).
34. D. Corbin. *Environ. Protect.* **14(1)** 26–27 (2003).
35. H. E. Hesketh, Fabric filtration. *Handbook of Environmental Engineering, Volume 1, Air and Noise Pollution Control* (L. K., Wang, and N. C. Pereira, eds.), Humana, Totowa, NJ, 1979, pp. 41–60.
36. US EPA. Fabric Filtration Design and Baghouse Components. APTI Virtual Classroom, Lesson 1. http://yosemite.epa.gov. US Environmental Protection Agency, Washington DC. Jan. 23, 2004.
37. L. K. Wang, N. C. Pereira, and Y.-T. Hung (eds.). *Advanced Air and Noise Pollution Control.* Humana Press, Totowa, NJ, 2005.

APPENDIX 1

HAP EMISSION STREAM DATA FORM*

Company ______________ Plant contact ______________
Location (Street) ______________ Telephone No ______________
(City) ______________ Agency contact ______________
(State, Zip)______________ No. of Emission Streams Under Review __________

A. Emission Stream Number/Plant Identification ______________
B. HAP Emission Source (a)__________ (b)__________ (c)__________
C. Source Classification (a)__________ (b)__________ (c)__________
D. Emission Stream HAPs (a)__________ (b)__________ (c)__________
E. HAP Class and Form (a)__________ (b)__________ (c)__________
F. HAP Content (1,2,3)** (a)__________ (b)__________ (c)__________
G. HAP Vapor Pressure (1,2) (a)__________ (b)__________ (c)__________
H. HAP Solubility (1,2) (a)__________ (b)__________ (c)__________
I. HAP Adsorptive Prop. (1,2) (a)__________ (b)__________ (c)__________
J. HAP Molecular Weight (1,2)(a)__________ (b)__________ (c)__________
K. Moisture Content (1,2,3) __________ P. Organic Content (1)***__________
L. Temperature (1,2,3) __________ Q. Heat/O_2 Content (1)__________
M. Flow Rate (1,2,3) __________ R. Particulate Content (3)__________
N. Pressure (1,2) __________ S. Particle Mean Diam.(3)__________
O. Halogen/Metals (1,2) __________ T. Drift Velocity/SO_3 (3)__________
U. Applicable Regulation(s) ______________
V. Required Control Level ______________
W. Selected Control Methods ______________

*The data presented are for an emission stream (single or combined streams) prior to entry into the selected control method(s). Use extra forms if additional space is necessary (e.g., more than three HAPs) and note this need.

**The numbers in parentheses denote what data should be supplied depending on the data on lines C and E:

1 = organic vapor process emission
2 = inorganic vapor process emission
3 = particulate process emission

***Organic emission stream combustibles less HAP combustibles shown on lines D and F.

APPENDIX 2

METRIC CONVERSIONS

Nonmetric	Multiplied by	Yields metric
MMBtu/h	1054.35	MM J/h
°F	0.555556(°F-32)	°C
ft	0.3048	m.
acfm	0.028317	acmm
dscfm	0.028317	dscmm
gal	3.78541	L
hp	746	J/s
in.	2.54	cm
lb	0.453592	kg
mil	0.0254	mm
mile	1609.344	m.
ton	0.907185	Metric ton (1000 kg)
yd^3	0.76455	m^3

3
Cyclones

José Renato Coury, Reinaldo Pisani, Jr., and Yung-Tse Hung

CONTENTS

1. INTRODUCTION

The cyclone is a well-known device used primarily for solid–fluid separation. It has been extensively utilized and studied for more than a century, and much has been written about it in the technical and scientific literature.

The objective of this chapter is to provide a practical view on cyclone performance, presenting correlations useful for its evaluation and design. Theoretical aspects are kept to a minimum while emphasizing workable correlations leading to cyclones with predictable performances. This chapter targets the readers with adequate knowledge in physics and mathematics. Those seeking in-depth information on the fluid dynamics of this device are advised to look at some excellent texts available, such as the ones by Licht (1), Leith and Jones (2), Ogawa (3), Bohnet and colleagues (4,5), and Boysan et al. (6). This list is by no means exclusive.

The chapter is divided in three parts. The first part deals with the cyclones utilized in industrial applications, intended to perform solid–gas separation of relatively large volumes of effluents. This is the aspect for which the cyclones are mostly known and deals with the search for a configuration capable for removing the solid as efficiently as possible, with a minimum of power consumption. The second part addresses cost analysis of cyclones and auxiliary process equipment, such as fans, ductwork, dampers, and stacks. The third part addresses a more recent application for cyclones: its use as a sampler for environmental and occupational monitoring. In this case, the main objective is to have a device with a performance that can be related to the current criteria adopted by the legislation. In all parts, correlations are presented and discussed, and examples of calculations are given.

From: *Handbook of Environmental Engineering, Volume 1: Air Pollution Control Engineering*
Edited by: L. K. Wang, N. C. Pereira, and Y.-T. Hung © Humana Press, Inc., Totowa, NJ

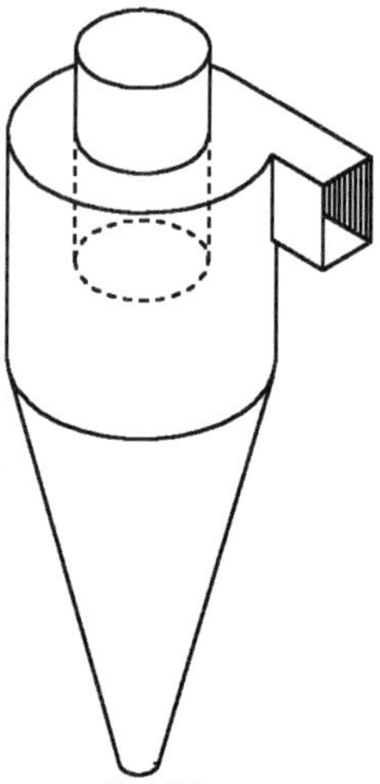

Fig. 1. Reverse flow cyclone, with tangential entry.

2. CYCLONES FOR INDUSTRIAL APPLICATIONS

2.1. *General Description*

Cyclones are one of the most utilized devices for solid–gas separation. It works by forcing the gaseous suspension to flow spirally (thus the name cyclone) within a confined space, so that the particles are expelled toward the walls of the vessel by centrifugal force. Once on the walls, the particles move downward, mainly by gravity, and are removed from the cyclone, whereas the gas spins out, usually upward. Several geometries are in use, but the conical–cylindrical reverse-flow type, illustrated in Fig. 1, predominates. Also, a variety of cyclone entries can be utilized, as the ones shown in Fig. 2, depending on the application.

Cyclones have a wide range of industrial applications either in product recovery or in gas cleaning and can be found in virtually every site where powder handling takes place. It is relatively inexpensive and easy to construct, requires little maintenance, and can, in principle, work at high temperatures and pressures. Depending on the process, it can be used as a precollector, for removing larger particles before bag filters or electrostatic precipitators. If well designed, the cyclone collects particles larger than 10 μm with good efficiency. For smaller particles, the efficiency drops considerably. It cannot be used in the processing of sticky particles or with solids with a high moisture content, as caking and clogging may occur.

The range of gas flow rates that can be treated by a cyclone is wide, spanning from 50 to 50,000 m^3/h. However, it is common practice to divide the flow in parallel cyclones when the total flow rate exceeds 20,000 m^3/h, to avoid scale problems associated with blowers and other ancillary items. Moreover, smaller cyclones are usually more efficient and work at smaller pressure drops (2,7).

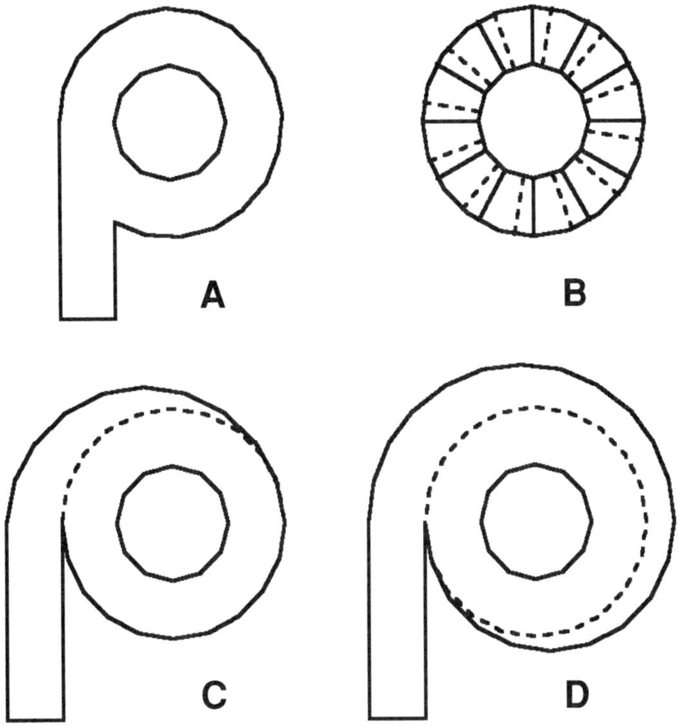

Fig. 2. Cyclone entries: (A) tangential; (B) swirl vane; (C) half scroll; (D) full scroll.

Attention here will be focused on the reverse-flow cyclone, with tangential entry, which is by far the most common geometry found in practice. The main dimensions of the device are illustrated in Fig. 3 and will be referred to in the remaining of this text: the cylindrical body has a diameter D_c and a height h; the conical section has a height Z and ends in the solids outlet of diameter B; the gas entry can be either circular (diameter D_{in}) or rectangular (height a, width b); the exit duct, also called vortex finder, has a diameter D_e and starts at a distance S from the top of the cylindrical body; the cyclone has a total height H.

In the cylindrical body, where the entrance and the exit of the gaseous current are placed, the gas begins a spiral movement down, creating a centrifugal field, intensified in the conical part, that impels the particles in the direction of the wall of the equipment, where they are collected. There is a stagnated region close the wall, as a result of the laminar sublayer, which allows the fall of the collected particles into the reservoir situated in the base of the equipment. In opposition to the centrifugal force, there is the drag force caused by the radial movement of the gas stream toward the central axis of the cyclone, and the turbulence of the gaseous current, whose combined effect is to carry the noncollected particles to the exit duct (8,9).

The movement of the gas is constituted by an external downward vortex and an internal upward vortex (10). The intermediate region between these vortexes defines the central axis of the cyclone, with height and diameter Z_c or h^m and d_c, respectively, delimited by the diameters of the exits B and D_e. In the external vortex, the tangential

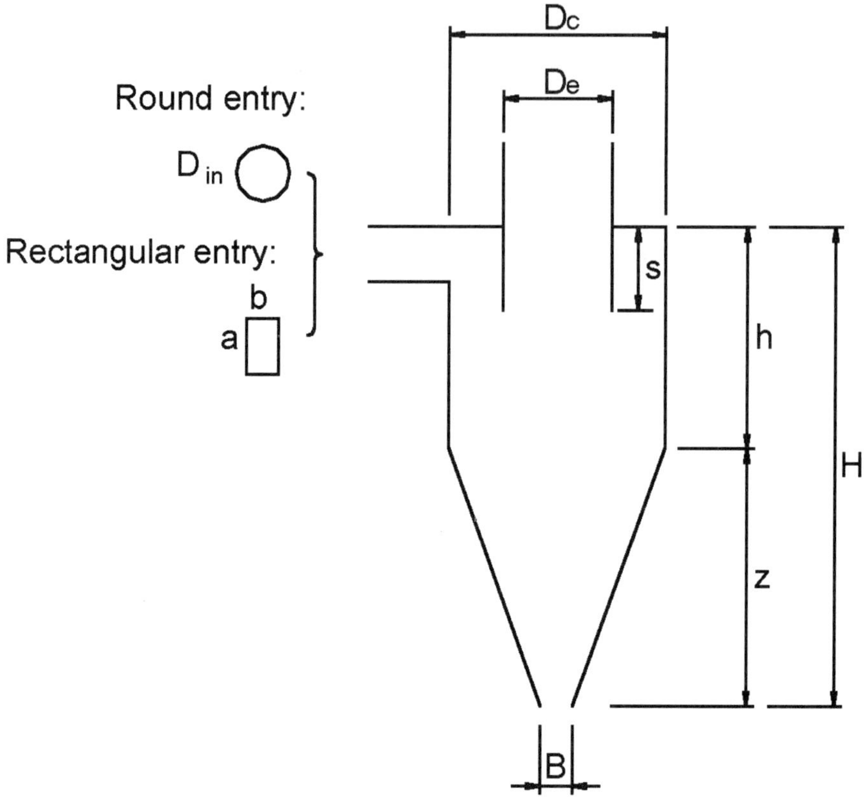

Fig. 3. Cyclone dimensions.

velocity of the gas (v_t) increases with the decrease of the radial position to a maximum value ($v_{t\max}$), and in the internal vortex the tangential velocity decreases toward the center (6,11).

Several theories were proposed in the past 50 yr to predict the performance of cyclones in terms of collection efficiency and pressure drop associated with the operation. These parameters are related to the cyclone dimensions, to the physical properties of the solid and of the gas, and to operational conditions such as gas velocity in the entrance, temperature, and pressure. The adopted concepts are based on distinct approaches such as the following: the trajectory of the particle derived from the balance of forces acting on it, as in the models of Barth (12) and Iozia and Leith (10,13); the residence time of the particles inside the device, as in Lapple (14), Leith and Licht (9), and Dietz (8); or the derivation of dimensionless numbers that correlate the collection efficiency with the operational conditions, as the one of Buttner (15) and Moore and McFarland (16).

Next, the calculation procedure for designing a cyclone based on the models of Barth (12), Leith and Licht (9), and Iozia and Leith (13) will be presented, followed by an example of application and by the comparison of the results with the experimental data obtained by Dirgo and Leith (17). The models were chosen based on their objectivity and also with the intention of giving a chronological overview of the subject.

Table 1
Dimensions for the Design of Standard Cyclones

Family:	Lapple	Swift	Stairmand	Swift	Stairmand	Swift
Use:	General purpose	General purpose	High efficiency	High efficiency	High flow rate[a]	High flow rate[a]
Q/D_c^2 (m^3/h)	6,860	6,680	5,500	4,940	16,500	12,500
a/D_c	0.5	0.5	0.5	0.44	0.75	0.8
b/D_c	0.25	0.25	0.2	0.21	0.375	0.35
H/D_c	4.0	3.75	4.0	3.9	4.0	3.7
h/D_c	2.0	1.75	1.5	1.4	1.5	1.7
D_e/D_c	0.5	0.5	0.5	0.4	0.75	0.75
B/D_c	0.25	0.4	0.375	0.4	0.375	0.4
S/D_c	0.625	0.6	0.5	0.5	0.875	0.85
ΔH	8.0	7.6	6.4	9.2	7.2	7.0

[a]Half-scroll entry.
Source: ref. 2.

2.2. *Correlations for Cyclone Efficiency*

The design of cyclones is usually based on seven geometrical relations of the above-mentioned dimensions. These dimensions, when expressed as fractions of the body diameter D_c, are fixed for a given cyclone "family" that can also be subdivided, according to their use (high efficiency, high flow rate, general purpose). Table 1, taken from Leith and Jones (2), lists the dimensional relations for the classical Lapple, Stairmand, and Swift cyclone families and includes a criterion for the adoption of D_c based on the gas volumetric flow rate (in m^3/h).

2.2.1. *The Barth Model*

The Barth model (12) predicts the collection efficiency as a function of the relation between the terminal velocities of a particle of a given diameter and the particle collected with 50% efficiency, which has equal probability of being or not collected (17,18). For a particle of diameter D_i, the efficiency is thus given by

$$\eta_i = \frac{1}{\left[1+\left(v_{ts}/v_{ts}^m\right)^{-3.2}\right]} \tag{1}$$

where v_{ts} and v_{ts}^m are the terminal velocities for the particle and for the one with 50% collection, respectively.

The ratio v_{ts}/v_{ts}^m can be related to the mean radial velocity of the gas in the central axis of the cyclone, where the maximum tangential velocity, v_{tmax}, occurs. Assuming Stokes law and negligible gas density, this ratio can be expressed as

$$\frac{v_{ts}}{v_{ts}^m} = \frac{\pi h^m \rho_p {v_{t\max}}^2 D_i^2}{9\mu Q} \tag{2}$$

The height of the cyclone central axis, h^m, is limited by the gas exit duct diameter, D_e, and by the dust exit diameter, B, and can be estimated from Eq. (3) or (4):

$$h^m = H - S, \text{ for } D_e \leq B \tag{3}$$

$$h^m = \frac{(H - h)(D_c - D_e)}{(D_c - B)} + (h - S), \text{ for } D_e \geq B \tag{4}$$

According to Barth, the maximum tangential velocity, $v_{t\max}$, can be obtained by the following correlation:

$$v_{t\max} = v_0 \left[\frac{(D_e/2)(D_c - b)\pi}{2ab\alpha + h^m (D_c - b)\pi\lambda} \right] \tag{5}$$

where v_0 is gas velocity at the cyclone exit, given by

$$v_0 = \frac{4Q}{\pi D_e^2} \tag{6}$$

The parameter λ is the friction factor, for which the value 0.02 is suggested. The parameter α can be related to the dimensions b and D_c by

$$\alpha = 1 - 1.2(b/D_c) \tag{7}$$

2.2.2. *The Leith and Licht Model*

The Leith and Licht model (9,58) is based on the assumption that the noncollected particles are fully mixed in the radial direction at a given point of the axial position, because of turbulence. Therefore, the residence time of the particle inside the device can be associated with the time it needs to move in the radial and axial directions in order to reach the wall (1,19). This principle is semitheoretically treated in a number of equations and it results in the following expression for the particle collection efficiency:

$$\eta_i = 1 - \exp\left\{ -2 \left[\frac{G\tau_i Q}{D_c^3} (n + 1) \right]^{\frac{1}{2n+2}} \right\} \tag{8}$$

where G is a dimensionless geometry parameter, n is the vortex exponent, and τ_i is the relaxation time.

The geometry parameter G is expressed in terms of the dimensions of the cyclone families and can be written as

$$G = \frac{D_c}{a^2 b^2} \left\{ 2\left[\pi (S - a/2)(D_c^2 - D_e^2) \right] + 4V_{nl,H} \right\} \tag{9}$$

where $V_{nl,H}$ is an annular volume related to the vortex penetration inside the cyclone. Alexander (20) defines as "natural length" the distance below the bottom of the exit duct where the vortex turns. Depending on the value of Z_c, the volume to be considered in Eq. (9) is either V_{nl} or V_H, as follows:

$$V_{nl} = \frac{\pi D_c^2}{4}(h - S) + \left(\frac{\pi D_c^2}{4} \right) \left(\frac{Z_c + S - h}{3} \right) \left(1 + \frac{d_c}{D_c} + \frac{d_c^2}{D_c^2} \right) - \frac{\pi D_e^2 Z_c}{4} \tag{10}$$

$$\text{if } (H - S) > Z_c$$

or

$$V_H = \frac{\pi D_c^2}{4}(h-S) + \left(\frac{\pi D_c^2}{4}\right)\left(\frac{H-h}{3}\right)\left(1+\frac{B}{D_c}+\frac{B^2}{D_c^2}\right) - \frac{\pi D_e^2 (H-S)}{4} \quad (11)$$
$$\text{if } (H-S) < Z_c$$

The diameter of the cyclone central axis, d_c, is given by

$$d_c = D_c - (D_c - B)\left(\frac{S + Z_c - h}{H - h}\right) \quad (12)$$

The vortex natural length, Z_c, and the vortex exponent, n, can be estimated by the following expressions (20):

$$Z_c = 2.3 D_e \left(\frac{D_c^2}{ab}\right)^{\frac{1}{3}} \quad (13)$$

$$n = 1 - \left[1 - 0.67\left(D_c^{0.14}\right)\right]\left(\frac{T}{283}\right)^{0.3} \quad (14)$$

with D_c given in meters and the gas temperature T in degrees Kelvin.

Equation (13), although often used, is not entirely satisfactory, because it does not include the dependence of Z_c on the gas velocity at the entrance, which was experimentally verified by Hoffmann et al. (21).

The relaxation time, τ_i, is given by

$$\tau_i = \frac{\rho_p (D_i)^2}{18\mu} \quad (15)$$

The parameter G is sometimes expressed in terms of K_a, K_b, and K_c, as

$$G = \frac{8K_c}{K_a^2 K_b^2} \quad (16)$$

with

$$K_a = \frac{a}{D_c} \quad (17)$$

$$K_b = \frac{b}{D_c} \quad (18)$$

and

$$K_c = \frac{(2V_s + V_{nl,H})}{D_c^3} \quad (19)$$

V_s is the annular volume between the central plane of the inlet duct and the bottom of the exit duct, S, and is given by

$$V_s = \pi(S - a/2)\frac{(D_c^2 - D_e^2)}{4} \quad (20)$$

Therefore, the average residence time of the gas inside the cyclone, θ, can be estimated by

$$\theta = \frac{\left(V_s + V_{nl,H}/2\right)}{Q} = \frac{K_c D_c^3}{Q} \tag{21}$$

A criterion for choosing the appropriate gas velocity at the entrance, v_i, is to compare it with the saltation velocity, v_s, which is the minimum velocity capable of "peeling off" particles from the wall. In principle, velocities larger than the saltation cause resuspension of the collected particles and a decrease in efficiency. According to Kalen and Zenz (22) and Koch and Licht (23), the recommended ratio is

$$1.20 \leq \frac{v_i}{v_s} \leq 1.35$$

in which v_s is given by

$$v_s = 2.055\left(\frac{4g\mu\rho_p}{3\rho^2}\right)^{1/3}\left\{\frac{\left(b/D_c\right)^{0.4}}{\left[1-\left(b/D_c\right)\right]^{1/3}}\right\}D_c^{\,0.067}v_i^{\,2/3} \tag{22}$$

for the variables in English units (lb_m, ft, and s). In general, $v_i/v_s = 1.25$ [valid for $T = 38°C$ and $\rho_p = 2580$ kg/m^3 (18)] is adopted for maximizing collection efficiency.

2.2.3. The Iozia and Leith Model

The Iozia and Leith model (10,13) departs from the Barth (12) model by proposing new equations for estimating the diameter and length of the cyclone central axis, the maximum tangential velocity, and the dependence of these variables on the device dimensions, based on 26 experiments performed at ambient temperature. The correlation proposed for the collection efficiency is based on the particle Stokes diameter, which is defined as the diameter of the sphere that has the same terminal velocity of the particle. The proposed expression can be written as

$$\eta_i = \frac{1}{\left[1+\left(D_{50}/D_i\right)^{\beta}\right]} \tag{23}$$

where D_{50} is the Stokes diameter of the particle with 50% collection efficiency, D_i is the Stokes diameter of the particle whose collection efficiency is being determined, and β is an exponent dependent of the cut diameter D_{50}.

The cut diameter can be estimated by

$$D_{50} = \left(\frac{9\mu Q}{\pi\rho_p Z_c v_{t\max}^2}\right)^{0.5} \tag{24}$$

The natural length Z_c in this model is estimated as a function of the diameter of the central axis, d_c, according to the expression

$$\frac{d_c}{D_c} = 0.47\left(\frac{ab}{D_c^2}\right)^{-0.25}\left(\frac{D_e}{D_c}\right)^{1.4} \tag{25}$$

For $d_c > B$,

$$Z_c = (H - S) - \left[\frac{(H - h)}{(D_c/B) - 1}\right]\left[\left(\frac{d_c}{B}\right) - 1\right] \tag{26}$$

For $d_c < B$,

$$Z_c = (H - S) \tag{27}$$

The maximum tangential gas velocity, $v_{t\max}$, is given by

$$v_{t\max} = 6.1 v_i \left(\frac{ab}{D_c^2}\right)^{0.61} \left(\frac{D_e}{D_c}\right)^{-0.74} \left(\frac{H}{D_c}\right)^{-0.33} \tag{28}$$

where v_i is the gas velocity at the cyclone entry:

$$v_i = \frac{Q}{ab} \tag{29}$$

The exponent β in Eq. (23) is dependent on the cut diameter, and a correlation was derived from 11 experiments at ambient temperature (13,24,25):

$$\ln\beta = 0.62 - 0.87\ln(D_{50}) + 5.21\ln\left(\frac{ab}{D_c^2}\right) + 1.05\left[\ln\left(\frac{ab}{D_c^2}\right)\right]^2 \tag{30}$$

with D_{50} in centimeters.

2.3. *Correlations for Cyclone Pressure Drop*

The pressure drop, ΔP, in a particle-free cyclone can be estimated by

$$\Delta P = \frac{\rho v_i^2}{2}\Delta H \tag{31}$$

where ΔH is a dimensionless parameter that depends on the cyclone geometry (1,23) and can be calculated by the following correlation proposed by Shepherd and Lapple (26):

$$\Delta H = 16\left(\frac{ab}{D_e^2}\right) \tag{32}$$

Alternatively, Casal and Benet (27), after performing a number of experimental tests, adjusted (with a standard deviation of 1.61 against the 2.58 of Shepherd and Lapple) the following expression:

$$\Delta H = 11.3\left(\frac{ab}{D_c^2}\right)^2 + 3.33 \tag{33}$$

Ramachandran et al. (24), based on 98 cyclone configurations, statistically determined the best correlation for predicting ΔH that had the following form:

$$\Delta H = 20\left(\frac{ab}{D_e^2}\right)\left[\frac{S/D_c}{(H/D_c)(h/D_c)(B/D_c)}\right]^{\frac{1}{3}} \tag{34}$$

2.4. Other Relations of Interest

2.4.1. The Effect of Particle Loading

The pressure drop in cyclones decreases with increasing particle load in the gas, and collection efficiency increases. This is attributed to the impact of the larger particles against the smaller ones, forcing them toward the stagnation region near the wall (17,28,29). This effect can be quantified by the following expressions:

$$\frac{\Delta P_{\text{at } c_{\text{in}}}}{\Delta P_{\text{at } c_{\text{in}}=0}} = 1 - 0.013 c_{\text{in}}^{0.5} \quad (35)$$

$$\frac{100 - \eta_{01}}{100 - \eta_{02}} = \left(\frac{c_{\text{in2}}}{c_{\text{in1}}} \right)^{0.182} \quad (36)$$

with c_{in} in grains per cubic feet.

2.4.2. Cyclones in Parallel (Multicyclones)

The pressure drop in multicyclones (dozens, or even hundreds of cyclones associated in parallel) can be estimated by Eq. (37), which is a function of the total volumetric flow rate, Q, the geometrical parameters K_a and K_b, and the diameter of the cyclone body, D_c (7):

$$\Delta P = \left(\frac{\Delta H \rho Q^2}{2 K_a^2 K_b^2 N_c^2 D_c^4} \right) \quad (37)$$

Multicyclones present a ΔP considerably smaller than a single cyclone, for the same collection efficiency. The division of the total volumetric flow rate among the number of N_c cyclones in Eq. (8) gives for collection efficiency of multicyclones:

$$\eta_i = 1 - \exp\left\{ -2 \left[\frac{G \tau_i Q}{N_c D_c^3} (n+1) \right]^{\frac{1}{2n+2}} \right\} \quad (38)$$

Nevertheless, the difficulties arising from distributing the dust-laden gas uniformly among the cyclones results, in practice, in smaller collection efficiencies than that predicted by Eq. (38).

2.4.3. Cyclones in Series

In case a second cyclone in series is needed, it is necessary to calculate the size distribution in the exit of the first cyclone, which will be the feed of the next. This size distribution is easily obtained from a mass balance for each mass fraction that can be expressed in terms of mass flow rate of particles with diameter D_i at the exit by the relation

$$m_{0i} = c_{\text{in}} Q x_i (1 - \eta_i) \quad (39)$$

Therefore, the size fraction of a particle of diameter D_i at the first cyclone exit is

$$x_{0i} = \frac{m_{0i}}{\sum m_{0i}} \quad (40)$$

Table 2
Grade efficiency (%) of a High-Efficiency Stairmand Cyclone as a Function of the Gas Velocity at the Entrance, v_i

D_i (μm)	v_i = 5 m/s	10 m/s	15 m/s	20 m/s	25 m/s
1.4	0.0	2.2	4.4	8.9	16.7
2.1	0.0	6.7	22.2	50.0	75.6
2.9	0.0	5.6	24.4	61.1	80.0
3.7	2.2	37.8	77.8	93.3	97.8
4.4	3.3	57.8	86.7	94.4	96.6
5.1	12.2	84.4	96.7	95.6	
5.8	33.3	93.3			
6.6	70.0				
7.4	77.8				

Source: ref. 17.

2.4.4. Overall Efficiency

Once a designer has adopted a given collection efficiency model and decided upon a given cyclone configuration, the designer is in a position to evaluate the response of that arrangement as far as the process at hand.

The overall efficiency of the cyclone η_o (i.e., the collected fraction of the total mass entering it) can be calculated from the grade efficiency η_i by

$$\eta_o = \sum(\eta_i x_i) \tag{41}$$

where x_i is the mass fraction of particles with diameter D_i collected with efficiency η_i.

The overall concentration at the cyclone exit can therefore be obtained as:

$$c_o = (1 - \eta_o)c_{\text{in}} \tag{42}$$

where c_{in} is the overall concentration at the entrance.

2.5. Application Examples

The three models presented above will be used for the prediction of the grade efficiency of a cyclone and the results will be compared to experimental measurements from the literature.

Dirgo and Leith (17) utilized a high-efficiency Stairmand cyclone, with a 0.305-m body diameter, at ambient temperature. They determined the grade efficiency and the pressure drop as a function of the gas velocity at the entrance, for spherical oil droplets with a density of 860 kg/m^3. Their results are shown in Tables 2 and 3.

Example 1

Design a high-efficiency Stairmand cyclone for a volumetric flow rate of 500.4 m^3/h and determine the grade efficiency curves utilizing the models of Barth, and Iozia and Leith for the entry gas velocities of 10, 15, and 20 m/s. Also, cstimate the maximum power consumption of a fan, neglecting the loss in the external ducts and assuming 55% efficiency for the fan motor. The particle density is 860 kg/m^3 and the temperature is 27°C. The air density and viscosity at 27°C are 1.18 kg/m^3 and 1.8×10^{-5} kg/m s, respectively.

Table 3
Pressure Drop in a High-Efficiency Stairmand Cyclone as a Function of the Gas Velocity at the Entrance, v_i

ΔP (Pa)	v_i (m/s)
87	5.1
336	10.0
785	15.0
1407	20.0
2205	25.0

Source: ref. 17.

Solution

Table 1 gives the high-efficiency Stairmand cyclone configuration, and the body diameter can be calculated as follows:

$$\frac{500.4 \text{ m}^3/\text{h}}{D_c^2} = 5500$$

$$D_c = 0.302 \text{ m}$$

The seven remaining dimensions are obtained from the other relations as

$$\frac{a}{0.302 \text{ m}} = 0.5 \rightarrow a = 0.151 \text{ m}$$

$$\frac{b}{0.302 \text{ m}} = 0.2 \rightarrow b = 0.060 \text{ m}$$

$$\frac{H}{0.302 \text{ m}} = 4.0 \rightarrow H = 1.208 \text{ m}$$

$$\frac{h}{0.302 \text{ m}} = 1.5 \rightarrow h = 0.453 \text{ m}$$

$$\frac{D_e}{0.302 \text{ m}} = 0.5 \rightarrow D_e = 0.151 \text{ m}$$

$$\frac{B}{0.302 \text{ m}} = 0.375 \rightarrow B = 0.113 \text{ m}$$

$$\frac{S}{0.302 \text{ m}} = 0.5 \rightarrow S = 0.151 \text{ m}$$

Barth Model

For $v_i = 10$ m/s, the gas volumetric flow rate [Eq. (29)] is given by

$$Q = \left(10\frac{\text{m}}{\text{s}}\right)(0.151 \text{ m})\ (0.060 \text{ m})$$

$$Q = \left(0.0906\frac{\text{m}^3}{\text{s}}\right)\left(\frac{3600\text{s}}{1 \text{ h}}\right) = 326.2 \text{m}^3/\text{h}$$

The gas velocity at the exit [Eq. (6)] is

$$v_o = \frac{4(0.0906\ \text{m}^3/\text{s})}{\pi(0.151\ \text{m})^2}$$

$$v_o = 5.1\ \text{m/s}$$

The values of α [Eq. (7)] and h^m [Eq. (4), as $D_e > B$] are given by

$$\alpha = 1 - 1.2(0.060\ \text{m}/0.302\ \text{m})$$

$$\alpha = 0.76$$

$$h^m = \frac{(1.208\ \text{m} - 0.453\ \text{m})(0.302\ \text{m} - 0.151\ \text{m})}{0.302\ \text{m} - 0.113\ \text{m}} + (0.453\ \text{m} - 0.151\ \text{m})$$

$$h^m = 0.905\ \text{m}$$

Once α and h^m only depend on the cyclone dimensions, their values remain constant for the entry velocities of 10, 15, and 20 m/s, as well as the value of λ (0.02).

The maximum tangential velocity [Eq. (5)] is

$$v_{t\max} = 5.1\,\text{m/s}\left[\frac{(0.151\,\text{m}/2)(0.302\ \text{m} - 0.060\ \text{m})\pi}{2[(0.151\ \text{m})(0.060\ \text{m})(0.76)] + 0.905\ \text{m}(0.302\ \text{m} - 0.060\ \text{m})(\pi)0.02}\right]$$

$$v_{t\max} = 10.6\ \text{m/s}$$

Now v_{ts}/v_{ts}^m can be calculated as a function of particle diameter [Eq. (2)]:

$$\frac{v_{ts}}{v_{ts}^m} = \frac{\pi(0.905\ \text{m})(860\,\text{kg/m}^3)(10.6\,\text{m/s})^2 D_i^2}{9(1.8\times10^{-5}\,\text{kg/ms})(0.0906\,\text{m}^3/\text{s})}$$

$$\frac{v_{ts}}{v_{ts}^m} = 1.8718\times10^{10}\ D_i^2$$, with D_i in meters.

The collection efficiency [Eq. (1)] can therefore be written as

$$\eta_i = \frac{1}{\left[1 + (1.8718\times10^{10}\text{D}_i^2)^{-3.2}\right]}$$, with D_i in meters.

For the other entry velocities, the results are as follows:

For v_i= 15 m/s,

$$Q = 0.139\,\frac{\text{m}^3}{\text{s}}\,\frac{3600\text{s}}{1\ \text{h}} = 500.4\ \text{m}^3/\text{h}$$

$$v_o = 7.8\ \text{m/s}$$

$$v_{t\max} = 16.3\ \text{m/s}$$

$$\frac{V_{ts}}{V_{ts}^{m}} = 2.8850 \times 10^{10} D_i^2, \text{ with } D_i \text{ in meters.}$$

$$\eta_i = \frac{1}{\left[1 + \left(2.885 \times 10^{10} D_i^2\right)^{-3.2}\right]}, \text{ with } D_i \text{ in meters.}$$

For $v_i = 20.0$ m/s,

$$Q = 0.1812 \frac{\text{m}^3}{\text{s}} \frac{3600\text{s}}{\text{h}} = 652.3 \text{ m}^3/\text{h}$$

$$v_o = 10.1 \text{ m/s}$$

$$v_{t\max} = 21.1 \text{ m/s}$$

$$\frac{V_{ts}}{V_{ts}^{m}} = 3.7083 \times 10^{10} D_i^2, \text{ with } D_i \text{ in meters.}$$

$$\eta_i = \frac{1}{\left[1 + \left(3.7083 \times 10^{10} D_i^2\right)^{-3.2}\right]}, \text{ with } D_i \text{ in meters.}$$

Figure 4 shows the efficiency curves obtained with the Barth model compared to the experimental results of Dirgo and Leith (17), for the same operating conditions (note that the difference in D_c is of only 3 mm).

Leith and Licht Model

For $v_i = 10$ m/s, the length and diameter of the vortex are calculated from Eqs. (13) and (12), respectively:

$$Z_c = 2.3(0.151 \text{ m})\left(\frac{(0.302 \text{ m})^2}{(0.151 \text{ m})(0.060 \text{ m})}\right)^{\frac{1}{3}}$$

$$Z_c = 0.750 \text{ m}$$

$$d_c = 0.302 \text{ m} - (0.302 \text{ m} - 0.113 \text{ m})\left(\frac{0.151 \text{ m} + 0.750 \text{ m} - 0.453 \text{ m}}{1.208 \text{ m} - 0.453 \text{ m}}\right)$$

$$d_c = 0.190 \text{ m}$$

Because $H–S = 1.208$ m–0.151 m = 1.057 m > Z_c, the volume of the cyclone natural vortex is given by Eq. (10):

$$V_{nl} = \frac{\pi(0.302 \text{ m})^2}{4}(0.453 \text{ m} - 0.151 \text{ m}) + \left(\frac{\pi(0.302 \text{ m})^2}{4}\right)$$
$$\left(\frac{0.750 \text{ m} + 0.151 \text{ m} - 0.453 \text{ m}}{3}\right)$$
$$\left(1 + \frac{0.190 \text{ m}}{0.302 \text{ m}} + \frac{(0.190 \text{ m})^2}{(0.302 \text{ m})^2}\right) - \frac{\pi(0.151 \text{ m})^2 0.750 \text{ m}}{4}$$

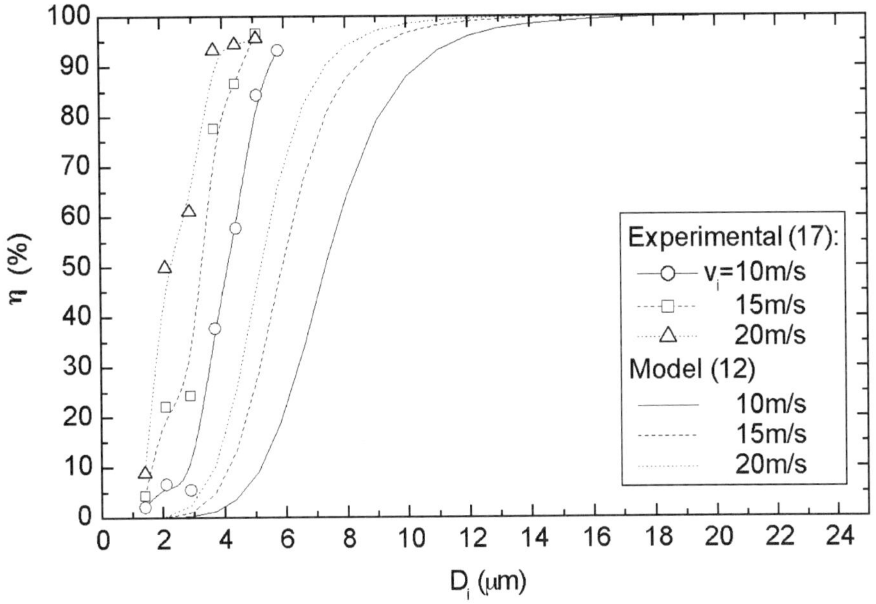

Fig. 4. Collection efficiency as a function of particle diameter, for three entry velocities: curves obtained from the Barth model (12) and experimental points from Dirgo and Leith (17).

$$V_{nl} = 0.0299 \text{ m}^3$$

Once V_{nl} is known, the geometry parameter is calculated from Eq. (9):

$$G = \frac{0.302 \text{ m}}{(0.151 \text{ m})^2(0.060 \text{ m})^2}\left\{2\left[\pi(0.151 \text{ m} - 0.151 \text{ m}/2)\left((0.302 \text{ m})^2 - (0.151 \text{ m})^2\right)\right] + 4\left(0.0299 \text{ m}^3\right)\right\}$$

$$G = 559.4$$

The vortex exponent [Eq. (14)] is obtained from D_c and the operation temperature as

$$n = 1 - \left[1 - 0.67\left(0.302^{0.14}\right)\right]\left(\frac{(27 + 273)}{283}\right)^{0.3}$$

$$n = 0.56$$

The relaxation time [Eq. (15)] can now be calculated as a function of particle diameter:

$$\tau_i = \frac{860 \text{ kg/m}^3 (D_i)^2}{18\left(1.8 \times 10^{-5} \text{ kg/ms}\right)}$$

$$\tau_i = 2654321 D_i^2$$, with D_i in meters.

Therefore, the grade collection efficiency [Eq. (8)] results in

$$\eta_i = 1 - \exp\left\{-2\left[\frac{559.4(2654321 D_i^2)0.0906\,\mathrm{m^3/s}}{(0.302\ \mathrm{m})^3}(0.56+1)\right]^{\frac{1}{2(0.56)+2}}\right\}$$

$$\eta_i = 1 - \exp\left(-2905.6 D_i^{0.64}\right),\ \text{with } D_i \text{ in meters.}$$

If only Eq. (8) is dependent on the flow rate, the procedure applied to the other velocities is as follows:

For $v_i = 15$ m/s,

$$Q = 0.139\,\frac{\mathrm{m^3}}{\mathrm{s}}\,\frac{3600\ \mathrm{s}}{1\ \mathrm{h}} = 500.4\ \mathrm{m^3/h}$$

$$\eta_i = 1 - \exp\left(-3332.2 D_i^{0.64}\right)\ \text{with } D_i \text{ in meters.}$$

For $v_i = 20$ m/s,

$$Q = 0.1812\,\frac{\mathrm{m^3}}{\mathrm{s}}\,\frac{3600\mathrm{s}}{1\ \mathrm{h}} = 652.3\,\mathrm{m^3/h}$$

$$\eta_i = 1 - \exp\left(-3627.2 D_i^{0.64}\right)\ \text{with } D_i \text{ in meters.}$$

Figure 5 shows the efficiency curves obtained with the Leith and Licht model compared to the experimental results of Dirgo and Leith (17), for the same operating conditions.

Iozia and Leith Model

For $v_i = 10$ m/s, we can start by calculating the diameter of the cyclone's central axis [Eq. (25)]:

$$\frac{d_c}{0.302\ \mathrm{m}} = 0.47\left(\frac{0.151\ \mathrm{m}\ 0.060\ \mathrm{m}}{(0.302\ \mathrm{m})^2}\right)^{-0.25}\left(\frac{0.151\ \mathrm{m}}{0.302\ \mathrm{m}}\right)^{1.4}$$

$$d_c = 0.096\ \mathrm{m}$$

Because $d_c < B$, Eq. (27) is utilized for the calculation of Z_c:

$$Z_c = (1.208\ \mathrm{m} - 0.151\ \mathrm{m})$$

$$Z_c = 1.057\ \mathrm{m}$$

The maximum tangential velocity is calculated from Eq.(28):

$$v_{t\max} = 6.1(10\,\mathrm{m/s})\left(\frac{(0.151\ \mathrm{m})(0.060\ \mathrm{m})}{(0.302\ \mathrm{m})^2}\right)^{0.61}\left(\frac{0.151\ \mathrm{m}}{0.302\ \mathrm{m}}\right)^{-0.74}\left(\frac{1.208\ \mathrm{m}}{0.302\ \mathrm{m}}\right)^{-0.33}$$

$$v_{t\max} = 15.8\,\mathrm{m/s}$$

Knowing Q, Z_c, $v_{t\max}$, and the properties of the gas and particles, the diameter is calculated from Eq. (24):

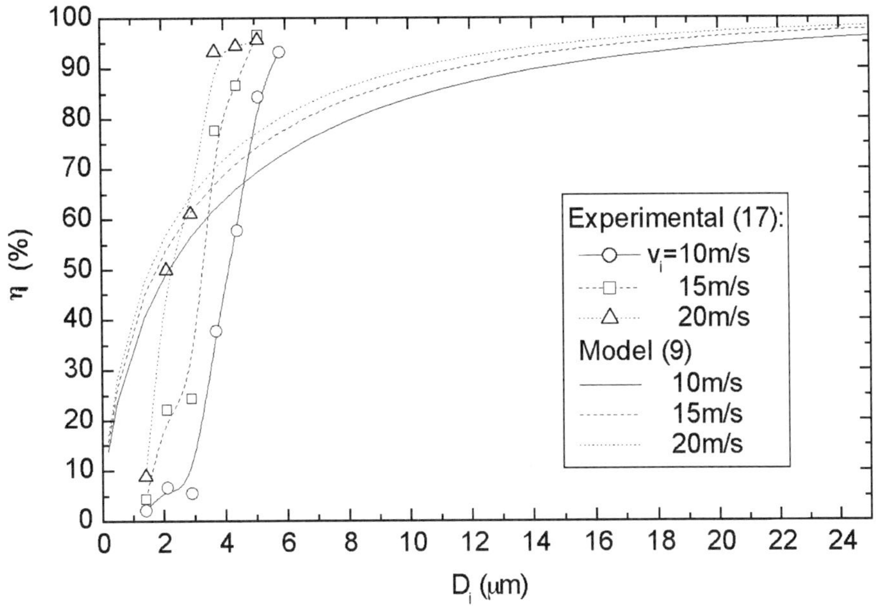

Fig. 5. Collection efficiency as a function of particle diameter, for three entry velocities: curves obtained from the Leith and Licht model (9) and experimental points from Dirgo and Leith (17).

$$D_{50} = \left(\frac{9(1.8 \times 10^{-5}\,\mathrm{kg/ms})(0.0906\,\mathrm{m^3/s})}{\pi(860\,\mathrm{kg/m^3})(1.057\,\mathrm{m})(15.8\,\mathrm{m/s})^2} \right)^{0.5}$$

$$D_{50} = 4.5 \times 10^{-6}\,\mathrm{m} = 4.5\ \mu\mathrm{m}$$

The exponent β is a function of the cut diameter (in centimeters) and of the cyclone dimensions [Eq. (30)]. It is therefore implicitly dependent on the gas entry velocity:

$$\ln\beta = 0.62 - 0.87\ln\left(4.5\times10^{-4}\right) + 5.21\ln\left(\frac{(0.151)(0.060)}{(0.302)^2}\right) + 1.05\left[\ln\left(\frac{(0.151)(0.060)}{(0.302)^2}\right)\right]^2$$

$$\beta = 2.44$$

Therefore, the grade efficiency [Eq. (23)] at 10.0 m/s can be written as

$$\eta_i = \frac{1}{\left[1 + (4.5/D_i)^{2.44}\right]}, \text{ with } D_i \text{ in micrometers.}$$

For v_i = 15 m/s, the vortex diameter and length, d_c and Z_c, do not depend on the entry velocity. Thus, the calculated values hold:

$$d_c = 0.096 \text{ m}$$

$$Z_c = 1.057 \text{ m}$$

$$v_{t\max} = 23.6 \text{ m/s}$$

$$D_{50} = 3.7 \times 10^{-6} \text{ m} = 3.7 \text{ µm}$$

$$\beta = 2.89$$

The collection efficiency is

$$\eta_i = \frac{1}{\left[1 + (3.7/D_i)^{2.89}\right]} \quad \text{with } D_i \text{ in micrometers.}$$

For $v_i = 20$ m/s

$$d_c = 0.096 \text{ m}$$

$$Z_c = 1.057 \text{ m}$$

$$v_{t\max} = 31.5 \text{ m/s}$$

$$D_{50} = 3.2 \times 10^{-6} \text{ m} = 3.2 \text{ µm}$$

$$\beta = 3.28$$

and the efficiency is

$$\eta_i = \frac{1}{\left[1 + (3.2/D_i)^{3.28}\right]} \quad \text{with } D_i \text{ in micrometers.}$$

Figure 6 shows the efficiency curves obtained with the Iozia and Leith model compared to the experimental results of Dirgo and Leith (17), for the same operating conditions.

By looking at Figs. 4–6, it can be verified that the Iozia and Leith model provided the best prediction in the studied conditions. However, it is worth noting that the model underestimated the collection efficiency of the larger particles (see Fig. 6). The Barth model provided an efficiency curve with an adequate slope, but displaced to the right of the experimental points, as can be seen in Fig. 4. This is probably the result of the calculated values of $v_{t\max}$, underestimated by Eq. (5). The Leith and Licht model provided the worst prediction of the three, overestimating the efficiency of the smaller particles and underestimating the larger ones (see Fig. 5).

2.5.1. Pressure Drop and Power Consumption

The pressure drop in the cyclone can be estimated by Eq. (31), with ΔH given by Eq. (32), (33), or (34). Note that, in SI units, the resulting ΔP is in Pascals (Pa).

The power, W_c, consumed by the fan in order to maintain the required volumetric flow rate in the cyclone can be estimated by a correlation given by Cooper and Alley (30):

$$W_c = \frac{Q\Delta P}{E_f} \tag{43}$$

where E_f is the fan efficiency.

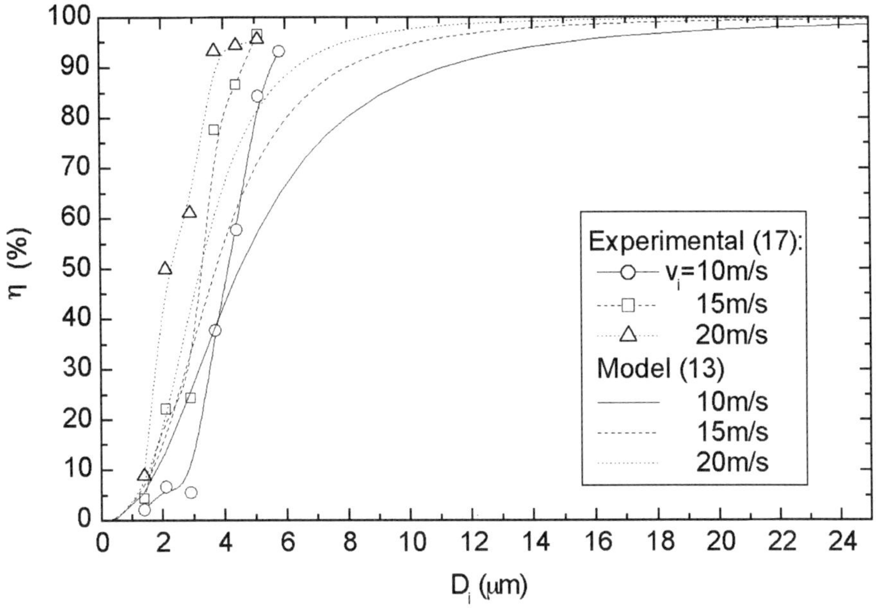

Fig. 6. Collection efficiency as a function of particle diameter, for three entry velocities: curves obtained from the Iozia and Leith model (13) and experimental points from Dirgo and Leith (17).

In the conditions of this example, the case of higher power consumption occurs for the gas entry velocity of 20 m/s and particle-free gas. In this case, Eqs. (32)–(34) give

$$\Delta H = 16\left(\frac{(0.151\ \text{m})(0.060\ \text{m})}{(0.151\ \text{m})^2}\right)$$

(a) $\Delta H = 6.4$

$$\Delta H = 11.3\left(\frac{(0.151\ \text{m})(0.060\ \text{m})}{(0.151\ \text{m})^2}\right)^2 + 3.33$$

(b) $\Delta H = 5.1$

$$\Delta H = 20\left(\frac{(0.151\ \text{m})(0.060\ \text{m})}{(0.151\ \text{m})^2}\right) \times \left[\frac{0.151\ \text{m}/0.302\ \text{m}}{(1.208\ \text{m}/0.302\ \text{m})(0.453\ \text{m}/0.302\ \text{m})(0.113\ \text{m}/0.302\ \text{m})}\right]^{\frac{1}{3}}$$

(c) $\Delta H = 4.8$

The pressure drop [Eq. (31)] calculated utilizing the above values of ΔH above are respectively

$$\Delta P = \left(\frac{(1.18\,\mathrm{kg/m^3})(20\,\mathrm{m/s})^2(6.4)}{2} \right)$$

$$\text{(a) } \Delta P = 1510\,\frac{\mathrm{kg}}{\mathrm{ms^2}} = 1510\ \mathrm{Pa}$$

$$\text{(b) } \Delta P = 1204\,\frac{\mathrm{kg}}{\mathrm{ms^2}} = 1204\ \mathrm{Pa}$$

$$\text{(c) } \Delta P = 1133\,\frac{\mathrm{kg}}{\mathrm{ms^2}} = 1133\ \mathrm{Pa}$$

Table 3 shows that the experimental value found by Dirgo and Leith was 1407 Pa. Therefore, the result given by Eq. (32) is the closest and provides some safety margin. In this case, the calculated fan power [Eq. (43)] with 0.55 efficiency is

$$W_c = \frac{(0.1812\ \mathrm{m^3/s})(1510\ \mathrm{N/m^2})}{0.55}$$

$$W_c = 497\,\frac{\mathrm{Nm}}{\mathrm{s}} = 497\ \mathrm{W}$$

Example 2

Use the grade efficiency results given by the Iozia and Leith model to calculate the overall efficiency for the entry velocities of 10, 15, and 20 m/s. Assume that the size distribution of the particles is that listed in Table 4 and that the powder concentration is 0.02 kg/m^3. Determine the concentration and size distribution at the equipment exit as well as the pressure drop for the velocity of 20 m/s. The physical properties of the particles and the gas are the same as in the previous example.

Table 4
Particle Size Distribution Utilized in Example 2

Size range (μm)	x_i (in mass basis)
0–1	0.01
1–2	0.02
2–4	0.04
4–6	0.06
6–8	0.08
8–10	0.10
10–20	0.13
20–30	0.15
30–40	0.12
40–50	0.10
50–60	0.07
60–70	0.05
70–80	0.04
80–90	0.02
90–100	0.01

Table 5
Results from the Calculation of the Overall Efficiency in the Cyclone

Range (μm)	D_i (μm)	x_i (in mass)	η_i (10 m/s)	$\eta_i x_i$ (10 m/s)	η_i (15 m/s)	$\eta_i x_i$ (15 m/s)	η_i (20 m/s)	$\eta_i x_i$ (20 m/s)
0–1	0.5	0.01	0.00467	0.0000467	0.00307	0.0000307	0.00226	0.0000226
1–2	1.5	0.02	0.06413	0.0012825	0.06854	0.0013709	0.07690	0.0015380
2–4	3	0.04	0.27104	0.0108417	0.35295	0.0141180	0.44728	0.0178910
4–6	5	0.06	0.56392	0.0338351	0.70479	0.0422872	0.81212	0.0487270
6–8	7	0.08	0.74613	0.0596903	0.86326	0.0690605	0.92874	0.0742990
8–10	9	0.10	0.84439	0.0844392	0.92883	0.0928833	0.96745	0.0967446
10–20	15	0.13	0.94968	0.1234582	0.98280	0.1277633	0.99374	0.1291862
20–30	25	0.15	0.98499	0.1477489	0.99602	0.1494024	0.99882	0.1498233
30–40	35	0.12	0.99334	0.1192009	0.99849	0.1198188	0.99961	0.1199531
40–50	45	0.10	0.99638	0.0996382	0.99927	0.0999269	0.99983	0.0999828
50–60	55	0.07	0.99778	0.0698446	0.99959	0.0699713	0.99991	0.0699938
60–70	65	0.05	0.99852	0.0499261	0.99975	0.0499874	0.99995	0.0499974
70–80	75	0.04	0.99896	0.0399583	0.99983	0.0399933	0.99997	0.0399987
80–90	85	0.02	0.99923	0.0199846	0.99988	0.0199977	0.99998	0.0199996
90–100	95	0.01	0.99941	0.0099941	0.99992	0.0099992	0.99998	0.0099999
		$\Sigma x_i = 1.0$		$\Sigma \eta_i x_i = 0.87$		$\Sigma \eta_i x_i = 0.91$		$\Sigma \eta_i x_i = 0.93$

Solution

Table 5 can be easily constructed from Eq. (41) and the collection efficiencies calculated from Eq. (23) for the velocities of 10, 15, and 20 m/s. Therefore, the overall collection efficiencies are:

$\eta_o = 87\%$ for $v_i = 10$ m/s
$\eta_o = 91\%$ for $v_i = 15$ m/s
$\eta_o = 93\%$ for $v_i = 20$ m/s

For $v_i = 20$ m/s the particle concentration in the cyclone exit [Eq. (42)] is calculated as

$$c_o = (1 - 0.93)0.02\,\text{kg/m}^3$$

$$c_o = 1.4 \times 10^{-3}\,\frac{\text{kg}}{\text{m}^3} = 1.4\,\text{g/m}^3$$

The particle size distribution at the exit is obtained from Eqs. (39) and (40), utilizing the results listed in Table 6.

The results show that there are few particles larger than 10 μm in the cyclone exit, where particles below 4 μm predominate. This is very useful information for defining of a downstream particle collector that might be needed.

The pressure drop in the cyclone operating with particle loaded gas can be estimated by Eq. (35), where the entry concentration c_{in} in grains per cubic feet is needed:

$$c_{\text{in}} = 0.02\,\frac{\text{kg}}{\text{m}^3}\,\frac{1\text{ gr}}{6.48 \times 10^{-5}\,\text{kg}}\,\frac{(0.3048\text{ m})^3}{(1\text{ ft})^3}$$

$$c_{\text{in}} = 8.7\,\text{gr/ft}^3$$

Table 6
Results from the Calculation of the Particle Size Distribution in the Cyclone Exit.

Range (μm)	D_i (μm)	η_i (20 m/s)	$\eta_i x_i$ (20 m/s)	m_{oi} (kg/s)	x_{oi} (in mass)
0–1	0.5	0.00226	0.0000226	3.6158E-05	0.138877
1–2	1.5	0.07690	0.0015380	6.6906E-05	0.256976
2–4	3	0.44728	0.0178910	8.0123E-05	0.307739
4–6	5	0.81212	0.0487270	4.0853E-05	0.156911
6–8	7	0.92874	0.0742990	2.0661E-05	0.079354
8–10	9	0.96745	0.0967446	1.1798E-05	0.045312
10–20	15	0.99374	0.1291862	2.9493E-06	0.011328
20–30	25	0.99882	0.1498233	6.4035E-07	0.002459
30–40	35	0.99961	0.1199531	1.7004E-07	0.000653
40–50	45	0.99983	0.0999828	6.2154E-08	0.000239
50–60	55	0.99991	0.0699938	2.2529E-08	8.65E-05
60–70	65	0.99995	0.0499974	9.3040E-09	3.57E-05
70–80	75	0.99997	0.0399987	4.6551E-09	1.79E-05
80–90	85	0.99998	0.0199996	1.5439E-09	5.93E-06
90–100	95	0.99998	0.0099999	5.3597E-10	2.06E-06
			$\Sigma \eta_i x_i = 0.93$	$\Sigma m_{oi} = 2.604\text{E-}4$	$\Sigma x_{oi} = 1.0$

From Example 1, the pressure drop in cyclone operating with a particle-free gas at 20 m/s was estimated using

$$\Delta P_{\text{at } c_{\text{in}}=0} = 1510 \text{ Pa}$$

Therefore, Eq. (35) gives

$$\frac{\Delta P_{\text{at } c_{\text{in}}}}{1510 \text{ Pa}} = 1 - 0.013(8.7)^{0.5}$$

$$\Delta P_{\text{at } c_{\text{in}}} = 1452 \text{ Pa}$$

3. COSTS OF CYCLONE AND AUXILIARY EQUIPMENT*

3.1. Cyclone Purchase Cost

Cyclones are used upstream (60) of particulate control devices (e.g., fabric filters, ESPs) to remove larger particles entrained in a gas stream. Equation (44) yields the cost of a carbon steel cyclone with support stand, fan, and motor, and a hopper or drum to collect the dust:

$$P_{\text{cyc}} = 6250 A_{\text{cyc}}^{0.9031} \tag{44}$$

*This subject is also presented in the chapter "Technical, Energy and Cost Evaluation of Air Pollution Control Technologies", by L. K. Wang et al. in Volume 2 of the *Handbook of Environmental Engineering* series. The examples given here were taken from that chapter.

where P_{cyc} is the cost of the cyclone (August 1988 US$) and A_{cyc} is the cyclone inlet area (ft^2 [$0.200 \le A_{\text{cyc}} \le 2.64$ ft^2]).

The cost of a rotary air lock for hopper or drum is given by

$$P_{\text{ral}} = 2730 A_{\text{cyc}}^{0.0985} \tag{45}$$

where P_{ral} is the cost of a rotary air lock (August 1988 US$) and A_{cyc} is the cyclone inlet area (ft^2 [$0.350 \le A_{\text{cyc}} \le 2.64$ ft^2]).

The cost of the complete cyclone unit is given by the sum of P_{cyc} and P_{ral}.

3.2. *Fan Purchase Cost*

In general, fan costs are most closely correlated with fan diameter (see Chapter 7 for a detailed fan design). Equations (46)–(48) can be used to obtain fan prices. Costs for a carbon steel fan motor ranging in horsepower from 1 to 150 hp are provided in Eqs (49) and (50). Equation (47) or (48) is used in conjunction with Eqs. (49) or (50), respectively.

The cost of a fan is largely a function of the fan wheel diameter, d_{fan}, which, in turn, is related to the ductwork diameter. The fan wheel diameter can be obtained for a given ductwork diameter by consulting the appropriate manufacturer's multirating tables or by calling the fan manufacturer.

For a centrifugal fan consisting of backward-curved blades including a belt-driven motor and starter and a static pressure range between 0.5 and 8 in. of water, the cost as a function of fan diameter (d_{fan}) in July 1988 dollars is provided by

$$P_{\text{fan}} = 42.3 d_{\text{fan}}^{1.20} \tag{46}$$

where P_{fan} is the cost of the fan system (July 1988 US$) and d_{fan} is the fan diameter (in. [$12.25 \le d_{\text{fan}} \le 36.5$ in.]).

The cost of a fiber-reinforced plastic (FRP) fan, not including the cost of a motor or starter, is provided by Eq. (47). The cost of a motor and starter as obtained in Eq. (49) or (50) should be added to the fan cost obtained in Eq. (47):

$$P_{\text{fan}} = 53.7 d_{\text{fan}}^{1.38} \tag{47}$$

where P_{fan} is the cost of the fan without motor or starter (April 1988 US$) and d_{fan} is the fan diameter (in. [$10.5 \le d_{\text{fan}} \le 73$ in.]).

A correlation for a radial-tip fan with weld, carbon steel construction, and an operating temperature limit of 1000°F without a motor or starter is provided by Eq. (48). The values for the parameters a_f and b_f are provided in Table 7.

$$P_{\text{fan}} = a_f d_{\text{fan}}^{b_f} \tag{48}$$

where P_{fan} is the cost of the fan without motor or starter (July 1988 US$), a_f and b_f are obtained from Table 7, and d_{fan} is the fan diameter (in.).

The cost of fan motors and starters is given in Eq. (49) or (50) as a function of the horsepower (hp) requirement (W_c). The cost obtained from either of these equations should be added to the fan cost obtained in Eq. (47) or (48). For low horsepower requirements,

$$P_{\text{motor}} = 235 \text{hp}^{0.256} \tag{49}$$

Table 7
Equation (48) Parameters

Parameter	Group 1	Group 2
Static pressure (in.)	2–22	20–32
Flow rate (acfm)	700–27000	2000–27000
Fan wheel diameter (in.)	19.125–50.5	19.25–36.5
a_f	6.41	22.1
b_f	1.81	1.55

where P_{motor} is the cost of the fan motor, belt, and starter (February 1988 US$) and hp is the motor horsepower ($1 \leq \text{hp} \leq 7.5$). For high-horsepower requirements,

$$P_{\text{motor}} = 94.7\text{hp}^{0.821} \tag{50}$$

where P_{motor} is the cost of the fan motor, belt, and starter (February 1988 US$) and hp is the motor horsepower ($7.5 \leq \text{hp} \leq 250$).

3.3. *Ductwork Purchase Cost*

The cost of ductwork for a HAP control system is typically a function of material (e.g., PVC, FRP), diameter, and length. To obtain the duct diameter requirement as a function of the emission stream flow rate at actual conditions (Q_a), use Eq. (51). This equation assumes a duct velocity (U_{duct}) of 2000 ft/min.

$$d_{\text{duct}} = 12\left[\frac{4Q_a}{\pi U_{\text{duct}}}\right]^{0.5} = 0.3028Q_a^{0.5} \tag{51}$$

The cost of PVC ductwork in US$/ft for diameters between 6 and 24 in. is provided using

$$P_{\text{PVCd}} = a_d d_{\text{duct}}^{bd} \tag{52}$$

where P_{PVCd} is the cost of PVC ductwork (US$/ft [August 1988 US$]), d_{duct} is the duct diameter (in.) (factor of 12 in./ft), $a_d = 0.877$ ($6 \leq d_{duct} \leq 12$ in.) or 0.0745 ($14 \leq d_{duct} \leq 24$ in.), and $b_d = 1.05$ ($6 \leq d_{duct} \leq 12$ in.) or 1.98 ($14 \leq d_{duct} \leq 24$ in.).

For a FRP duct having a diameter between 2 and 5 ft, Eq. (53) can be used to obtain the duct cost. Note that the duct diameter is in units of feet for this equation.

$$P_{\text{FRPd}} = 24D_{\text{duct}} \tag{53}$$

where P_{FRPd} is the cost of the FRP ductwork (US$/ft [August 1988 US$]) and D_{duct} is the duct diameter (ft).

It is more difficult to obtain ductwork costs for carbon steel and stainless-steel construction because ductwork of this material is almost always custom fabricated. For more information on these costs, consult refs. 31 and 32.

3.4. *Stack Purchase Cost*

Because stacks are usually custom fabricated, it is also difficult to obtain stack cost correlations. Smaller stacks are typically sections of straight ductwork with supports.

Table 8
Equation (54) Parameters for Costs of Large Stacks

Lining	Diameter (ft)	a_s	b_s
Carbon steel	15	0.0120	0.811
316 L stainless	20	0.0108	0.851
Steel in top	30	0.0114	0.882
Section	40	0.0137	0.885
Acid resistant	15	0.00602	0.952
Firebrick	20	0.00562	0.984
	30	0.00551	1.027
	40	0.00633	1.036

Source: ref. 31.

However, the cost of small (e.g., 50–100 ft) FRP stacks can be roughly estimated as 150% of the cost of FRP ductwork for the same diameter and length. Similarly, the cost of small carbon steel and stainless-steel stacks is also approx 150% of the cost of corresponding ductwork (31,32).

For larger stacks (200–600 ft), the cost is typically quite high, ranging from US$ 1,000,000 to US$ 5,000,000 for some applications. Equation (54) and Table 8 can be used to obtain costs of large stacks:

$$P_{\text{stack}} = a_s H_{\text{stack}}^{b_s} \tag{54}$$

where P_{stack} is the total capital cost of large stack (10^6 US$), H_{stack} is the stack height (ft), and a_s and b_s refer to Table 8.

3.5. Damper Purchase Cost

Dampers are commonly used to divert airflow in many industrial systems. Two types of damper are discussed: backflow and two-way diverter valve dampers. The cost of backflow dampers for duct diameters between 10 and 36 in. is given by

$$P_{\text{damp}} = 7.4 d_{\text{duct}}^{0.944} \tag{55}$$

where P_{damp} is the cost of the damper (February 1988 US$) and d_{duct} is the ductwork diameter (in.).

The cost of a two-way diverter valve for ductwork diameters between 13 and 40 in. is given by

$$P_{\text{divert}} = 4.846 d_{\text{duct}}^{1.50} \tag{56}$$

where P_{divert} is the cost of the two-way diverter valve (February 1988 US$) and d_{duct} is the ductwork diameter (in.).

3.6. Calculation of Present and Future Costs

For the purposes of this handbook, auxiliary equipment is defined to include the cost of fans, ductwork, stacks, dampers, and cyclones (if necessary) that commonly accompany

Table 9
CE Equipment Index

Date	Index	Date	Index	Date	Index
Feb. 1990	389.0	May 1988	369.5	Aug. 1986	334.6
Jan. 1990	388.8	Apr. 1988	369.4	July 1986	334.6
Dec. 1990	390.9	Mar. 1988	364.0	June 1986	333.4
Nov. 1989	391.8	Feb. 1988	363.7	May 1986	334.2
Oct. 1989	392.6	Jan. 1988	362.8	Apr. 1986	334.4
Sept. 1990	392.1	Dec. 1987	357.2	Mar. 1986	336.9
Aug. 1989	392.4	Nov. 1987	353.8	Feb. 1986	338.1
July 1989	392.8	Oct. 1987	352.2	Jan. 1986	345.3
June 1989	392.4	Sept. 1987	343.8	Dec. 1985	348.1
May 1989	391.9	Aug. 1987	344.7	Nov. 1985	347.5
Apr. 1989	391.0	July 1987	343.9	Oct. 1985	347.5
Mar. 1989	390.7	June 1987	340.4	Sept. 1985	347.2
Feb. 1989	387.7	May 1987	340.0	Aug. 1985	346.7
Jan. 1989	386.0	Apr. 1987	338.3	July 1985	347.2
Dec. 1988	383.2	Mar. 1987	337.9	June 1985	347.0
Noc. 1988	380.7	Feb. 1987	336.9	May 1985	347.6
Oct. 1988	379.6	Jan. 1987	336.0	Apr. 1985	347.6
Sept. 1988	379.5	Dec. 1986	335.7	Mar. 1985	346.9
Aug. 1988	376.3	Nov. 1986	335.6	Feb. 1985	346.8
July 1988	374.2	Oct. 1986	335.8	Jan. 1985	346.5
June 1988	371.6	Sept. 1986	336.6	Dec. 1984	346.0

Source: ref. 33.

control equipment. These costs must be estimated before the purchase equipment cost (PCE) can be calculated. Costs for auxiliary equipment were obtained from ref. 31.

If the equipment costs must be escalated to the current year, the *Chemical Engineering* (CE) equipment index can be used (33). Monthly indices for 5 yr are provided in Table 9.

The following equation can be used for converting the past cost to the future cost or vice versa.

$$\text{cost}_b = \text{cost}_a \frac{(\text{index}_b)}{(\text{index}_a)} \qquad (57)$$

where cost_a is the cost in the month–year of a (US$), cost_b is the cost in the month–year of b (US$), index_a is the CE equipment cost index in the month–year of a, and index_b is the CE equipment cost index in the month–year of b. It should be noted that although the CE equipment cost indices are recommended here for index_a and index_b, the ENR cost indices (34) can also be adopted for updating the costs.

3.7. Cost Estimation Examples

Example 3

Assume an emission stream actual flow rate of 1000 acfm, a particle density of 30 $\text{lb}_\text{m}/\text{ft}^3$, an emission stream density of 0.07 $\text{lb}_\text{m}/\text{ft}^3$, an emission stream viscosity of 1.41×10^{-5}

lb_m/ft s, and the cyclone inlet area has been calculated to be 2.41 ft^2; determine the following:

A. The August 1988 cost of the cyclone body
B. The August 1988 cost of the total cyclone system
C. The August 1990 cost of the total cyclone system
D. The present cost of the total cyclone system assuming that the present CE equipment index obtained from Chemical Engineering (33) is 410.

Solution

A. The August 1988 cost of a cyclone is then obtained from Eq. (44) as follows:

$$P_{\text{cyc}} = 6250A_{\text{cyc}}^{0.9031}$$

$$P_{\text{cyc}} = 6250(2.41)^{0.9031}$$

$$P_{\text{cyc}} = \text{US\$ } 13{,}832$$

B. The August 1988 cost of a rotary air lock for this system is given by Eq. (45):

$$P_{\text{ral}} = 2730A_{\text{cyc}}^{0.0985}$$

$$P_{\text{ral}} = 2730(2.41)^{0.0985}$$

$$P_{\text{ral}} = \text{US\$ } 2{,}977$$

The August 1988 cost of a cyclone system is the sum of these two costs, or US$ 16,809.

C. The February 1990 cost of a cyclone system is given by Eq. (57). The CE equipment indices for August 1988 and February 1990 are 376.3 and 389.0, respectively, as shown in Table 9:

$$\text{cost}_b = \text{cost}_a \frac{\text{index}_b}{\text{index}_a}$$

$$\text{cost}_a = \text{US\$ } 16{,}809 \frac{389.0}{376.3}$$

$$\text{cost}_a = \text{US\$ } 17{,}376.0$$

D. The present cost (when the index is 410) is

$$\text{cost}_b = \text{US\$ } 16{,}809 \frac{410.0}{376.3}$$

$$\text{cost}_b = \text{US\$ } 18{,}314.35$$

Example 4

Determine the fan costs in July 1988 and in the future when the future CE equipment cost index is projected to be 650.0. Assume the required static pressure equal 8 in. of water with a fan diameter of 30 in.

Solution

Equation (46) can be used to obtain the fan cost as follows:

$$P_{\text{fan}} = 42.3d_{\text{fan}}^{1.20}$$

$$P_{\text{fan}} = 42.3(30)^{1.20}$$

$$P_{\text{fan}} = \text{US\$ } 2,505$$

The future fan cost when the CE equipment cost index will reach 650.0. The July 1988 index is known to be 374.2. Equation (57) can be used for calculation:

$$\text{cost}_b = \text{cost}_a \frac{\text{index}_b}{\text{index}_a}$$

$$\text{cost}_b = \text{US\$ } 2,505 \frac{650.0}{374.2}$$

$$\text{cost}_b = \text{US\$ } 4,351.28 \text{ in the future}$$

Example 5

Determine the required FRP duct diameter assuming a duct velocity (U_{duct}) of 2,000 ft/min, and an actual air emission rate (Q_a) of 15,300 acfm.

Solution

d_{duct} is obtained using Eq. (51):

$$d_{\text{duct}} = 12\left(\frac{4Q_a}{\pi U_{\text{duct}}}\right)^{0.5}$$

$$d_{\text{duct}} = 12\left(\frac{4(15,300)}{\pi(2,000)}\right)^{0.5}$$

$$d_{\text{duct}} = 37.4 \text{ in. or } 3.12 \text{ ft.}$$

Example 6

Determine the cost of a 50-ft FRP duct ($D_{\text{duct}} = 3.12$ ft) in the future when the CE equipment cost index reaches 700.0.

Solution

The August 1988 cost of FRP ductwork can be calculated using Eq. (53):

$$P_{\text{FRPd}} = 24D_{\text{duct}}$$

$$P_{\text{FRPd}} = 24(3.12)$$

$$P_{\text{FRPd}} = \text{US\$ } 74.88/\text{ft}$$

Thus, for a 50-ft duct length, the August 1988 cost of ductwork equals 50(US\$ 74.88) = US\$ 3,744.
The future cost when the CE equipment cost index reaches 700.0 is

$$\text{cost}_b = \text{cost}_a \frac{\text{index}_b}{\text{index}_a}$$

$$\text{cost}_b = \text{US\$ } 3,744\,\frac{700.0}{376.3}$$

$$\text{cost}_b = \text{US\$ } 6,964.66$$

4. CYCLONES FOR AIRBORNE PARTICULATE SAMPLING

Because of the complex flow pattern of the gas within the cyclone, its removal efficiency tends to increase substantially for small body dimensions. Cyclones with some centimeters, or even millimeters, in body diameter can reach high collection efficiencies for fine particles (diameters below 5 μm), unlike their "grown-up" relatives.

For these reasons, cyclones have been widely utilized as sampler devices for particulate matter (PM) monitoring in environmental and occupational applications (59). Their size and geometry, allied to suitable collection efficiencies, make possible the design of portable monitors, very flexible in their use. They can, for example, be carried around, clipped on a person's body, continuously sampling the ambient air one is breathing.

The design criteria, in this case, are centered on the cyclone collection efficiency performance and its comparison with the standards for monitoring devices, defined by the governmental agencies and/or legislation. These standards constantly change, and a brief overview of their present status is given below.

4.1. *Particulate Matter in the Atmosphere*

It is well known that the inhalation of particles is harmful to people's health. Studies carried out mainly in the second half of the last century also verified that the degree of penetration of particles in the respiratory system is a function of the particle size. These findings led to the establishment of criteria for aerosol monitoring, which are normally presented in the form of precollectors acting as parts of the respiratory system: the particles that penetrate through these precollectors are equivalent to those that penetrate through the corresponding part of the human body. The increasingly rigorous standards for air quality, which include definitions of particle size fractions in relation to their penetration through standard precollectors, fall into four categories, according to the American Conference of Governmental Industrial Hygienists (ACGIH) (35):

- The ***inhalable fraction*** (IPM) is the mass fraction of total airborne particles inhaled through the nose and mouth and is given by

$$\text{IPM} = 0.5\left[1 + \exp\left(-0.06D_{\text{ae}}\right)\right] \tag{58}$$

where D_{ae} is the particle aerodynamic diameter (in μm), defined as the diameter of an equivalent spherical particle of unit density and the same terminal velocity as the particle in question. The International Standards Organization (ISO) expects to adopt a similar definition that includes the ambient wind speed, U (in m/s), and can be written as (36)

$$\text{IPM} = 0.5\left[1 + \exp\left(-0.06D_{\text{ae}}\right)\right] + 10^{-5}U^{2.75}\exp\left(0.05D_{\text{ae}}\right) \tag{59}$$

- The ***thoracic fraction*** (TPM) is the mass fraction of inhaled particles penetrating the respiratory system beyond the larynx and is given by

$$\text{TPM} = \text{IPM} * \left[1 - F(x)\right] \tag{60}$$

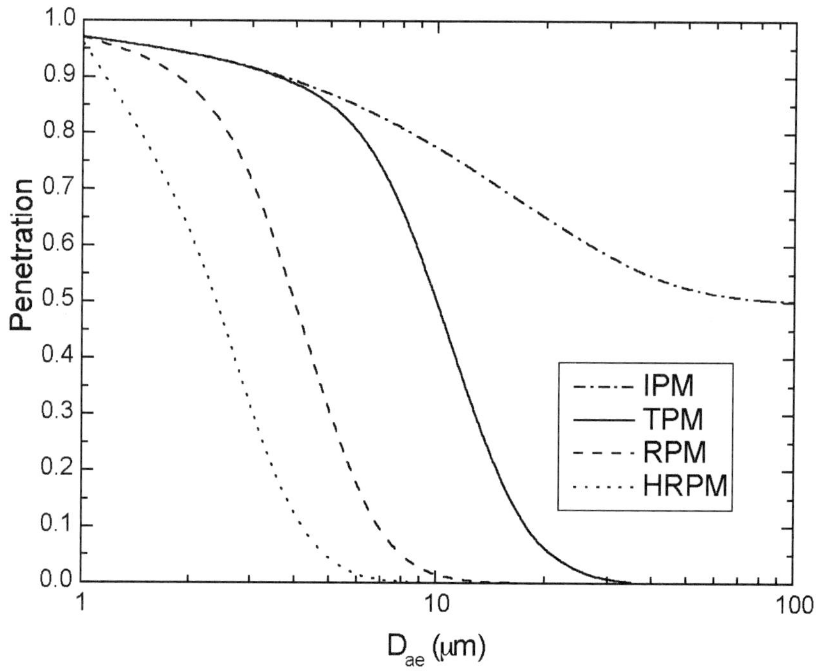

Fig. 7. The sampling conventions for the IPM, TPM, RPM, and HRPM size fractions.

where $F(x)$ is the cumulative probability function of the standardized normal variable x given by

$$x = \ln\left(D_{ae}/\Gamma\right)/\ln\sigma \tag{61}$$

where $\Gamma = 11.64$ μm and $\sigma = 1.5$.

- The ***respirable fraction*** (RPM) is the mass fraction of inhaled particles that penetrates to the alveolar region of the lung and is given by

$$\text{RPM} = \text{IPM} * \left[1 - F(x)\right] \tag{62}$$

where x is given by Eq. (61), with $\Gamma = 4.25$ μm and $\sigma = 1.5$.

- The ***high-risk respirable fraction*** (HRPM) is the mass fraction of inhaled particles more sensitive to the sick and infirm, or children, and is currently in the process of adoption by the international standards. The HRPM mass fraction is given by

$$\text{HRPM} = \text{IPM} * \left[1 - F(x)\right] \tag{63}$$

where x is also given by Eq. (61), with $\Gamma = 2.50$ μm and $\sigma = 1.5$.

Figure 7 shows the curves resulting from Eqs. (58), (60), (62), and (63).

The legislation concerning maximum levels of PM in the atmosphere has followed these findings closely. The US Environmental Protection Agency (EPA) has, since 1987, adopted a convention very similar to the thoracic fraction, known as PM_{10}, for the validation of data from precollectors for ambient air monitoring. More recently, the EPA

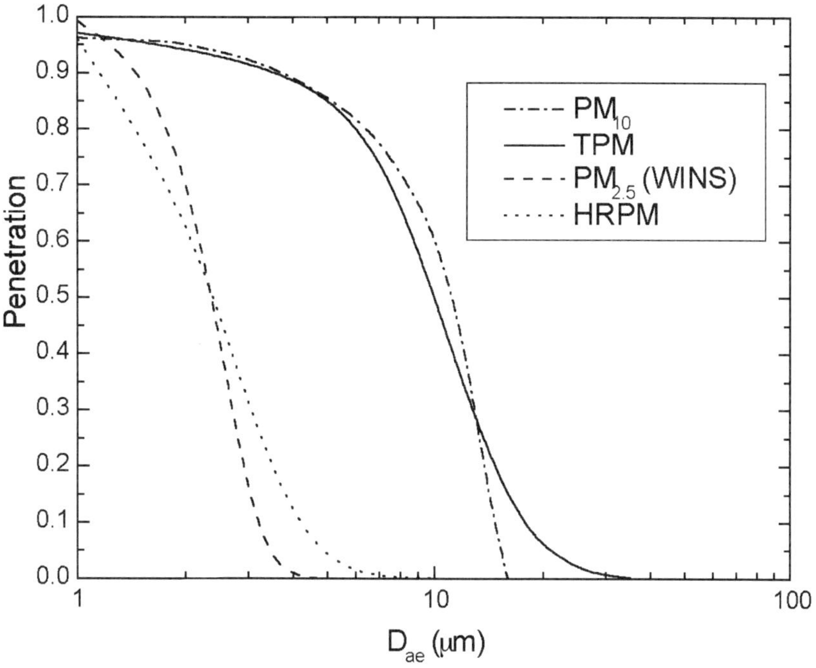

Fig. 8. Comparison between the PM_{10} and TPM curves, and between the $PM_{2.5}$ and HRPM curves.

has included the fine fraction of the PM_{10} in their criteria (37). This fraction, known as $PM_{2.5}$, is referenced by the penetration curve through a cut-type impactor called Well Impactor Ninety Six (WINS). The resulting curve is similar to the HRPM curve. Figure 8 compares the TPM and HRPM curves with the PM_{10} and $PM_{2.5}$, respectively [the $PM_{2.5}$ curve was taken from Adams et al. (38)].

A great deal of work has been spent in developing and validating precollectors to conform to these conventions, and a few reviews are available (39,40). Virtually all of the published material are constituted of empirical approaches, and most of them are centered on one specific equipment or case (41–45). However, some generalized correlations for predicting cyclone performance can be found, and four examples are presented next. It is interesting to note that, as a rule, little attention is given to pressure drop in all of these publications.

4.2. General Correlation for Four Commercial Cyclones

Chan and Lippmann (46) conducted an extensive experimental investigation on the performance of four commercially available portable cyclones, namely the 10-mm Nylon (Dorr–Oliver), the Unico 240, and the Aerotec 2 and 3/4 cyclones. The full description of their geometry was not given. Some dimensional characteristics of these cyclones are listed in Table 10.

The cyclones were tested in a calibration apparatus specially developed for this purpose (48), consisting of a test chamber with multiple cyclone sampling ports. Filter holders were mounted after each cyclone with precalibrated flowmeters downstream of

Table 10
Dimensional Characteristics of the Cyclones Tested

Cyclone	Overall length (mm)	Body diameter (mm)	Outlet tube diameter (ID) (mm)	Ref.
10-mm Nylon	50	10	5.6	47
Unico 240	133	50.8	11	48
Aerotec 2	280	114	3.5	48
Aerotec 3/4	120	41.3	7.5	48

Source: ref. 46.

each filter. The test aerosols greater than 1 μm used were γ-tagged monodispersed iron oxide particles, with a density of 2560 kg/m^3. The submicron particles were produced with an atomizer–impactor aerosol generator (ERC model 7300). The particle size distribution for the larger particles (>1 μm) was determined by optical microscopy, and radiometric counting was used for the counting efficiencies of the submicron particles.

Comparisons were made between the experimental grade efficiency curves and the predictions calculated from existing correlations for industrial size cyclones. The conclusion was that no theory could adequately describe the cyclone performance—all of them underestimating particle collection. An empirical equation for the grade efficiency was developed, with the following form:

$$\eta = 0.5 + 0.5 \tanh\left[B\left(\frac{D_{ae}}{KQ^n}\right)^2 + (A - 2B)\left(\frac{D_{ae}}{KQ^n}\right) + B - A \right] \tag{64}$$

$$\text{valid for } \frac{D_{ae}}{KQ^n} < 1 - \frac{A}{2B} \tag{65}$$

where η is the cyclone efficiency for a particle of aerodynamic diameter D_{ae} (in μm), operating at a gas flow rate Q (in L/min). The empirical parameters K, n, A, and B are listed in Table 11 for the studied configurations and ranges. This equation fit the results of all tests within 95% accuracy.

Figures 9–12 show the estimated performance of the four cyclones, operating within their tested ranges, compared with the $PM_{2.5}$ (WINS) and HRPM curves. It can be noted that the Unico 240 (Fig. 9) and the Aerotec 3/4 (Fig. 10) cyclones have performance curves more adjustable to the HRPM convention. The Aerotec 2 cyclone (Fig. 11) was tested in a range that does not include the two criteria for D_{ae} = 2.5 μm. The 10-mm Nylon cyclone (Fig. 12) adapts very closely to the $PM_{2.5}$ (WINS) curve.

4.3. *A Semiempirical Approach*

Lidén and Gudmundsson (49), based on a thorough review of the available theories, developed a semiempirical model for predicting the cutoff size and slope of the collection efficiency curve of a cyclone as a function of the operating conditions and dimensional ratios. They based their study on previously published data for four different cyclone designs (Stairmand, Lapple, Z, and SRI), with sizes varying from industrial (D_c > 20 cm)

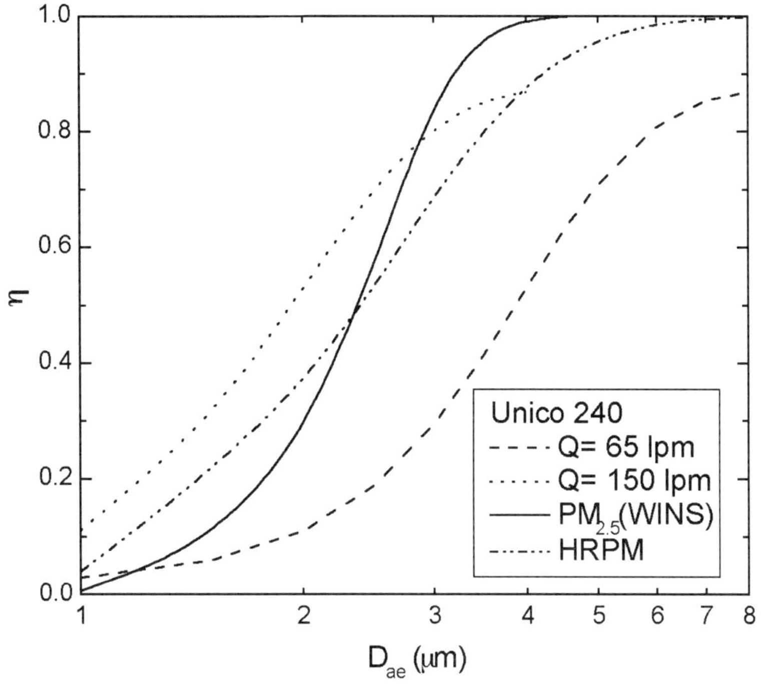

Fig. 9. Comparison between two performance curves for the Unico 240 cyclone, and the $PM_{2.5}$ (WINS) and HRPM curves.

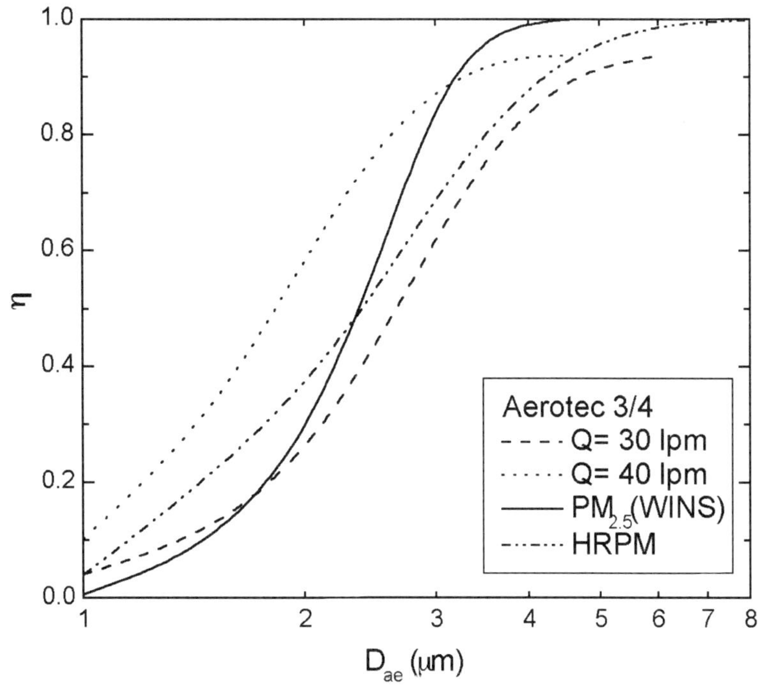

Fig. 10. Comparison between two performance curves for the Aerotec 3/4 cyclone, and the $PM_{2.5}$ (WINS) and HRPM curves.

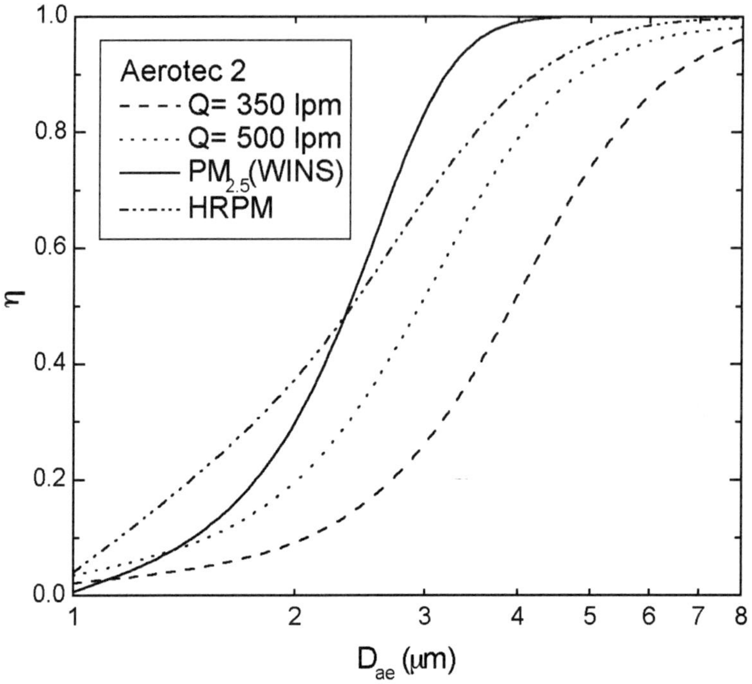

Fig. 11. Comparison between two performance curves for the Aerotec 2 cyclone, and the $PM_{2.5}$ (WINS) and HRPM curves.

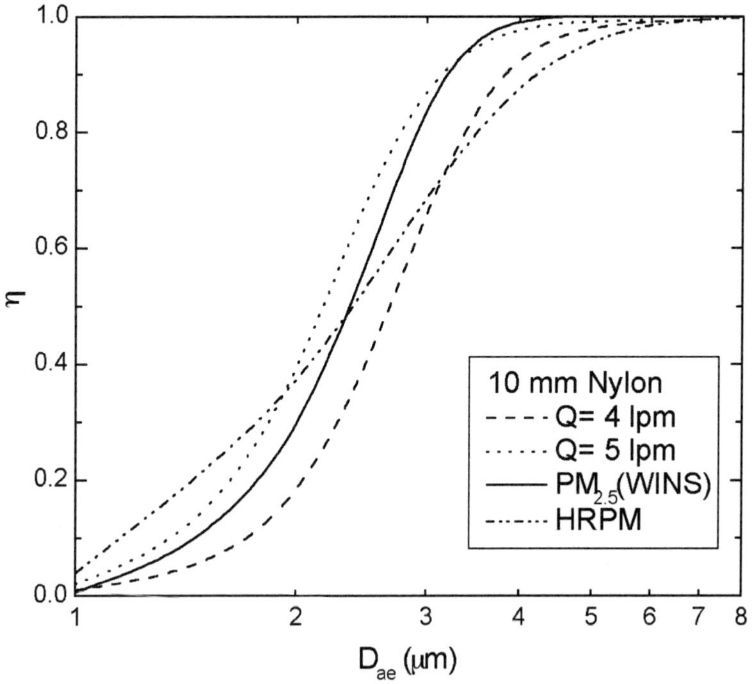

Fig. 12. Comparison between two performance curves for the 10-mm Nylon cyclone, and the $PM_{2.5}$ (WINS) and HRPM curves.

Table 11
Empirical Constants of Eq. (64) for Four Commercial Sampling Cyclones

Cyclone	D_{pc}[a] (μm)	Q[b] (L/min)	K	n	A	B
Aerotec 2	2.5–4.0	350–500	468.01	–0.80	2.02	–0.68
UNICO 240	1.0–5.0	65–350	123.68	–0.83	1.76	–0.82
Aerotec 3/4	1.0–5.0	22–65	214.17	–1.29	2.04	–0.77
10-rnm Nylon	1.8–7.0	0.9–5.0	6.17	–0.75	3.07	–0.93
	1.0–1.8	5.8–9.2	16.10	–1.25	1.19	–0.59
	0.1–1.0	18.5–29.6	178.52	–2.13	0.74	–0.07

[a] Range of cut diameters, D_{pc}.
[b] Range of operation flow rates, Q.
Source: ref. 46.

to sampling scale ($1.9 < D_c < 3.7$ cm). Tables 12 and 13 list the dimensional characteristics of the cyclones investigated. The dimensions are defined in Fig. 13.

Based on the work of Dring and Suo (50), a cutoff size parameter ψ_{50} is defined as:

$$\Psi_{50} = \frac{\sqrt{F_s}\,D_{50}}{D_c} \tag{66}$$

where D_{50} is the aerodynamic diameter of the particle corresponding to 50% collection efficiency. The parameter F_s is the Cunningham slip factor, which accounts for the reduced viscous drag on a particle whose size is comparable to the mean free path of the gas. This correction factor is a strong function of particle size and becomes very important in the submicron range, as can be seen in Fig. 14.

The cutoff size D_{50} was estimated by adjusting the efficiency curves to

$$\eta = \frac{e^{f(\theta)}}{1+e^{f(\theta)}} \tag{67}$$

where $f(\theta)$ is given by

$$f(\theta) = a_1 + a_2\theta + a_3\theta^2 \tag{68}$$

The parameters a_1, a_2, and a_3 were determined by fitting, and θ was the normalized collection efficiency, defined as

$$\theta = \frac{\sqrt{F_s(D_{ae})}\,D_{ae}}{\sqrt{F_s(D_{ae})}\,D_{50}} - 1 \tag{69}$$

By examining the work of Moore and McFarland (51,52), it could be verified that the cutoff size parameter ψ_{50} was highly correlated to the Reynolds number based on the cyclone annular dimension, Re_{ann}, defined as

$$\mathrm{Re}_{ann} = 0.5\left(1 - \frac{D_t}{D_c}\right)\frac{\rho v_i D_c}{\mu} \tag{70}$$

where D_t is the vortex finder external diameter (usually taken as D_e), v_i is the gas velocity at the cyclone entrance, and ρ and μ are the gas density and viscosity, respectively.

Table 12
Range of Body Diameters and Cyclone Reynolds Number Studied

Cyclone	D_c(cm)	Re_{cyc}
Stairmand	1.9, 20.3, 25.2, 30.5	5,000–500,000
Lapple	3.8–14.0	2,000–64,000
Z	12–300	12,000–300,000
SRI II and III	3.1–3.7	6,000–32,000

Source: (Adapted from Lidén and Gudmundsson (49), with permission from Taylor and Francis; http://www.tandf.co.uk/journals.)

Table 13
Dimensional Ratios for the Cyclones Studied

Cyclone	D_e/D_c	a/D_c	b/D_c	D_{in}/D_c	S/D_c	H/D_c	h/D_c
Stairmand	0.50	0.50	0.20		0.50	4.0	1.5
Lapple	0.50	0.50	0.25		0.75	4.0	2.0
Z	0.25	0.45	0.11		~0.82	~3.4	~0.86
SRI II	0.286			0.286	0.43	1.9	1.3
SRI III	0.27			0.27	0.35	1.6	0.45

Source: (Adapted from Lidén and Gudmundsson (49), with permission from Taylor and Francis; http://www.tandf.co.uk/journals.)

A multiple regression analysis was performed on ψ_{50} versus the annular Reynolds number and several ratios of cyclone dimensions, and at the 5% level of significance, the result was

$$\psi_{50} = \exp(F_0)\mathrm{Re}_{ann}^{F1}\left(\frac{S}{D_c}\right)^{F2} \tag{71}$$

The results of the multiple regression for the parameters F_0, F_1, and F_2 are given in Table 14, and are valid for the four cyclone types studied. Comparison between the model and experiments were given for the Stairmand and SRI cyclones. From the respective curves, it was possible to extract the parameters a_1, a_2, and a_3 [Eq. (68)], which are listed in Table 15. Therefore, by calculating D_{50} from Eq. (71), the efficiency curve for the cyclones can be derived from Eq. (67). The results for D_{50} = 2.5 μm are shown in Figs. 15 and 16 for the Stairmand and SRI cyclones, respectively, and are compared to the $PM_{2.5}$ (WINS) curve. It can be noted that the SRI cyclone has a much better fit than the Stairmand, which is not surprising, as the former was developed for sampling, and the latter is a classical design for industrial use. The final fitting between model and experiments, presented by Lidén and Gudmundsson, show a considerable scatter for the Stairmand data and much less for the SRI data. In any case, these relations are useful guidelines for cyclone scaling, but caution is advised before using the equations for accurate cyclone design.

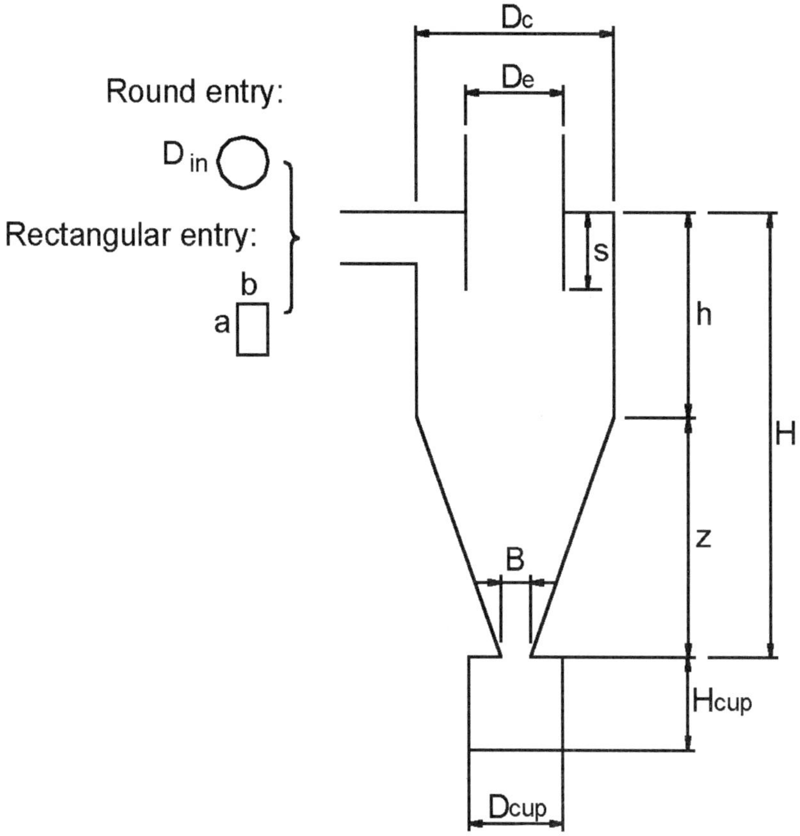

Fig. 13. Dimensions of the monitoring cyclone.

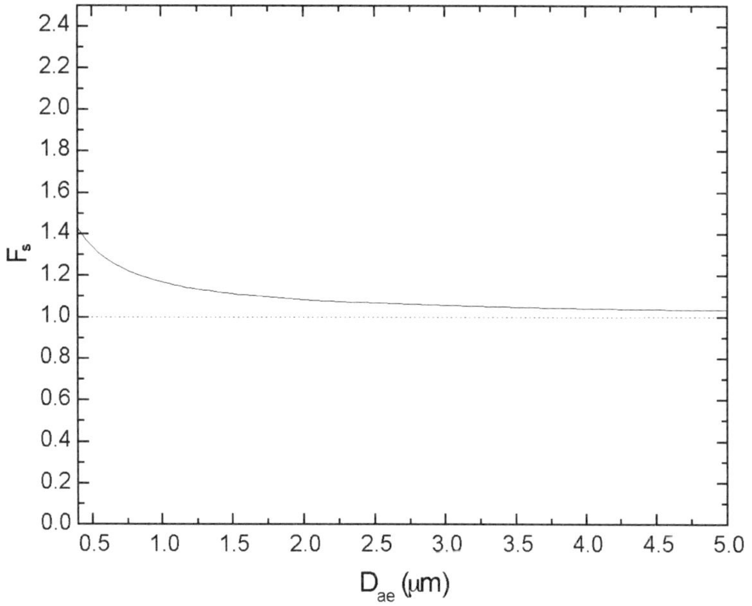

Fig. 14. The Cunningham slip factor F_s as a function of particle size.

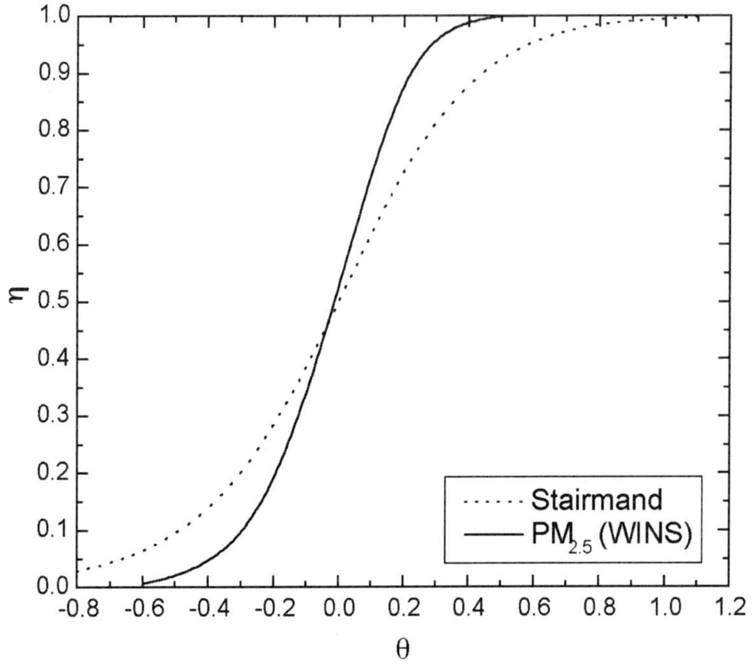

Fig. 15. Comparison between the performance of the Stairmand cyclone (D_c = 1.9 cm; Q = 30 L/min) and the $PM_{2.5}$ (WINS) curve.

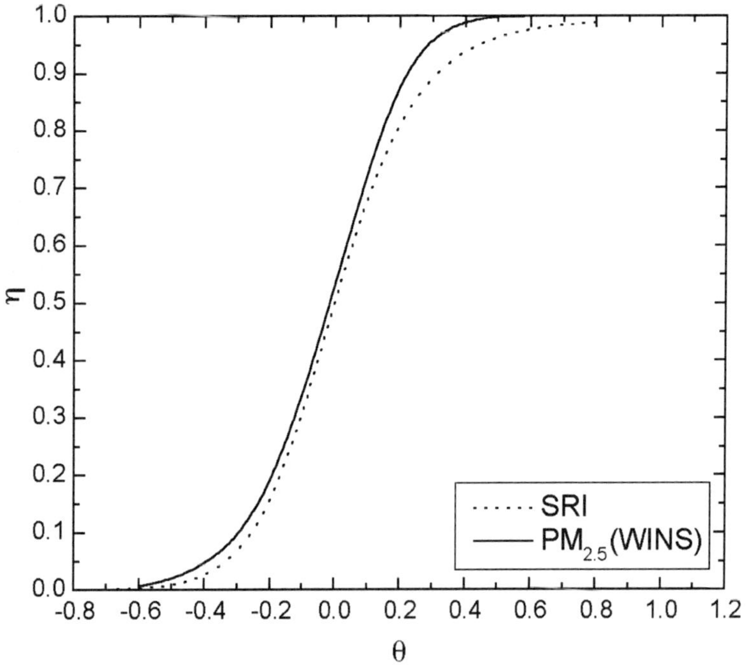

Fig. 16. Comparison between the performance of the SRI II cyclone (D_c = 3.1 cm; Q = 45 L/min) and the $PM_{2.5}$ (WINS) curve.

Table 14
Multiple Regression Results for ψ_{50}

Parameter	Coefficient	Standard deviation
exp (F_0)	0.0414	
F_1	–0.713	0.008
F_2	–0.172	0.033

$R^2 = 0.981$; GSD = 1.12

Source: (Adapted from Lidén and Gudmundsson (49), with permission from Taylor and Francis; http://www.tandf.co.uk/journals.)

Table 15
Parameters a_1, a_2, and a_3 for Eq. (68)

Cyclone	a_1	a_2	a_3
Stairmand	0	4.848	0.627
SRI	0	8.196	–3.239

4.4. *The "Cyclone Family" Approach*

Kenny and Gussman (53) recently presented a detailed experimental study based on three cyclone "families" (whose relative dimensions are in fixed proportions to the body diameter) named Extra Sharp Cut Cyclone (ESCC), Sharp Cut Cyclone (SCC), and Gussman Kenny (GK). The authors refer to the ESCC cyclone as designed to give a very sharp penetration curve, intended for ambient air monitoring. The base of the cyclone is cylindrical rather than conical. The SCC family is based on the design of the SRI–III cyclone (54) and, according to Kenny and Gussman, provides both a sharp cut and stable performance under loading. The GK cyclone has a wider penetration curve and was designed for use in the workplace and indoor air monitoring. The detailed geometrical characteristics of the families are not clearly stated in the article. However, the authors give indications that led to the dimensions listed in Table 16, which may contain deviations.

The experimental work, described in detail by Maynard and Kenny (55), included the testing of 36 diverse cyclone combinations, resulting from modular combinations of cyclone parts, all with a body diameter D_c of 17.5 mm. At the first stage, the cyclone tested comprised more than the three families. In a second phase, the experiments were focused on the three families: seven ESCC, seven SCC, and eight GK cyclones were tested. The test aerosol was made of poly-dispersed glass microspheres, with nominal diameters up to 25 μm and density of 2450 kg/m^3. The size distribution was measured with an aerodynamic particle sizer (APS 3310 or APS 3320, manufactured by TSI Inc.). Penetration curves were determined, as a function of particle size and gas flow rates. Each test was characterized by the cut diameter D_{50} (the diameter with 50% collection efficiency) and by the sharpness of the penetration curve, given by

Table 16
Dimensions of the Cyclone Families Studied

Family	D_{in}/D_c	D_e/D_c	B/D_c	H/D_c	Z/D_c	S/D_c	H_{cup}/D_c	D_{cup}/D_c
ESCC1[a]	0.13	0.24	?	1.69	0	0.38	?	?
SCC[b]	0.24	0.27	0.24	0.45	1.13	0.35	0.71	1.00
GK[b]	0.20	0.23	0.20	0.40	0.90	0.23	0.87	1.03

[a]From ref. 53, Fig.5, p. 1416.
[b]From ref. 56, Table.1, p. 678.
Source: (Adapted from Kenny and Gussman (53), with permission from Elsevier Science.)

$$\sigma = \left(\frac{D_{16}}{D_{84}} \right)^{1/2} \tag{72}$$

which is equivalent to the geometric standard deviation of a log-normal curve fitted to the penetration data. This parameter was used for characterizing the penetration curve sharpness, even if it was not log-normal.

In the first stage, results from all 36 combinations tested in two flow rates (2 and 4 L/min) were statistically fitted to a multiple-regression model. The fittings for the D_{50}, D_{16}, and D_{84} (in μm) are listed in Table 17.

In the second stage, the results for the three families were analyzed, and the cut diameter D_{50} was well represented by the correlation:

$$D_{50} = \frac{aD_c^b}{Q^{b-1}} \tag{73}$$

where D_{50} is the cut diameter (in μm), D_c is the cyclone body diameter (in cm), and Q is the flow rate (in L/min). Table 18 lists the values for the empirical parameters a and b as well as the regression coefficient, R^2. Table 19 lists the range of body diameters and flow rates tested in this stage.

Cut diameters, D_{50}, calculated from the multiple regression model (*see* Table 17) were compared to the respective ones calculated from Eq. (73), and some discrepancy was found. The authors therefore recommend Eq. (73) for the specific family it has been deduced to fit.

Figures 17–19 show the cut diameters D_{50} (in μm) as a function of cyclone body diameter D_c (in cm) for the ESCC, SCC, and GK families, respectively. The curves cover the range of flow rates tested and show that a large number of cut diameters in the region of interest (e.g., for PM_{10} and $PM_{2.5}$) can be achieved by the proper combination of cyclone configuration with flow rate. Nevertheless, the lack of information on the penetration curves steepness constitutes a problem for the designer. Some qualitative information can be drawn from the multiple-regression models (*see* Table 17), which allows the calculation of σ for two flow rates ($Q = 2$ and 4 L/min).

4.5. $PM_{2.5}$ Samplers

Special attention is directed nowadays to the establishment of reliable samplers for the high-risk respirable fraction of aerosols. Peters et al. (57) conducted a series of tests

Table 17
Multiple-Regression Models for D_{50} (μm), D_{16} (μm), and D_{84} (μm) at 2 and 4 L/min

Variable (mm)	Parameter value for 2 L/min data			Parameter value for 4 L/min data		
	D_{50} (μm)	D_{16} (μm)	D_{84} (μm)	D_{50} (μm)	D_{16} (μm)	D_{84} (μm)
Constant	0	1.705	−1.538	0	0.629	0.125
D_{in}	0.172104	0	0.856	−0.170	−0.305	−0.219
D_e	0	0.832	0	0	0	0
h	0	0.05498	0.04676	0	0	0
B	0	0	0.06395	0	−0.02728	0
H_{cup}	0	0	0.01976	0	0	−0.02269
$D_{in}D_e$	0.155059	0.368	0	0.07873	0.122	0.09303
$D_{in}S$	0.043678	0.03636	0.02756	0.01636	0.01394	0.01337
$D_{in}h$	0.012937	0	0	0.00969	0.009032	0.009583
$D_{in}Z$	0.017697	0.01969	0.0143	0.00765	0.0143	0.006529
$D_{in}D_{cup}$	0.018444	0.02134	0	0.01773	0.01785	0.1257
D_eZ	−0.011925	0	0	0	−0.00591	0
SZ	−0.004054	−0.00337	−0.002958	0	0	0
ZZ	0.001225	0	0	0	0	0.00047
BH_{cup}	0.003311	0.003325	0	0.00240	0.005152	0.003827
R^2	0.998	0.986	0.938	0.994	0.972	0.971

Source: Adapted from Kenny and Gussman (53), with permission from Elsevier Science.

Table 18
Parameters a and b and Regression Coefficient R^2 Obtained from Experimental Fittings for the Cyclone Families

Cyclone family	a	b	R^2
ESCC	2.6538±1.014	1.837±0.019	0.9886
SCC	4.2503±1.018	2.131±0.017	0.9753
GK	2.5492±1.017	2.105±0.017	0.9907

Source: Adapted from Kenny and Gussman (53), with permission from Elsevier Science.

Table 19
Summary of Family Cyclones Tested

Cyclone designation	Body diameter (cm)	Flow rates tested (L/min)
SRI–IIIa (SCC)	3.45	10–20
SCC	1.25–3.495	1–16.7
ESCC	0.816–3.0	1–16.7
GK	1.299–3.45	1–18

Source: Adapted from Kenny and Gussman (53), with permission from Elsevier Science.

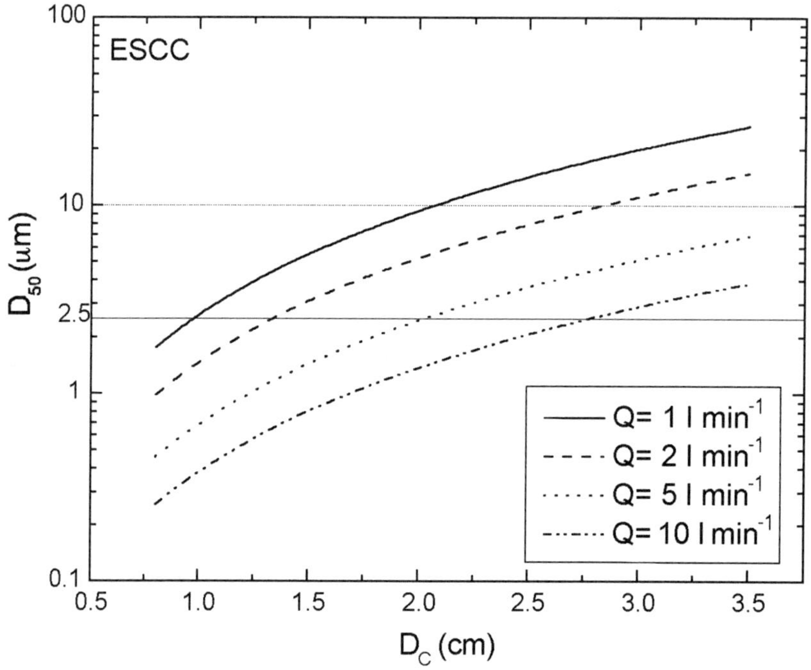

Fig. 17. Cut diameter D_{50} as a function of the cyclone body diameter D_c and gas flow rate, for the ESCC family.

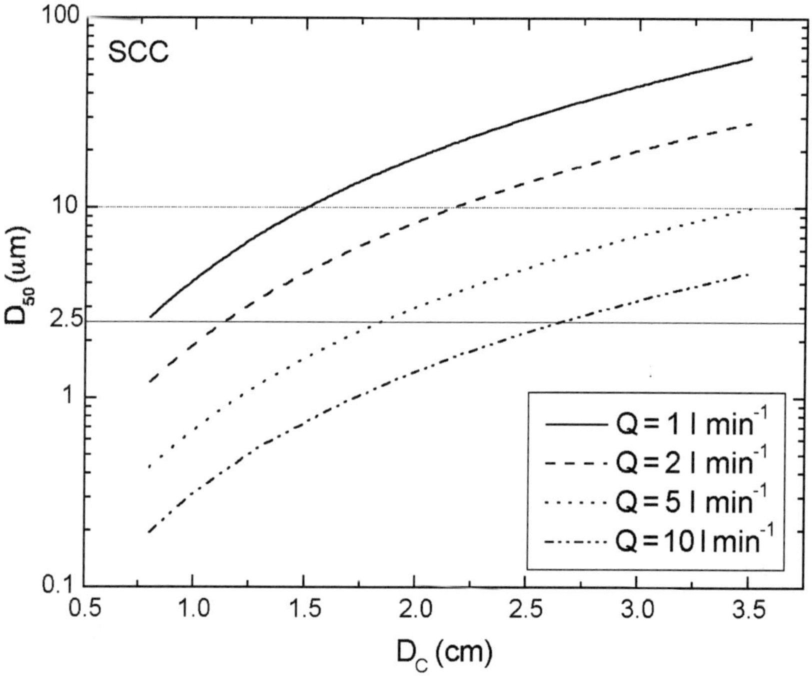

Fig. 18. Cut diameter D_{50} as a function of the cyclone body diameter D_c and gas flow rate, for the SCC family.

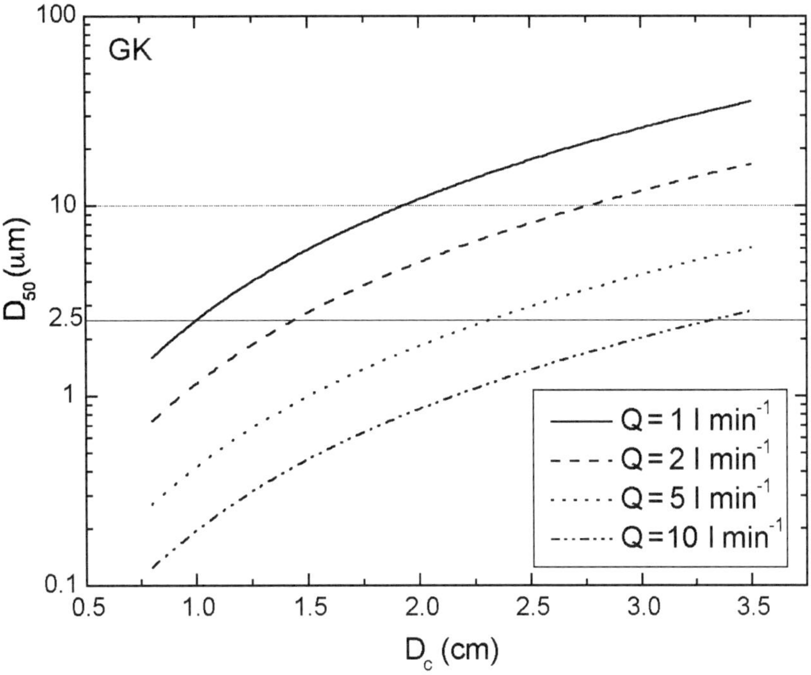

Fig. 19. Cut diameter D_{50} as a function of the cyclone body diameter D_c and gas flow rate, for the GK family.

for evaluating the collection characteristics of three cyclones used in $PM_{2.5}$ speciation samplers. The cyclones investigated were the SCC 1.829, SCC 2.141, and AN 3.68 (*Note*: The number following the letters refers to the body diameter, in centimeters.) The detailed dimensions of the SCC family were already presented in Table 16, and those referring to the AN 3.68 are shown in Table 20. The experimental technique used was the same as described in the previous subsection.

The penetration curves were fitted by a reverse asymmetric sigmoid equation, expressed, for D_{ae} (in μm), as

$$P = 1 - \eta = a + b\left\{1 - \left[1 + \exp\left(\frac{\left(D_{ae} + d\ln\left(2^{\frac{1}{e}} - 1\right) - c\right)}{d}\right)\right]^{-e}\right\} \tag{74}$$

where P is the penetration (= 1− η) and *a–e* are the curve parameters. The geometric standard deviation is given by Eq. (72).

Table 21 shows the adjusted parameters for the three cyclones and the one obtained for the WINS impactor, adopted by the EPA as the descriptor penetration curve for the $PM_{2.5}$. Figures 20 and 21 show the penetration curves for the three cyclones investigated, in comparison with the PM2.5 (WINS) standard. It can be seen that the AN 3.68

Table 20
Detailed dimensions of the AN 3.68 cyclone

Dimension (cm)	AN 3.68
D_c	3.68
D_{in}	1.01
D_e	1.09
B	1.28
H	7.07
h	2.33
Z	4.74
S	1.55
H_{cup}	2.26
D_{cup}	3.10

Source: Adapted from Peters et al. (57), with permission from Elsevier Science.

can be better adjusted to the $PM_{2.5}$ criteria; however, it has a D_{50} cut size of 2.7 μm at its design flow rate. The SCC curves show a longer tail in the coarse particles region. Field experiments were carried out by monitoring the ambient air of a US city. The authors had some concern about the effect of the particle loading variations and with the performance of the separators after becoming dirty from use.

4.6. *Examples*

Example 7

Deduce the flow rate and derive the performance curve for the SRI II cyclone working at 23°C at atmospheric pressure for a cut diameter of 2.5 μm. Compare it with the $PM_{2.5}$ (WINS) criterion.

Table 21
Adjusted Parameters for Eq. (74), Referring to the SCC 2.141, SCC 1.829, and AN 3.68 Cyclones and the WINS Impactor

Cyclone:	SCC 2.141		SCC 1.829	AN 3.68	WINS	
Flow rate:	6.7 L/min	7.0 L/min	5.0 L/min	24.0 L/min	28.1 L/min	16.7 L/min
a	1	1	1	1	1	1
b	−1	−1	−1	−1	−1	−1
c^a	2.52	2.35	2.44	2.72	2.33	2.48
d	0.1823	0.1534	0.1270	0.1926	0.1945	0.3093
e	0.3005	0.2640	0.2160	0.6318	0.6688	3.3683
D_{84} (μm)	2.09	1.96	2.08	2.38	1.98	2.05
D_{16} (μm)	3.22	3.04	3.16	3.17	2.76	2.85
GSD	1.24	1.24	1.23	1.15	1.18	1.18

[a] $c = D_{50}$.
Source: Adapted from Peters et al. (57), with permission from Elsevier Science.

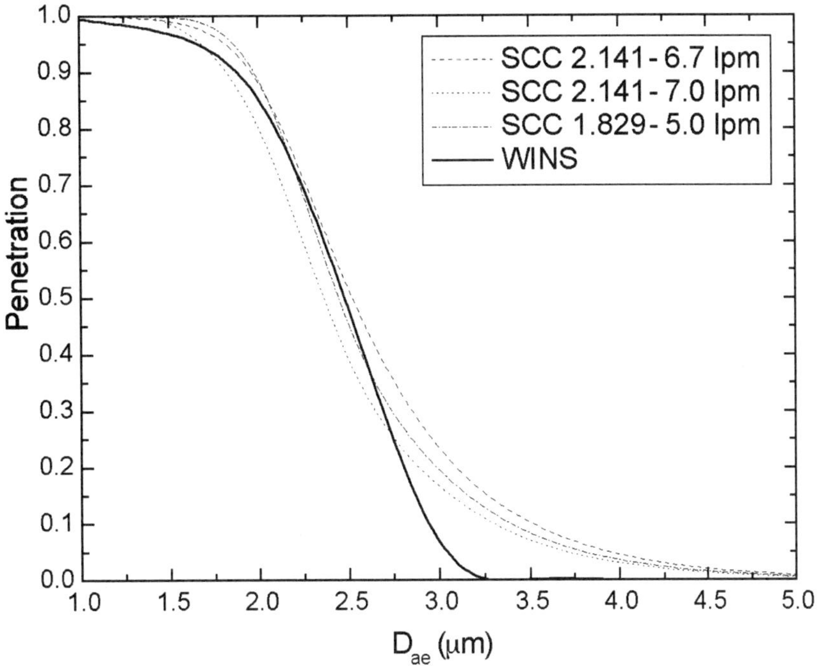

Fig. 20. Comparison between two performance curves for the SCC 2.141 (for $Q = 6.7$ and 7.0 L/min) and one for the SCC 1.829 (for $Q = 5.0$ L/min) cyclones, with the $PM_{2.5}$ (WINS) curve.

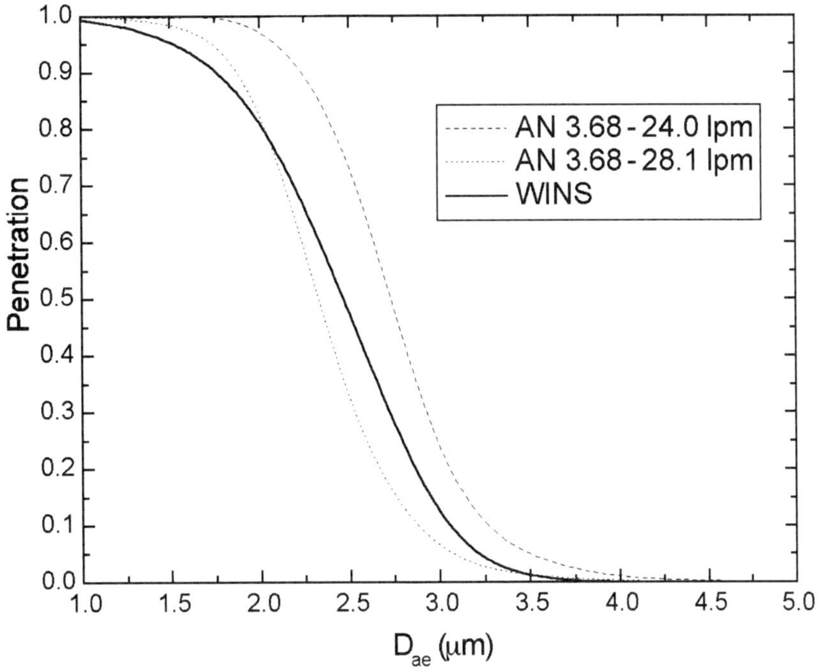

Fig. 21. Comparison between two performance curves for the AN 3.68 cyclone (for $Q = 24.0$ and 28.1 L/min), with the $PM_{2.5}$ (WINS) curve.

Solution

The first step consists in calculating the gas flow rate that gives the desired cut diameter for the SRI II cyclone, whose dimensions can be found in Table 13. For this, it is necessary to calculate F_s, the Cunningham slip factor, which is given by

$$F_s = 1 + (\lambda/D_{ae})[2.514 + 0.8 \exp(-0.55D_{ae}/\lambda)] \tag{75}$$

where λ is the gas mean free path. For air, at atmospheric pressure and 23°C, $\lambda = 6.702\times 10^{-8}$ m. Thus,

$$F_s(D_{50}) = 1.0674$$

From Eq. (66), taking $D_c = 3.1$ cm, comes

$$\psi_{50} = \frac{\sqrt{1.0674}\times 2.5\times 10^{-6}\,\text{m}}{3.1\times 10^{-2}\,\text{m}} = 8.332\times 10^{-5}$$

From Eq. (71) and Tables 13 and 14,

$$\psi_{50} = 0.0414\text{Re}_{\text{ann}}^{-0.713}(0.43)^{-0.172} = 8.332\times 10^{-5}$$

Thus,

$$\text{Re}_{\text{ann}} = 7412.73$$

From Eq. (70) and Table 13, for ρ_g=1.17 kg/m^3 and μ=1.80$\times$10^{-5} kg/ms,

$$\text{Re}_{\text{ann}} = 0.5(1-0.286)\frac{1.17\text{kg/m}^3\times 3.1\times 10^{-2}\,\text{m}}{1.8\times 10^{-5}\,\text{kg/ms}}v_i = 7412.73$$

Thus:

$$v_i = 10.305/\text{ms}$$

Finally, the gas flow rate in the cyclone can be calculated as

$$Q = \frac{\pi}{4}(0.286D_c)^2 v_i = 6.362\times 10^{-4}\,\text{m}^3/\text{s} = 38.2\text{ L min}^{-1}$$

For this flow rate, the efficiency curve can be constructed as follows: For a given, D_{ae}, the slip factor F_s is calculated as above. Then, the factor θ comes from Eq. (69) as

$$\theta = \frac{\sqrt{F_s(D_{ae})D_{ae}(\mu\text{m})}}{\sqrt{1.0674}\times 2.5(\mu\text{m})}$$

Next, $f(\theta)$ comes from Eq. (68) and Table 15 as

$$f(\theta) = 8.196\theta - 3.239\theta^2$$

Finally, the efficiency η is calculated from Eq. (67). Table 22 lists the values of these steps for a number of diameters. Figure 22 shows η as a function of D_{ae} for the SRI II cyclone with a cut diameter of 2.5 µm (for which a flow rate of 38.2 L/min is necessary). The $PM_{2.5}$ curve is also plotted; some deviation can be seen in the larger-particle range. It is worth noting that the predicted flow rate for the SRI cyclone (38.2 L/min) is somewhat higher than the range reported by Smith et al. (54).

Table 22
Results of the Grade Efficiency of the SRI II Cyclone, Operating at 38.2 L/min

D_{ae} (μm)	F_s	θ	$f(\theta)$	η_i
1.1	1.15318	–0.5427	–5.40153	0.00449
1.2	1.14041	–0.5039	–4.95194	0.00702
1.4	1.12035	–0.4263	–4.08240	0.01659
1.6	1.10531	–0.3487	–3.25220	0.03725
1.8	1.09360	–0.2712	–2.46122	0.07862
2.0	1.08424	–0.1937	–1.70929	0.15326
2.2	1.07659	–0.1162	–0.99636	0.26966
2.4	1.07020	–0.0387	–0.32246	0.42008
2.5	1.06740	0	0	0.5
2.6	1.06480	0.0387	0.31252	0.57750
2.8	1.06017	0.1162	0.90857	0.71271
3.0	1.05616	0.1936	1.46572	0.81241
3.2	1.05265	0.2711	1.98397	0.8791
3.4	1.04956	0.3486	2.46336	0.92153
3.6	1.04680	0.4260	2.90382	0.94804
3.8	1.04434	0.5035	3.30545	0.96462
4.0	1.04212	0.5809	3.66818	0.97511

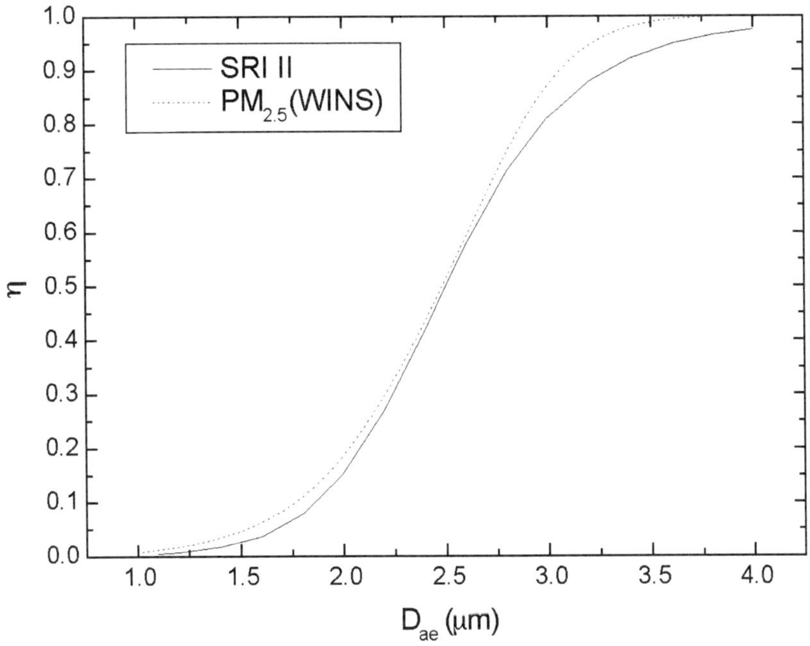

Fig. 22. Collection efficiency curve for the the SRI II cyclone operating at 38.2 L/min compared to the $PM_{2.5}$ (WINS) criterion.

Example 8

An initial survey of the air inside the production building at an industrial site revealed that the ambient was loaded with the particulate matter they produced, which has a density of 2750 kg/m^3. The TSP was determined and a concentration of 298 mg/m^3 was detected. Also, the particle size distribution was measured with an apparatus that utilized Stokes diameter as the operating principle. The distribution is presented in Table 23. After analysis of the results, it was decided that a systematic monitoring of the $PM_{2.5}$ concentration would be carried out, and the AN 3.68 minicyclone was purchased, to work at a flow rate of 28.1 L/min. Considering that the balance available has the sensibility of 10^{-5} g, estimate the minimum sampling time for the cyclone.

Solution

The first step is to transform the given diameter (Stokes) into the aerodynamic diameter utilized in all monitoring equations. By definition, the Stokes diameter is the equivalent diameter of a sphere that has the same terminal velocity as the particle. Therefore, the two diameters can be related by

$$D_{\text{ae}} = \sqrt{\frac{\rho_p}{\rho_{\text{unit}}}} D_{\text{St}} \tag{76}$$

where ρ_{unit} is the unit density. Therefore, in this case, the aerodynamic particle diameter is

$$D_{\text{ae}} = \sqrt{2.75} D_{\text{St}} = 1.658 D_{\text{St}}$$

Table 24 lists the mean aerodynamic diameters for each particle size range. It can be noted that the particles smaller than 2.5 μm correspond to the four smaller ranges. This means that, according to Table 23, 15.6% of the TSP is constituted of particles smaller than 2.5 μm, which corresponds to a concentration of 46.5 μg/m^3, well above the maximum of 15 μg/m^3 recommended by the EPA for an healthy ambient. This implies that some protection measures need to be taken.

As far as monitoring is concerned, the next step consists of determining the mass concentration in the cyclone exit, once this is the mass of $PM_{2.5}$ collected by the membrane placed after it. This membrane is assumed to have 100% collection efficiency. The procedure here is similar to the one adopted in Example 2 in Section 2.5.1: the collection efficiency of each range, η_i, is determined and multiplied by the size fraction x_i. Here, the efficiency is calculated utilizing Eq. (74) that, for the AN 3.68 minicyclone working at 28.1 L/min, can be written as (see Table 24)

$$\eta = \left\{1 - \left[1 + \exp\left(\frac{\left(D_{\text{ae}} + 0.1945\ln\left(2^{\frac{1}{0.6688}} - 1\right) - 2.33\right)}{0.1945}\right)\right]^{-0.6688}\right\}$$

Table 24 lists the calculated efficiencies as well as the total mass collected, which was 91%. The next steps are straightforward:

$$\text{Concentration of } PM_{2.5} \text{ in the cyclone exit} = (1 - 0.91) \times 298\,\text{mg/m}^3 = 26.82\,\text{mg/m}^3$$

Table 23
Particle Size Distribution, in Stokes Diameter

Range (μm)	x_i (mass basis)
0.3–0.5	0.0095
0.5–0.7	0.0095
0.7–1.0	0.039
1.0–3.0	0.098
3.0–5.0	0.245
5.0–10.0	0.296
10.0–20.0	0.228
20.0–50.0	0.075

Table 24
Results from the Calculation of the Particle Size Distribution in the Cyclone Exit

Range (μm)	D_{St} (μm)	D_{ae} (μm)	η_i	$\eta_i x_i$
0.3–0.5	0.4	0.66	0.00023	0.000002
0.5–0.7	0.6	0.99	0.00127	0.000012
0.7–1.0	0.85	1.41	0.01060	0.000413
1.0–3.0	1.5	2.49	0.66270	0.064945
3.0–5.0	4.0	6.63	0.99999	0.244998
5.0–10.0	7.5	12.43	1.00000	0.296000
10.0–20.0	15.0	24.87	1.00000	0.228000
20.0–50.0	35.0	58.03	1.00000	0.075000
				$\Sigma \eta_i x_i = 0.91$

Table 25
Determination of the Particle Size Distribution in Example 9

D_{50} (μm)	Q (L/min)	c_{in} (μg/m^3)	% in range	accum. %
1.0	2.73	0.195	0.1	0.1
1.5	1.68	1.562	0.8	0.9
2.0	1.19	7.027	3.6	4.5
2.3	1.01	21.667	11.1	15.6
2.8	0.80	38.064	19.5	35.1
3.2	0.68	33.574	17.2	52.3
3.6	0.59	38.259	19.6	71.9
4.0	0.52	26.352	13.5	85.4
5.0	0.40	18.739	9.6	95
7.0	0.27	9.565	4.9	99.9
10.0	0.17	0.195	0.1	100
		$\Sigma c_{in} = 195.2$		

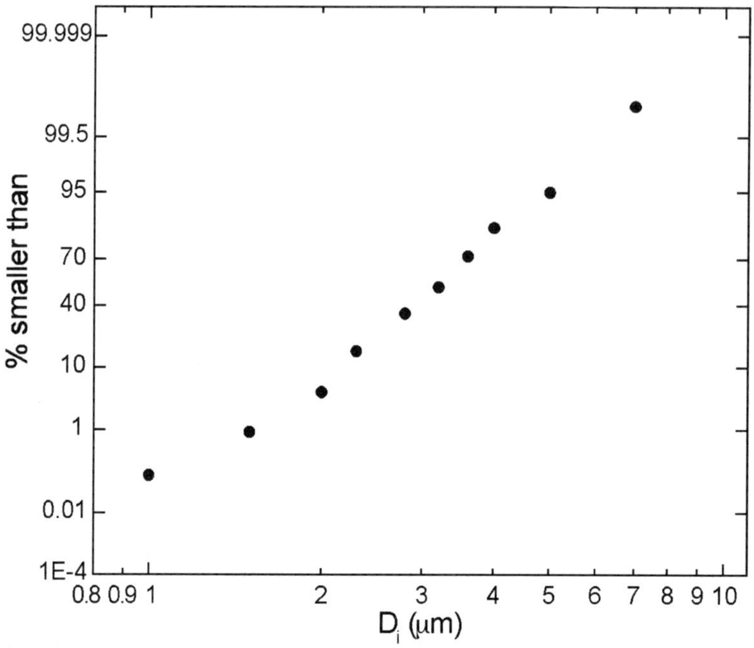

Fig. 23. Particle size distribution calculated in Example 9.

$$\text{Volumetric flow rate of air} = 28.1\ \text{L/min} = 4.683 \times 10^{-4}\ \text{m}^3/\text{s}$$

$$\text{Rate of deposition of PM}_{2.5}\ \text{in the membrane} = 26.82 \times 4.683 \times 10^{-4}\ \mu\text{g/s}$$
$$= 1.25 \times 10^{-2}\ \mu\text{g/s}$$

If the balance weighs 10^{-5} g, the minimum mass that can reliably handled is one order of magnitude higher. Thus

$$\text{Minimum weight measured} = 100\mu\text{g}$$

$$\text{Minimum monitoring time} = \frac{100\mu\text{g}}{1.256\times10^{-2}\ \mu\text{g/s}} = 7962\text{s} = 2\text{h}13\ \text{min}$$

Example 9

Define a strategy for measuring the particle size distribution of a given site, utilizing a minicyclone.

Solution

If a proper cyclone is chosen, the particle size distribution in a given environment can be determined with reasonable accuracy. Let us choose the ESCC cyclone, which has a sharp efficiency curve and a well-defined dependence of the cut diameter on the gas flow rate given by Eq. (73). Within the range of body diameters tested (53) let us take $D_c = 1.25$ cm. Thus, rearranging Eq. (73) and taking a and b from Table 18, we obtained:

$$Q = \left(\frac{aD_c}{D_{50}}\right)^{\frac{1}{b-1}} = \left(\frac{2.6538\text{x}1.25}{D_{50}}\right)^{\frac{1}{1-1.837}}$$

Table 25 lists the calculated gas flow rates for cut diameters ranging from 1 to 10 μm. For the sake of illustration, let us suppose that the measurements of particle concentrations for each flow rate are given by c_{in} in Table 25. In this case, the percentage of particles with average diameter D_{50} can be estimated as

$$\% \text{ in range} \approx \frac{c_{\text{in}}}{\sum c_{\text{in}}}$$

Figure 23 shows the calculated particle size distribution. It is worth mentioning that the use of a microgram balance (10^{-6} g sensibility) in this study is likely to occur.

NOMENCLATURE

a	Height of the cyclone entry duct (m)
a	Empirical parameter in Eq. (73) or (74), given in Table 18 or 21
a_d	Constant in Eq. (52)
a_f	Constant in Table 7
a_s	Constant in Table 8
a_1, a_2, a_3	Parameters in Eq. (68), listed in Table 15
A	Empirical parameter listed in Table 11
A_{cyc}	Cyclone inlet area (ft^2)
b	Width of the cyclone entry duct, (m)
b	Empirical parameter in Eq. (73) or (74), given in Table 18 or 21
b_d	Constant in Eq. (52)
b_f	Constant in Table 7
b_s	Constant in Table 8
B	Diameter of the dust exit, at the base of the cyclone (m)
B	Empirical parameter listed in Table 11
c	Parameter in Eq. (74)
c_{in}	Overall particle concentration at the cyclone entry [kg/m^3 or gr/ft^3 in Eq. (35)]
c_o	Particle concentration at the cyclone exit (kg/m^3)
cost_a	The cost in the month–year of a (US$)
cost_b	The cost in the month–year of b (US$)
d	Parameter in Eq. (74)
d_c	Diameter of the cyclone central axis (m)
d_{duct}	Duct diameter (in).
d_{fan}	Fan diameter (in).
D_{ae}	Particle aerodynamic diameter [μm, or *m* in Eq. (75)]
D_c	Diameter of the cyclone cylindrical body [m, or cm in Eq. (73)]
D_{cup}	Diameter of the dust receiver cup (m)
D_{duct}	Duct diameter (ft)
D_e	Diameter of the exit cyclone duct (vortex finder) (m)
D_i	Particle diameter (m)

D_{in}	Diameter of cyclone entry duct (m)
D_{St}	Stokes diameter μm
D_t	Vortex finder external diameter (m)
D_{16}	Diameter of the particle collected with 16% efficiency (μm)
D_{50}	Diameter of the particle collected with 50% efficiency [m, or μm in Eqs. (69) and (73)]
D_{84}	Diameter of the particle collected with 84% efficiency (μm)
e	Parameter in Eq. (74)
E_f	Fan efficiency
$f(\theta)$	Parameter in Eq. (67)
F_s	Cunningham slip factor
$F(x)$	Cumulative probability function of the standardized normal variable x
F_0, F_1, F_2	Parameters in Eq. (71), given in Table 14
g	Gravity, (m/s^2)
G	Geometric parameter in the Leith and Licht model
h	Height of the cyclone cylindrical body (m)
hp	Motor horsepower
h^m	Height of the central axis in the Barth model (m)
H	Cyclone total height (m)
H_{cup}	Height of the dust receiver cup (m)
HRPM	High-risk respirable fraction
H_{stack}	Stack height (ft)
IPM	Inhalable fraction
index_a	CE equipment cost index in the month–year of a
index_b	CE equipment cost index in the month–year of b
K	Empirical parameter, listed in Table 11
K_a	Dimensionless parameter in Eq. (17)
K_b	Dimensionless parameter in Eq. (18)
K_c	Dimensionless parameter in Eq. (19)
m_{oi}	Mass flow rate of particles with diameter D_i in the exit (kg/s)
n	Vortex exponent
n	Empirical parameters listed in Table 11
N_c	Number of cyclones in parallel
P	Penetration
P_{cyc}	Cost of cyclone (US$)
P_{damp}	Cost of damper (US$)
P_{divert}	Cost of two-way diverter valve (US$)
P_{fan}	Cost of fan system (US$)
P_{FRPd}	Cost of FRP ductwork (US$/ft)
P_{motor}	Cost of fan motor, belt and starter (US$)
$PM_{2.5}$	Particle matter smaller than 2.5 μm
PM_{10}	Particle matter smaller than 10 μm
P_{PVCd}	Cost of PVC ductwork (US$/ft)
P_{ral}	Cost of rotary air lock (US$)
P_{stack}	Total capital cost of large stack (US$)

Q	Gas volumetric flow rate [m^3/s, or L/min in Eqs. (64) and (73)]
Q_a	Emission stream flow rate at actual condition (ft^3/min)
$\mathrm{Re}_{\mathrm{ann}}$	Reynolds number based on the cyclone annular dimension
RPM	Respirable fraction
S	Internal height of the cyclone exit duct (vortex finder) (m)
T	Temperature (K)
TPM	Thoracic fraction
U	Ambient wind speed (m/s)
U_{duct}	Duct velocity (ft/min)
v_i	Gas velocity at the cyclone entry (m/s)
v_o	Gas velocity at the cyclone exit (m/s)
v_s	Saltation velocity (ft/s)
v_t	Gas tangential velocity (m/s)
$v_{t\max}$	Gas maximum tangential velocity (m/s)
v_{ts}	Particle terminal velocity (m/s)
v_{ts}^m	Terminal velocity of the particle collected with 50 % efficiency in the Barth model (m/s)
V_H	Volume below the exit duct, excluding the central axis (m^3)
V_{nl}	Annular volume between S and the end of the vortex length (Z_c), excluding the central axis (m^3)
V_s	Annular volume between S and the medium height of the entry duct (m^3)
x	Standardized normal variable
x_i	Particle mass fraction of diameter D_i in the entry
x_{oi}	Particle mass fraction of diameter D_i in the exit
W_c	Fan power (W)
Z_c	Height of the cyclone central axis (m)
α	Constant in Eq. (7)
β	Exponent related to the cut diameter in the Iozia and Leith model
Γ	Constant in Eq. (61) (μm)
ΔH	Dimensionless parameter
ΔP	Pressure drop (Pa)
η	Cyclone efficiency for a particle of aerodynamic diameter D_{ae}
η_i	Collection efficiency of a particle of diameter D_i
η_0	Overall particle collection efficiency
θ	Gas average residence time (s)
θ	Normalized collection efficiency
λ	Friction coefficient
λ	Gas mean free path (m)
μ	Gas viscosity (kg/ms)
ρ	Gas density (kg/m^3)
ρ_p	Particle density (kg/m^3)
ρ_{unit}	Unit density (kg/m^3)
σ	Geometric standard deviation
τ_i	Relaxation time (s)
ψ_{50}	Cutoff size parameter

REFERENCES

1. W. Licht, *Air Pollution Control Engineering* 2nd ed., Marcel Dekker, New York, 1988, pp. 277–330.
2. D. Leith and D. L Jones, in *Handbook of Powder Science and Technology* 2nd ed. (M.E. Fayed and L. Otten, eds.), Chapman & Hall, New York, 1997, pp. 727–752.
3. A. Ogawa, *Separation of Particles from Air and Gases*. CRC, Boca Raton, FL, 1984, Vol. II, pp. 1–49.
4. M. Bohnet, *Chem. Eng. Process.* **34**, 151–156 (1995).
5. M. Bohnet, O. Gottschalk, and M. Morweiser, *Adv. Powder Technol.* **8**, 137–161 (1997).
6. F. Boysan, W. H. Ayers, and J. Swithenbank, *Trans. Inst. Chem. Eng.* **60**, 222–230 (1982).
7. J. Benítez, *Process Engineering and Design for Air Pollution Control*. PTR Prentice-Hall, Englewood Cliffs, NJ, 1993.
8. P. W. Dietz, *AIChE J.* **27**, 888–892 (1981).
9. D. Leith and W Licht, *Am. Inst. Chem. Eng. Symp. Ser.* **68**, 196–206 (1972).
10. D. L. Iozia and D. Leith, *Aerosol Sci. Technol.* **10**, 491–500 (1989).
11. A. J. ter Linden, *Proc. Inst. Mech. Eng.* **160**, 233–245 (1949).
12. W. Barth, *Brennstoff-Warme-Kraft*. **8**, 1–9 (1956).
13. D. L. Iozia, and D. Leith, *Aerosol Sci Technol.* **12**, 598–606 (1990).
14. C. Lapple, *Chem. Eng.* **58**, 144–151 (1951).
15. H. Buttner, *J. Aerosol Sci.* **30**, 1291–1302 (1999).
16. M. E. Moore, and A. R. McFarland, *Am. Ind. Hyg. Assoc. J.* **51**, 151–159 (1990).
17. J. Dirgo, and D. Leith, *Aerosol Sci. Technol.* **4**, 401–415 (1985).
18. F. Dullien, *Industrial Gas Cleaning*, Academic, San Diego, CA, 1989.
19. A. K. Cocker, *Chem. Eng. Prog.* **89**, 51–55 (1993).
20. R. M. Alexander, *Proc. Austrl. Inst. Mining Met. New. Series* **152(3)**, 203–208 (1949).
21. A. C. Hoffmann, R. Dejonge, H. Arends, et al., *Filtrate Sep.* **32**, 799–804 (1995).
22. B. Kalen, and F. A. Zenz, *AIChE Symp. Ser.* **70**, 388–396 (1974).
23. W. H. Koch, and W. Licht, in *Industrial Air Pollution Engineering* (V. Cavaseno, ed.), McGraw-Hill, New York, 1980, pp. 175–183.
24. G. Ramachandran, D. Leith, J. Dirgo, et al., *Aerosol Sci. Technol.* **15**, 135–148 (1991).
25. Z. B. Maroulis, and C. Kremalis, *Filtrate Sep.* **32**, 969–976 (1995).
26. C. B. Shepherd, and C. E. Lapple, *Ing. Eng. Chem.* **32**, 972–984 (1940).
27. J. Casal, and J. M. M. Benet, *Chem. Eng.* **24**, 99–100 (1983).
28. L. C. Whiton, *Chem. Met. Eng.* **39**, 150 (1932).
29. J. G. Masin, and W.H. Koch, *Environ. Prog.* **5**, 116–122 (1986).
30. C. D. Cooper, and F. C. Alley, *Air Pollution Control: A Design Approach*, Waveland, Prospect Heights, IL, 1994.
31. W. M. Vatavuk, *Estimating Costs of Air Pollution Control*, Lewis, Chelsea, MI, 1990.
32. B. G. Liptak, *Environmental Engineers' Handbook, Volume II: Air Pollution*, Chilton, Radnor, PA, 1974.
33. Chemical Engineering, *Equipment Indices*, McGraw-Hill, New York, 2001.
34. Engineering News Record, *ENR Indices*, McGraw-Hill, New York, 2001.
35. O. G. Raabe, and B. O. Stuart, in *Particle Size-Selective Sampling for Particulate Air Contaminants* (J.H. Vincent, ed.), ACGIH, Cincinnati, OH, 1999, pp. 73–95.
36. QUARG—Quality of Urban Air Review Group. Airborne Particulate Matter in the United Kingdom, Third Report, Department of Environment, UK, 1996.
37. S. Marquardt, in *Air Quality Control Handbook* (E. R. Alley, ed.), McGraw-Hill, New York, 1998 pp. 4.1–4.33.
38. H. S. Adams, L. C. Kenny, M. J. Niewenhuijsen, et al., *J. Expos. Anal. Environ. Epidemiol.* **11**, 5–11 (2001).

39. V. A. Marple, K. L. Rubow, and B. A. Olson, in *Aerosol Measurement: Principles, Techniques and Applications* (K. Willeke, and P. A. Baron, eds.), Van Nostrand Reinhold, New York, 1993, pp. 206–232.
40. J. H. Vincent, *Aerosol Science for Industrial Hygienists*, Pergamon, Oxford, 1995.
41. W. John, and G. Reischl, *JAPCA* **30**, 873–876 (1980).
42. D. L. Bartley, and G. M. Breuer, *Am. Ind. Hyg. Assoc. J.* **43**, 520–528 (1982).
43. W. D. Griffiths, and Boysan, *J. Aerosol Sci.* **27**, 281–304 (1996).
44. M. Gautam, and A. Sreenath, *J. Aerosol Sci.* **28**, 1265–1281 (1997).
45. Y. F. Zhu, and K. W. Lee, *J. Aerosol Sci.* **30**, 1303–1315 (1999).
46. T. Chan, and M. Lippmann, *Environ. Sci. Technol.* **11**, 377–382 (1977).
47. T. T. Mercer, *Aerosol Technology in Hazard Evaluation, US*, Academic, New York, 1973, pp. 297–299.
48. M. Lippmann, and T. Chan, *Am. Ind. Hyg. Assoc. J.* **35**, 189–200 (1974).
49. G. Lidén, and A. Gudmundsson, *J. Aerosol Sci.* **28**, 853–874 (1997).
50. R. P. Dring, and M. Suo, *J. Energy* **2**, 232–237 (1978).
51. M. E. Moore, and A. R. McFarland, *Environ. Sci. Technol.* **27**, 1842–1848 (1993).
52. M. E. Moore, and A. R. McFarland, *Environ. Sci. Technol.* **30**, 271–276 (1996).
53. L. C. Kenny, and R. A. Gussman, *J. Aerosol Sci.* **31**, 1407–1420 (2000).
54. W. B. Smith, R. R. Wilson, and D. B. Harris, *Environ. Sci. Technol.* **13**, 1387–1392 (1979).
55. A. D. Maynard, and L. C. Kenny, *J. Aerosol Sci.* **26**, 671–684 (1995).
56. L. C. Kenny, and R. A. Gussman, *J. Aerosol Sci.* **28**, 677–688 (1997).
57. T. M. Peters, R. A. Gussman, L. C. Kenny, et al., *Aerosol Sci. Technol.* **34**, 422–429 (2001).
58. L. K. Wang and N. C. Pereira (eds.). *Handbook of Environmental Engineering, Volume 1, Air and Noise Pollution Control.* Humana Press, Totowa, NJ, 1979.
59. US EPA. Cyclones. APTI Virtual Classroom. Lesson 2. http//search.epa.gov. US Environmental Protection Agency, Washington DC, (2004).
60. L. K. Wang, N. C. Pereira, and Y.-T. Hung (eds.). *Advanced Air and Noise Pollution Control.* Humana Press, Totowa, NJ, 2005.

4
Electrostatistic Precipitation

Chung-Shin J. Yuan and Thomas T. Shen

CONTENTS

1. INTRODUCTION

Electrostatic precipitation (ESP) is defined as the use of electrostatic forces to remove charged solid particles or liquid droplets from gas streams in which the particles or droplets are carried in suspension. It is one of the most popular and efficient particulate control devices and accounts for about 95% of all utility particulate controls in the United States (1). The first commercial electrostatic precipitator was designed by Walker and Hutchings and installed at a lead smelter works at Baggily, North Wales in 1885. However, this first attempt was not successful owing to inadequate power supply and poor properties of lead fume for electrostatic precipitation (i.e., small particle sizes, high temperature, and high resistivity of the particles) (2).

The principle of electrostatic precipitation was first developed by Dr. Frederick G. Cottrell, an American chemistry instructor at the University of California in Berkley. Cottrell also developed the first successful commercial electrostatic precipitator in 1906, which was installed at an acid manufacturing plant near Pinole, California (3). The first US electrostatic precipitation patent was then issued in 1908 for which the original ESP was a single-stage, cylindrical shape with a high-voltage electrode rod suspended in the center of the cylinder. Since then, electrostatic precipitators have been used extensively to remove both solid particles and liquid droplets from stationary combustion sources and a variety of industrial processes.

The ESP that we are most familiar with is based on the two-stage precipitator principle and developed in the 1930s. This allowed for reduction in ozone by utilizing the

From: *Handbook of Environmental Engineering, Volume 1: Air Pollution Control Engineering*
Edited by: L. K. Wang, N. C. Pereira, and Y.-T. Hung © Humana Press Inc., Totowa, NJ

very fine tungsten wires 5–10 mils in diameter with which everyone is familiar. The thin wires operated at very low voltages (12-kV ionizer and 6-kV collector) and utilized currents of positive polarity. The compact size and lower cost for the collector were achieved by using light aluminum plates spaced about 0.25 in. apart. These basic design elements were incorporated in the "Precipitation" first marketed by Westinghouse in the late 1930s. In general, the removal efficiencies of modern electrostatic precipitators can approach 99.9% or higher (4). However, if not properly designed and/or operated, small changes in the properties of particles/droplets or the gas stream can significantly affect the removal efficiency of the electrostatic precipitators (5,6).

The electrical mechanisms for the precipitation of particles or droplets are provided by discharge electrodes, which charge the particles or droplets in a corona discharge and create the electrostatic field that causes the charged particles or droplets to migrate toward the collecting electrodes. The essential components of the electrode system consist of one or more discharge electrodes of relatively small diameter (such as wires) as well as collecting electrodes (such as plates or tubes). In general, the discharge electrodes are of negative polarity, whereas the collecting discharges are at ground potential and considered positive polarity.

Electrostatic precipitation differs fundamentally from the fabric filtration and scrubbing processes in that the separation forces are electrical and are applied directly to the particles/droplets themselves, rather than indirectly through the gas stream. The electrical process has the inherent capability of capturing submicron particles or droplets at high efficiency with relatively low energy consumption and small pressure drop through the gas cleaning system. In comparison with other commercial particulate control devices, electrostatic precipitators have the following advantages and disadvantages (7):

A. Advantages
   1. High removal efficiency of fine particles/droplets
   2. Handling of large gas volumes with low pressure drop
   3. Collection of either dry powder materials or wet fumes/mists
   4. Sustenance of a wide range of gas temperature up to approx 700°C
   5. Low operating costs, except at very high removal efficiencies

B. Disadvantages
   1. High capital costs
   2. Unable to collect gaseous pollutants
   3. Large space requirement
   4. Inflexibility of operating conditions
   5. Variation of removal efficiency with particle/droplet properties (e.g., resistivity of particles/droplets)

This chapter is intended to serve as a guide to the understanding of electrostatic precipitation. It covers principles of operation, types of precipitator, design methodology, major field of application, limitations, and future developments. The reader is referred to refs. 4–13 for further reading on the subject of electrostatic precipitation.

2. PRINCIPLES OF OPERATION

Compared to other particulate control devices, electrostatic precipitators are as elegant as they are efficient. Instead of performing work on the entire gas stream in the cleaning

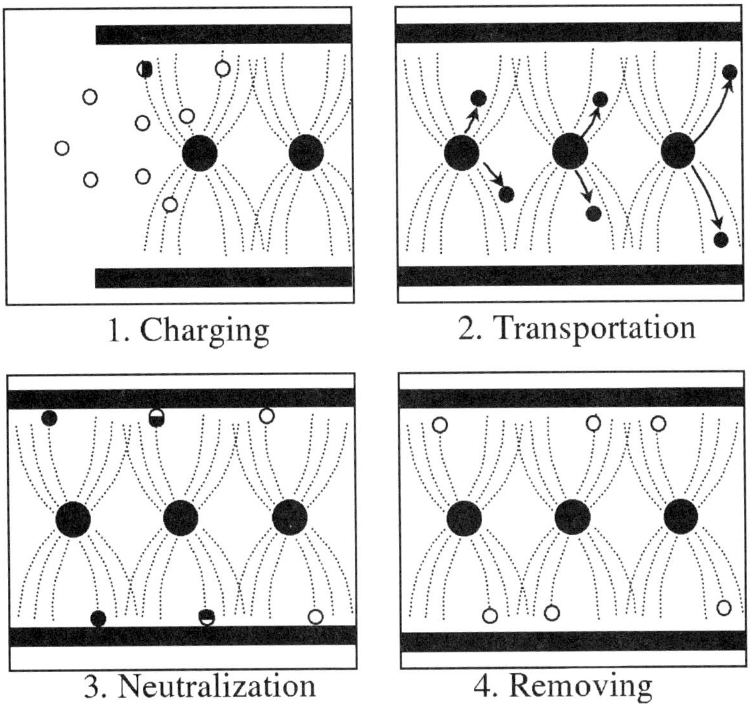

Fig. 1. Steps for charging, transportation, neutralization, and removing of particles/ droplets in the electrostatic fields.

process, the electrostatic forces are applied directly to the suspended particles in the electrostatic fields (4). Current knowledge states that particles/droplets in the precipitation process are charged, transported, neutralized, and removed as briefly described by the following steps and illustrated in Fig. 1.

1. The particles/droplets are charged in passing through an ionized electrostatic field
2. The charged particles/droplets are transported by the electrostatic force onto the surfaces of grounded collecting electrodes of opposite polarity
3. The charged particles/droplets are neutralized while arriving at the surfaces of collecting electrodes
4. The collected particles/droplets are removed from the surfaces of collecting electrodes by rappers, or other means, to a hopper beneath the electrostatic precipitator.

Electrostatic precipitators are built in either a single stage or two stages. Single-stage precipitators are designed for the combination of discharge electrodes and collecting electrodes together in a single section and are of two basic forms. The flat surface type (also called plate–wire precipitator) consists of several grounded parallel plates that serve as collecting electrodes, together with an array of parallel high-potential wires mounted in a plane midway between each pair of plates; these wires are the corona discharge electrodes (*see* Fig. 2A). The alternative single-stage precipitator design consists of an array of grounded cylinders or tubes that serves as collecting electrodes; coaxial to each cylinder is a high-potential wire, which is the corona discharge electrode (*see*

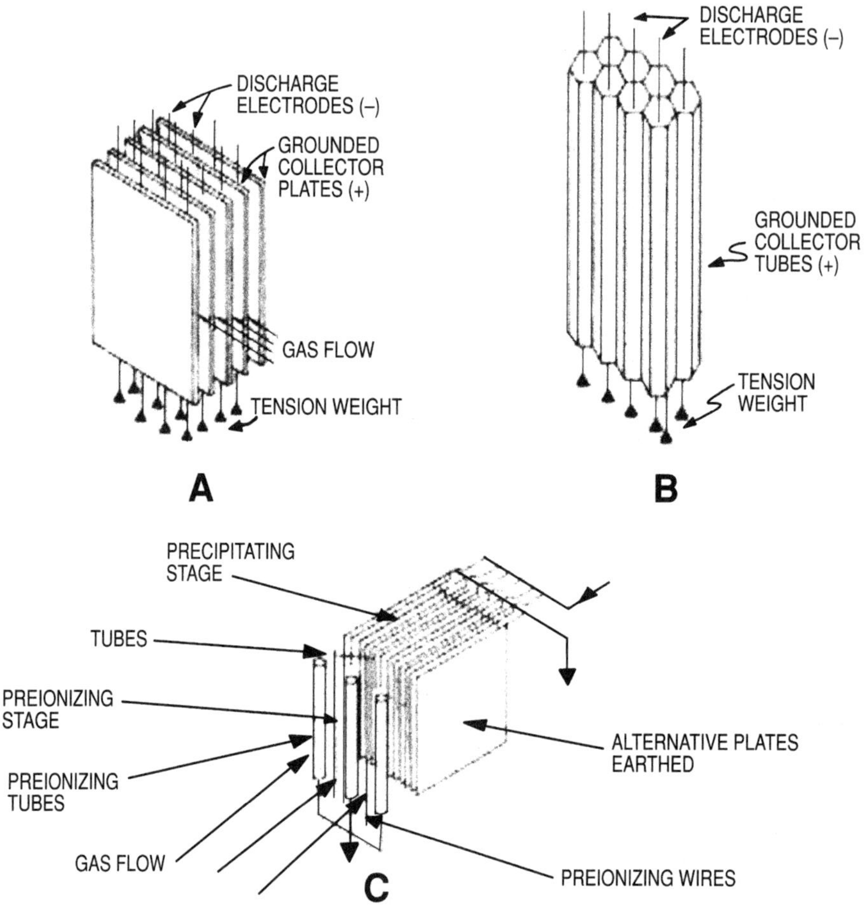

Fig. 2. Types of electrostatic precipitators: (**A**) single-stage flat surface type; (**B**) single-stage tubular type; (**C**) two-stage type.

Fig. 2B). In both forms of single-stage precipitator, the ionization and the collection of particles/droplets are achieved in a single stage; that is to say, the corona discharge and precipitating field extend over the full length of the apparatus. The two-stage precipitators differ in the sense that the ionization of particles/droplets is carried out in the first stage confined to the region around the corona discharge wires, followed by particle collection in the second stage, which provides an electrostatic field whereby the previously charged particles are migrated onto the surface of collecting electrodes (*see* Fig. 2C).

A gas stream with suspended particles/droplets is passed between the parallel plates or through the cylinders. Assuming that a sufficient potential difference exists between the discharge and collecting electrodes, a corona will form around the wires. As a result, large numbers of negative and positive ions are formed in the corona zone near the wires. With the discharge electrodes at negative polarity, the negative ions are attracted to the wires. The particles/droplets moving with the gas stream in passing through the interelectrode space are subjected to intense bombardment by the negative ions and become highly

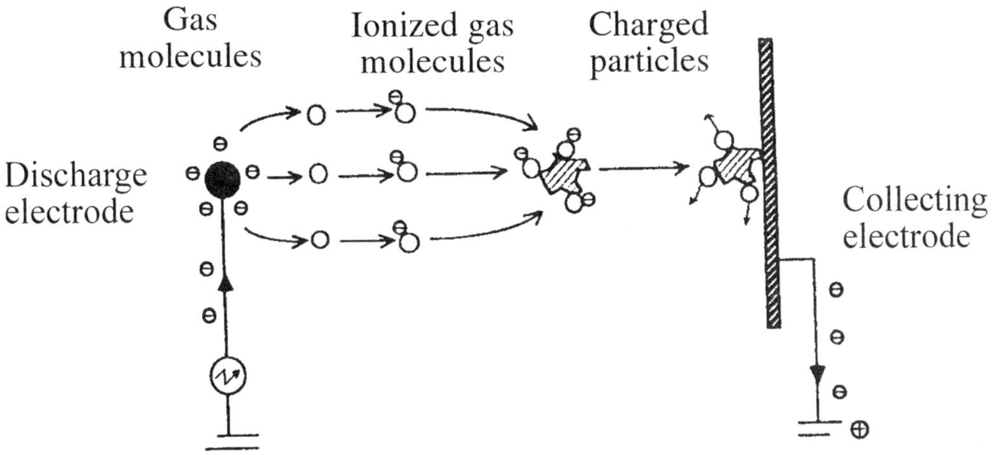

Fig. 3. Ionization of particles/droplets during the corona discharge.

charged in a short time (0.1 s or less). Typically, 1-μm particles/droplets will carry about 300 electron charges, whereas a 10-μm particle will carry about 30,000 electron charges (12). The charged particles/droplets, in turn, being under the influence of the high potential difference maintained between the discharge and collecting electrodes, are attracted to the collecting electrodes and thus are separated from the gas stream. Solid particles build up a layer on the collecting surface, from which the accumulated deposit has to be periodically removed by rapping or flushing, and are allowed to collect into a hopper. Liquid droplets form a film on the collecting surface, which then drips off into a sump. Single-stage precipitators have proved to be universally applicable in the cleaning of contaminated industrial gases, and two-stage precipitators are generally used for domestic and commercial indoor air cleaning, especially when low ozone generation is essential.

In the following subsections, some of the fundamental aspects of precipitator operation, such as corona discharge, electrical field, particle charging, and particle collection, are analyzed.

2.1. Corona Discharge

The high-voltage direct corona utilized in the ionization stage of electrostatic precipitation is a stable self-maintaining gas discharge between a discharge electrode and a collecting electrode. As the potential difference between the discharge and collecting electrodes is raised, the gas in the vicinity of the more sharply curved electrode breaks down at a voltage less than the spark-breakdown value for the gap length in question. This incomplete breakdown, known as the corona, appears in air as a highly active region of glow, extending into the gas a short distance beyond the discharge electrode. Ionization processes of the corona discharge are confined to or near this glow region. Most of the ionization is produced by free electrons, which then migrate and attach to gas molecules, forming ionized gas molecules. The ionization process of corona discharge is illustrated in Fig. 3.

The type of corona that is produced depends on the polarity of the discharge electrode. If the discharge electrode is positive, negative ions are accelerated toward the electrode, causing the breakdown of gas molecules, with the result that positive ions are repelled outward from the discharge electrode in the form of a corona glow. Conversely, if the discharge electrode is negative, positive ions are accelerated toward the discharge electrode and negative ions are repelled from the discharge electrode to produce a corona discharge.

One has the choice of applying either a positive or a negative potential to the discharge electrode. The negative potential generally yields a higher current at a given voltage than the positive, and the sparkover voltage (voltage at which complete breakdown of the gas dielectric occurs), which sets the upper limit to the operating potential of the precipitator, is also usually higher; in addition, a positive corona tends to be sporadic and unstable. A negative corona is usually used in industrial precipitators, whereas a positive corona, because of its lower ozone generation properties, is used in domestic and commercial air conditioning.

When the corona-starting voltage is reached, the ions repelled from the discharge electrodes toward the collecting electrodes constitute the only current in the entire space outside the corona glow. This interelectrode current increases slowly at first and then more rapidly with increasing voltage. As sparkover is approached (i.e., as complete breakdown of the gas dielectric occurs), small increments in voltage give sizable increases in current. Typical corona currents are of the order of 0.1–5.0 mA/m^2 of collecting electrode area. Sparkover does not occur infrequently; in general, a frequency of sparkover less than 100 times per minute is acceptable.

2.2. *Electrical Field Characteristics*

The fundamental differential equation that describes the field distribution between two electrodes is Poisson's equation, which expressed vectorially is

$$\nabla \cdot \mathrm{E} = \sigma_i / K_0 \tag{1}$$

where σ_i is the ion space-charge density (i.e., the quantity of electrical charge per unit volume of the space between the discharge and collecting electrodes) and K_0 is the permittivity of free space (8.85×10^{-12} F/m). Applying Eq. (1) to the simplest precipitator geometry, the coaxial wire–cylinder combination, the following is obtained:

$$(1/r)(d/dr)(rE) = \sigma_i / K_0 \tag{2}$$

where r is the radial distance from the cylinder axis. Prior to the onset of the corona, σ_1 is zero. Taking V_0 as the potential at the wire surface r_0 and grounding the cylinder (i.e., $V = 0$ at $r = r_1$), integration of Eq. (2) yields the electrostatic field

$$E = \frac{V_0}{r \ln(r_1/r_0)} \tag{3}$$

Equation (3) shows that the finer the discharge electrode, the greater will be the field strength at the surface of that electrode. Furthermore, as r increases, i.e., at point distant from the discharge electrode, the field strength decreases.

Above the corona threshold the interelectrode space charge becomes significant and must be included in Eq. (2). Under these conditions, the solution of Eq. (2) yields

$$E = \left[\frac{i_1}{2\pi K_0 m_i} + \left(\frac{r_0}{r} \right)^2 \left(E_c^{\ 2} - \frac{i_1}{2\pi K_0 m_i} \right) \right]^{1/2} \quad (4)$$

where m_i is the ionic mobility, i_1 is current per unit length of wire, and E_c is the minimal field intensity required to form a corona and also the field strength at the wire ($r - r_0$). This is the maximum field strength, and as r increases, E decreases until for sufficiently large r,

$$E = \left(\frac{i_1}{2\pi K_0 m_i} \right)^{1/2} \quad (5)$$

The electric field in terms of the potential V can be expressed as $E = -dV/dr$. Therefore, integrating the right-hand side of Eq. (4) with respect to r gives the corona current–voltage relationship

$$V_0 = V_c + E_c r_0 \left[(1+\phi)^{1/2} - 1 - \ln \frac{1 + (1+\phi)^{1/2}}{2} \right] \quad (6)$$

where V_c is the corona-starting potential related to E_0 by Eq. (3) as follows:

$$E_c = \frac{V_0}{r_0 \ln(r_1/r_0)} \quad (7)$$

ϕ is a dimensionless current given by

$$\phi = \left(\frac{r_1}{E_c r_0} \right)^2 \frac{i_1}{2\pi K_0 m_i} \quad (8)$$

Thus far, the space charge considered is the ionic space charge with no consideration given to the presence of charged particles in the interelectrode space. Such particles do contribute to the total space charge, and their contribution to the total field may be approximated as follows. The particle space charge density σ_p and ionic space charge density σ_i are assumed to be independent of position, (i.e., r). Inclusion of these two densities into Poisson's equation and solving for E gives

$$E = \frac{V_c}{r \ln(r_1/r_0)} + \frac{(\sigma_i + \sigma_p)}{2K_0} \frac{(r^2 - r_0^{\ 2})}{r} \quad (9)$$

Integration of Eq. (10) gives the potential of the wire as

$$V_0 = V_c + \left(\frac{\sigma_i + \sigma_p}{4K_0} \right) r_1^{\ 2} \quad (10)$$

where the cylinder is assumed to have zero potential and r_0 is assumed negligible with respect to r_1. Using Eq. (5) to eliminate σ_i in Eq. (10) yields

$$V_0 = V_c + \frac{\sigma_p}{4K_0} r_1^2 + \frac{i_1 \ln(r_1/r_0)}{8\pi K_0 m_1 V_0} \quad (11)$$

The principal effect of additional space charge as a result of particles is that, for a fixed voltage between the discharge and collecting electrodes, there is a reduction in current. Particle space charge may be calculated using Eq. (21). In recapitulating the various expressions for field strength, note that the simplest expression, Eq. (3), neglects any form of space charge; this is tantamount to negligible corona current. The next expression, Eq. (4), considers the presence of ionic space charge. The final expression, Eq. (9), recognizes both ionic and particle space charge.

The above results hold true for the coaxial wire–cylinder geometry only. In trying to obtain similar results for the duct or wire–plate geometry, the solution of Poisson's equation poses some mathematical difficulties. One way around this is to find approximate solutions for low corona current cases. It can be shown that (14,15)

$$i_1 = \frac{4\pi K_0 m_I}{b^2 \ln(d'/r_0)} V_0(V_0 - V_c) \quad (12)$$

where b is the wire-to-plate spacing and d' is a parameter given by

$$d' = 4b/\pi \quad \text{for} \quad \text{b/c} \leq 1.0$$

where c is the wire-to-wire spacing. The wire–plate geometry where $b/c \leq 1.0$ covers most of the practical duct-precipitation cases. The corona-starting voltage V_c in Eq. (12) is given by

$$V_c = r_0 E_c \ln(d'/r_0) \quad (13)$$

The average plate-current density is given by

$$i_a = i_1/2c \quad (14)$$

The case of uniformly distributed particle space charge can be expressed by

$$V = V_c + \frac{\sigma_p b^2}{2K_0} + i_1 \frac{\ln(d'/r_0)}{4pK_0 m_i V_0} \quad (15)$$

Up to this point no mention has been made of calculating the corona-starting field intensity E_c or the sparkover field. It is very difficult to calculate reliable values of these parameters by using atomic data. Practical values are best established by observing various similar installations. The reader is referred to Table 1 in Section 3.1 for some typical electrical parameter of practical electrostatic installations.

An empirical evaluation of E_c for round wires and outer electrodes of arbitrary shape can be obtained from (15)

$$E_c = \delta' \left[A' + B'/(r_0 \delta')^{1/2} \right] \quad (16)$$

where A' and B' are constants for a specific gas and δ' is the relative gas density taken with respect to 1 atm and 25°C. For air, the values $A' = 32.2 \times 10^5$ V/m and $B' = 8.46 \times 10^4$ V/m$^{1/2}$ are recommended. Values for other gases are reported in the literature (16).

Example 1

Find the corona-starting voltage for a duct precipitator of a 28-cm plate-to-plate spacing and a 10-cm wire-to-wire spacing and a 109-mil diameter wire. Assume that the gas is air at 40°C and 2 atm. Compare with a 109-mil-diameter wire in a 28-cm-diameter cylinder.

Solution

The corona-starting voltage for duct geometries is given by Eq. (13), for which E_c and d' are required. E_c can be determined from Eq. (16) for which δ' is required:

$$\delta' = \frac{T_0}{T}\frac{P}{P_0} = \frac{298}{313} \times \frac{2}{1} = 1.90$$

$$2r_0 = 109 \text{ mils} = 2.77 \times 10^{-3}\,\text{m}$$

Thus, from Eq. (16),

$$E_c = 1.90\left[32.2 \times 10^5 + \frac{8.46 \times 10^4}{\left(1.39 \times 10^{-3} \times 1.9\right)^{1/2}}\right]$$
$$= 9.25 \times 10^6 \text{ V/m}$$

Now, $d' = 4b/\pi$, where b is the wire-to-plate spacing and is assumed to be one-half of the plate-to-plate spacing (i.e., the wires are assumed to be placed midway between the plates); thus, $b = 0.14$ m. Therefore, $d' = (4)(0.14)/\pi = 0.178$ m; thus, from Eq. (13),

$$V_0 = \left(1.39 \times 10^{-3}\right)\left(9.25 \times 10^6\right) \ln\left(\frac{0.178}{1.39 \times 10^{-3}}\right)$$
$$= 62.4 \times 10^{-3}\,\text{V}$$

The starting voltage for the cylinder is given by Eq. (8). Thus,

$$V_0 = \left(9.25 \times 10^6\right)\left(1.39 \times 10^{-3}\right) \ln\left(\frac{0.14}{1.39 \times 10^{-3}}\right)$$
$$= 59.3 \times 10^{-3}\,\text{V}$$

(Note that $r_1 = 0.14$ m.) For equal duct width (parallel-plate spacing) and cylinder diameter and identical wire size, the cylinder-starting voltage will always be lower than the duct-starting voltage. In industrial precipitators, corona-starting voltages are somewhat lower than calculated estimates because of irregular electrode spacing and extraneous discharges from dust films, nicks, and the like on corona wires. In ducts, this is also attributable to the lower starting voltage of the end wires.

Corona discharge is accompanied by a relatively small flow of electric current. Sparking usually involves a considerably larger flow of current, which cannot be tolerated. However, with suitable controls, precipitators have been operated continuously with a small amount of sparking to make certain that the voltage is in the correct range to ensure corona.

2.3. Particle Charging

Particle-charging mechanisms generally considered relevant in electrostatic precipitation are (1) field charging, also known as impact charging or ion bombardment, wherein particles are bombarded by ions moving under the influence of the applied electric field, and (2) diffusion charging, wherein particles are charged as a result of the motion of the ions produced by the thermal motion of surrounding gas molecules. The field-charging mechanism predominates for particles larger than 0.5 μm, whereas the diffusion-charging mechanism predominates for particles smaller than 0.1 μm. However, both mechanisms are important in the size range 0.1–0.5 μm.

2.3.1. Field Charging

Assume that a spherical particle of radius a is placed in a uniform corona discharge field E_0 in a gas and that the particle bears no charge initially. As soon as a charge is acquired by the particle as a result of ion bombardment, an electric field is created that repels similarly charged ions. Some ions continue to strike the particle, but the rate at which they do so diminishes until the charge acquired by the particle is sufficient to prevent further ions striking it. This is the limiting charge that can be acquired by the particle. Assume that the particle has acquired a uniform surface charge q. The particle charge distorts the field E_0 and imparts to it a radial component, which can be shown (11) to be

$$E_r = E_0 \cos\theta \left[2\left(\frac{D-1}{D+2}\right)\frac{a^3}{r^3} + 1 \right] + \frac{q}{4\pi K_0 r^2} \qquad r \geq a \tag{17}$$

where D is the particle dielectric constant, r is the radial direction from the center of the particle, and θ is the polar angle between r and the undistorted field E_0. An ion of charged q_i is attracted to the particle if the ion approaches from an angle θ for which the radial force $F_r = q_i E_r$ is negative. Particle charging ceases at $F_r = 0$. Setting $\theta = \pi$ and $r = a$ and defining

$$P = \frac{3D}{D+2} \tag{18}$$

the saturation or limiting surface charge acquired by the particle turns out to be

$$q_{\max} = 4\pi K_0 P a^2 E_0 \tag{19}$$

Pauthenier and Moreau-Hanot (17) found that the particle charge as a function of time is given by

$$q_f = q_{\max}\left(\frac{t}{t+\tau}\right) \tag{20}$$

where τ, in seconds, is the particle-charging time constant, given by

$$\tau = \frac{4K_0}{q_i m_i N} \tag{21}$$

where N is the ion concentration and τ is the time required for 50% of maximum charge to be acquired.

The factor P in Eq. (18) is a measure of field distortion as a result of particle charge. It reduces to unity (i.e., no distortion) for $D = 1$, and it approaches 3 for large values of D (i.e., for conducting particles). For most dielectric substances, D is less than 10.

2.3.2. *Diffusion Charging*

Field charging becomes less important as particle size decreases, and, subsequently, diffusion charging begins to play a more important role. The role of diffusion charging can be examined by the following analysis based on the kinetic theory of gases. It is known from kinetic theory that the density of gas in a potential field varies according to

$$N = N_0 \exp(-U/KT) \tag{22}$$

where N_0 is the initial ion density, U is the ion potential energy, K is the Boltzmann constant, and T is the absolute temperature. The potential energy of an ion with charge q_i, in the vicinity of a uniformly charged spherical particle, is

$$U = \frac{qq_i}{4\pi K_0 r} \tag{23}$$

where r is the distance from the center of the particle to the ion. The number of ions that strike the particle per second is, from kinetic theory,

$$(Nv_i/4)(4\pi a^2) = \pi a^2 N v_i$$

where v_i is the root mean square velocity of the ion. Thus, the time interval associated with the ion–particle collisions is

$$t(s) = \frac{1}{\pi a^2 N v_i} \tag{24}$$

Assuming that every ion that makes contact with the particle is captured, the particle charging can be described by

$$\frac{dq}{dt} = \pi\, a^2 N v_i \tag{25}$$

The solution of Eq. (25) with the initial condition $q = 0$ is

$$q_d = \frac{4\pi K_0 akT}{q_i} \ln\left[\frac{aN_0 q_i^2 v_i t}{4K_0 kT} + 1\right] \tag{26}$$

Example 2

Estimate the number of electronic charges acquired by a spherical conducting particle ($P = 3$) under conditions of field charging and diffusion charging. Assume a particle radius of 1 μm and a charging time of 0.1 s.

Solution

Repeat calculations for a particle radius of 0.1 μm. Use the following conditions: $E_0 = 1\times10^6$ V/m, $N_0 = 1\times10^{14}$ ions/m^3, $m_i = 4\times10^{-4}$ m^2/V s, $K_0 = 8.85\times10^{-12}$ C^2/N m^2, $q_i = 1.602\times10^{-19}$ C (electronic charge), $k = 1.38\times10^{-23}$ N m/K, $T = 313$ K, and $v_i = 500$ m/s.

In the case of ion bombardment, combining Eqs. (19)–(21) gives

$$\begin{aligned} q_f &= 4\pi K_0 P a^2 E_0 \left(\frac{t}{t + 4K_0 / q_i m_i N_0} \right) \\ &= 4\pi(8.85\times10^{-12})(3)(10^{-12})(10^6) \\ &\quad \times \left[\frac{0.1}{0.1 + 4(8.85\times10^{-12})/(1.602\times10^{-19})(4\times10^{-4})(10^{14})} \right] \\ &= 3.162\times10^{-16}\,\mathrm{C}(\text{or } 1{,}974 \text{ electronic charges}) \end{aligned}$$

Note that q_{max} for this particular case happens to be 2082 electronic charges. Thus, in 0.1 s, the particle has already acquired 95% of its maximum charge.

For the case of diffusion charging, Eq. (26) provides the answer:

$$\begin{aligned} q_d &= \frac{4\pi(8.85 \times 10^{-12})(10^{-6})(1.38 \times 10^{-23})(313)}{1.602 \times 10^{-19}} \\ &\quad \times \ln\left[\frac{(10^{-8})(10^{14})(2.566 \times 10^{-38})(500)(0.1)}{4(8.85 \times 10^{-12})(1.38 \times 10^{-23})(313)} + 1 \right] \\ &= 2.02 \times 10^{-17}\,\mathrm{C}(\text{or } 126 \text{ electronic charges}) \end{aligned}$$

Comparing the charges from two mechanisms illustrates the significance of field charging as opposed to diffusion charging for a particle of 1 mm radius. Diffusion charging contributes less than 6% of the total charge.

Repeating the above calculations for a particle of 0.1 μm radius and charging time of 0.1 s gives field charging of 20 electronic charges and diffusion charging of 8 electronic charges. Being a contributor of almost 30% of the total charge, it obviously shows that the diffusion-charging mechanism cannot be neglected for smaller particles.

Previously, we made use of particle space charge density σ_p without actually defining it; see Eqs. (9), (11), and (15). We are now in a position to do so. Assuming that field charging is the more dominant charging mechanism, Eq. (19) gives the maximum charge acquired by a particle:

$$q_{\max} = 4\pi K_0 P a^2 E$$

Assuming that the particle concentration is N_p per unit volume, the total charge acquired by all particles (i.e., particle space-charge density) is

$$\sigma_p = 4\pi \mathrm{K}_0 P a^2 E N_p$$

which may be also be expressed as

$$\sigma_p = K_0 PES \tag{27}$$

where S is the total particle surface per unit volume of gas,

$$S = 4\pi a^2 N_p$$

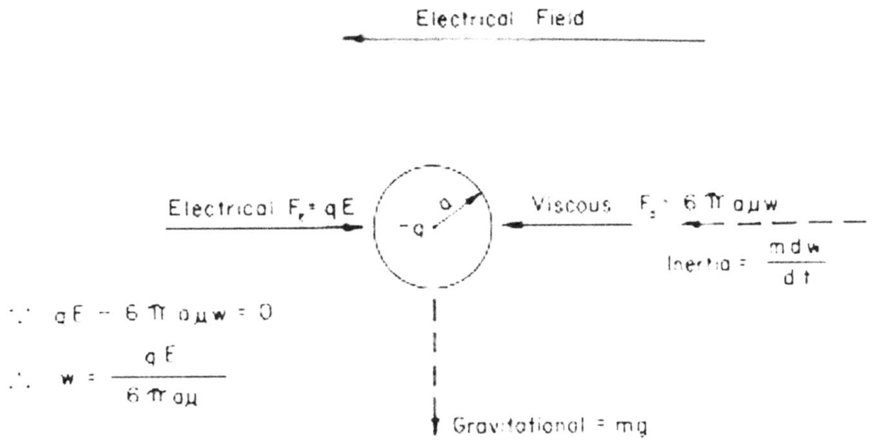

Fig. 4. Free-body diagram of a charged particle.

The above expression for particle space-charge density can be utilized to obtain a more rigorous expression—more rigorous than Eq. (11)—for the effect of particle space charge on the cylindrical corona field. Using σ_p, as given in Eq. (28), in Poisson's equation, Eq. (2), and solving for E gives

$$E = \left\{ \left[\left(\frac{r_0}{r} \right)^2 \left(E_c^{\,2} - \frac{i_1}{2\pi K_0 m_i} \right) + \frac{i_1}{4\pi K_0 m_i (PSr)^2} \right] e^{-2PSr} - \frac{i_1}{4\pi K_0 m_i} \left[\frac{2}{PSr} + \frac{1}{(PSr)^2} \right] \right\}^{1/2} \tag{28}$$

A plot of field strength E versus radial distance from the above expression will show that particle space charge works to lower the field near the wire surface (r_0) and to raise it at the cylinder wall (r_1). The field reduction at the wire occurs because the space charge tends to shield the discharge wire from the cylinder.

2.4. Particle Collection

Particle collection in electrical precipitators occurs when the charged particles move to the surface of the collecting electrodes and are trapped by the electrostatic field. The particles are accelerated toward the collecting electrodes by Coulomb forces, but inertial and viscous forces resist the motion. Consequently, a particle in the precipitation field attains a velocity, known as particle migration velocity, which is a fundamental parameter important to all theories of particle precipitation.

2.4.1. Particle Migration Velocity

The motion of a charged particle is governed by the dynamics of the force system acting on the particle, as illustrated in Fig. 4. The various forces acting in the precipitation system are gravitational, inertial, viscous, and electrical forces. For fine particles of interest in electrostatic precipitation, gravitation forces are quite insignificant and, therefore, may be neglected. A balance between the remaining forces on the particle yields

$$m\frac{dw}{dt} = F_e - F_d \tag{29}$$

where m and w are particle mass and migration velocity, respectively, and F_e and F_d are the electrostatic and viscous drag forces acting on the particle, respectively. For a collecting field of E_p, F_e is given by qE_p. Also, assuming laminar flow exists, the viscous drag force on a spherical particle is given by Stoke's law, $F_d = 6\pi\mu aw$, where μ is the gas viscosity. Thus, Eq. (29) becomes

$$m\frac{dw}{dt} = qE_p - 6\pi\,\mu\,aw \tag{30}$$

whose integration velocity yields

$$w = \left(\frac{qE_p}{6\pi\mu a}\right)\left[1 - \exp\left(-\frac{6\pi\mu at}{m}\right)\right] \tag{31}$$

The exponential term is quite negligible for $t > 0.01$ s. Dropping this term is equivalent to ignoring the acceleration term on the left-hand side of Eq. (31). Consequently, Eq. (30) reduces to

$$w = \frac{qE_p}{6\pi\mu a} \tag{32}$$

which is a result of equating electrostatic and viscous drag forces. For particles charged by ion bombardment, the maximum charge attained by the particle is given by Eq. (20). Thus, Eq. (32) becomes

$$w = \frac{2K_0PaE_cE_p}{3\mu} \tag{33}$$

where E is the electric strength of charging field. For single-stage precipitators, the charging field E_c and the collecting field E_p are approximately the same.

The migration velocity is therefore seen to be proportional to the discharge and collecting fields and also the particle radius a, but inversely proportional to gas viscosity μ. In practice, laminar flow is seldom achieved, and Eq. (32) would overestimate migration velocities for nonlaminar flow. If the particle size approaches the mean free path of gas molecules ($\lambda = 6.8\times10^{-8}$ m in atmospheric air at 25°C), then Eq. (33) must be multiplied by the Cunningham correction factor,

$$C = 1 + \left(\frac{\lambda}{a}\right)\left[1.26 + 0.40\exp\left(-1.10\frac{a}{\lambda}\right)\right] \tag{34}$$

This means an increase in migration velocity; for example, for a particle of 0.5 μm radius in atmospheric air at 25°C, the Cunningham correction factor is 1.17, an increase of 17% in migration velocity.

Assume that a particle at the discharge electrode must move a distance d to be collected at the collecting electrode. Let the particle velocity in the direction of gas flow be the same as the gas velocity v, whereas the transverse velocity from the discharge electrode to the collecting electrode be given by the migration velocity w. The particle will move to the collecting electrode in time $t' = d/w$, and the duct length for collection efficiency of 100% is given by

$$L = vt' = v\left(\frac{d}{w}\right) \tag{35}$$

where L is in the direction of gas flow.

Example 3

Find the minimum length of a collecting electrode for a single-stage wire–plate-type precipitator with a 8-in. (0.2032-m) plate-to-plate spacing and an applied voltage of 600,000 V. Air velocity through the precipitator is 3 ft/s (0.9144 m/s) and the minimum particle diameter is 1.0 μm.

Solution

Assume that E_c and E_p are the same; that is,

$$E_c = E_p = \frac{60{,}000}{0.1016} = 590{,}550 \text{ V/m}$$

Furthermore, let P =1, and μ for air at 25°C is 1.8×10^{-5} N s/m^2. Thus, from Eq. (32),

$$\begin{aligned} w &= \frac{2(8.85 \times 10^{-12})(0.5 \times 10^{-6})(590{,}550)^2}{3 \times 1.8 \times 10^{-5}} \\ &= 0.057 \text{ m/s} \end{aligned}$$

which, after multiplying by the Cunningham correction factor, C=1.17, becomes

$$w = (0.057)(1.17) = 0.0667 \text{ m/s}$$

Therefore, the length of electrode from Eq. (34) is

$$L = (0.9144)(0.2032 / 2)(0.0667) = 1.4 \text{ m}$$

Example 3 shows that 100% collection efficiency should result from a precipitator about 1.4 m in length. This value may be representative of controlled laboratory conditions. However, in practice, a precipitator for the conditions in Example 3 may well be two to three times that length because the migration velocity w may, in practice, be two or three times smaller than the idealized value given by Eq. (34). Such a discrepancy arises because the migration velocity under realistic precipitator conditions is subject to several factors such as uneven gas flow, re-entrainment of collected particles, and "effective" values of field intensity or space-charge density, which cannot be included in the idealized theory. In engineering design, it is practical to use modified values of w that are determined from actual field experience or are established by pilot-plant tests. The theoretical equations therefore serve as a basis for analyzing field-precipitator performance and for calculating a new design in which previous practical values for w exist. The reader is referred to Tables 1 and 2 (Section 3.1) for typical values.

2.4.2. Particle Collection Efficiency

The particle collection efficiency of electrostatic precipitators was first developed empirically by Evald Anderson in 1919 and then theoretically developed by W. Deutsch in 1922. Thus, the collection efficiency equation of electrostatic precipitators is usually

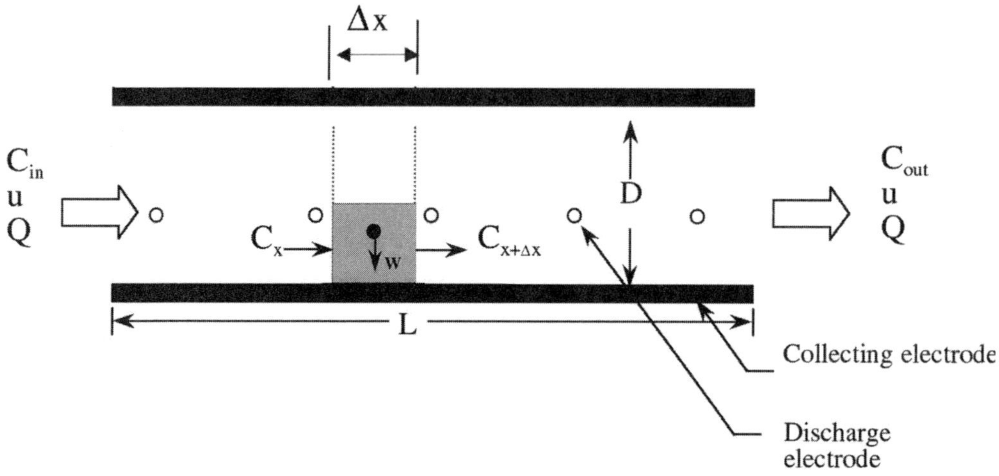

Fig. 5. Schematic of gas flow passing through two plates in an electrostatic precipitator.

known as the Deutsch–Anderson equation. The derivation of the Deutsch–Anderson equation is based on the following assumptions:

1. The particle concentration at any cross-sectional area normal to the gas flow is uniform
2. Gases move downstream at constant velocity with no longitudinal mixing
3. The charging and collecting electrical fields are constant and uniform
4. Particles move toward the collecting electrodes with a constant migration velocity
5. Re-entrainment of collected particles on the surface of collecting electrodes is negligible.

Consider a dusty gas flow in a rectangular channel confined by two parallel collecting plates in an electrostatic precipitator, as illustrated in Fig. 5. The concentration of particles decreases gradually with distance because of the migration of particles toward the collecting plates. A material balance on particles flowing into and out of a control volume shows that the difference between the mass of particles flowing through the slice Δx must equal the mass of particles collected at the collecting plates:

$$vH(D/2)C_x - vH(D/2)C_{x+\Delta x} = w\, C_{x+\Delta x/2} H\Delta x \tag{36}$$

where v is the gas flow velocity, H is the height of the plate, D is the width of the plate, C is the particle concentration, and w is the migration velocity of particles. Dividing through Eq. (36) by Δx and taking the limit as Δx approaches to zero

$$(-vHD/2)(dC/dx) = wHC \tag{37}$$

Integrate Eq. (37) for distance x from 0 to L and for particle concentration C from inlet particle concentration C_{in} to outlet particle concentration C_{out},

$$\ln(C_{\text{out}}/C_{\text{in}}) = -2wHL/(vHD) \tag{38}$$

Define the particle collection efficiency of an electrostatic precipitator η in terms of inlet and outlet particle concentrations,

$$\eta = 1 - C_{\text{out}}/C_{\text{in}} = 1 - \exp(-wA/Q) \tag{39}$$

which is the Deutsch–Anderson equation. A is the area of collecting plates and Q is the volumetric gas flow rate.

This equation can be applied to both wire–cylinder and wire–plate (duct-type) precipitators. For a cylinder of radius R and length L and gas flow at velocity v,

$$A = 2\pi RL$$

$$Q = \pi R^2 v$$

hence,

$$\eta = 1 - \exp(-2wL/Rv) \tag{40}$$

Similarly, for a duct precipitator of plate length L, plate height H, and the wire-to-plate spacing b,

$$A = 2LH$$

$$Q = DHv$$

hence,

$$\eta = 1 - \exp\left(-wL/bv\right) \tag{41}$$

Comparison of Eqs. (40) and (41) shows that for a given collection efficiency, a cylindrical precipitator may be operated at twice the gas velocity of a duct precipitator of equal length and electrode spacing.

Equation (39) holds for conductive spherical particles in the size range for which Stokes' law is valid (i.e., laminar flow). In practice, laminar flow is rarely achieved. However, at the boundary layer, the gas flow is laminar, and particles entering the boundary layer will be collected. Nonspherical shapes and dielectric factors may very well change the numerical coefficients but not the basic form of the equation. The Deutsch–Anderson equation indicates clearly that for a given particle size, the collection efficiency increases with increasing particle migration velocity or collecting surface area, whereas the collection efficiency decreases with increasing gas flow rate. For a constant volume of gas passing through the electrostatic precipitator, the maximum collection efficiency occurs when the velocity is uniform. Collection efficiency decreases as gas viscosity increases. The density and concentration of the particle do not appear in the Deutsch–Anderson equation; however, they may exert a secondary influence. For example, light and fluffy particles on the collecting surface are harder to remove; they tend to fall more slowly to the hoppers and are subject to re-entrainment. High particle concentrations mean a greater mass of materials to be collected and disposed of. Thus, particulate buildup on the collecting electrodes will be greater and rapping of the electrodes may become critical.

The temperature and pressure of the flue gas have several important effects on the performance of an electrostatic precipitator. First, gas viscosity increases with temperature. An increase in gas viscosity would reduce migration velocity proportionally, as defined previously in Eq. (32). Second, the gas volume V is directly proportional to the absolute gas temperature T and inversely proportional to pressure P, as expressed in the universal gas law $PV = nRT$, where n and R are the number of moles and universal gas volume, respectively, resulting from higher temperatures. Finally, the gas temperature and pressure also influence the voltage–current relationship as a result of changing gas density because ion mobility decreases with gas density.

The particle size of the dust is also very important in the determination of the value of migration velocity for design purposes. The variations in migration velocity result largely from particle size variations. With improved particle sizing techniques, the Deutsch–Anderson equation may be modified as

$$\eta_i = 1 - \exp\left(-\frac{w_i A}{Q}\right) \tag{42}$$

where η_i is the fractional collection efficiency for the *i*th particle size, w_i is the migration velocity of the *i*th particle size, and the overall collection efficiency η is the summation of fractional collection efficiency η_i times mass fraction f_i:

$$\eta = \sum_{i=1}^{n} \eta_i f_{mi} \tag{43}$$

where f_{mi} is the mass fraction of the *i*th particle size.

Penney commented that the Deutsch–Anderson equation neglects the adhesion problem (18). In a two-stage precipitator in which the electrical force reverses and tends to pull the particle off, adhesive forces can still hold the particle. Adhesion is essential in the collection of the lower-receptivity particle. Also, it is of importance in the transfer of the particle from the collecting electrodes to the hopper. The effective transfer of particles to the hopper is mainly dependent on the function of chunks or agglomeration of particle, which can effectively fall with a minimum re-entrainment.

Adhesion resulting from differences in contact potential appears to be effective for particles of a few micrometers or less, but ineffective for large particles. More basic research on the adhesive behavior is required in parallel with precipitator performance tests so that proper rapping mechanisms can be designed to balance the various effects on changes in the adhesive behavior.

2.4.3. Particle Re-entrainment

Once captured in an electrostatic precipitator, particles remain captured only in the case of liquid droplets. Dry, solid particles are only lightly held onto the collecting electrode and can be easily dislodged and re-entrained into the gas stream. Re-entrainment may occur as a result of (1) low particle resistivity, (2) erosion of the particle from collecting electrodes, and (3) rapping.

The dominant force holding particles on the collecting surface results from the flow of current through the particles; if the particle resistivity is too low, not enough charge is retained by the particle. In this case, the negative charge may leak off the particle, which, in turn, acquires a positive charge from the collecting electrode and is forcefully accelerated away from that electrode. Large fly ash particles and carbon black particles exhibit low resistivity. In the case of fly ash, the re-entrainment problem can be reduced by using high-efficiency cyclones preceding the electrostatic precipitator. Carbon black particles are too small to be separated by cyclones; nevertheless, the electrostatic precipitator helps to agglomerate the carbon particles into coarser particles, which then can be removed by cyclones that follow the electrostatic precipitator.

Nonuniform gas velocity can result in excessively high gas velocity through some sections of the electrostatic precipitator. The excessive turbulence from such high velocities produces re-entrainment of the collected particles on the electrodes owing to the scouring action of the gas. To prevent such erosion, various special designs of collecting electrodes are used. The objective in all of these designs is to provide quiescent zones to prevent or reduce erosion. The various designs increase collection efficiency by (1) providing baffles to shield deposited particles from the re-entraining forces of the gas stream, (2) providing catch pockets that convey precipitated particles into a quiescent gas zone behind the collecting electrode, and (3) minimizing protrusions from the plate surface in order to raise sparkover voltage. Furthermore, gas flows through the hoppers can sweep collected particles back into the gas stream. This can be minimized by installing baffles in the hoppers to reduce the circulation of gas bypassing the electrodes through the hoppers.

The frequency and intensity of the rapping cycle have an important effect on collection efficiency because the collected particles may falls as much as 12 m (40 ft) through a transverse gas stream before reaching the hoppers. High collection efficiency requires that the particles, when rapped loose from the collecting plate, should fall as coarse aggregates so that they are not redispersed into the gas stream. This is achieved by frequent, gentle rapping. Rapping cycles are determined experimentally after the electrostatic precipitator is placed in operation. Typically, a rapping frequency of one impact per minute may be used.

3. DESIGN METHODOLOGY AND CONSIDERATIONS

Electrostatic precipitator design involves (1) the determination of precipitator size and electrical energization equipment required to achieve a given level of collection efficiency, (2) the selection of the electrode systems, (3) the design of a gas flow system to provide acceptable gas flow quality, (4) the structural design of the precipitator housing, and (5) the selection of means to remove the collected particles. The overall design must result in a completely integrated system. The essential components and partial cross-sectional views of a typical single-stage electrostatic precipitator of the flat surface type are given in Fig. 6.

Over the past 40 yr, significant improvements have been made in the design and construction of electrostatic precipitator (EPS) components; however, in terms of design practice, the present methodology is still based on empirical relations. The values for current design variables were obtained mostly from experiences with similar ESP applications. Unfortunately, the records of these accumulated field experiences, often regarded as proprietary, are unavailable to the public. Therefore, the designer will face many decisions for which there are no clear-cut solutions.

In a plate–wire ESP, gases flow between parallel plates of sheet metal and high-voltage electrodes. The electrodes consist of long weighted wires hanging between the plates and supported by rigid frames. The gases must pass through the wires as they traverse the ESP unit. This configuration allows many parallel lanes of flow and is well suited for handling large volumes of gas. The cleaning and power supplies for this type are often sectioned, to improve performance. The plate–wire ESP is the most popular type.

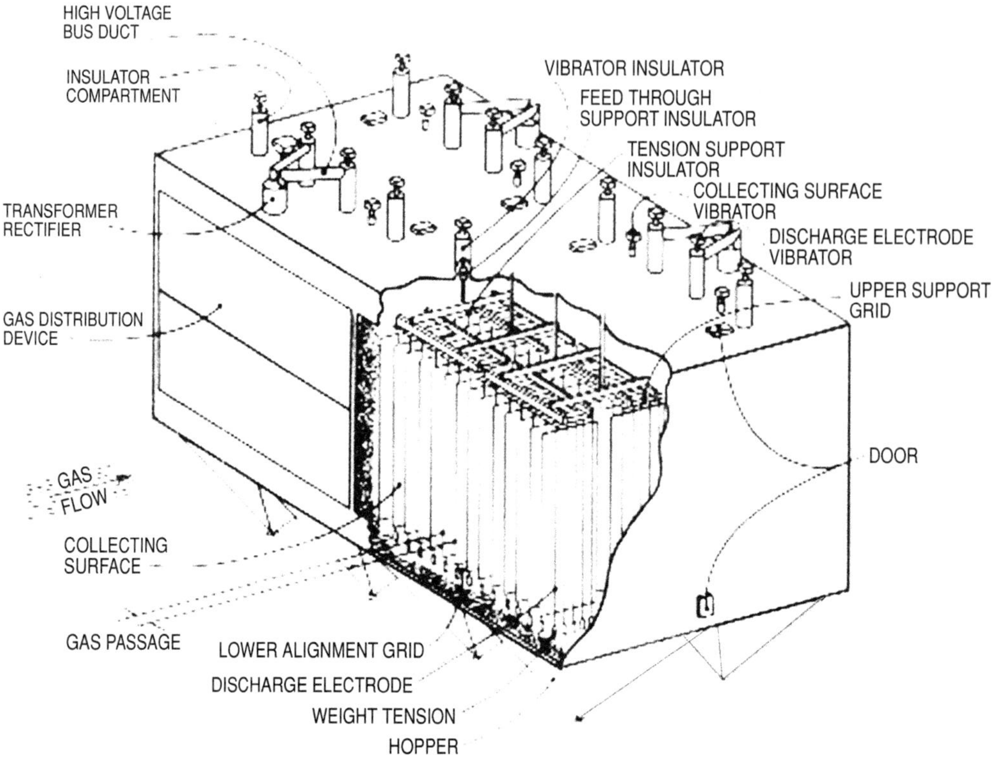

Fig. 6. Components of a single-stage electrostatic precipitator of the flat surface type.

Flat–plate ESPs differ from plate–wire types in that the electrons consist of flat plates rather than wires. A number of smaller precipitators use flat plates instead of wires. These plates increases the average electric field used to collect particles, and they provide increased surface collection area, relative to plate wires. A flat-plate ESP operates with little or no corona (a region of gaseous ions), which leads to high rapping losses, particularly if the emission stream velocity is high. These ESPs perform well with small, high-resistivity particles, provided the velocity is low.

Tubular ESPs are the oldest type and the least common. Tubular ESPs are typically used in sulfuric acid plants, Coe oven byproduct gas cleaning (tar removal), and iron and steel sinter plants. The tube is usually a circular, square, or hexagonal honeycomb with gas flowing lengthwise through the system. The tubular ESP is most commonly applied where the particles are wet or sticky.

The two-stage ESP is a series device where the first unit is responsible for ionization and the second is responsible for collection. This results in more time for particle changing and economical construction for smaller (less or equal 50,000 acfm) applications. Two-stage units are often used to collect oil mists, smokes, fumes, and other sticky particulates because there is little electrical for to hold the collected particles onto the plates.

Ionizing wet scrubbers (IWSs) also may be used as a particulate control device. An IWS combines the principle of wet electrostatic particle charging with packed-bed scrubbing into a two-stage collection system. A constant DC voltage is applied to the ionizing section, which the emission stream passes through before introduction to the scrubbing section. The electrostatic plates in the ionizing section are continually flushed with water to prevent resistive layer buildup. The cleaned gas exiting the ionizing section is further scrubbed in a packed-bed section. Unlike dry ESPs, IWSs are fairly insensitive to particle resistivity. For best performance of IWSs, monitoring of plate voltage and packed-bed-scrubbing water is recommended.

A rigorous design of a given ESP system can become quite complex, as it normally includes consideration of electrical operating points (voltages and currents), particle charging, particle collection, sneakage, and rapping re-entrainment. The most important variable considered in the design of an ESP is the specific collection plate area assuming that the ESP is already provided with an optimum level of secondary voltage and current. Secondary voltage or current is the voltage or current level at the plates themselves, and this voltage and current are responsible for the electric field. The collection plate area is a function of the desired collection efficiency gas stream flow rate and particle drift velocity.

Pretreatment of the emission stream temperature should be within 50–100°F above the stream dew point. If the emission stream temperature does not fall within the stated range, pretreatment (i.e., emission stream preheat or cooling) is necessary. The primary characteristics affecting ESP sizing are drift velocity of the particles and flow rate. Therefore, after selecting a temperature for the emission stream, the new stream flow rate must be calculated. The calculation method depends on the type of pretreatment performed. The use of pretreatment mechanical dust collectors may also be appropriate. In the emission stream (20–30 μm), pretreatment with mechanical dust collectors is typically performed.

3.1. Precipitator Size

Although there are many variations in the details of determining the size of an ESP to handle a given volumetric flow of gas, the Deutsch–Anderson equation or its modified form is generally used. Other design approaches are the use of tests in a pilot-scale electrostatic precipitator to arrive at the design conditions or theoretical analysis to extrapolate known conditions to those corresponding to the new requirements.

The Deutsch–Anderson equation provides the basis for the development of quantitative relationship (i.e., η, w, A, and Q) in spite of the fact that other variables and conditions must be included. These variables are discussed later. In engineering design practice, however, the modified Deutsch–Anderson equation based on the empirical data has been found to be practical for developing approximate solutions, which are sufficiently accurate for determining the size of an ESP. Sometimes, the overall shape and size of the ESP is governed by the space available, particularly in retrofit installations. The ranges of design variables for ESPs (12,13,19) are summarized in Table 1. The values of these variables vary with particulate and gas properties, with gas flow, and with required collection efficiency. The typical values of migration velocity for various applications are listed in Table 2.

The quantitative rclationships of migration velocity, collecting plate area, gas flow rate, and collection efficiency, as indicated in the Deutsch–Anderson equation can be best illustrated by the following simple examples. It should be noted, however, that the

Table 1
Ranges of Design Variables for Electrostatic Precipitators

Design variable	Range of values
Migration velocity	3.1–21.4 cm/s (0.1–0.7ft/s)
Specific collection area (plate surface area/gas flow rate)	19–95 $m^2/(m^3/s)$ (100–500ft^2/1000ft^3/min)
Gas velocity	0.6–2.4 m/s (2–8 ft/s)
Aspect ratio (duct length/height)	0.5–2.0
Corona power ratio (corona power/gas flow)	100–1000 W/(m^3/s)
Applied voltage	30–75 kV
Electrical field strength	6–15 kV/cm (15–40 kV/in)
Corona current/plate area	50–700 μA/m^2 (5–65 μA/ft^2)
Corona current/wire length	0.03–30.0 μA/m (0.01–10 μA/ft)
Plate area/electrical set	500–8000 m^2 (5380–86,000 ft^2)
Space between plates	5–30 cm (2–12 in.)
Horizontal length of plate/vertical height of plate	0.5–1.0
Vertical height of plate	8–15 m (26–50 ft)
No. of high tension sections in gas flow direction	2–8
Degree of high-tension sectionalization	0.01–0.10 high-tension bus section/(m^3/s)

Source: Data from refs. 12, 13, and 19.

equation applies only to a very narrow particle size range and relatively constant migration velocity.

Example 4

Find the collecting plate area of a horizontal-flow, single-stage electrostatic precipitator handling an average gas flow of 2.5 m^3/s from a pulverized coal-fired boiler. The required collection efficiency of ESP is 98%.

Solution

Given: $\eta = 98\%$ and $Q = 2.5$ m^3/s

From the data given in Table 2, select $w = 12$ cm/s (or 0.12 m/s).

$$\eta = 1 - \exp(-wA/Q)$$

$$0.98 = 1 - \exp(-0.12A / 2.5)$$

$$A = \ln(1 - 0.98)(2.5) / 0.12$$
$$= 81.5 \text{ m}^2 \text{ (required)}$$

Reference is made to the ranges of design variables shown in Table 1. The width of the plates is generally between 0.5 and 1.0 times the height. Therefore, use two sections formed by plates of 4.0 m wide × 5.2 m high on 25-cm centers.

$$\text{Total collecting plate area} = 2 \times 4.0 \times 5.2 \times 2 = 83.2 \text{ m}^2 \text{ (provided)}$$

Table 2
Typical Values of Migration Velocity for Various Applications

	Migration velocity (cm/s)	
Application	Average	Range
Pulverized coal (fly ash)	13.0	4.0–20.3
Paper mills	7.6	6.4–9.4
Open-hearth furnace	5.2	4.9–5.8
Secondary blast furnace (80% foundry iron)	9.1	15.7–19.4
Gypsum	17.0	2.2–3.2
Hot phosphorus	2.7	6.1–8.5
Acid mist (H_2SO_4 or TiO_2)	7.3	6.1–8.5
Flash or multiple hearth roaster	7.6	6.6–9.2
Cement plant (wet process)	10.6	9.1–12.2
Cement plant (dry process)	6.4	5.8–7.0
Catalyst dust	7.6	6.9–8.9
Gray iron cupola (iron/coke = 10:1)	3.3	3.0–3.7

Source: Data from refs. 9 and 13.

Example 5

Find the collection efficiency of a horizontal-flow, single-stage electrostatic precipitator consisting of two sections formed by plates 4.0 m wide and 6.0 m high on 25-cm centers, handling a gas flow of 2.5 m^3/s. Assume that the migration velocity is 12 cm/s.

Solution

Given: The plate area of each section $A = 4 \times 6 \times 2 = 48\ m^2$
The average flow rate per section $Q = 2.5/2 = 1.25\ m^3/s$
The migration velocity w = 12 cm/s = 0.12m/s

For uniform gas velocity,

$$\eta = 1 - \exp(-wA/Q) = 1 - \exp[-(0.12)(48.0)/1.25] = 99\%$$

In the Deutsch–Anderson equation, the most sensitive variable among others is migration velocity, which is closely associated with the particulate collection efficiency. Migration velocity in reality differs for different applications and often differs considerably within the same application field (20–22). This mainly results from the variations in particle characteristics (e.g., resistivity, particle size) and gas conditions (e.g., gas temperature, moisture content, and sulfur oxides content). These variables can change migration velocity by as much as a factor of 3 (9). Thus, it is very difficult to select a proper migration velocity for a specific application in the design of electrostatic precipitators based on the Deutsch–Anderson equation. For example, an electrostatic precipitator design on the basis of migration velocity of 12 cm/s for a collection efficiency of 98% would give a collection efficiency only around 75% if the migration velocity of 4 cm/s were used. Typical values of migration velocity for various applications are given in Table 2.

In operation, migration velocity also depends strongly on factors such as accuracy of electrode alignment, uniformity and smoothness of gas flow through the precipitator, rapping of the electrodes, and the size and electrical stability of the rectifier sets. Migration velocity can be estimated from a pilot-scale or an existing ESP system by using known values for the collection plate surface area, volumetric gas flow rate, and particulate collection efficiency in the Deutsch–Anderson equation.

Example 6

Find the migration velocity for an existing electrostatic precipitator, which the collection plate area is 110 m^2, gas flow rate is 2.5 m^3/s, and collection efficiency is 99.5%.

Solution

Given: $\eta = 99.5\%$, $Q = 2.5$ m^3/s, and $A = 110$ m^2

$$\eta = 1 - \exp(-wA/Q)$$

$$0.995 = 1 - \exp[-w(100) / 2.5]$$

$$w = \ln(1 - 0.995)(2.5)/(110) = 0.12 \text{ m/s (or 12 cm/s)}$$

3.2. Particulate Resistivity

Particulate resistivity, a measure of a particle's resistance to electrical conduction, is a fundamental indicator of migration velocity of the particles. Resistivity is of extreme importance not only because it varies widely but also because it strongly influences the collection efficiency of the precipitator. It could influence the electrostatic charges exerted on the particles as well as the re-entrainment of collected particles from the collecting plates. Once collected, the particles would release their charges to the collecting plates depending on the particulate resistivity. The transfer of electrostatic charges completes the electrical circuit, produces current flow, and allows maintenance of voltage drop between the discharge and collecting electrodes.

The resistivity of a material can be determined experimentally by establishing a current flow through a slab of the material. It is of importance to make resistivity measurements of freshly collected particles in actual gas stream. In general, the measurements should be made in the field rather than in the laboratory. Resistivities measured in the laboratory on the same particles can be 100–1000 times greater than field resistivities (23). The resistivity is defined as the resistance times the cross-sectional area normal to the current flow divided by the path length (7):

$$p = \frac{Ra}{l} = \frac{Va}{il} \tag{44}$$

where p is the particulate resistivity, R is the particulate resistance, a is the cross-sectional area normal to the current flow, l is the path length in the direction of current flow, V is the potential, and i is the current.

The resistivity of materials generally ranges from 10^{-3} to 10^{14} Ω-cm, whereas the best range of the resistivity for particle collection in an ESP is 10^{7}–10^{10} Ω-cm. In general, ESP design and operation are difficult for particulate resistivities above 10^{11} Ω-cm.

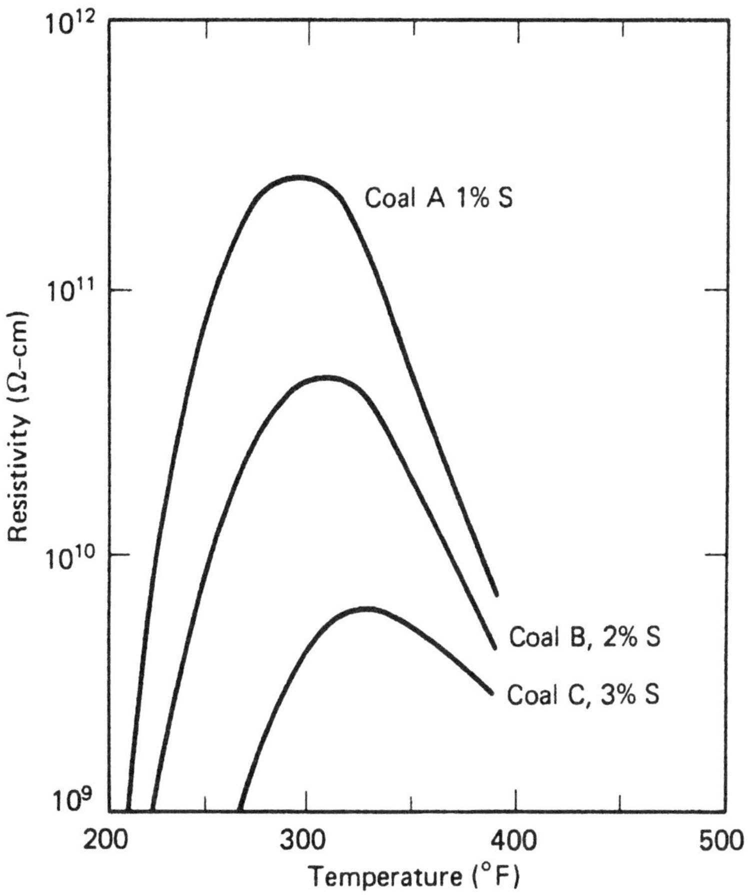

Fig. 7. Variation of particulate resistivity with moisture content and gas temperature for three different coals. (From ref. 24.)

For lower resistivity, the particles can be charged quickly to its saturated charge level, resulting in higher migration velocity, and, accordingly, achieve higher collection efficiency of ESP. However, if the resistivity of particles is too low (i.e., the particles are considered a good conductor for $p<10^7$ Ω-cm), the electrostatic charge could drain off quickly and the collected particles are then easily re-entrained back into the gas stream. On the contrary, if the resistivity of particles is too high (i.e., the particles are considered a good insulator for $p>10^{11}$ Ω-cm), the particles become difficult to be charged in the electrical field and the charges on collected particles do not easily drain off at the collecting plates. Under this condition, the particles remain strongly attracted to the collecting plates and are difficult to rap off. Moreover, a "back corona" phenomenon might develop and, accordingly, reduce the migration velocity of particles in the gas stream.

Major operating parameters influencing particulate resistivity include gas temperature, moisture levels, and chemical composition of particles (7). The gas temperature of the maximum resistivity is unfortunate because operators often cannot reduce ESP

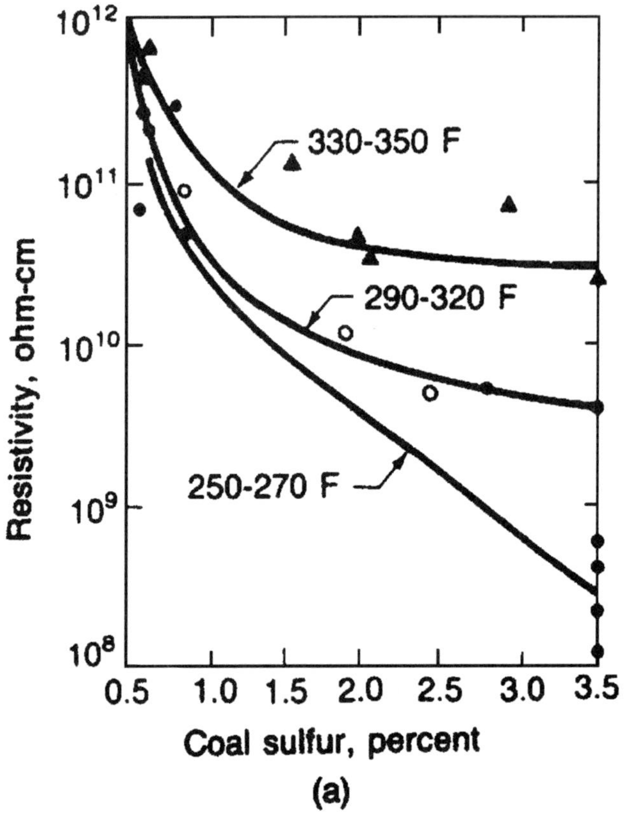

Fig. 8. Variation of fly ash resistivity with sulfur content and gas temperature. (From ref. 25.)

temperatures below 250°F without risking the condensation of sulfuric acid on cold surfaces. On the other hand, increasing the temperature above 350°F results in unnecessary loss of heat out the stack, which represents a monetary loss. Moreover, the operation of ESP at a gas temperature of approx 300°F would result in a maximum resistivity of the particles (*see* Fig. 7). Therefore, it is recommended to operate the precipitator at gas temperatures either below or above 300°F, which develops cool-side ESP and hot-side ESP, respctively. Between these two types of ESP, the hot-side ESP is currently more popular than the cool-side ESP.

Figure 7 also illustrates that particulate resistivity decreases with increasing moisture content of the gas stream. Therefore, injection moisture into the gas stream has been practically applied to reduce the particulate resistivity and thus enhance the collecting efficiency of ESP. However, increasing the moisture level might increase the sparkover ratio, which contrarily reduces the collecting efficiency of ESP. Also, the sulfur content of the fuel (e.g., coal) plays an important role in determining the particulate resistivity. The maximum resistivity of particles decreases significantly from 4×10^{11} to 8×10^{9} Ω-cm as the sulfur content of coal increases from 1% to 3% (*see* Fig. 8). The resistivity decreases with increasing coal sulfur content because of increased adsorption of conductive gases by fly ash.

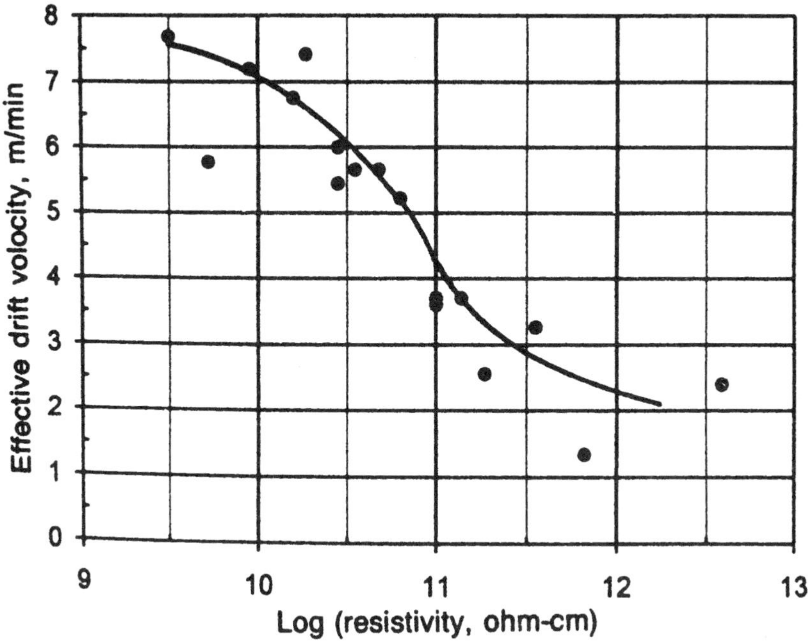

Fig. 9. Effects of fly ash resistivity on effective migration velocity in an ESP. (From ref. 23.)

A highly resistive particle increases the occurrence of sparking in an ESP and forces a lower operating voltage. A serious back corona can develop, which reduces both charging and collection of particles. The effects of resistivity are more significant above 10^{11} Ω-cm, but can be accounted by for the design for effective migration velocity. Figure 9 illustrates the effect of fly ash resistivity on effective migration velocity of particles.

Example 7

Estimate the collecting plate area required for an ESP that is applied for removing fly ash in the gas stream emitted from an utility power plant. Assuming that (1) the designed collection efficiency is 99.5%, (2) the volumetric gas flow rate is 9,600 m^3/min, and (3) the resistivity of fly ash is 8×10^9 Ω-cm.

Solution

Given: $p = 8 \times 10^9$ Ω-cm and Q = 9,600 m^3/min = 160 m^3/s
From Fig. 9, w = 7.2 m/min = 0.12 m/s. Applying the Deutsch–Anderson equation

$$\eta = 1 - \exp(-wA/Q)$$

$$0.995 = 1 - \exp(-0.12A/160)$$

$$A = \ln(1 - 0.995)(160)/(0.12) = 7,040 \text{ m}$$

3.3. *Internal Configuration*

The design of the internal configuration of an ESP is of great importance; however, it was usually ignored in most textbooks. The even distribution of gas flow through the ducts is very crucial for proper operation of an ESP, as are uniform plate spacing, proper discharge electrode arrangement, trueness of plates, slopes of hoppers, adequate numbers of electrical sections, and other features inside the ESP (26,27). Using the practical design parameters given in Table 1 and the basic understanding of ESP configuration, we can specify the geometry of an ESP. The overall width of the precipitator is virtually equal to the number of ducts for gas flow as follows (7):

$$N_d = Q/vDH \tag{45}$$

where N_d is the number of ducts, Q is the total volumetric gas flow rate, v is the linear gas velocity in the ESP, D is the width of ducts between two collecting plates, and H is the height of the plates (8–15 m). The overall length of the precipitator is given as follows (7):

$$L_t = N_s L_p + (N_s - 1)L_s + L_{\text{in}} + L_{\text{out}} \tag{46}$$

where L_t is the overall length, N_s is the number of electrical sections in the direction of gas flow (2–8), L_p is the length of the collecting plate (1.0–4.0 m), L_s is the spacing between electrical sections (0.5–2.0 m), L_{in} is the length of the inlet section (3–5 m), and L_{out} is the length of the outlet section (3–5 m).

The overall height of an ESP could be 1.5–3.0 times the plate height because of hoppers, superstructure, controls, and so forth. The number of electrical sections depends on the aspect ratio (the ratio of overall plate length to plate height) and plate dimensions. However, the number of electrical sections must be sufficient to provide the minimum collection area required but not a great excess of area. The number of electrical sections can be estimated by

$$N_s = r_a H/L_p \tag{47}$$

where N_s is the number of electrical sections and r_a is the aspect ratio.

When the numbers of ducts and sections are specified, the actual overall plate area can be calculated by

$$A_a = 2HL_p N_s N_d \tag{48}$$

In general, the performance of ESP can be improved with increasing sectionalization because of more accurate alignment and spacing for smaller sections and more stable rectifier sets operating at higher voltages. Large numbers of electrical sections allow for meeting the overall efficiency targets even if one or more sections are inoperable. However, adding extra sections increases the capital cost.

Example 8

Estimate the overall width and length of an ESP designed for treating 20,000 m^3/min of gas with total plate area of 14,000 m^2. Assume the plates are available in 8-15 meters high and 3 meters long.

Solution

Given: Q = 20,000 m^3/min and A = 14,000 m^2

From Table 1, we select H = 10 m, D = 0.25 m, L_p = 3 m, L_s = 0.3 m, L_{in} = 4 m,

L_{out} = 4 m, v = 100 m/min, and r_a = 0.9.

From Eqs. (45)–(48), we obtain

$$N_d = Q/vDH = (20,000) / (100)(0.25)(10) = 80 \text{ ducts}$$

$$N_s = r_a H/L_p = (0.9)(10) / (3) = 3 \text{ sections}$$

$$D_t = N_d D = (80)(0.25) = 20 \text{ m}$$

$$L_t = N_s L_p + (N_s - 1)L_s + L_{\text{in}} + L_{\text{out}} = (3)(3) + (3 - 1)(0.3) + 4 + 4 = 17.6 \text{ m}$$

$$A_a = 2HL_p N_s N_d = (2)(10)(3)(3)(80) = 14{,}400 \text{ m}^2 \ (>14{,}000 \text{ m}^2)$$

3.4. *Electrode Systems*

Good precipitator design provides for definite structural relationships between the electrode systems: discharge electrode and collecting electrode. The type and positioning of the discharge and collecting electrodes can be major factors in the operation and maintenance of an ESP. As illustrated in Fig. 6, the discharge electrode system consists of a high-voltage duct, feed-through support insulator, tension support insulator, upper support grid, discharge electrode vibrator and wires, lower alignment grid, and weight tension. The discharge wires energized negatively are usually designed of round, 12-gage steel spring wires with sharp edges to facilitate the formation of a corona around them. They are reinforced at the top and bottom to ensure good electrical contact and to resist mechanical and electrical erosion. The discharge wires are taut by weights and positioned through guides to prevent excess swaying. The wires tend to be high-maintenance items. Corrosion can occur near the top of the wires because of air leakage and acid condensation. Moreover, long weighted wires tend to oscillate. The middle of the wire can approach the collecting plates quite closely, causing increased sparking and wear. Some types of discharge wire are illustrated in Fig.10.

The collecting electrode system is designed to have maximum collecting surface, high-sparkover voltage characteristics, no tendency to buckle or warp, resistance to corrosion, and aerodynamic shielding of collecting surfaces to minimize re-entrainment of the collected particles. Standard planar electrodes are usually made of cold-rolled steel sheets to ensure flatness. Collecting electrode panels are grouped within the precipitator housing to form independently suspended and independently rapped collecting electrode modules, and they are rapped periodically by electromechanical means. Because the collecting electrodes are generally cleaned by the rapping and dropping of collected particles by gravity, they have to be separated at a sufficient distance for the free fall of the particles. This leads to widening the distance between electrodes, which, in turn, requires a higher voltage to produce the desired corona discharge. The dimensions of the electrode systems are fixed largely by the required voltage and area of electrode surface per unit volume of gas. Some types of collecting plate are illustrated in Fig.10.

3.5. *Power Requirements*

The power requirements for an electrostatic precipitator vary with collection efficiency. It is important for the power supply to deliver a unidirectional current to the

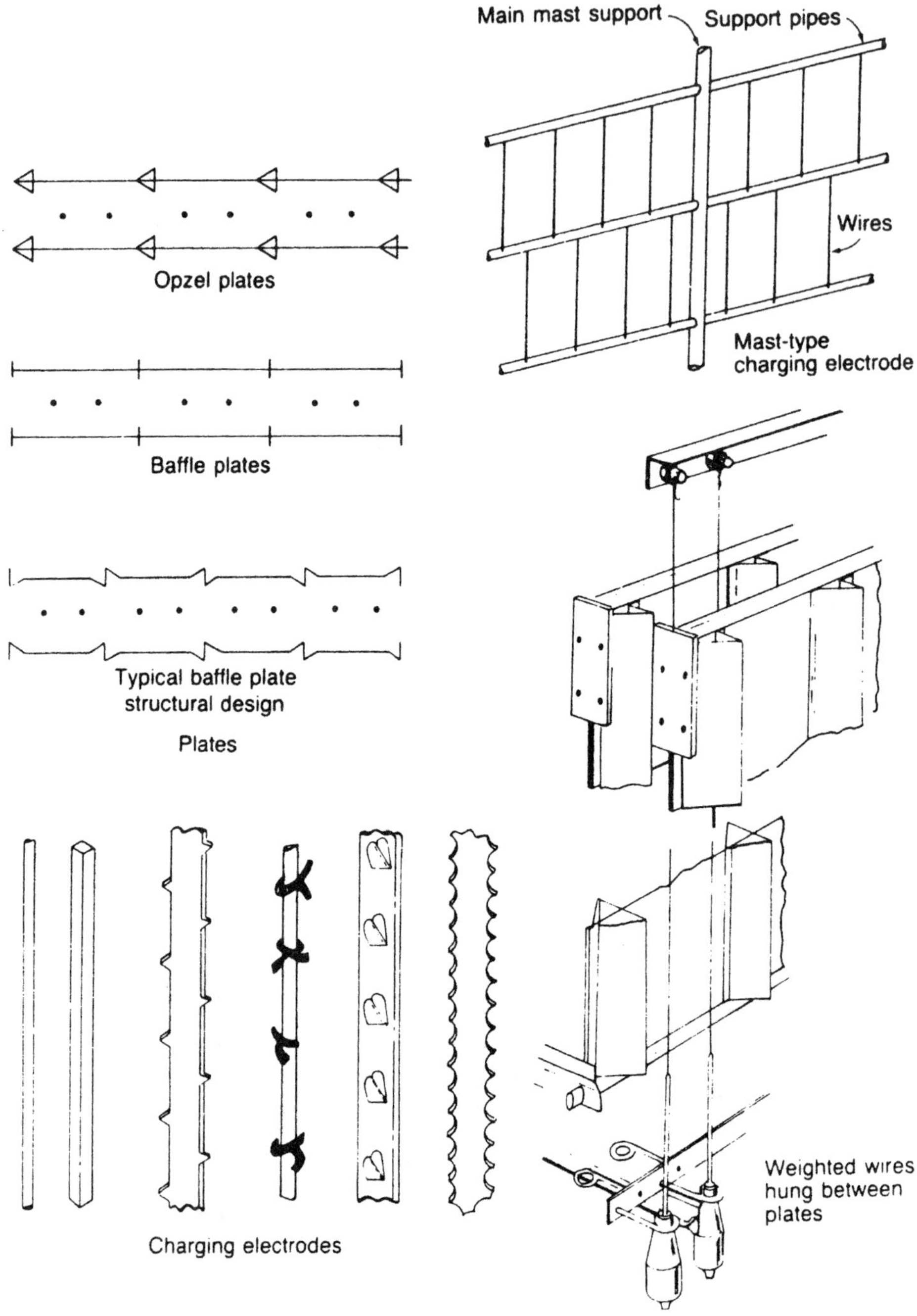

Fig. 10. Schematic of various discharge wires and collecting plates. (From ref. 7.)

electrodes at a potential very close to that which produces arcing. Selection of power requirements has been generally based on data relating efficiency to corona power per unit volume of gas flow (9,10). These data are experimentally developed for each type of application and for varying particulate properties. The recommended

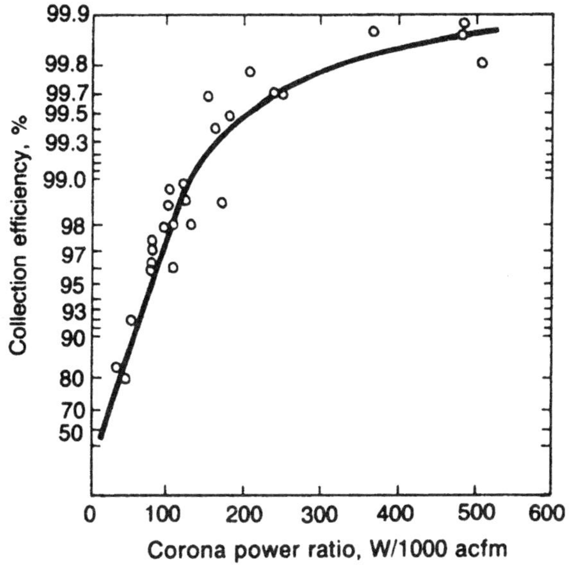

Fig. 11. Collection efficiency as a function of corona power ratio. (From ref. 23.)

values of energy requirement range from 100 to 1000 W/(m^3/s of gas flow) (*see* Table 1). To ensure continuous maximum collection efficiency, automatic voltage regulators are generally used for adjusting voltages automatically, even under widely varying operating conditions.

Modern power packs are equipped with a system of automatic voltage regulation to maintain an optimum, average precipitator sparking rate, usually in the range of 5 or 10 to about 75 sparkovers per minute, according to the application and dust concentration proceeding from outlet to inlet (27). Because maximum collection efficiency depends predominantly on maximum voltage, power packs should have substantial overcapacity both with regard to voltage as well as current.

Operating power consumption in an ESP mainly comes from corona power and pressure drop, with corona power being the main source. The corona power can be approximated by

$$P_c = I_c V_{avg} \tag{49}$$

where P_c is the corona power, I_c is the corona current, and V_{avg} is the average voltage. Even though voltages in ESPs are very high, the current flow as a result of gas ion migration is low. Thus, the power consumption is not unreasonably high. The effective migration velocity of particles can be related to the corona power:

$$w = \frac{k' P_c}{A} \tag{50}$$

where w is the migration velocity of particles, k' is simply an adjustable constant (0.5–0.7), and A is the surface area of the collecting plates. The ratio of P_c to A (P_c/A) is also known as the power density. Although the power density often increases gradually from the inlet of the gas flow to the outlet, the overall power density is a fairly stable and representative parameter. The typical range of the overall power density is 10–20 W/m^2 (27). By substituting Eq. (50) ($wA = k'P_c$) into the Deutsch–Anderson equation, the corona power can be related to the collection efficiency:

$$\eta = 1 - \exp\left(-\frac{k'P_c}{Q}\right) \tag{51}$$

Equation (51) with $k' = 0.015$ for P_c/Q in units of W/(m^3/s) [or $k' = 0.55$ for P_c/Q in units of W/(ft^3/s)] is reasonably accurate for particulate collection efficiencies up to approx 98.5% (28). For collection efficiencies above 98.5%, the required corona power increases rapidly for an increase in collection efficiency, as shown in Fig. 11.

Example 9

An electrostatic precipitator is designed to treat 9000 m^3/min of gas stream to remove 99.8% of particles.

Solution

Given: Q = 9000 m^3/min = 333,333 ft^3/min and η= 99.8%
From Fig. 11, P_c/Q = 330 W/(1000 acfm) and

$$P_c = (P_c/Q)(Q) = (330/1{,}000)(333{,}333)(1/1{,}000) = 110 \text{ kW}$$

3.6. Gas Flow Systems

The design of the gas flow system is commonly based on model studies with large systems. The best operating conditions for an electrostatic precipitator occur when the gas velocity ranges between 0.6 and 2.4 m/s (2–8 ft/s) and is uniformly distributed (13, 19). Because conveying velocities within flue systems are too high for particle precipitation, it is necessary to reduce the gas flow rate by careful design of the connecting systems. In practice, it is almost impossible to achieve a completely uniform velocity distribution in a practical duct system. However, it is possible to approach an acceptable quality of flow at the precipitator inlet by the use of splitters, strengtheners, vanes, baffles, and diffusion plates. The purpose of these designs is to achieve a nearly uniform gas flow rate, using the best duct design procedures plus field corrective measures as required (13).

A nonuniform gas flow rate can result in excessive deposits of particles as well as variation of flow pattern in the ducts and gas flow system. It changes the velocity distribution across the ESP and consequently alters the designed residence time of the gas stream in each duct. Although the increasing of residence time in some ducts might enhance the collection efficiency of particles, the decreasing of residence time in other ducts would definitely reduce the collection efficiency of particles. The combination of these two effects, however, results in the reduction of particulate removal efficiency in an ESP.

3.7. Precipitator Housing

The precipitator housing is built of weatherproof gastight materials, suitable for outdoor or indoor installations. Major housing parts include the shell, hopper, inlet and outlet duct connections, inspection doors, and insulator casings. A variety of construction materials are used to meet specific gas and particulate properties, operating practices, and other factors. Construction materials for the electrostatic precipitator can be either reinforced concrete supports, frame with brick wall, or steel throughout. Steel construction permits shop fabrication before on-site installation. Concrete shells are not recommended when the operating temperature exceeds 260°C.

The insulation of an ESP is also one of the basic design requirements in order to prevent the ESP from the condensation of water vapor and acidic gases. Condensation of moisture and acidic gases on the outer shell walls results mainly from the contact of cool ambient air, which could cause potential corrosion of construction materials. Some electrically controlled heating blankets are usually used for this design. In addition, corrosion-resistant materials are generally used when the effluent gas stream of an industrial process contains corrosive gases.

The most common material used in ESP construction is carbon steel, in cases where the gas stream contains high concentrations of SO_3 or where liquid–gas contact areas are involved, stainless steel may be required. However, by keeping the emission stream temperature above the dew point and by insulating the ESP (the temperature drop across an insulated ESP should not exceed 20°F), the use of stainless steel should not be necessary.

3.8. Flue Gas Conditioning

As mentioned earlier in this section, gas temperature, moisture content, and chemical composition of flue gas have strong influence on particulate resistivity and, thus, on the collection efficiency of particles. However, the gas temperature often cannot be decreased owing to the possibility of acid and moisture condensation. Ducting can sometimes be arranged such that the gas temperature entering the ESP can be raised, but this procedure is usually quite costly. Therefore, chemical conditioning of flue gas becomes one of the most practical approaches. Several flue gas conditioning systems are commercially available and work quite well with reasonably small expense.

Flue gas conditioning is extremely important for the improvement of particulate resistivity. Flue gas can be conditioned by adjusting gas temperature and moisture content, by adding sulfur trioxide and ammonium to the gas stream, and by varying sulfur content of fuel. Among them, moisture conditioning can be accomplished by stream injection or by liquid water spray into the dusty gas stream. Proper spray nozzle design, spacing, and careful temperature control are crucial. If too much water is injected, the particles will cake on the interior of the ESP (27) and increase the sparkover rate, which results in the decrease of particulate collection efficiency in an ESP (7). Figure 7 illustrated the effectiveness of adding moisture to cement kiln dust exhaust streams. In addition to the temperature and humidity of the gas stream, chemical compounds such as SO_3, NH_3, and NaCl are commonly used as conditioning agents.

3.9. Removal of Collected Particles

Particles accumulating on the collecting plates must be removed periodically. In wet ESPs, the liquid flowing down the collector surface removes the particles. In dry ESPs, the particles are removed by vibrating or rapping the collector plates. For dry ESPs, this is a critical step in the overall performance because improperly adjusted or operating rappers can cause re-entrainment of collected particles or sparking because of excessive particulate buildup on the collection plates or discharge electrodes. In normal operation, dust buildup of 6–25 mm is allowed before rapping of a given intensity is initiated. In this way, collected material falls off in large clumps that would not be re-entrained. If rapping is initiated more frequently or if the intensity of rapping is lowered, the resulting smaller clumps of particulate matter are more likely to be re-entrained, reducing the collection efficiency of the ESPs. Optimal adjustment of the ESP can best be made by direct visual inspections through sight ports.

In the ESP, the collected particles or droplets on collecting plates are generally removed by rapping or by washing. The design of electromechanical rapping includes an electric motor drive, pneumatic drive, or magnetic impulse drive. The selection of the type and number of rappers varies among manufacturers and with the characteristics of particles being collected. Modern precipitators are designed for two independent electromechanical rapping systems: one for keeping the high-voltage discharge electrodes continuously clean, and the other for sequential rapping of the collecting electrode modules.

If rapping does not provide for complete cleaning of the electrodes, particulate collection efficiency of the precipitator may decrease in the course of operation owing to particulate buildup. As a consequence, the condition of inadequate rapping may require an increase of corona power input in order to maintain the level of particulate collection efficiency. Successful rapping depends mainly on a certain range of particulate resistivity at various temperatures. For certain particles, the application of an adhesive to the collecting electrodes is necessary. In this case, the removal of the accumulated material can only be accomplished by a washing procedure. After washing is completed, the adhesive fluid again is applied before the unit is put back in operation.

In most American designs, the collecting plates are rapped by a falling weight. The intensity of the rap can be easily adjusted by varying the height from which the weight is dropped or by adjusting the acceleration field strength. In a typical European design, rapping is accomplished by hammers connected to a motor rotating at a constant speed. Thus, to adjust the rapping intensity, the hammers must be changed physically. Generally, one rapping unit is designed and provided for every 110–150 m^2 (or 1200–1600 ft^2) of collection area (28). Both designs allow for convenient adjustment of the rapping interval varying from 1 to 10 min.

Hoppers are designed to catch the falling particles as well as to provide space for temporary storage. Most hoppers have a pyramidal shape that converges to either a round or square discharge. Hopper walls must be steeply sloped (at least 60% slope) to prevent dust caking and bridging. In addition, hoppers are often heat traced because warm dust flows much better than cold dust. In general, approx 60–70% of the collected dust can be removed through the first inlet set of the hoppers. However, in the case of failure of the first electrical set, the dust load is then transferred to the next downstream hopper. Therefore, liberal sizing of the hoppers is recommended. Proper support structure must be provided so that a hopper will not collapse when filled with dust. The discharge of dust from the hoppers with regular frequency is also crucial for avoiding dust bridging and hopper collapsing.

3.10. Instrumentation

Instrumentation is of major importance in electrostatic precipitation and falls under the following two categories: (1) process instrumentation and (2) instrumentation for electrical variables. Process instrumentation provides the measurement of process variables such as gas flow rate, gas temperature, relative humidity, and gas pressure. Variations in these process conditions can affect precipitator performance and it is therefore necessary to monitor them during normal operation. Conventional instruments such as Pitot-tube meters, thermocouples, hygrometers, and manometers are used for this purpose. Various analytical instruments may also monitor specific compounds in the gas stream. Other process instrumentation consists of sensors to measure the dust level in the collection hoppers and to detect the intensity of the rappers for rapping control.

Electrical variables that are measured are high voltage, current, and sparkover rate for the discharge electrodes and readings for the rectifier equipment. Kilovoltmeters and conventional milliammeters provide information whereby current to the discharge electrodes may be set to provide the maximum voltage. Direct-reading sparkover-rate meters are used to obtain the optimum sparkover rate for a given precipitator.

Oscilloscopes (CRT) are especially useful for studying sparking characteristics and for troubleshooting electrical faults. Furthermore, oscilloscopes aid in the monitoring of current and voltage waveforms. The optimum voltage wave shape is one that has a balance between the peak voltage and the average current or voltage, because the charging field is determined by the peak voltage and the collecting field is a function of the average current or voltage.

4. APPLICATIONS

Electrostatic precipitators have been used not only for collecting solid particles and liquid droplets to comply with air pollution control regulations, but also for removing particles in office buildings and stores and in manufacturing and process operations, in which particle-free air is essential. Precipitators have also been used in industrial processes to recover valuable materials such as copper, lead, or gold in the fluidized catalyst process and soda ash in Kraft paper mills (9,12). Other applications pertain to purifying fuel and chemical process gases for quality improvement, collecting partially condensable vapors for chemical product or byproduct recovery, and separating contaminant gases and vapors from gas streams by sorption on solid particles for later removal (9,12,13). The major fields for the application of electrostatic precipitator are summarized in the following subsections.

4.1. Electric Power Industry

Electrostatic precipitators in the electric power industry are used principally for collection of solid particles from coal-fired power plants. The application constitutes the largest single use of precipitators in the United States—about 75% of the total application in terms of gas volume treated (19,25). The ash content of the coals being burned varies from 5% to 25%; typical particulate emissions range from 4.6 to 16.1 g/m^3 of stack gas [or 2–7 grains/standard cubic foot (SCF)]. The particle size distribution of fly ash varies with the type of boiler and the characteristics of coal. The median diameter of fly ash is around 5–15 μm (9). At gas temperatures of 232°C or above, the particulate resistivity is likely to be below the critical value of 10^{10} Ω-cm (16).The particulate collection efficiency has been rated better than 99%. Newer installations can handle gases up to 370°C, particularly for high-resistivity particles generated from low-sulfur coals and residual fuel oils.

4.2. *Pulp and Paper Industry*

Precipitators are used in recovering salt cake from the flue gases of Kraft mill recovery boilers and in collecting acid mist from paper mills. Particulate emissions from the recovery boiler are extremely fine and hygroscopic. They are composed principally of sodium sulfate and sodium carbonate with small quantities of sodium chloride, sodium sulfide, and sodium sulfite. Because of its hygroscopic nature, sampling the gas to determine particle size distribution is quite difficult. The median particle size for recovery is approx 1.9 μm (9). The particulate collection efficiencies of ESP range from 90% to 98%.

4.3. *Metallurgical Industry*

Applications in the ferrous industry have been in the cleaning of gaseous effluent from steel-making furnaces, blast furnaces, foundry cupolas, sinter machines, and byproduct coke ovens. The use of precipitators in the nonferrous industry has been standard practice for copper, lead, and zinc smelters in cleaning the off gases from the extraction process. Precipitators are also used in cleaning gases from electrolytic cells in the reduction of bauxite to produce aluminum (9). The particulate collection efficiencies of ESPs range from 85% to 99%. The particulate collection efficiencies are relatively low when applied to electric arc furnaces because of large quantities of high-temperature gas.

4.4. *Cement Industry*

Precipitator applications to cement kilns have been particularly favorable because they permit the recovery of cement as well as the control of particulate emissions. Precipitators have also been used for the cleaning of ventilating gases and dryer gases. The particulate emission rate for a cement kiln is highly variable because of variation in the raw feed and kiln design. Particulate matter from cement kilns generally has high resistivity. Early applications in the cement industry were hampered by the resistivity problem, but newer installations have successfully overcome the problem of resistivity by controlling gas temperature, by conditioning with moisture, and by improving electrical energization. In the wet process, particulate resistivity is less of a problem. The trend of precipitator designs is toward higher collection efficiency, current precipitators being designed for collection efficiencies of ESP over 99.5% (9).

4.5. *Chemical Industry*

Precipitators have been used to collect sulfuric and phosphoric acid mists and to remove particulates from elemental phosphorus in the vapor phase. In the manufacture of sulfuric acid, gases from the smelter contain approx 3–10% sulfur dioxide and contaminated particulates. They must be removed before being introduced into the converter to prevent fouling of the catalyst. Precipitators designed for the chemical industry are generally of the tubular type with vertical gas flow. The particulate collection efficiencies of ESP range from 97% to 99.5% (9).

4.6. Municipal Solid-Waste Incinerators

The use of precipitators on municipal solid-waste incinerators is a relatively new application. Emission of particulates ranges from 50 to 300 g/kg of refuse or from 1.2 to 5.7 g/m^3 of gas (5). The properties and composition of the particulate matter vary greatly because the composition of refuse is highly variable. Particle size varies from a median diameter of 15–30 μm (9). The resistivity of fly ash varies with temperature, moisture content, and particle size. Gases from municipal solid-waste incinerators are at temperatures of 655–900°C and must be cooled before entering the precipitator (9). The particulate collection efficiencies of ESP range from 90% to 99%.

4.7. Petroleum Industry

Principal uses in the petroleum industry are for the collection of particulate emitted from fluidized-bed catalytic cracking units (FCC), for the removal of tar from gas streams, such as fuel gases, acetylene, and shale oil distillation gases, and for the collection of particulates emitted from fluidized-bed waste sludge incinerators. The median particle size is approx 10–12 μm (9).

4.8. Others

Two-stage precipitators are applied for aerosol sampling, food processing, asphalt saturating, high-speed grinding machines, galvanizing kettles, rubber-curing ovens, and radioactive particle collection (5,9,12). They are also used at hospitals as well as in office buildings, where particle-free air is essential.

5. PROBLEMS AND CORRECTIONS

Despite many successful installations in various industrial operations, electrostatic precipitators in many cases have failed to meet performance requirements by somewhat large margins. Even the best available precipitator cannot handle all situations. The problems for the precipitator are classified in the four major categories: fundamental, mechanical, operational, and chemical.

5.1. Fundamental Problems

Fundamental problems are associated with (1) the assumptions in the derivation of the Deutsch–Anderson equation, (2) the high resistivity of particles, (3) nonuniform gas flow, (4) improperly designed electrode systems, (5) insufficient high-voltage electrical equipment, (6) inadequate rapping equipment, and (7) re-entrainment of collected particles.

Much effort has been spent via theoretical, empirical, and statistical methods at refining the Deutsch–Anderson equation and to make the resulting collection efficiency expression more useful for design purposes. In a study dealing with precipitation of fly ash from coal-fired electric power plants, Selzler and Watson (29) suggested that in the determination of particulate migration velocity the important factors are electrical power input to the precipitator, the particle size distribution in the entering gases, and the sulfur-to-ash ratio of the coal burned. Based on this approach, a proposed empirical collection efficiency equation is,

$$\eta = 1 - \exp\left[(-0.57)(203)(A/Q)^{1.4}(kW/Q)^{0.6}(S/AH)^{0.22}\right] \tag{52}$$

where A is the surface area of collecting plate (1000 ft^2), Q is the flue gas volumetric flow rate (1000 actual ft^3/min), kW is the power input to the discharge electrodes, and *S/AH* is the sulfur-to-ash ratio of the coal burned (by weight).

Selzler and Watson derived the numerical parameters by the use of least-squares regression techniques. The data used for the development of the above equation were obtained from the US Environmental Protection Agency (US EPA) and questionnaires sent to utility companies and the Federal Power Commission.

Frisch and Coy (30) considered Selzler and Watson's approach as a good attempt at a more systematic method for sizing the ESPs. However, they commented that it is meaningless to use the sulfur-to-ash ratio as an independent variable to describe precipitator performance at elevated temperatures and that it is erroneous to assume power density as an unconstrained independent variable. The following empirical efficiency equation was developed (30):

$$\eta = 1 - \exp\left[-k(P_c/A)^{a'}(A/Q)^{b'}(\bar{v})^{c'}(\bar{x})^{d'}\right] \tag{53}$$

where P_c/A is the power density (W/ft^2), A/Q is the specific collecting area (ft^2/1000 actual ft^3/min), $\bar{v}$ is the average treatment velocity (ft/s), $\bar{x}$ is the mass median particle diameter (μm), and k, a', b', c', and d' are empirical constants.

On the basis of theoretical considerations and a comparison of observed versus predicted collection efficiency, the use of the Frisch–Coy equation [Eq. (53)] showed better results than the Selzler–Watson equation [Eq. (52)] in estimating the size of hot-side precipitators. The hot-side precipitator is recently applied because it reduces particulate resistivity and prevents acid condensation by means of placing the precipitator in the front of an air preheater, where flue gas temperature is much higher than that at the usual downstream location.

Cooperman (31) and Robinson (32) have modified the Deutsch–Anderson equation and brought the theoretical and empirical aspects of precipitation phenomena into closer agreement. The modification takes into account the erosion of collected particles and the nonuniformity of particle concentration over the precipitator cross-section. Soo (33) rationalized precipitator design by providing knowledge of the equilibrium position of a corona wire and the effect of turbulent diffusion in the electrostatic field. Soo (34) also introduced the concept of particle–gas–surface interactions applied to the electrostatic precipitators. Potter (35) has utilized the concept of an extended Deutsch–Anderson equation: A semilogarithmic plot of particulate collection efficiency against the product of specific collection area and the square of operating voltage generates a "performance line" for a particulate–precipitator combination. Such performance lines enable interpolations of precipitator performance to be reliably made for combinations of operating conditions other than those used in the original pilot tests. Important effects of particle size and of carrier gas additives readily emerge through the performance line.

High particulate resistivity is one of the principal causes of poor performance by precipitators. Particle deposits on the surface of collecting electrodes must possess at least a small degree of electrical conductivity in order to allow transportation of ions through the dust layer. If the dust is a good conductor, there is little or no disturbance of the corona discharge. However, as the particulate resistivity increases, a point is reached at which the corona ions begin to be impeded. A further increase in particulate resistivity causes the voltage across the dust layer to increase and corona discharge sets in, which severely reduces the particulate collection efficiency of ESPs.

Most dusts and fumes have dielectric breakdown strengths of about 10 kV/cm; thus, with a typical corona current density of 1 $\mu A/cm^2$, the critical resistivity appears to be around 10^{10} Ω-cm. Loss of precipitator performance increases with increasing particulate resistivity above the critical value of 10^{10} Ω-cm. Here, sparkover voltages are reduced, back corona may form, and corona currents are disturbed or disrupted; the effects are limited to reduce operating voltages and currents. When particulate resistivity exceeds 10^{11} Ω-cm, it becomes difficult to achieve reasonable collection efficiencies with precipitators of conventional design. Above 10^{12} Ω-cm, precipitator performance drops to such low levels as to become impracticable for most applications.

Methods for overcoming the high resistivity of dusts can be classified under several categories and include the following (22):

1. Keeping collecting plate surfaces as clean as possible. Numerous schemes have been proposed toward this objective, such as moving brushes, scrapers, and belts. The most commonly used method to keep the collecting plate surface clean is by high-impact rapping, using accelerations at the plate surfaces of as high as 50*g* to 100*g*.
2. Improving the electrical energization of the precipitator. Experience shows that precipitator performance improves considerably with higher operating voltages and currents. In practice, sparkover voltages limit the maximum operating voltage. Practical methods for improving electrical energization include greater sectionalization of corona electrodes, use of pulsating voltages, fast-acting spark-quenching circuits, and automatic control systems.
3. Conditioning of flue gas. Control of particulate resistivity by varying the moisture and chemical conditioning of the carrier gases is achieved by increasing particle conductivity as a result of adsorption of moisture and the chemical substances from the gas. Adsorption is a surface effect and is greater at lower temperatures. Moisture (steam) conditioning is effective at 120–150°C. For chemical conditioning, SO_3, NH_3, and NaCl have been commonly used as conditioning agents.
4. Changing operating temperatures of the precipitators. The particulate resistivity depends on temperature according to $P = A'\exp(-E_a/kT)$, where A' is a constant, E_a is an activation energy, k is the Boltzmann constant, and T is absolute temperature. This curve passes through a maximum as gas temperature is increased; thus, at low temperatures (100–150°C) and at high temperatures (300–370°C), the particulate resistivity is below the level at which the precipitation problems will be encountered.
5. Temperature-controlled electrodes. It is similar to category 4 except that only the temperature of the deposited dust layer is changed rather than the whole gas steam. The dust layer is temperature controlled by heating or cooling the collecting electrodes.

6. Changing raw materials. Changes in the raw materials used in a plant process can have a profound effect on particulate resistivity. The basic factors that govern particulate conductivity can be used as guidelines in finding better raw materials from the viewpoint of improving precipitation.
7. Graded-resistance electrodes. Making the electrodes of semiconductor material, which will counteract the adverse effects of the high-resistivity deposits, alters the basic electrical properties of the collecting electrodes.

It must be mentioned in passing that low particulate resistivity can also be a problem, as was pointed out in Section 2.4.3. If the particulate resistivity is below 10^4 Ω-cm, the collected particles are so conductive that their charges leak to ground faster than they are replenished by the corona. With no charge to hold them, the particles are either reentrained in the exit gas or they pick up positive charges and are repelled.

Nonuniform gas flow through a precipitator lowers performance in two ways. First, uneven treatment of the gas lowers collection efficiency in the high-gas-velocity zones to a degree not compensated for in the low-velocity zones. Second, particles already captured may be blown off the plate surfaces in high-gas-velocity regions and be lost from the precipitator. The second loss predominates where gas flow is especially bad. Techniques available for controlling and correcting gas flow patterns include the use of guide vanes to change gas flow direction, flue transitions to couple flues of different sizes and shapes, and various types of diffusion screen and device to reduce turbulence.

The remainder of the fundamental problems can usually be corrected and even avoided by sound engineering design and judicious selection of precipitator components, as discussed in Section 3.

5.2. Mechanical Problems

Mechanical problems are associated with (1) poor alignment of electrodes and sectionalization design, (2) vibrating or swinging discharge wires, (3) bowed or distorted collecting plates, (4) excessive dust deposits on electrodes, (5) full or overflowing with collected dust in hoppers, (6) air leakage in hoppers, gas ducts, or shell, and (7) dust piles in connecting gas ducts. A sound maintenance program based on routine measurements and on-site observations is most effective and highly recommended.

5.3. Operational Problems

Operational difficulties are associated with (1) process changes, (2) poor electrical settings, (3) mismatched power supply to load, (4) failure to empty hoppers, (5) overloading precipitator equipment by excessive gas flow rate and/or particle loading in gas stream, and (6) upsets in operation of the furnace or process equipment to which the precipitator is connected. To overcome these difficulties, it is essential to have a set of simple but complete operation instructions or standard operation procedure (SOP) for the electrostatic precipitator.

5.4. Chemical Problems

A large number of physical components of the electrostatic precipitator are exposed to the potential attack of corrosive atmospheres, mainly acidic gases and moisture content in the gas stream. Critical zones of the precipitator that are most vulnerable to metal corrosion include the outer shell walls, the roof plate, the collecting plate surfaces, the high-voltage system, the hoppers, the access doors, the expansion joints, and the test ports. Good design and proper maintenance with an understanding of common corrosion processes and preventive measures as outlined in Hall and Katz (36) can provide viable equipment with long life in service.

6. EXPECTED FUTURE DEVELOPMENTS

Among the candidates for upgrading conventional ESPs are advanced digital voltage controls, flue gas conditioning, intermittent energization, temperature-controlled precharging, wide plate spacing, and positive energizition of corona electrodes for hot-side ESPs. For future developments, the following expected emphases in ESP development are forecast over the next few years:

1. Use of computer models for precipitator design and performance analysis
2. Derivation of a valid theory whereby the relative importance of the different factors is reflected directly in the precipitator equations
3. Reduction of the size and cost of a precipitator required for a specific duty
4. Use of wet precipitators in controlling fine particles
5. Design toward even higher efficiency, particularly in the collection of small particles, which is generally the main justification for using an electrostatic precipitator
6. Use of the electrostatic precipitation process for newer industrial processes such as coal gasification, gas turbine, and magneto-hydrodynamic (MHD) power generation

NOMENCLATURE

a	Cross-sectional area normal to the current flow (cm^2)
a''	Empirical constant
A	Total collecting electrode surface area (m^2)
A_a	Actual overall collecting plate surface area (m^2)
A	Particle radius (m)
A'	Constant for gas (V/m)
B'	Constant for gas ($V/m^{1/2}$)
b	Wire-to-plate spacing (m)
b''	Empirical constant
C	Cunningham correction factor (dimensionless)
C_{in}	Inlet particulate concentration (kg/m^3)
C_{out}	Outlet particulate concentration (kg/m^3)
C_{p}	Particulate concentration (kg/m^3)
c	Wire-to-plate spacing (m)
c''	Empirical constant
d	Electrode duct width (m)
d'	Dimension variable, Eq. (13)
d''	Empirical constant
D	Width of ducts
E	Electric field (V/m)
E_a	Activation energy
E_e	Corona-starting field (V/m)

E_p	Precipitating or collecting field (V/m)
E_r	Radial component of electrical field near particle (V/m)
F_0	Electrostatic force (N)
F_d	Viscous drag (N)
H	Height of plate
i	Current (A)
i_a	Average current density at plate (A/m^2)
i_i	Linear current density (A/m)
I_c	Corona current
J	Power requirement [kW/(m^3/s)]
k	Boltzmann's constant, 1.38×10^{-23} (N m/K)
k'	Adjustable constant
k''	Empirical constant
K	Power cost ($/kWh)
l	Path length in the direction of current flow (cm)
L	Electrode duct length (m)
L_{in}	Length of inlet section (m)
L_{out}	Length of outlet section (m)
L_p	Length of collecting plate (m)
L_s	Spacing between electrical section (m)
L_t	Overall length of precipitator, m plate (A/m^2)
m	Particle mass (kg)
m_i	Ion mobility (m^2/V s)
M	Maintenance cost, $/(cm^3/s)
N	Ion concentration in potential Field m^{-3}
N_0	Initial ion concentration (m^3)
N_d	Number of ducts (m^2/V)
N_p	Particle number density (m^3)
p	Particle resistivity (Ω-cm)
P	Pressure (atm)
P_0	Standard pressure (1 atm)
P_c	Corona power
P	Dimensionless, Eq. (18)
q	Particle charge (C)
q'	Ion charge 1.602×10^{-19}
q_d	Diffusion charging of particle (C)
q_f	Field charging of particle (C)
q_i	Ion charge (C)
q_{max}	Limiting particle charge (C)
Q	Gas flow rate (m^3/s)
r	Radius, radial distance from particle center, radial distance from cylinder axis (m)
r_0	Wire radius (m)
r_1	Cylinder or tube radius (m)
r_a	Aspect ratio

R	Particulate resistivity (Ω)
S	Particle surface area per unit gas volume (m^{-1})
t	Time (s)
T	Absolute temperature (K)
T_0	Standard temperature (298 K)
T'	Annual operating time (h)
U	Ion potential energy (N m)
v	Gas velocity (m/s)
v_1	Root mean square ion velocity (m/s)
$\bar{v}$	Average treatment velocity (ft/s)
V	Potential (V)
V_a	Corona-starting potential (V)
V_a'	Potential at wire surface (V)
V_{avg}	Average voltage (V)
w	Particle migration velocity (m/s)
$\bar{x}$	Mass median particle diameter (μm)
δ'	Gas density relative to 25°C (298 K) and 1 atm (dimensionless)
η	Efficiency fraction (dimensionless)
θ	Polar angle (rad)
μ	Gas viscosity (N s/m^2)
ρ	Particle resistivity (Ω-cm)
σ_e	Ion space-charge density (C/m^3)
σ_p	Particle space-charge density (C/m^3)
τ	Charging time constant (s)
ϕ	Dimensionless current, Eq. (7)

REFERENCES

1. G. R. Offen and R. F. Altman, *J. Air Pollut. Control Assoc.* **41**, 222 (1991).
2. H. J. White, *J. Air Pollut. Control Assoc.* **7**, 166–177 (1957).
3. F. Cameron, *Cottrell—Samaritan of Science*, Doubleday, Garden City, NY, 1952.
4. P. R. Bibbo, in *Electrostatic Precipitators* (L. Theodore and A.J. Buonicore, eds.), EST International Inc., Roanoke, VA, 1992, pp. 283–354.
5. US DHEW PHS, *Control Techniques for Particulate Air Pollutants*, NAPCA Publication No. AP-51, US Public Health Service, Washington, DC, 1969, pp. 81–102.
6. R. D. Ross (ed.), *Air Pollution and Industry*, Van Nostrand-Reinhold, Princeton, NJ, 1972.
7. C. D. Cooper and F. C. Alley, *Air Pollution Control —A Design Approach*, 2nd ed. Waveland Press, 1996, pp. 152–153 and 172.
8. C. F. Gottschlich, in *Air Pollution*, 2nd ed. (A. C. Stern, ed.), Academic, New York, 1968, Vol. **3**, pp. 437–455.
9. S. Oglesby *et al.*, *A Manual of Electrostatic Precipitator Technology, Part 1*, PB-196-380, APTD 0610, PHS CPA 22-69-73 Southern Research Institute, Birmingham, AL, 1970.
10. M. Robinson, in *Electrostatic Precipitation in Air Pollution Control, Part I* (W. Strauss, ed.), Wiley–Interscience, New York, 1971, pp. 256–281.
11. G. B. Nichols, *Proc. Symp. Control Fine-Particulate Emissions Ind. Sources*, 1974, pp. 142–144.
12. H. J. White, *Industrial Electrostatic Precipitation*, Addison-Wesley, Reading, MA, 1963.

13. J. A. Danielson (ed.), *Air Pollution Engineering Manual*, 2nd ed., US EPA Publication No. AP-40, US Environmental Protection Agency, Washington, DC, 1973, pp. 135–166.
14. P. Cooperman, *Trans. Am. Inst. Elec. Eng.* **79**, 47 (1960).
15. M. Robinson, in *Electrostatics and Its Applications* (A. D. Moore, ed.), Wiley–Interscience, New York, 1973.
16. W. M. Thornton, Phil. Mag. **28**, 666 (1939).
17. M. Pauthenier and M. Moreau-Hanot, *Electrician* **113**,187 (1934).
18. G. W. Penney, J. *Air Pollut. Control Assoc.* **25(2)**, 113–117 (1975).
19. H. J. White, *Proc. Symp. Control Fine-Particulate Emissions Ind. Sources*, 1974, pp. 58–76.
20. J. Katz, *J. Air Pollut. Control Assoc.* **15(11)**, 525–528 (1965).
21. J. T. Reese and J. Greco, *J. Air Pollut. Control Assoc.* **18(8)**, 523–528 (1968).
22. H. J. White, *J. Air Pollut. Control Assoc.* **24(4)**, 314–338 (1974).
23. H. J. White, in *Handbook of Air Pollution Technology* (S. Calvert and H. M. England, eds.), Wiley, New York, 1984.
24. L. Theodore and A.J. Buonicore (eds.), *Air Pollution Control Equipment: Selection, Design, Operation and Maintenance*, ETS, Inc., 1992, p. 298.
25. H. J. White, *J. Air Pollut. Control Assoc.* **27(1)**, 15; **27(2)**, 115; **27(3)**, 206; **27(4)**, 309 (1977).
26. G. G. Schneider, T. I. Horzella, J. Cooper, et al., Striegl, *J. Chem. Eng.* **82** (1975).
27. H. J. Hall, *J. Air Pollut. Control Assoc.* **25(2)**, 134 (1975).
28. US EPA, *Operation and Maintenance Manual for Electrostatic Precipitators*, EPA-625/1-85-017, US Environmental Protection Agency, Research Triangle Park, NC, 1985.
29. D. R. Selzler and W. D. Watson, *J. Air Pollut. Control Assoc.* **24(2)**, 115–121 (1974).
30. N. W. Frisch and D. W. Coy, *J. Air Pollut. Control Assoc.* 24(**9**), 872–875 (1974).
31. P. Cooperman, *Air Pollution Control Association* Preprint No. 69-4, APCA Annual Meeting, 1969.
32. M. Robinson, *Atmos. Environ.* **1(3)**, 193–204 (1967).
33. S. L. Soo, AIChE Symp. Ser. **68**, 185–193 (1972).
34. S. L. Soo, *Int. J. Multiphase Flow* **1**, 89–101 (1973).
35. E. C. Potter, *J. Air Pollut. Control Assoc.* **28(1)**, 40–16 (1978).
36. H. J. Hall and J. Katz, *J. Air Pollut. Control Assoc.* **26(4)**, 312 (1976).
37. J. R. Benson and M. Corn, *J. Air Pollut. Control Assoc.* **24(4)**, 340–348 (1974).
38. J. R. O'Connor and J. F. Citarella, *J. Air Pollut. Control Assoc.* **20(5)**, 283–286 (1970).
39. *Survey of Current Business*, US Dept. of Commerce, Bureau of Economic Analysis, Washington, DC (monthly publication).
40. 2002 Buyer's guide: electrostatic precipitators. *Environ. Protect.*, **13(3)**, 114 (2002).
41. US EPA. Air Pollution Control Technology Series Training Tool: Electrostatic Precipitators. www.epa.gov. US Environmental Protection Agency, Washington, DC, 2004.
42. US EPA. Electrostatic Precipitator. APTI Virtual Classroom, Lessons 1–10. http://search.epa.gov. US Environmental Protection Agency, Washington, DC, 2004.

5
Wet and Dry Scrubbing

Lawrence K. Wang, Jerry R. Taricska, Yung-Tse Hung, James E. Eldridge, and Kathleen Hung Li

CONTENTS

1. INTRODUCTION

1.1. General Process Descriptions

The scrubbing process is a unit operation in which one or more components of a gas stream are selectively absorbed into an absorbent. The term "scrubbing" is used interchangeably with "absorption" when describing this process. In wet scrubbing, water is the most common choice of absorbent liquor. In special cases, another relatively nonvolatile liquid may be used as the absorbent. In dry scrubbing, a dry powder or semidry slurry are also possible absorbents, depending on the requirements of a given situation.

Scrubbing is commonly encountered when treating flue gas (or some other polluted gas stream) to control acid gases, particulates, heavy metals, trace organics, and odors. Often, a scrubbing system is composed of two or more scrubbers in series. This is done so that an individual scrubber stage can utilize an absorbent specific to a targeted pollutant or pollutants. Higher total removal efficiencies are often possible in a multistage scrubber system than would otherwise be possible with a single-stage scrubber. An example of this is commonly found in the rendering industry, where both ammonia and hydrogen sulfide are produced during normal operations. A scrubber using an acid-based absorbent liquid is used to remove ammonia from the air. The hydrogen sulfide is then scrubbed using a caustic solution, sometimes with an oxidizing agent added to the liquid. In this example, physical or chemical absorption (or both) could occur in the scrubbing process. If a pollutant is simply trapped (e.g., particulates impinging on water) or dissolved, then it is a physical absorption process. If the pollutant being absorbed also undergoes a

From: *Handbook of Environmental Engineering, Volume 1: Air Pollution Control Engineering*
Edited by: L. K. Wang, N. C. Pereira, and Y.-T. Hung © The Humana Press, Inc., Totowa, NJ

chemical reaction (e.g., HCl being absorbed into and then reacting with lime-based slurry of $CaCl_2$), then the process is a chemical absorption process. In dry scrubbing, an alkaline reagent is injected into the gas stream while preventing the gas from being saturated with water vapor.

Although the most common name for such a unit operation installation is a scrubber or absorber, other names commonly used to reference such installations in industry are spray towers and packed or plate columns. It should be noted that the latter three unit operations may operate slightly differently from the wet and dry scrubbers (absorbers) defined here. These terms are mentioned here, because they are sometimes used interchangeably with mass transfer unit operations.

1.2. Wet Scrubbing or Wet Absorption

The physical criteria in designing a wet scrubber are simple:

1. Use a liquid for absorption that offers a high solubility of the pollutant in the gas stream being treated
2. Maximize gas–liquid contact surfaces

When both conditions are met, the pollutant will readily diffuse out of the gas phase and be absorbed into the liquid phase.

Theoretically, absorption of a pollutant in a gas phase into a contacting liquid phase occurs when the liquid contains less than the equilibrium concentration of the pollutant. In other words, the pollutant in the gas phase must have some solubility in the liquid phase. For absorption into the liquid phase to occur, the maximum concentration of the same pollutant in the liquid phase must be avoided initially. This is because the concentration difference across the phase boundary is the driving force for absorption to occur between the two phases. Additionally, absorption (mass transfer) from gas into liquid (or vice versa) is dependent on the physical properties of the gas–liquid matrix (e.g., diffusivity, viscosity, density) as well as the conditions of the scrubber system (e.g., temperature, pressure, gas and liquid mass flow rates). Absorption of a pollutant is enhanced by lower temperatures, greater liquid–gas contact surfaces, higher liquid–gas ratios, and higher concentration of the pollutant in the gas phase (or, alternately, lower concentration of the pollutant in the liquid phase). In some instances, elevated pressures are used to give added driving force of the pollutant into the liquid stream as well (1–21).

Wet scrubbers are often the technology of choice if high removal efficiencies of acid gases are required. An HCl removal efficiency* greater than 99% is easy to obtain in wet scrubbers. SO_2 is a more difficult pollutant to wet scrub; traditionally, wet scrubber designs call for 90–95% removal efficiency for SO_2. Scrubber designs have been challenged by new regulations regarding SO_2 removal efficiency. In 1998, the US EPA (Environmental Protection Agency) instituted new air pollution regulations known as NESHAP (National Emission Standards for Hazardous Air Pollutants) that call for the retrofitting of existing SO_2 wet scrubbers to achieve 98% removal efficiency and for all new wet scrubbers used to control SO_2 emissions to achieve 99% removal efficiency.

*This efficiency is possible only for HCl that is not in aerosol form. Possible formation of aerosols when scrubbing HCl must always be accounted for, as such aerosols will not be treated in a wet scrubber. The presence of HCl in aerosol form will form a distinctive white plume when exiting the stack of a wet scrubber.

Carbon dioxide gas can be effectively controlled with a wet scrubber. Unfortunately, initial capital costs as well as subsequent operating and maintenance costs of such a wet scrubber limit the use of such scrubbers.

The wet absorption of particulate matter (PM) from a gas stream involves the use of specially designed particulate scrubbers. A Venturi scrubber captures PM by impingement and agglomeration of the PM with liquid droplets.

1.3. Dry Scrubbing or Dry Absorption

Two principle methods of dry absorption systems are currently being used in industry: dry–dry absorption and semidry absorption. A dry–dry system injects a powdered alkali absorption agent into the polluted gas stream. The semidry method injects concentrated slurry into the polluted gas stream and then removes the liquid by evaporation, leaving the active, dry alkali absorption agent. Both methods remove any alkali agent not consumed or other solid wastes with an electrostatic precipitator or a fabric filter. All dry scrubbers contain a chemical injection zone followed by a reaction zone where the pollutant in the gas being treated reacts with the dry alkali. The process is completed with the removal of residual PM by an electrostatic precipitator or a fabric filter.

2. WET SCRUBBERS

2.1. Wet Absorbents or Solvents

2.1.1. Absorption of Gaseous Pollutants

As previously discussed, absorption is either physical or chemical. Physical absorption occurs when the pollutant compound dissolves into the solvent (absorbent). If there is a subsequent reaction between the pollutant and the solvent or chemicals present in the solvent, then the absorption is said to be chemical. Commonly used liquid absorbents are most often water or water-based solutions. Less commonly encountered, but nevertheless significant, are wet scrubbing systems using mineral oil or nonvolatile hydrocarbon oils as the absorbing liquid.

It is important to note here that when a pollutant is physically absorbed in a wet scrubber, no destruction of the pollutant species has occurred. The pollutant has simply moved from the gas phase into the liquid phase. In a subsequent chemical reaction, the pollutant may be neutralized or otherwise altered but still not destroyed. As a result, a wet scrubber often produces a liquid stream that must be treated to achieve final destruction of the given pollutant. Such secondary treatment requires an additional cost that needs to be included when considering the economics of a wet scrubber project.

The pH of the scrubbing liquor is often an important process parameter. Low-pH liquor is required for ammonia scrubbing, neutral or high pH is needed for acid gas scrubbing. When scrubbing trace organics, liquor with alkaline pH is often used as the absorbent. Common alkali liquors used in scrubbing acid gases and CO_2 are lime and caustic solutions.

Sodium-based salts are always preferable to calcium or other group II (periodic table) metal salts for adjusting liquor chemistry. This is because almost all sodium compounds are soluble, whereas deposits from hard water (Ca^{2+} and Mg^{2+} salts) are often observed to foul wet scrubber internal components. Lime, $Ca(OH)_2$, is less expensive than sodium reagents, but the latter normally offer higher removal efficiency of the pollutants in the gas stream. If sodium-based solids do form in a wet scrubber, they present a greater disposal problem than calcium-based solids. Therefore, the possible formation of solids in a wet scrubber must be considered, as well as possible use or disposal of such byproducts when considering a wet scrubber.

2.1.2. *Absorption of Particulate Matter*

A typical wet scrubber is a vertical tower in which liquid, normally water, enters from the top. The polluted gas being treated enters from the bottom of the scrubber so that the water flows down and the gas flows up. This is the classic "countercurrent" flow scheme. If the pollutant being removed from the gas stream is PM, no special chemical reagents are used. Simple water suffices as the absorbent liquor. If a gaseous pollutant is removed simultaneously with PM removal, the need for a chemical reagent will depend on the particular gaseous pollutant being controlled.

2.2. *Wet Scrubbing Systems*

Several methods are available for wet scrubbing. Figure 1 illustrates four common methods of wet scrubbing. A discussion of these and other methods of wet scrubbing follows. The method of wet scrubbing chosen to treat a given gaseous pollutant is always specific to the given pollutant or pollutants present in the gas stream being treated. A "standard" wet scrubber does not exist.

2.2.1. *High-Efficiency Venturi Scrubber*

As seen in Fig. 1c, a Venturi scrubber is often a primary control solution. This scrubber operates at low pH and will remove PM and HCl. Removal efficiency of such a Venturi scrubber should be 80–95% for particles greater than 2 μm (15).

The Venturi principle states that as gas enters a narrow constriction (the Venturi), the velocity of the gas increases. At this point of constriction, the absorbent liquor (scrubbant) is introduced. The high-velocity gas forces the liquor to atomize into small droplets, which offer a large total surface area of liquor into which the PM absorbs. After passing through the Venturi, the gas returns to near original velocity. At this lower velocity, the scrubbing liquor agglomerates back into the bulk liquid phase, containing the PM.

Suppliers of Venturi scrubbers commonly provide prefabricated units capable of treating gaseous streams of up to 80,000 cubic feet per minute (cfm). These units normally operate at a high pressure drop that increases the power costs of the unit. Such prefabricated units nevertheless have a considerable initial capital cost advantage when compared to electrostatic precipitators (ESPs) and fabric filters.

2.2.2. *Jet Venturi Scrubber Systems*

A slightly different type of Venturi scrubber is the Jet Venturi scrubber. In this type of scrubber, energy from a flow of pressurized liquid forces a draft to form. This draft captures PM with an efficiency of greater than 90%. Normal process installations use a quencher ahead of a Jet Venturi scrubber. However, the Jet Venturi also sometimes accepts a gas stream directly from a combustion chamber.

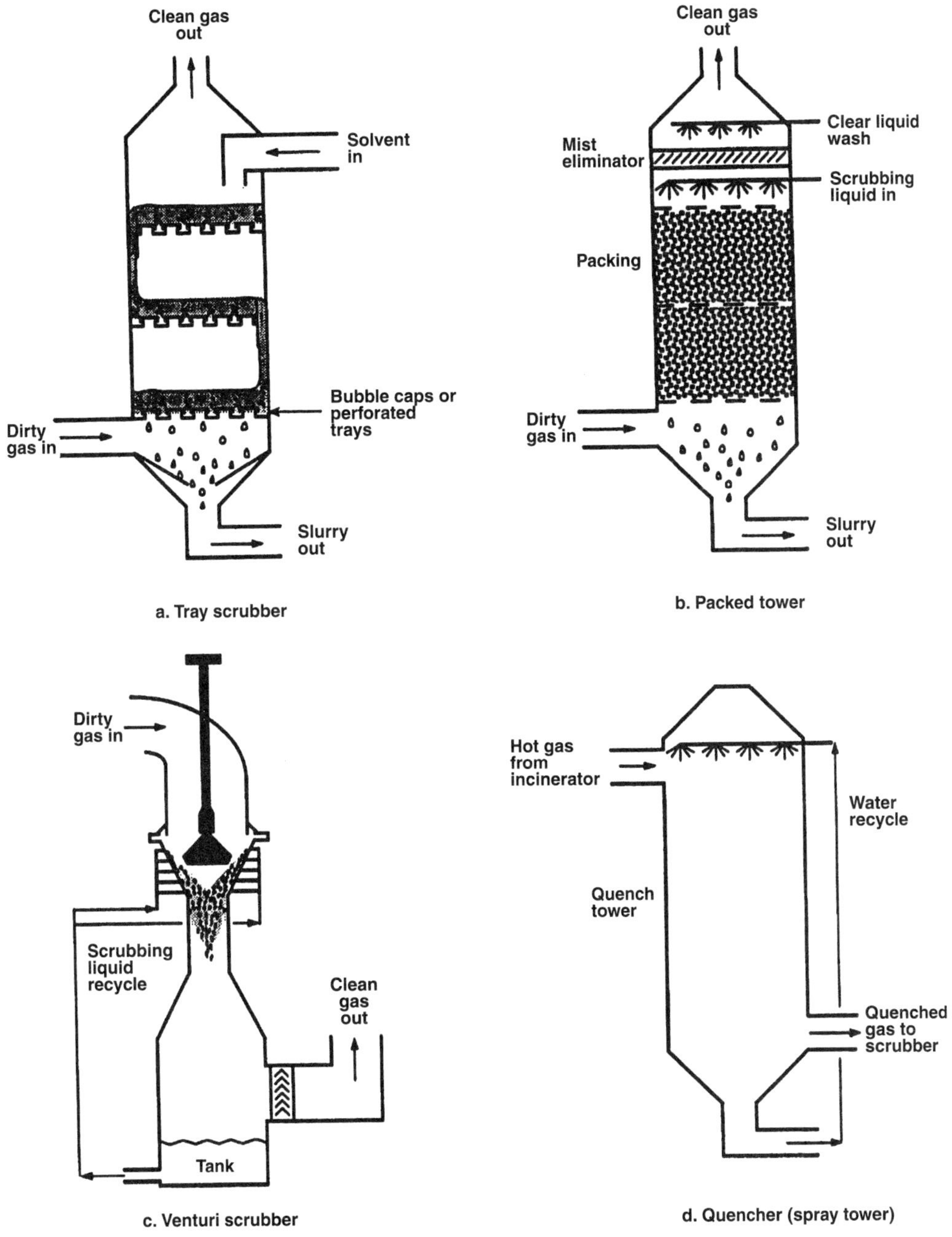

Fig. 1. Four common types of wet scrubbing systems. (From ref. 15.)

2.2.3. *Packed Towers (Scrubbers)*

A packed tower (scrubber), as seen in Fig. 1b, is commonly used to absorb pollutant(s) present in a gas stream. As discussed previously, the classic scrubber tower takes advantage of the countercurrent flow of gas and liquid; the gas passes up through the tower as the liquid passes downward. The actual mass transfer from gas to liquid occurs in the packed bed of the tower. Packing may either be random dumped or structured, depending on the given situation. Regardless of which type of packing is used, the purpose is to promote gas–liquid contact so that the pollutant(s) being removed from the gas stream is absorbed into the liquid stream. The packed bed is held in place by a packing support grid at the base of the packed bed. A bed limiter may be needed to hold down the top of the packing.

In extremely large towers, intermediate packing support may also be required. Excellent liquid distribution of the scrubbing liquor onto the packed bed is always required in a scrubbing tower. Several types of distributor are available, as are full cone spray nozzles, to ensure adequate wetting of the packing. Above the liquid inlet in the tower is the mist eliminator. As with liquid distribution, various methods of forcing droplets to coalesce from the gas stream exist. The most common types of mist eliminator are mesh pads and chevron blades. Above the mist eliminator, the scrubber tower narrows to the exit stack, where treated gas is released into the atmosphere. This type of scrubber is used when extremely high removal efficiency of a pollutant(s) from a gas stream is required, typically 99%.

2.2.4. *Spray Tower Scrubbers*

This is another option used to treat polluted gas streams. Three configurations of spray towers are employed in industry:

1. Polluted gas flows upward as liquid spray flows downward (countercurrent flow pattern).
2. Polluted gas flow and liquid spray flow are both downward (cocurrent flow pattern).
3. Polluted gas flows laterally as liquid spray flows downward (perpendicular or cross-flow pattern).

In flow pattern 1, if packing is added, the spray tower becomes a packed tower. Packing is also sometimes used in both configurations 2 and 3 to enhance gas–liquid contact. In these two situations, a hybrid of packed and spray towers is used.

When designing a strictly spray tower (no packing), the critical design parameters to consider are tower height and diameter, gas and liquid flows, as well as liquid-to-gas ratio, gas velocity, droplet size and liquid chemical composition. Spray towers often use a higher liquid-to-gas ratio than for packed towers. The higher ratio is needed to achieve high removal efficiency.

2.2.5. *Tray Towers (Scrubbers)*

This application uses a tower with numerous trays within (*see* Fig. 1a). As the scrubbing liquor passes the tower, a certain amount of liquid is collected (or held) on each tray. The trays have openings to give a specified open area, per the tower design, to allow for gas to pass upward through the tower. As the gas passes and the liquid flows downward, high gas–liquid contact occurs in the countercurrent flow scheme. Common types of tray are bubble cap, perforated, and valve types. The number of trays within the tower is a function of the needed removal efficiency of the tower.

2.2.6. *Quenchers*

As previously mentioned, quenchers are sometimes needed as a first step in conjunction with another gas treatment step. Similar to a spray tower, the quencher is used for temperature and humidification control of the polluted gas being treated. Often used just after an incineration process, the quencher cools the exhaust gas formed in the process to saturation, or near saturation, temperature. In so doing, the volume of the gas to be treated in the next step is greatly reduced. The approach to saturation temperature is a function of liquid rate, droplet size and gas residence time. Also, as the humidity of the gas is increased in the quencher, improved absorption (higher efficiency) is supported in the next step of the treatment process.

A quencher is also a scrubber, to a limited degree. With alkali liquor, the quencher can approach 50% removal efficiency of acid gases. This reduces the size of the high-efficiency packed tower or Venturi unit required after the quencher.

2.3. *Wet Scrubber Applications*

2.3.1. *General Downstream and Upstream Applications*

As previously described, a scrubber system moves a pollutant(s) from the gas phase into the liquid phase. Therefore, after scrubbing, a liquid separator is often required. Typical liquid separators are mist eliminators, cyclones (or sometimes called hydrocyclones to specify liquid vs air cyclones), and swirl vanes. All of these separators use impaction or centrifugal force to remove liquid droplets (coalesce) in the process stream. Mist eliminators, as previously discussed, are either mesh-pad or chevron-blade type.

A typical application example of a wet scrubber is the treatment of an acid gas stream with a quencher (cool, condense, some removal), then a Venturi for PM removal, followed by a scrubber (packed or spray tower type) to complete the removal of acid gas from the polluted airstream. Wet scrubbers also often follow an ESP or fabric-filter unit operation. This scenario is common when high PM removal is required and such removal cannot be accomplished with a single Venturi step. An example of this is a polluted airstream containing acid gas, heavy metals, and, possibly, organic residues. Several air pollution control process steps, each targeted for one of the above pollutants, are needed to fully cleanse the air.

2.3.2. *Incineration Pollution Control*

Incineration or combustion processes produce pollutants that must be removed from an airstream prior to atmospheric release. Possible pollutants formed are acid gases, carbon dioxide, carbon monoxide, nitric oxides (NO_x), heavy metals, and particulates.

If fine PMs (<10 μm) are not a concern and/or if total PM removal required is not needed, a wet scrubber will probably be used to treat the polluted air. All types of wet scrubber described here are a possible solution for air pollution control. Often, the ability of a wet scrubber to remove all of the pollutants mentioned earlier makes this option the easiest to use for a given air pollution control problem.

2.3.3. *Thermal Desorption*

A wet scrubber is sometimes useful for thermal desorptional, though its PM removal capacity is less than that of a fabric filter (or baghouse) or ESP. As flue gas from an incineration process is cooled to near-saturation temperature in a wet scrubber, the dispersion properties of the released flue gas may cause a plume to form. If this occurs, the gas will need reheating to eliminate the plume.

A quench chamber followed by a Venturi scrubber is often used to control PMs. This scenario is also possible when controlling acid gases and/or halogenated organic compounds.

2.3.4. General Remediation Applications

Wet scrubbers are simple to operate compared to other air pollution control options, making it popular among air pollution engineers. If used to control volatile organic compounds (VOCs), scrubbing liquor other than water may be required because of solubility concerns. Such solvents are often proprietary and are always more expensive to use than water. If VOC concentrations being treated are low, another control step is often needed to reach the desired removal efficiency, which entails added costs.

With the exception of mercury (Hg), volatile metals will condense at the normal operating temperature of the typical wet scrubber. Therefore, high-efficiency removal of heavy metals is possible in a wet scrubber. Unfortunately, the high vapor pressure of mercury prevents ready condensation of mercury in a wet scrubber. As a result, the removal efficiency of mercury vapors in wet scrubbers is not established in the literature.

If volumetric flows being treated are low, wet scrubbers do not have high removal efficiencies. Imparted turbulence in the scrubbing liquor will improve the removal efficiency achieved. Common scrubbing liquors are water, water solutions, and nonvolatile organic liquids. Two-stage scrubbing systems, first with water and then with an alkaline solution, are common as acid gas removal efficiency is improved at pH >7.

If PM removal is required, the actual particle size distribution and the required removal efficiency will determine what type of wet scrubber is used for control purposes. The various types of wet scrubbers dealt with in this discussion commonly achieve a removal efficiency of 99.5%. To further improve upon this, as well as to lower the costs of control operations, wet scrubbers are being developed in tandem with other technologies, such as ionization.

2.4. Packed Tower (Wet Scrubber) Design

2.4.1. General Design Considerations

The efficiency of an absorption process used to remove a pollutant or pollutants from an air flow will depend, in part on the following:

1. The solubility of the pollutant(s) in the chosen scrubbing liquor
2. Pollutant(s) concentration in the airstream being treated
3. Temperature and pressure of the system
4. Flow rates of gas and liquid (liquid/air ratio)
5. Gas–liquid contact surface area
6. Stripping efficiency of the liquor and recycling of the solvent

Of the above parameters, the ability to increase gas–liquid contact will always result in a higher absorption efficiency in a wet scrubber. If the temperature can be reduced and the liquid-to-air ratio increased, then the absorption efficiency will also be improved in the scrubber.

The actual design of the tower (diameter, height, depth of packed bed, etc.) will also depend on the given vapor–liquid equilibrium for the specific pollutant/scrubbing liquor. Additionally, the type of tower (packed vs tray, etc.) used will affect this equilibrium.

Such data are often not available for all pollutants encountered in industry today. If data are available, empirical data will always be superior to theoretical data for design purposes. If such empirical data are unavailable, a similar type of pollutant having available data, with an added safety factor built into the design, should be used to model the system.

2.4.2. *Packed Tower (Wet Scrubber) Design Variables*

As an in-depth analysis of design methods for all types of absorption tower is beyond the scope of this discussion, a design for a typical, common wet scrubber is given here. The example is a packed tower wet scrubber, as shown in Fig. 1b. This type of tower is commonly found in air pollution control installations. The configuration used is somewhat simplified. The tower is packed with 2 in. ceramic Raschig Rings (note: 1 in. = 2.54 cm) and the scrubbing liquor (absorbent) used is water. The water is sprayed from top and the slurry is collected at the bottom. The scrubbing liquor spray system is described as a once-through process with no recirculation. It should be noted that in a field installation, this once-through method has the consequence of sending a large flow of water to a treatment facility. This example is applicable for either organic or inorganic air pollutant control (1–3,14–17).

In any absorption process, possible removal efficiency is controlled by the concentration gradient of the pollutant being treated between the gas and the liquid phases. As previously defined, this concentration gradient is the driving force to mass transfer between the phases. Therefore, the solubility of the given pollutant in the gas and liquid phases will determine the equilibrium concentration of the pollutant in the given example.

If a pollutant is readily soluble in the scrubbing liquor, the slope m of the equilibrium curve is low. There is an inverse relationship between m and driving force; the smaller the slope, the more readily the pollutant will dissolve into the scrubbing liquor. This represents a high-driving-force system. The size of the tower in such a system will be minimal, as mass transfer (absorption) between the phases occurs readily. If the slope is relatively large, approx 50 or more, this represents limited solubility of pollutant in the scrubbing liquor. For absorption to occur with limited driving force, the contact time between the phases must be extended, so the needed tower size will increase. A high liquid-to-gas ratio requirement is also indicated by the limited solubility of the pollutant in the scrubbing liquor if high removal efficiency of the pollutant is desired. As a practical rule of thumb, if $m > 50$, a removal efficiency of the pollutant of 99% will most likely not be practical.

In normal circumstances for an air pollution wet scrubber design control project, the inlet concentration of pollutant, gas flow rate, temperature, and pressure are fixed. The removal efficiency (outlet concentration of pollutant) is also normally specified and the available scrubbing liquor is known. The challenge of the design is to determine the scrubber tower diameter, the depth of packed section, and the needed scrubbing liquor flow rate to accomplish the specified outlet concentration of pollutant. The total height of the tower will then be determined based on these results. A further consideration is that the total head loss through the tower will directly impact the cost of operating the scrubber system.

In this example, there are no heat effects caused by absorption in the tower and both the airstream and liquid stream are dilute solutions. Flow rates are constant and the equilibrium curve is linear. The needed data for this design are available in the literature (1–5,11–17).

A material balance determines the scrubbing liquor required flow, based on the liquid-to-gas ratio determined from the equilibrium curve. The absorption factor is widely accepted to range from 1.25 to 2.0 for best economics in a scrubber design project. The absorption factor determines the liquid–gas molar flow rates (10–17). For this example, an absorption factor of 1.6 is used.

$$L_{\text{mol}} = (\text{AF})(m)(G_{\text{mol}}) \quad (1)$$

where AF is the absorption factor (explained earlier), L_{mol} is the liquid (absorbent) flow rate (lb-mol/h), G_{mol} is the gas flow rate (lb-mol/h), and, m is the slope of the equilibrium curve.

Note that the value of m is temperature dependent for the given system (1,4,5). Other systems are defined elsewhere (1,3,6,14).

At the assumed value of AF, Eq. (1) yields

$$L_{\text{mol}} = (1.6)(m)(G_{\text{mol}}) \quad (2)$$

Defining the gaseous stream flow rate in scfm to be Q_e, it follows that

$$G_{\text{mol}} = 0.155\ Q_e \quad (3)$$

where Q_e is the emission stream flow rate (scfm).

Now, L_{mol} can be converted to gpm:

$$L_{\text{gal}} = [L_{\text{mol}} \times \text{MW}_{\text{solvent}} \times (1/D_L) \times 7.48] / 60 \quad (4)$$

where $\text{MW}_{\text{slovent}}$ is the molecular weight of the scrubbing liquor (solvent), L_{gal} is the liquid (solvent) flow (gpm), and D_L is the density of the liquid (solvent) (lb/ft^3).

The factor 7.48 is used to convert cubic feet to gallons. When water is used as the solvent, then D_L is equal to 62.43 lb/ft^3 and the $\text{MW}_{\text{solvent}}$ is equal to 18 lb/lb-mol. Then, Eq. (4) yields

$$L_{\text{gal}} = 0.036\ L_{\text{mol}} \quad (5)$$

2.4.3. Packed Tower (Wet Scrubber) Sizing

Once the gas and liquid streams entering and leaving the packed tower are identified along with pollutant and solvent concentrations, the flow rates are calculated and operational conditions determined. These data combined with the type of packing used will determine the actual size of the tower. The tower size must be sufficient to accept the gas and liquid flows without excessive head loss.

The determination of the tower (*see* Fig. 1b) diameter has traditionally been based on an approach to flooding. Normal operating range to achieve maximum efficiency has been to use 60–75% of the flooding rate for tower sizing purposes. (Note: With flooding, the upward flow of gas through the tower impedes the downward flow of liquid. The actual point of flooding is somewhat arbitrary in definition.) A common correlation to determine the tower diameter is given in Fig. 2.

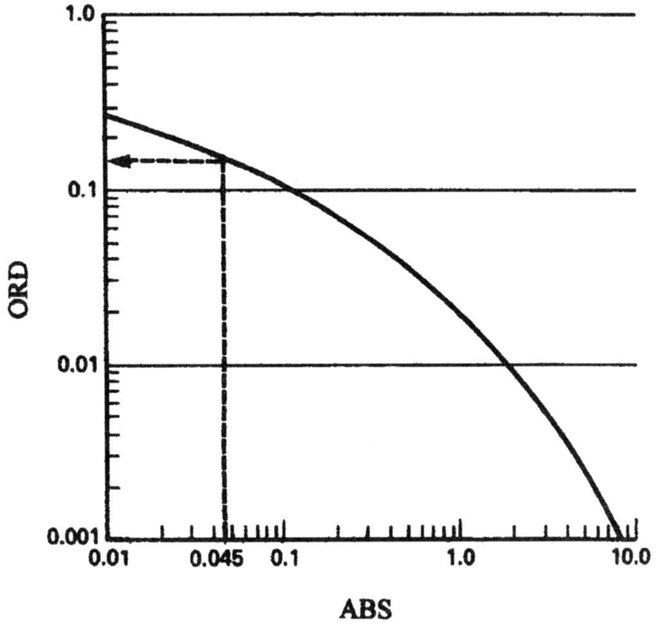

Fig. 2. Flooding correction in randomly packed towers.

The actual determination of the tower diameter is as follows:

Calculate the abscissa (ABS):

$$ABS = (L / G)(D_G / D_L)^{0.5} \tag{6}$$

where L is the solvent flow rate (lb/hr) = ($MW_{solvent}$)(L_{mol}), G is the gas flow rate (lb/hr) = (MW_e)(G_{mol}), D_L is the density of the liquid (lb/ft^3), and D_G is the density of the gas or emission stream (lb/ft^3).

D_G can be approximated as

$$D_G = PM/RT \tag{6a}$$

where P is the pressure (atm) (note, normally 1), M is the molecular weight of gas (lb/lb-mol), R is the gas constant (0.7302 ft^3 atm/lb-mol °R), and T is the temperature (°R).

The values for L and G are determined by multiplying L_{mol} and G_{mol} by their respective molecular weights. Next, the flooding line in Fig. 2 is used to read the ordinate (ORD). Now, the ordinate expression for $G_{area,f}$ at flooding is

$$ORD = [(G_{area,f})^2 (a/e^3) (\mu_L^{0.2})] / [(D_G D_L g_c)] \tag{7}$$

Solving for $G_{area,f}$, we have

$$G_{area,f} = \{(ORD\ D_G D_L g_c) / [(a/e^3)(\mu_L^{0.2})]\}^{0.5} \tag{8}$$

where $G_{area,f}$ is the gas stream flow rate based on tower cross-sectional area at the flooding point (lb/ ft^2-s), μ_L is the viscosity of the scrubbing (cP) (1 when using water), and g_c is the gravitational constant (32.2 ft/s^2). Note that a and e are packing factors (11).

Here, the assumption is made that f is the fraction of flooding appropriate for the given design. Using an f value of 0.6 in this example, the gas stream flow rate, G_{area}, for the cross sectional area determined above is

$$G_{area} = f\,G_{area,\,f} \tag{9}$$

As also previously mentioned, the normal operating range for fraction of flooding, f, is 0.60–0.75. Therefore, the column (tower) cross-sectional area is

$$A_{column} = G/(3{,}600G_{area}) \tag{10}$$

The diameter of the column (packed tower shown in Fig. 1b), D_{column}, may now be determined:

$$D_{column} = [(4/\pi)(A_{column})]^{0.5} = 1.13\,(A_{column})^{0.5} \tag{11}$$

where D_{column} is the column (tower) diameter (ft).

Now that the tower diameter is known, the height of the packed section, sufficient for the needed removal efficiency, is determined. This packed height is determined from the number of theoretical transfer units (NTU), which is multiplied by the height of transfer unit (HTU).

The HTU is dependent on the solubility of the pollutant being treated in the scrubbing liquor. Larger HTU values reflect more resistance to mass transfer by the pollutant into the scrubbing liquor. HTU is given in feet and is expressed as N_{og} or N_{ol}, depending on the limiting resistance to mass transfer in the system. In this example, where a pollutant is being scrubbed from a gaseous stream, the gas film resistance (as opposed to the liquid film) most likely controls mass transfer. So in this example, N_{og} is used.

The height of the column (packed tower) in Fig. 1b is determined by

$$\mathrm{Ht}_{column} = N_{og} \times H_{og} \tag{12}$$

where Ht_{column} is the column (packed tower) height(ft), N_{og} is the number of gas transfer units (based on overall gas film coefficients) (dimensionless), and H_{og} is the height of an overall gas transfer unit based on overall gas film coefficients (ft). The actual determination of N_{og} is beyond the scope of this text. Because the solutions here are dilute, the N_{og} is determined by

$$N_{og} = \ln\{[(\mathrm{HAP}_e / \mathrm{HAP}_o)(1 - (1/\mathrm{AF})) + (1/\mathrm{AF})]\} / (1 - 1/\mathrm{AF}) \tag{13}$$

where: HAP_e is the HAP (hazardous air pollutant) emission stream concentration (ppmv) and HAP_o is the HAP outlet concentration (ppmv).

This is a once-through system, so pure water (pollutant free) is used to scrub in this system. This makes the above expression possible.

Alternatively, Fig. 3 can be used to graphically determine N_{og}. Equation (13) determines the efficiency that will be realized in the scrubber tower. The inlet and outlet concentration of pollutant is related to the number of transfer units, N_{og}, through the absorption factor, AF, as shown in Equation (1).

The removal efficiency (RE) is determined from inlet and desired outlet concentration of the pollutant:

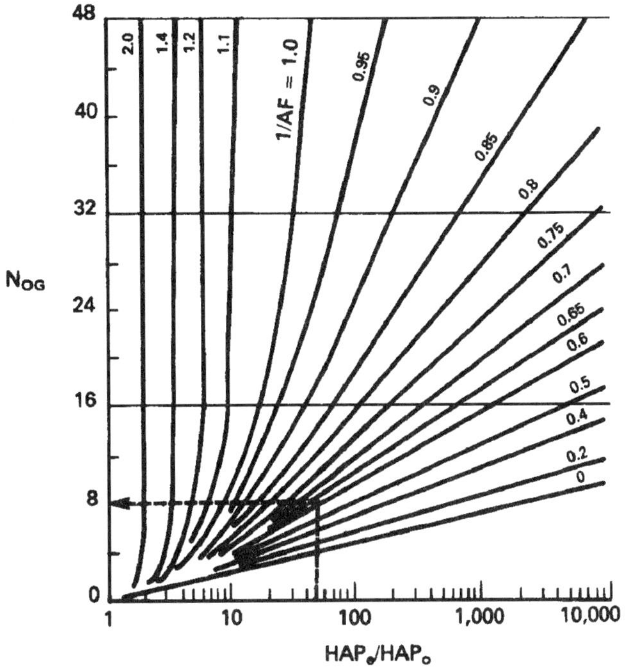

Fig. 3. Relationship among Nog, AF, and efficiency.

$$HAP_o = HAP_e \, [1 - (RE/100)] \tag{14}$$

The outlet concentration of the pollutant may now be substituted into Eq.(13) to obtain the depth of packing needed for the specified removal efficiency. A general statement is that a larger value of N_{og} yields a higher removal efficiency until the driving force (concentration gradient) is exhausted. At this point, no further transfer of pollutant between the two phases occurs.

Once the number of transfer units, N_{og}, required to meet the removal efficiency requirement is known, the height of each transfer unit, H_{og}, may be determined:

$$H_{og} = H_G + (1/AF)H_L \tag{15}$$

where H_G is the height of the gas transfer unit (ft) and H_L is the height of liquid transfer unit (ft).

Based on the packing chosen along with gas and liquid flow rates, generalized correlations to determine H_G and H_L are available:

$$H_G = [b(3{,}600\, G_{area})^c \,/\, (L'')^d]\, (Sc_G)^{0.5} \tag{16}$$

$$H_L = Y\, (L'' \,/\, \mu_L'')^s\, (Sc_L)^{0.5} \tag{17}$$

where b, c, d, Y, and s are empirical packing constants (11) from Tables 1 and 2, L'' is the liquid flow rate (lb/h-ft^2), μ_L'' is the liquid viscosity (lb/ft-hr), Sc_G is the Schmidt number for the gas stream (*see* Table 3), and Sc_L is the Schmidt number of the liquid stream (*see* Table 4).

Table 1
Constants for Use in Determining Height of a Gas Film Transfer Unit

Packing	*b*	*c*	*d*	Range of 3,600 G_{area} (lb/h-ft²)	Range of L'' (lb/h-ft²)
Raschig rings					
0.375 in.	2.32	0.45	0.47	200–500	500–1,500
1 in.	7.00	0.39	0.58	200–800	400–500
	6.41	0.32	0.51	200–600	500–4,500
1.5 in.	17.30	0.38	0.66	200–700	500–1,500
	2.58	0.38	0.40	200–700	1,500–4,500
2 in.	3.82	0.41	0.45	200–800	500–4,500
Bert saddles					
0.5 in.	32.40	0.30	0.74	200–700	500–1,500
	0.81	0.30	0.24	200–700	1,500–4,500
1 in.	1.97	0.36	0.40	200–800	400–4,500
1.5 in.	5.05	0.32	0.45	200–1,000	400–4,500
3-in Partition rings	650	0.58	1.06	150–900	3,000–10,000
Spiral rings (stacked staggered)					
3-in.Single spiral	2.38	0.35	0.29	130–700	3,000–10,000
3-in. Triple spiral	15.60	0.38	0.60	200–1,000	500–3,000
Drip-point grids					
No. 6146	3.91	0.37	0.39	130–1,000	3,000–6,500
No. 6295	4.56	0.17	0.27	100–1,000	2,000–11,500

Source: ref. 11.

Values of Sc_G and Sc_L for several pollutants are given in the literature (3,4). In this example, the effect of temperature on Sc is ignored. L'' is determined as the result of

$$L'' = L/A_{column} \tag{18}$$

Now, the total tower height, Ht_{total} using Ht_{column} determined in Eq. (12) is determined:

$$Ht_{total} = Ht_{column} + 2 + (0.25\,D_{column}) \tag{19}$$

The actual cost of packing is based on the volume of packing, $V_{packing}$ (ft³), needed to fill the tower:

$$V_{packing} = (\pi/4)\,(D_{column})^2\,(Ht_{column}) \tag{20}$$

$$V_{packing} = 0.785\,(D_{column})^2\,(Ht_{column}) \tag{20a}$$

The packing cost equals volume packing, $V_{packing}$, times the cost of packing ($/ft³ of packing). Note that now that the tower has been sized, if the tower design calls for a fractional foot diameter (i.e., 4.15 ft), the calculations must be repeated until an approximate 0.5 ft tower diameter is reached (i.e., 4.5 ft) or more preferably a whole foot diameter is reached (i.e., 4.0 ft). This is so because tower suppliers will quote a project based on their standard size of manufacture.

Table 2
Constants for Use in Determining Height of a Liquid Film Transfer Unit

Packing	*Y*	*S*	Range of L″ (lb/h-ft²)
Raschig rings			
0.375 in.	0.00182	0.46	400–15,000
0.5 in.	0.00357	0.35	400–15,000
1 in.	0.0100	0.22	400–15,000
1.5 in.	0.0111	0.22	400–15,000
2 in.	0.0125	0.22	400–15,000
Berl saddles			
0.5 in.	0.00666	0.28	400–15,000
1 in.	0.00588	0.28	4OO-15,000
1.5 in.	0.00625	0.28	400–15,000
3-in. Partition rings (stacked, staggered)	0.0625	0.09	3,000–14,000
Spiral rings (stacked, staggered)			
3-in. Single spiral	0.00909	0.28	400–15,000
3-in. Triple spiral	0.0116	0.28	3,000–14,000
Drip-point grids (continuous flue)			
Style 6146	0.0154	0.23	3,500–30,000
Style 6295	0.00725	0.31	2,500–22,000

Source: ref. 11.

Table 3
Schmidt Numbers for Gases and Vapors in Air at 77°F and 1 atm

Substance	Sc_G[a]	Substance	Sc_G[a]
Ammonia	0.66	Valeric acid	2.31
Carbon dioxide	0.94	*i*-Caproic acid	2.58
Hydrogen	0.22	Diethyl amine	1.47
Oxygen	0.75	Butyl amine	1.53
Water	0.60	Aniline	2.14
Carbon disulfide	1.45	Chloro benzene	2.12
Ethyl ether	1.66	Chloro toluene	2.38
Methanol	0.97	Propyl bromide	1.47
Ethyl alcohol	1.30	Propyl iodide	1.61
Propyl alcohol	1.55	Benzene	1.76
Butyl alcohol	1.72	Toluene	1.84
Amyl alcohol	2.21	Xylene	2.18
Hexyl alcohol	2.60	Ethyl benzene	2.01
Formic acid	0.97	Propyl benzene	2.62
Acetic acid	1.16	Diphenyl	2.28
Propionic acid	1.56	*n*-Octane	2.58
i-Butyric acid	1.91	Mesitylene	2.31

[a] $Sc_G = \mu_G / P_G D_G$, where D_G and μ_G are the density and viscosity of the gas stream, respectively, and P_G is the diffusivity of the vapor in the gas stream.

Source: ref. 13.

Table 4
Schmidt Numbers for Compounds in Water at 68°F and 1 atm

Solute[a]	Sc_L[b]	Solute[a]	Sc_L
Oxygen	558	Glycerol	1400
Carbon dioxide	559	Pyrogallol	1440
Nitrogen Oxide	665	Hydroquinone	1300
Ammonia	570	Urea	946
Bromine	840	Resorcinol	1260
Hydrogen	196	Urethane	1090
Nitrogen	613	Lactose	2340
Hydrogen chloride	381	Maltose	2340
Hydrogen sulfide	712	Mannitol	130
Sulfuric add	580	Raffinose	2720
Nitric acid	390	Sucrose	2230
Acetylene	645	Sodium chloride	745
Acetic acid	1140	Sodium hydroxide	665
Methanol	785	Carbon dioxide[c]	445
Ethanol	1105	Phenol[c]	1900
Propanol	1150	Chloroform[c]	1230
Butanol	1310	Acetic acid[d]	479
Allyl alcohol	1080	Ethylene dichloride[d]	301
Phenol	1200		

[a]Solvent is water except where indicated.

[b]$Sc_L = \mu_L/P_L D_L$, where μ_L and P_L are the viscosity and density of the liquid, respectively, and D_L is the diffusivity of the solute in the liquid.

[c]Solvent is ethanol.

[d]Solvent is benzene.

Source: ref. 13.

2.4.4. Packed Tower (Wet Scrubber) Operation and Maintenance

As previously mentioned, the pressure drop (head loss) through a packed tower (*see* Fig. 1b) has a major impact on the economics of a tower. When in the design phase, the most accurate pressure drop data for a given packing should be provided by the packing supplier. However, for the purpose of a general example, the following is a relatively accurate correlation:

$$P_a = \left(g \times 10^{-8}\right)\left[10^{(rL''/D_L)}\right]\left(3,600G_{\text{area}}\right)^2 / D_G \tag{21}$$

where P_a is the pressure drop (lb/ft^2-ft) and g and r are the packing constants from Table 5 (4).

The total pressure drop through a packed tower (*see* Fig. 1b) wet scrubber is

$$P_{\text{total}} = P_a \, \text{Ht}_{\text{column}} \tag{22}$$

The fan power requirement, F_p (in kWh/yr), is calculated as follows:

$$F_P = 1.81 \times 10^{-4} \, (Q_{e,a})(P_{\text{total}})(\text{HRS}) \tag{23}$$

where F_P is the fan power requirement (kWh/yr), $Q_{e,a}$ is the actual emission stream flow rate (acfm), P_{total} is the system pressure drop (in. H_2O), and HRS is the system operating hours per year (h/yr).

Tables 5
Pressure Drop Constants for Tower Packing

Packing	Nominal size (in.)	*g*	*r*	Range of L″ (lb/h-ft^2)
Raschig rings	0.5	139	0.00720	300–8,600
	0.74	32.90	0.0045	1,800–10,800
	1	32.10	0.00434	360–27,000
	1.5	12.08	0.00398	720–18,000
	2	11.13	0.00295	720–21,000
Berl saddles	0.5	60.40	0.00340	300–14,100
	0.74	24.10	0.00295	360–14,400
	1	16.10	0.00295	720–78,800
	1.5	8.10	0.00225	720–21,600
Intalox saddles	1	12.44	0.00277	2,520–14,400
	1.5	5.66	0.00225	2,520–14,400
Drip-point	No. 6146	1.045	0.00214	3,000–17,000
grid tiles	Continuous flue			
	Cross-flue	1.218	0.00227	300–17,500
	No. 6295	1.088	0.00224	850–12,500
	Continuous flue			
	Cross flue	1.435	0.00167	900–12,500

Source: ref. 13.

The value for $Q_{e,a}$ can be obtained from Q_e

$$Q_{e,a} = Q_e\,(T_e + 460) / 537 \tag{24}$$

where Q_e is the emission stream flow rate (scfm) and T_e is the emission stream temperature (°F). Equation (25) is used to determine annual electricity cost (AEC) of a packed tower wet scrubber. In January 1990, the UEC was \$ 0.059/kWh.

$$\text{AEC} = \text{UEC}\,(F_p) \tag{25}$$

where UEC is the unit electricity cost (\$/kWh).

The electric power needed to operate the fan feeding the gaseous stream to the scrubber tower is directly related to the total pressure drop of the fan. The Electric Power Institute of Palo Alto, California has provide a correlation between the acfm of gas being treated and the horsepower, hp, needed to drive the fan at a given pressure drop:

$$\text{hp} = Q_e\,P_{\text{total}} / 5{,}390 \tag{25a}$$

This correlation assumes an 80% efficient motor and 10% annual downtime. The cost of operating 1 hp (again from the Electric Power Institute), for 1 yr, at various electric power costs is

\$/kWh	0.04	0.06	0.08	0.10	0.12	0.14	0.16
Cost /yr	\$326	\$492	\$650	\$816	\$975	\$1, 134	\$1, 309

Thus, per this example, if the cost of power is \$0.10/kWh (during California's power crisis in the summer of 2001, this cost escalated to greater than \$0.30), and a plant has a

Table 6
Capital Costs Factors for Absorbers

Cost Item	Factor
Direct Costs (DC)	
Purchased equipment cost	
Absorber (tower & packing) + auxillary equipment	As estimated, EC
Instrumentation	0.10 EC
Sales tax	0.03 EC
Freight	0.05 EC
Purchased Equipment Cost (PEC)	1 PEC = 1.18 EC
Direct Installation Costs	
Foundation and supports	0.12 PEC
Erection and handling	0.40 PEC
Electrical	0.01 PEC
Piping	0.30 PEC
Insulation	0.01 PEC
Painting	0.01 PEC
Direct Installation Cost	0.85 PEC
Site preparation	As required. SP
Building	As required, Bldg.
Total Direct Costs	1.85 PEC + SP + Bldg.
Indirect Costs (IC)	
Engineering	0.10 PEC
Construction	0.10 PEC
Contractor fee	0.10PEC
Start–up	0.01 PEC
Performance test	0.01 PEC
Contingencies	0.03 PEC
Total Indirect Cost	0.35 PEC
Total Capital Costs	2.20 PEC + SP + Bldg.

Source: ref. 9.

scrubber tower treating 50,000 scfm with a pressure loss of 5 in. of H_2O, the fan is pulling approx 46 hp. This horsepower is costing the plant over $37,000 per year. Therefore, if a 50% reduction in pressure drop in the tower could be realized, the plant would net a power savings of over $18,000 per year.

Another consideration in scrubber tower operating cost is the cost of scrubbing liquor (absorbent), usually water that is consumed in normal operation of the tower. The annual solvent required (ASR) is expressed by

$$\mathrm{ASR} = 60\ (L_{gal})\ \mathrm{HRS} \tag{26}$$

Table 6 presents capital cost factors for several absorbents (9). The annual solvent cost (ASC) may now be calculated by multiplication of ASR by the unit solvent cost (USC):

$$\mathrm{ASC} = (\mathrm{USC})\ (\mathrm{ASR}) \tag{27}$$

Table 7
Annual Cost Factors for Absorbers Systems

Cost item	Factor
Direct Cost	
Utilities	
Electricity	\$0.059/kWh
Solvent (water)	\$0.20/10^3 gal
Operating Labor	
Operator labor	\$12.96/h
Supervisor	15% of operator labor
Maintenance	
Maintenance labor	\$14.96/h
Materials	100% of maintenance labor
Indirect Costs	
Overhead	0.60 (Operating labor and maintenance)
Administrative	2% of TCC
Property taxes	1% of TCC
Insurance	1% of TCC
Capital recovery[a]	0.1628 (TCC)

[a]The capital recovery cost is estimated as $i\,(1 + i)^n/[\,(1 + i)^n - 1]$, where i is interest (10%) and n is equipment life (10 yr).
Source: Data from refs. 9 and 12.

The USC is the unit solvent cost. The costs of various solvents are given in Table 7. As of January 1990, the solvent cost of water, on average, was \$0.20 per 1000 gal in the United States (1 US gal = 3.785 L).

2.5. *Venturi Wet Scrubber Design*

2.5.1. *General Design Considerations*

Venturi scrubbers provide excellent removal efficiency for particulate matter of 0.5 to 5 μm in diameter (*see* Fig. 1c). A general design criterion of Venturi scrubbers is that for any given pressure drop across a scrubber, the longer the constriction or "throat," the higher the removal efficiency. The throat cannot be made so long as to have frictional loses become significant, however. Suppliers of Venturi scrubbers also provide for variable throat sizes as a control mechanism of the scrubber. Changing the throat size will result in an adjustment of gas velocity, which affects pressure drop as well as efficiency.

In a typical Venturi scrubber, liquid, normally water, is introduced upstream of the Venturi or throat. As the water flows down the convergent sides of the throat, the sudden acceleration of gas velocity in the throat atomizes the water. This is referred to in industry as the wetted approach to Venturi scrubber design. As so implied, a nonwetted method is also possible. In this design, the water (or other liquid) is injected directly into the throat. The nonwetted scheme is used if the gaseous stream being scrubbed is near its saturation point. The nonwetted method requires that very clean water be used to avoid plugging of the injection nozzles. The wetted scheme is used if the gas being treated is hot, as this means that some amount of water must be evaporated (18).

As soon as the water is atomized, it collects particles by impaction of the particles on the water droplets. This impaction process is possible as the result of the difference in velocities between gas (high) and the water droplets (slow). As the gas–water droplet mixture passes out of the throat, the velocities of both gas and water droplets decelerate. At this point, further impaction causes the water droplets to reform into the bulk liquid phase. Particles that were captured by water droplets in the throat will remain in the bulk liquid phase. The water is then sent to a separator, where it is separated from the clean gas stream (18–21).

2.5.2. Venturi Scrubber Design Variables

The temperature of the gas stream entering a Venturi scrubber needs to be held to 50–100°F above the dew point of the gas stream. The determination of the dew point of a gaseous flow is explained elsewhere in this handbook. Cooling or heating is called for if the gaseous stream needing treatment is not within the temperature range stated. If such temperature adjustment of the gas stream is needed, then the physical properties of the stream will be altered. A Venturi scrubber's primary design parameters are the saturated gas flow rate, $Q_{e,s}$, which is a direct function of the temperature of the gas stream, the flow rate at actual conditions ($Q_{e,a}$), particle size distribution in the polluted gas stream, and throat size (18, 22). If the temperature of the gas stream is changed in a pre-heating or precooling step, the actual flow rate of the gas will also be changed. This new actual flow rate will, in turn, affect the saturated flow rate. Another pretreatment step that may be required is mechanical dust collection. If large particulate matter is present in the gas stream being cleansed, this pretreatment may be appropriate.

When designing a Venturi scrubber, three choices become apparent:

1. Rely on a previous design for a similar or identical application.
2. Do a pilot test on the air to be cleansed.
3. Collect empirical data about the air stream to be treated (particle size distribution, flow rate, temperature) to allow for use of published performance curves for a given Venturi system.

The first choice, although simple and direct, runs the risk of missing recent advances within the industry. The second option is time-consuming but will provide for a result that is based on timely data. The third option has the advantage of using industry data to reach an advanced design in a timely fashion. Therefore option 3 is discussed here. The primary consideration in this design will be pressure drop in the Venturi scrubber; a secondary consideration will be construction materials (9).

2.5.3. Venturi Scrubber Sizing

A Venturi scrubber must be sized after the decision is made that this technology is the best fit for the air pollution problem at hand (*see* Fig. 1c). A Venturi scrubber may be sized using the airflow at inlet conditions ($Q_{e,a}$) or the saturated airflow rate ($Q_{e,s}$) may be used for sizing calculations. Venturi scrubber original equipment manufacturers (OEMs) use either method, based on their own preferences. Cost data are generally based on emission stream flow rate at inlet conditions, $Q_{e,a}$. The saturated emission stream flow rate, $Q_{e,s}$, can be found as shown below (9,20). Psychometric charts (Fig. 4) are available to determine saturated air temperature ($T_{e,a}$).

$Q_{e,s}$ is then determined:

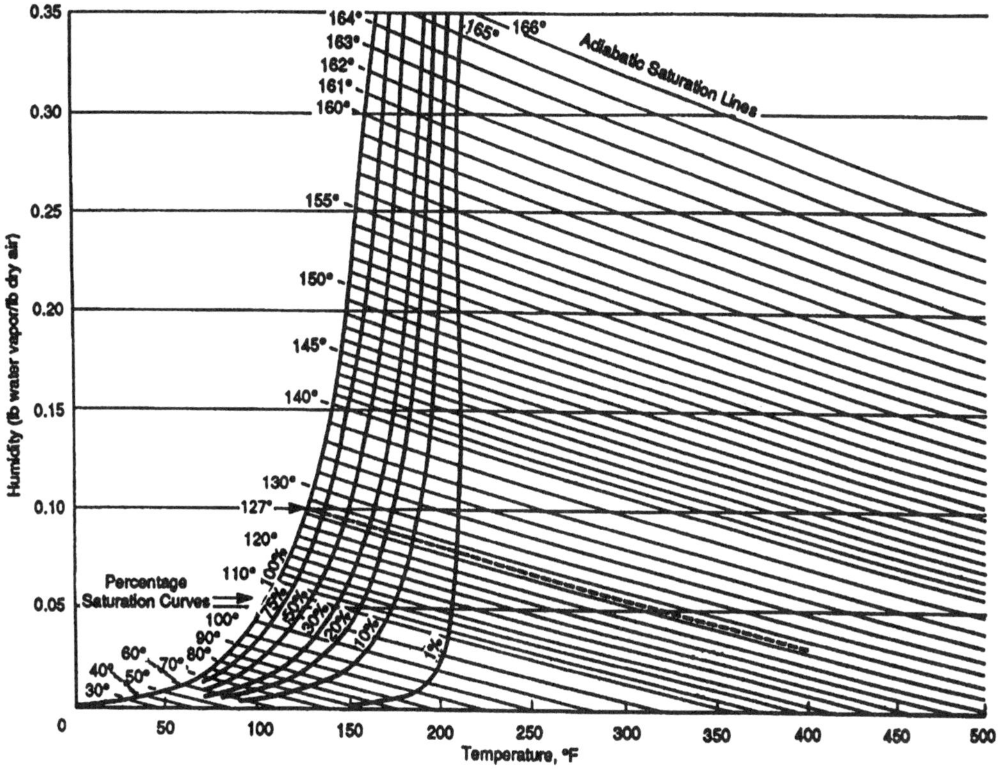

Fig. 4. Psychrometric chart.

$$Q_{e,s} = [Q_{e,a}\ (T_{e,s} + 460)\ /\ (T_e + 460)] + Q_w \tag{28}$$

where $Q_{e,s}$ is the saturated emission stream flow rate (acfm), $T_{e,s}$ is the temperature of the saturated emission stream (°F), T_e is the temperature of the emission stream at inlet air (°F), $Q_{e,a}$ is the actual emission stream flow rate from Eq. (24) (acfm), and Q_w is the volume of air added (ft^3/min). The volume of air added is determined using:

$$Q_w = Q_{e,\mathrm{ad}}\ (D_e)(L_{w,s} - L_{w,a})\ (1/D_w) \tag{29}$$

where $Q_{e,\mathrm{ad}}$ is the actual flow rate of dry air (acfm) (see example for considerations of density and moisture content of air at scrubber operating conditions), D_e is the density of the polluted air emission stream (lb/ft^3), $L_{w,s}$ is the saturated lb water/lb dry air (from Fig. 4), $L_{w,a}$ is the inlet lb water/lb dry air (from Fig. 4), D_w is the density of water vapor (lb/ft^3), and

$$Q_{e,\mathrm{ad}} = (1 - L_{w,a})\ Q_{e,a} \tag{29a}$$

Using the ideal gas law, an approximate density of any gas encountered in an air pollution control project can be made:

$$D = (PM)/(RT) \tag{30}$$

where D is the density (lb/ft^3), P is the pressure of the emission stream (atm), M is the molecular weight of the specific pollutant gas (lb/lb-mol), R is the gas constant (0.7302 atm-ft^3/lb-mol °R), and T is the temperature of the gas (°R).

2.5.4. Venturi Scrubber Operation and Maintenance

The performance of a Venturi scrubber, when plotted, normally yields a logarithmic curve. Such a curve relates collection efficiency in the Venturi scrubber to pressure drop and particle size (3,19–21,23). Standard industry practice has been to plot pressure drop versus mean particle diameter (D_p) for a specific Venturi size. An example of such a plot is given in Fig. 5. A Venturi vendor provided these data for a given removal efficiency at various pressure drops. With D_p data from a polluted airstream, one can estimate the removal efficiency possible for particulate matter from the airstream at various pressure drops across this Venturi scrubber. Figure 5 is typical of data supplied by Venturi scrubber system OEMs. Note that Fig. 5 is specific to one such OEM.

Also, because data are widely available from Venturi scrubber system OEM firms, it is used for most design purposes for Venturi scrubber projects. A fundamental understanding of the design equations presented here assists in understanding the design process for a Venturi scrubber; such equations, however, are generally not used by environmental engineers on a daily basis. It is important to note that the removal efficiency reported by OEM firms is a weighted average for each particle size in a known particle size distribution. The actual particle size distribution being treated in a polluted airstream may be, and most likely will be, different than the particle size distribution used by the Venturi scrubber OEM to generate Fig. 5 data. The D_p of the design (OEM) and the actual (field air pollution project) particle size distributions may also be the same or very similar, whereas the two particle size distributions are actually quite different. Thus, the removal efficiencies reported in Fig. 5 should be taken as approximations only.

Normal industry practice has been to use Venturi scrubbers that operate at pressure loss from 10 up to 80 in. of water. Above pressure loss of 80 in. of water, it has generally been found that particulate matter will not be removed efficiently within the Venturi scrubber.

A critical maintenance issue with any Venturi scrubber is that the spray nozzles where the liquid (normally water) is injected into the scrubber must be kept open. Routine inspection of nozzle openings and throat is good standard practice for any Venturi scrubber system. These measures, combined with normal pump maintenance, will help prevent both equipment failures as well as emission violations of a Venturi scrubber system (24,25).

Pressure drops from a variety of air pollution control applications using the Venturi principle are listed in Table 8 (9,18). These data are presented as typical of general industry applications. Specific applications, therefore, may have a pressure drop outside of this data range (9,18).

The capital expense of a Venturi scrubber system is straightforward. The system will have an initial capital expense at time of purchase. Additionally, there will be direct and indirect costs of site erection and commissioning of the scrubber system. Table 9 presents capital cost factors for typical Venturi scrubber systems (26,27).

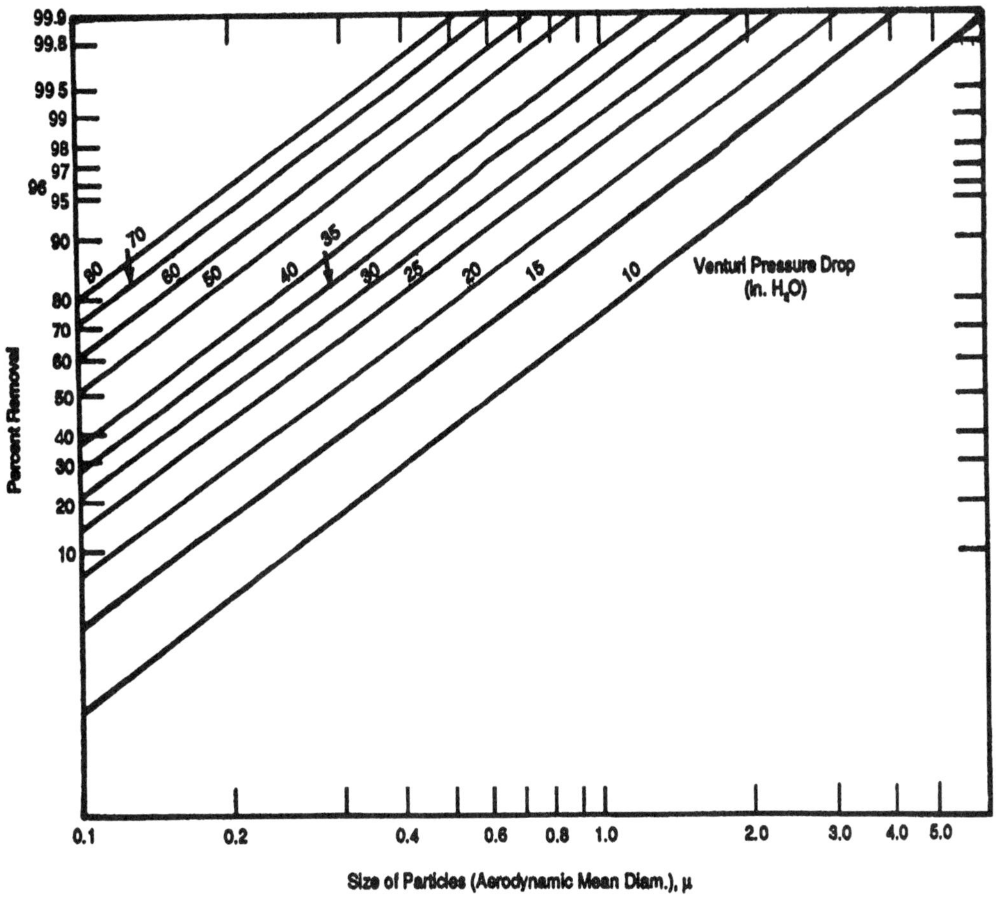

Fig. 5. Typical Venturi scrubber performance curve.

Annual operational costs of the system also reflect direct and indirect expenses. Such expenses are given in Table 7. It should be noted that the annual cost factors for Venturi and packed scrubbers are identical. Direct annual costs of a Venturi scrubber are also those of a packed scrubber: power, scrubbant liquor, labor, and maintenance costs.

Electric power cost is similar as for packed scrubbers: A certain amount of horsepower will be required to move the polluted air through the Venturi scrubber at a given pressure drop. Please refer to Section 2.4.4 for the formula from the Electric Power Institute of Palo Alto, California that may be used to estimate the horsepower needed to move the air volume in cfm if the pressure loss is known in inches of water column (WC). Annual electric costs of running 1 hp, at various power prices, are also given in Section 2.4.4.

The electrical costs are a function of the fan power required to move gas through the system. Equation (23) is used to estimate the fan power requirement assuming a fan-motor efficiency of 65% and fluid specific gravity of water (1.0).

Table 8
Pressure Drops for Typical Venturi Scrubber Application

Application	Pressure drop (in. H_2O)
Boilers	
Pulverized coal	15–40
Stoker coal	10–12
Bark	6–10
Combination	10–16
Recovery	30–40
Incinerators	
Sewage sludge	18–20
Liquid waste	50–55
Solid waste	
Municipal	10–20
Pathological	10–20
Hospital	10–20
Kilns	
Lime	15–25
Soda ash	20–40
Potassium chloride	30
Coal processing	
Dryers	25
Crushers	6–20
Dryers	
General spray	20–60
Food spray	20–30
Fluid bed	20–30
mining	
Crushers	6–20
Screens	6–20
Transfer points	6–20
Iron and steel	
Cupolas	30–50
Arc furnaces	30–50
BOFs	4–60
Sand systems	10
Coke ovens	10
Blast furnaces	2–30
Open hearths	20–30
Nonferrous metals	
Zinc smelters	20–50
Copper and brass smelters	20–50
Sinter operations	20
Aluminum reduction	60
Phosphorous	
Phosphoric acid	
Wet process	10–30
Furnace grade	40–80

Continued

Table 8 *(Continued)*

Application	Pressure drop (in. H_2O)
Asphalt	
Batch plants—dryers	10–15
Transfer points	6–10
Glass	
Container	25–60
Plate	25–60
Borosillicate	30–60
Cement	
Wet process kiln	10–15
Transfer points	6–12

Source: ref. 18.

Table 9
Capital Cost Factors for Venturi Scrubbers

Cost item	Factor
Direct Costs (DC)	
Purchased Equipment Costs (PEC)	
Venturl scrubber + auxiliary equipment	As estimated, EC
Instrumentation[a]	Included with EC
Sales tax	0.03 EC
Freight	0.05 EC
Purchased Equipment Cost, PEC	1.08 EC
Direct Installation Costs	
Foundation and supports	0.06 PEC
Erection and handling	0.40 PEC
Electrical	0.01 PEC
Piping	0.05 PEC
Insulation	0.03 PEC
Painting	0.01 PEC
Site preparation	As required, SP
Buildings	As required, Bldg.
Total Direct Costs	0.56 PEC + SP + Bldg.
Indirect Costs (IC)	
Engineering	0.10 PEC
Construction	0.10 PEC
Contractor fee	0.10 PEC
Start up	0.01 PEC
Performance test	0.01 PEC
Contingency	0.03 PEC
Total Indirect Cost	0.35 PEC
Total Capital Costs	1.91 PEC+ SP + Bldg.

[a]If instrumentation is not included with EC, estimate as 10% of the EC

Source: Data from refs. 9, 26, and 27.

$$F_p = 1.81 \times 10^{-4} (Q_{e,a}) (P) (\text{HRS}) \quad (23)$$

where F_p is the fan power requirement (kWh/yr), $Q_{e,a}$ is the actual emission stream flow rate (acfm), P is the system pressure drop (in. H_2O), and HRS is the system operating hours (h/yr).

The cost of scrubbing liquor (most likely water) is given by

$$\text{WR} = 0.6 (Q_{e,a}) \text{ HRS} \quad (31)$$

where WR is the water consumption (gal/yr).

A general assumption of needed operator labor is 2 h for each 8-h shift. Management labor costs are assumed to be 15% that of operator labor costs. Labor costs are presented in Table 7. Maintenance is normally estimated in industry as requiring 1 h from each 8-h shift. This cost is also provided in Table 7. Materials required for normal maintenance are assumed to equal cost of maintenance.

It should be noted that the cost of the wastewater generated by a Venturi scrubber is potentially quite high. Although not discussed here, such cost should be included when considering the use of a Venturi scrubber for an air pollution control project. Table 7 also includes such indirect costs as property taxes, capital recovery, administrative costs, and so forth. Table 10 presents both the US Environmental Protection Agency's conservative control efficiencies (CE) and the typical actual control efficiencies (CE) of a wet scrubber for removal of various hazardous air emissions. Either a fabric filter system or a four-field electrostatic precipitator is used as the pretreatment to the wet scrubber.

3. DRY SCRUBBERS

3.1. *Dry Absorbents*

A dry chemical absorbent scrubber will almost always use either a calcium- or sodium-based absorbent. These absorbents are classified as alkali absorbents, which have excellent to good absorbent properties for most of the acid gases as well as for some organic air pollutants. Additionally, the dry alkali absorbent most commonly used is slaked lime or $Ca(OH)_2$. In some instances, dry (powdered) activated carbon is also added to the dry absorbent. This is done so that the dry (or semidry) scrubber will also be able to remove heavy metals and/or trace (and often very toxic) organic pollutants. After the dry scrubbing, the solids (used and unused absorbent as well as particulate matter) are accumulated at the bottom of the scrubber tower or possibly directed to a baghouse (or other particulate collector). The removal efficiency of a wet scrubber is primarily dependent on the acid/alkali ratio used in the scrubber as well as outlet air temperature.

3.2. *Dry Scrubbing Systems*

Three dry scrubbers commonly found in air pollution control operations in industry are presented (15) in Fig. 6A,B. They are dry–dry, semi–dry, and spray dryer absorber systems.

3.2.1. *Dry–Dry Systems*

When injecting hydrated lime or pulverized limestone directly into a furnace (or other combustion chamber) or into the ducting downstream from a combustion process,

Table 10
Hazardous Waste Incinerator Emission Estimates

	US EPA conservative estimated efficiencies[a] (%)	Typical actual, control efficiencies (%)
Particulate matter	99+%	99.9+
Hydrogen chloride (HCl)	—	99+
Sulfur dioxide (SO_2)	—	95+
Sulfurlc Acid (H_2SO_4)	—	99+
Arsenic	95	99.9+
Beryllium	99	99.9
Cadmium	95	99.7
Chromium	99	99.5
Antimony	95	99.5
Barium	99	99.9
Lead	95	99.8
Mercury	85–90	40–90+
Silver	99	99.9+
Thallium	95	99+
PCDD/PCDF[b]	—	90–99+

[a]Based on spray dryer fabric-filter system or four field electrostatic precipitator followed by a wet scrubber.
[b]Total of all cogeners.

this air pollution control scheme is referred to as dry sorbent injection (DSI) (*see* Fig. 6A). Removal of about 50% of the SO_2 present is possible using DSI technology. When using DSI, a pneumatic device will inject the dry alkali directly into the hot air. An advantage to downstream injection is that this allows for the hot air to be humidified before dry alkali injection. This improves the removal efficiency of SO_2 achieved. In addition to higher moisture content of the air, lower temperature (20–30°F above saturation), increased alkali usage, and increased contact time between polluted air and dry alkali all work to increase SO_2 removal efficiency. Humidification before dry alkali injection is also useful, as this lowers the temperature of the air, and separation of other pollutants may be made easier as well. Additionally, at a lower gas temperature, formation of chlorinated dioxins and furan compound as well as other products resulting from incomplete combustion may be avoided. However, because of limiting factors of high alkali requirement and fatigue (corrosion) of the physical installation, dry–dry scrubbing systems are seldom used alone to control hazardous air pollutants.

3.2.2. Semidry Systems

The semidry scrubbing process is seen in Fig. 6B. In this process, adiabatic evaporation of water conditions the polluted air to a temperature somewhat above its saturation point. The alkali absorbent is injected directly or as slurry of water. The cooling of the gas will reduce the air volume of the gas, and chemical reaction between alkali and acid gas pollutant will occur. Downstream of the semidry scrubber, one will find either fabric filters or an electrostatic precipitator (ESP) to collect particles. This downstream pro-

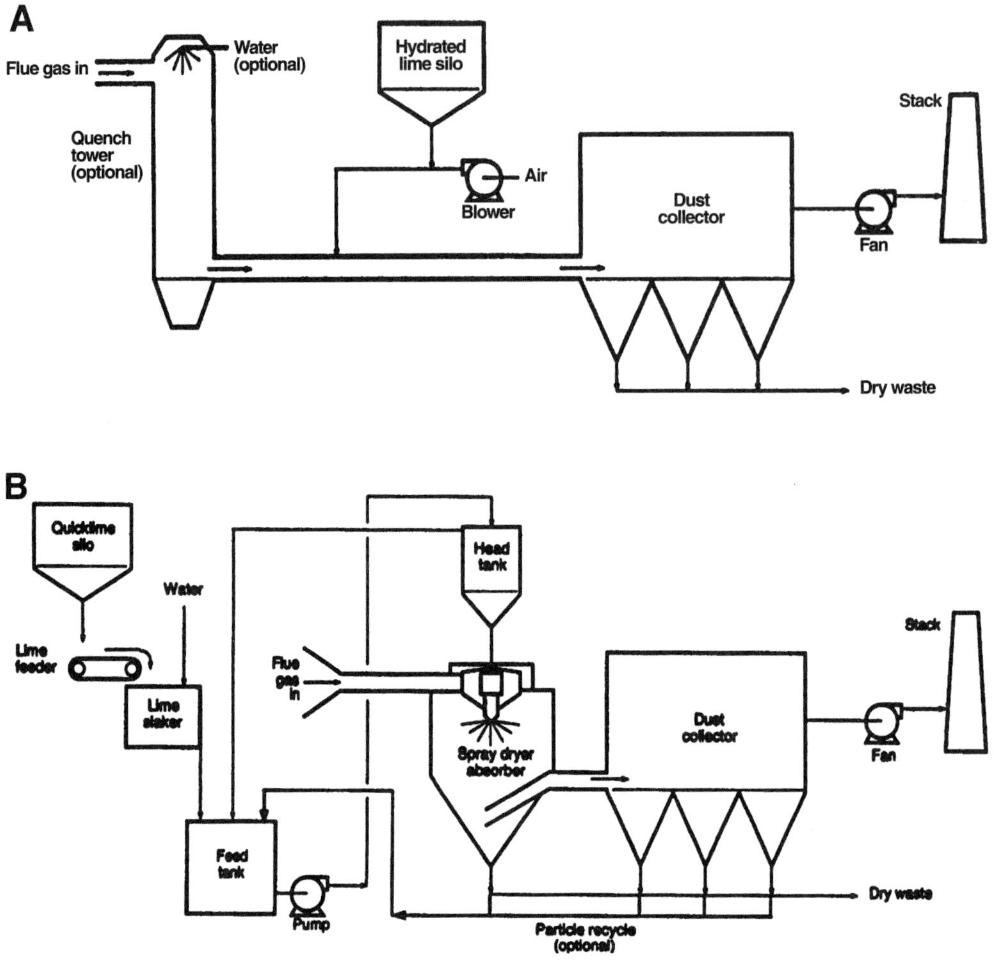

Fig. 6. (A) Dry sorbent injection process; (B) Spray dryer absorption (semidry) process. (From US EPA.)

cess may also act as a secondary reaction bed between the alkali absorbent and the acid gas being treated, with subsequent improvement in removal efficiency. The semidry scrubber offers the advantage of not producing a liquid waste stream that requires further treatment but avoids the problems of dry–dry scrubbers discussed in the previous subsection. Also, as liquid droplets are present in the semi-dry scrubbing process, higher acid gas removal efficiency is possible than if using the dry–dry scrubber process.

3.2.3. Spray Dryer Absorber Systems

Also seen in Fig. 6B is the spray dryer absorber (SDA) system. In the SDA, again, the most common alkali absorbent used is hydrated lime. The lime is slurried with water to approx 15% by weight of the total mixture. The dosage of slurry is controlled according to pollutant gas concentration in the airstream being treated based on needed removal efficiency. The SDA system is commonly found treating airstreams produced at power plants and solid-waste incinerators. SDA is a widely accepted and well-documented air pollution control technology. Alkali slurry is often injected in SDA by atomizers (single or multiple) or with dual-fluid spray nozzles. The optimum slurry droplet size in SDA is 50–90 μm. All types of flow schemes are found in SDA equipment: downflow, upflow, upflow using cyclonic precollection, as well as with single or multiple gas inlets. Typical removal efficiencies achieved in a SDA system are 95% for SO_2 and HF and 99%+ for HCl and SO_3. Polluted gas streams having temperatures as high as 1000°C may be treated with SDAs. This is possible as the result of the rapid cooling of gas that will take place as the result of evaporative cooling. Evaporation of water will bring the temperature of the hot gas below 200°C. Needed residence time for the polluted gas stream being treated is 10–18 s. Approximately 25% of the reaction products are recovered in the SDA unit as ash (15).

3.3. *Dry Scrubbing Applications*

3.3.1. *General Downstream and Upstream Dry Scrubbing Applications*

After dry scrubbing (downstream), a fabric filter or an ESP will be used to collect particulates. The efficiency of acid gas removal is enhanced by the presence of the particulate collection step. Generally, fabric filters improve such efficiency better than an ESP. Collected solids are also often suitable for recycling, which may recover some of the operation cost of the pollution control system.

Power plant air emissions are commonly treated and controlled with dry scrubber technology. Standard designs and operational data are therefore available for such applications. Less well understood is the use of dry scrubbers to control air emissions from waste incinerators. Numerous control schemes are possible, but no industry standard for treating air emissions from waste incinerators has emerged to date. Combinations of various types of scrubber is possible, as well as fabric filter, ESP, and so forth, to solve a particular air pollution problem. Example solutions are discussed below.

3.3.2. *Incineration Pollution Control*

The pollutants in an incinerator exhaust gas flow may be controlled using semidry scrubbing, with varying removal efficiencies. Particles that are sticky, gummy, and/or corrosive are processed easily in dry scrubbers. Much of the control process occurs in the dryer section, including chemical and moisture addition and, subsequently, most of the absorption as well. Some additional absorption will also occur in the dust collection stage. Dry absorption is best suited for control of trace metals, acid gas, and trace organic pollutants. Dry scrubbers often have high power costs because fluid nozzles operate at high pressure. Power consumption is lower for rotary atomizers than fluid nozzles. Wet absorbers generate lower chemical and disposal costs compared with dry scrubbers. A major advantage of dry scrubbers is lower initial capital expenditures compared to a wet scrubber system.

The final product of semidry scrubbing (*see* Fig. 6B) will be hygroscopic and contain a large soluble fraction. This end product also tends to cling more (or stick) than fly ash, so it is more difficult to handle than the latter. In addition to some fly ash, the final product of dry scrubbing will contain some trace heavy metals along with trace amounts of organic compounds. Consequently, the final product will always be classified as a hazardous waste material and, as such, will incur additional costs. Use of a fabric filter as a collection step after the dry scrubber is limited by possible clogging problems and/or temperature limitations. If either concern exists, an ESP must be used for the final collection step.

3.3.3. Thermal Desorption

A thermal desorber is typically a semidry scrubber followed by a baghouse. The scrubber removes acid gases, and the baghouse removes particulate matter. Such a control scenario will normally provide needed removal efficiencies for acid gases as well as heavy metals and trace organics.

3.3.4. General Remediation Applications

Two distinct limiting features of semidry absorption technology should be considered:

1. Very high removal efficiencies, as per wet scrubbing, are not possible.
2. As the absorbent is slurried, the dry waste produced in dry scrubbing will add to solid-waste disposal problems and costs.

The advantages of dry absorption are as follows:

1. Heavy metal contaminants can be removed in dry scrubbing. This is not possible in wet scrubbing.
2. Acid gases and trace organics are also removed in dry scrubbing, as well as some chlorinated dioxin and furan compounds.
3. Addition of activated carbon will further boost heavy metal removal efficiency.

3.4. Dry Scrubber Design

3.4.1. General Design Considerations

When properly designed and operated, a SDA system followed by a baghouse will achieve a removal efficiency as high as 99%+ of pollutants commonly present in an incinerator exhaust stream (acid gases and heavy metals). In older systems, removal rates of 70–80% of the pollutants present are common. A spray dryer with an ESP is capable of dioxin and furan removals of 98%. Additionally, SDA systems are used to treat hot polluted gas streams. A maximum temperature of 1000°C is possible for a gas stream being treated. Designs exist for a wide range of gas flow rates. SDA technology, however, is not useful for the control of gas streams with low concentrations of contaminants (28).

3.4.2. Semidry Scrubber Sizing

When designing a SDA system, the following concerns will govern the design process:

1. Polluted residence time, which is a function of the following:
   (a) Vessel volume/size
   (b) Polluted gas flow rate
   (c) Polluted gas inlet and outlet temperatures
2. Reagent slurry flow, which will determine the following:
   (a) Size of slaking equipment
   (b) Pump sizes
   (c) Atomization equipment requirements

Furthermore, the required reagent slurry flow will be dependent on, but not limited to, gas temperature, pollutant concentrations, types of pollutant, type of reagent slurry, and so forth.

3.4.3. Dry Scrubber Costs

When considering both dry and semidry scrubbers, the capital cost of a system is approximated with the sizing exponent (n):

$$n = 0.73 \tag{32}$$

Now, using the sizing equation

$$\text{Ib} = \text{Ia}\ (\text{Cb}/\text{Ca})^n \tag{33}$$

where Ib is the cost of the system being sized, Cb is the size of system being considered, Ia is the known cost of the existing system, and Ca is the size of the existing system. Therefore, Eq. (33) uses a known system (a) of a given size and given cost as a reference to estimate the cost of a differently sized system (b). Installed cost-to-purchase ratio is estimated to be 2.17 (22).

As for all pollution control technologies, several solutions will normally present themselves for a given pollution control challenge. As such, "average" cost estimation of a dry scrubbing project is difficult at best. Also complicating the estimation is the fact that OEM firms generally prefer to sell an entire scrubber system, not just one part. A total system cost may be less if it is a "standard" package from a given supplier as opposed to an individual design. However, the demands of the given pollution control project may dictate the need for a custom design.

As of 1991 (28), a dry scrubber system treating 278,000 actual cubic feet per hour (acfh) or 4633 acfm from an incineration source was reported to be \$66/acfm of air being treated or \$38 per actual cubic meter of air per hour (am^3/h). This cost is below that of a semidry system. However, a caveat to keep in mind is that a dry system will most likely require more alkali and will return a lower removal efficiency of acid gases than if the semidry scrubber option is chosen.

The installed cost of a SDA lime sprayer system (as of 1992) is \$49/acfm or \$28 per am^3/h. Using the same gas flow figures just mentioned (278,000 acfh or 4633 acfm), the installed capital cost of the SDA system is \$59/acfm (\$41 per am^3/h).

4. PRACTICAL EXAMPLES

Example 1

Absorption is the most widely used control process when dealing with inorganic vapor emission control situations. A typical emission stream is presented in Table 11, which contains inorganic vapors that are easily removed from the stream using absorption technology. This example will demonstrate the operating data that an industrial installation must compile to support an air emissions permit application.

Solution

From Table 11, the most important characteristics of the polluted emission stream are as follows:

Table 11
Effluent Characteristics for Emission Stream #5

HAP EMISSION STREAM DATA FORM*

Company: Glaze Chemical Company	Plant Contact: Mr. John Leake
Location (Street): 87 Octane Drive	Telephone No.: (999) 555-5024
(City): Somewhere	Agency Contact: Mr. Efrem Johnson
(State, Zip):	No. of Emission Streams Under Review: 7

A. Emission Stream Number/Plant Identification: #5 / Urea Evaporator Off-gas Exhaust

	(a)	(b)	(c)
B. HAP Emission Source	evaporator off-gas		
C. Source Classification	process point		
D. Emission Stream HAPs	ammonia		
E. HAP Class and Form	inorganic vapor		
F. HAP Content (1,2,3)**	20,000 ppmv		
G. HAP Vapor Pressure (1,2)	8.46 atm at 68° F		
H. HAP Solubility (1,2)	provided		
I. HAP Adsorptive Prop. (1,2)	not given		
J. HAP Molecular Weight (1,2)	17 lb/lb-mole		

K. Moisture Content (1,2,3): 2% vol	P. Organic Content (1)***:
L. Temperature (1,2,3): 85° F	Q. Heat/O_2 Content (1):
M. Flow Rate (1,2,3): 3,000 scfm (max)	R. Particulate Content (3):
N. Pressure (1,2): atmospheric	S. Particle Mean Diam. (3):
O. Halogen/Metals (1,2): none / none	T. Drift Velocity/SO_3 (3):

U. Applicable Regulation(s):
V. Required Control Level: assume 98% removal
W. Selected Control Methods: absorption

* The data presented are for an emission stream (single or combined streams) prior to entry into the selected control method(s). Use extra forms if additional space is necessary (e.g., more than three HAPs) and note this need.

** The numbers in parentheses denote what data should be supplied depending on the data on lines C and E:
1 = organic vapor process emission
2 = inorganic vapor process emission
3 = particulate process emission

*** Organic emission stream combustibles less HAP combustibles shown on lines D and F.

Emission stream flow rate (Q_e) = 3000 acfm
Emission stream temperature (T_e) = 85°F
Hazardous air pollutant (HAP) = ammonia, NH_3
HAP emission stream concentration (HAP_e) = 20,000 ppmv
Pressure (P_e) = 760 mm Hg

When a permit review is submitted for an absorption control solution, such data must be supplied by the applicant to the appropriate regulatory authority. In this example, calculations are presented which will be used to confirm the applicant's data. The following is a typical data sheet used for an absorption project permit review:

A. Emission Stream Data
   Hazardous Air Pollutant
   1. Emission stream flow rate (maximum possible), Q_e = ____________ acfm
   2. Temperature, T_e = ____________ °F
   3. HAP = ____________ (chemical name and formula)
   4. HAP_e, HAP emission concentration = ____________ ppmv
   5. Pressure, P_e = ____________ mm Hg or inches of water
   6. RE, required removal efficiency = ____________ %

B. Review Data for Permit Review (supplied by applicant)
   Absorption System Operating Conditions (at 77°F, 1 atm standard conditions)

1. Reported removal efficiency, $RE_{reported}$ = ______________%
2. Emission stream flow rate, Q_e = ______________ scfm
3. Temperature of emission stream, T_e = ______________ °F
4. Molecular weight of emission stream, MW_e = ______________lb/lb mol
5. Hazardous air pollutant, HAP =______________(chemical name and formula)
6. HAP concentration, HAP_e = ______________ ppmv
7. Solvent used (absorbent or scrubbing agent) = ______________
8. Slope of the solubility equilibrium curve (pollutant in absorbent), m = ______________
9. Solvent (absorbent) flow rate, L_{gal} = ______________ gal/min
10. Density of emission stream, D_e = ______________lb/ft^3
11. Schmidt number, for HAP/emission stream and HAP/solvent systems:
    a. Sc_G = ______________
    b. Sc_L = ______________
12. Solvent (absorbent) properties:
    a. Density, D_L = ______________ lb/ft^3
    b. Viscosity, μ_L = ______________ cP (centipoise)
13. Type of packing = ______________ (trade name, size, supplier)
14. Packing constants:
    a = ______________ b = ______________ c = ______________
    d = ______________ e = ______________ Y = ______________
    s = ______________ g = ______________ r = ______________
15. Column (absorber tower) diameter, D_{column} = ______________ft
16. Tower height (packed depth), Ht_{column} = ______________ft
17. Pressure drop (tower total including packed bed), P_{total}=______________ in. of water

Example 2

A step-by-step procedure for determining solvent flow rate required in a packed tower wet scrubbing system follows.

Solution

A. Assume a value for the absorption factor, AF = 1.6.
Determine m from published equilibrium data for the HAP/absorbent (solvent) system. (1,3,6)
m = ______________

Using Eq. (3) with
Q_e = ______________ scfm

$$G_{mol} = 0.155\, Q_e \tag{3}$$

G_{mol} = ______________ lb-mol/h

B. Using Eq. (2),

$$L_{mol} = (1.6)(m)(G_{mol}) \tag{2}$$

L_{mol} = ______________ lb-mol/h

C. Using Eq. (5)

$$L_{gal} = 0.036 L_{mol} \tag{5}$$

L_{gal} = ______________ gal/min

Example 3

An air emission stream #5 containing a regulated pollutant (HAP) is described in Table 11. A packed wet scrubber tower is a good choice to use to control in this situation (Fig. 1b). The slope of the equilibrium curve for this HAP/solvent (absorbent) system is very low, as $m = 1.3$ (note that this is well under 50, per previous discussion). As such, this indicates that the pollutant present is highly soluble in the solvent used as the absorbent in this scrubber. Therefore, a high driving force for absorption will be present in the proposed wet scrubber. The flow rate of solvent needed for the desired removal efficiency can now be determined.

Solution

$m = 1.3$
$Q_e = 3{,}000$ scfm

Using the above information and following Example 2, and using Eq. (2)–(5), the results are as follows:

$$G_{\text{mol}} = 0.155Q_e \quad (3)$$
$$G_{\text{mol}} = 0.155(3000)$$
$$G_{\text{mol}} = 465 \text{ lb-mol/h}$$

$$L_{\text{mol}} = (1.6)(m)(G_{\text{mol}}) \quad (2)$$
$$L_{\text{mol}} = (1.6)(1.3)(465)$$
$$L_{\text{mol}} = 967 \text{ lb-mol/h}$$

Equation (4) now becomes Eq. (5) when the factor 7.48 is used to convert from cubic feet to the US gallon basis for water flow, as water is the solvent (absorbent) being considered for this scrubber. Note that D_L is 62.43 lb/ft^3 and $\text{MW}_{\text{solvent}}$ is 18 lb/lb-mol for water.

Therefore,

$$L_{\text{gal}} = [L_{\text{mol}} \times \text{MW}_{\text{solvent}} \times (1/D_L) \times 7.48] / 60 \quad (4)$$
$$L_{\text{gal}} = [970 \times 18 \times (1/62.43) \times 7.48] / 60$$

$$L_{\text{gal}} = 0.036\, L_{\text{mol}} \quad (5)$$
$$L_{\text{gal}} = 0.036\, (967)$$
$$L_{\text{gal}} = 35 \text{ gal/min}$$

Example 4

A step-by-step procedure for determination of column diameter of packed tower wet scrubber (Fig. 1b) follows.

Solution

A. Using Fig. 2, calculate the abscissa (ABS) with Eq. (6):

$\text{MW}_{\text{solvent}} =$ ______________ lb/lb-mol
$L = (L_{\text{mol}})\, \text{MW}_{\text{solvent}}$
$L =$ ______________ lb/h
$\text{MW}_e =$ ______________ lb/lb-mol
$G = (G_{\text{mol}})\, \text{MW}_e$
$G =$ ______________ lb/h
$D_G =$ ______________ lb/ft^3
$D_L =$ ______________ lb/ft^3 from ref 1

$$\text{ABS} = (L/G)\,(D_G / D_L)^{0.5} \quad (6)$$

ABS = _______________

B. Again, using Fig. 2, determine the point of flooding ordinate (ORD).

ORD = _______________

C. Based on the chosen packing, use published data (11) to find the packing constants.

a = _______________, e = _______________

Using ref. 1, the viscosity, μ_L, is

μ_L = _______________ centipoise or cp

D. Using Eq (8), the $G_{\text{area},f}$ (mass flow of tower at flooding) is now determined:

$$G_{\text{area},f} = \{[(\text{ORD})\ D_G D_L\ g_c] / [(a/e^3)(\mu_L^{0.2})]\}^{0.5} \quad (8)$$

$G_{\text{area},f}$ = _______________ lb/s-ft²

E. As previously discussed, a fraction of flooding for design purposes is typically $0.6 < f < 0.75$.

f = _______________(chosen value)

So Eq. (9) can be used to determine G_{area}.

$$G_{\text{area}} = f\,G_{\text{area},f} \quad (9)$$

G_{area} = _______________ lb/h-ft²

F. The cross-sectional area of the column (tower) is now determined from Eq (10):

$$A_{\text{column}} = G / (3600 G_{\text{area}}) \quad (10)$$

A_{column} = _______________ ft²

G. Equation (11) will now yield the column (tower) diameter.

$$D_{\text{column}} = [(4/\pi(A_{\text{column}})]^{0.5} = 1.13(A_{\text{column}})^{0.5} \quad (11)$$

D_{column} = _______________ ft

Note, as previously discussed, a design should always be made to a readily available standard size of tower, normally a whole foot diameter. So this exercise is normally, in real practice, repeated for several iterations until such a tower diameter is calculated.

Example 5

Again, using Emission Stream 5 from Table 11, a packed tower (wet scrubber; Fig 1b) is determined to be the most effective control solution for the given pollutant/gas system. Determine the column (tower) diameter using the calculated results from Example 3 and the procedures in Example 4.

Solution

Using Eqs. (6a), (6), and (8)–(11),

$$L = (\text{MW}_{\text{solvent}})(L_{\text{mol}}) = (18)(967) = 17{,}410 \text{ lb/h}$$
$$G = (G_{\text{mol}})\ \text{MW}_e = (465)(28.4) = 13{,}200 \text{ lb/h}$$
$$D_G = PM/RT \quad (6a)$$
$$D_G = (1.0)(28.4) / [(0.7302)(460 + 85)] = 0.071 \text{ lb/ft}^3$$

From ref. 1 at 85°F,

$$D_L = 62.18 \text{ lb/ft}^3 \text{ at } 85°\text{F}$$

Then the abscissa (ABS) is determined from Eq. (6):

$$\text{ABS} = (L/G)\,(D_G / D_L)^{0.5} \quad (6)$$
$$\text{ABS} = (17{,}410 / 13{,}200)\,(0.071/62.18)^{0.5} = 0.045$$

Using Fig. 2, ABS = 0.045, so the ORD for flooding is equal to 0.15. If 2-in. ceramic Raschig rings are used to pack the column, the packing constants are (11)

$a = 28$ and $e = 0.74$

Also,

$g_c = 32.2$ ft/s^2 and $\mu_L = 0.85$ cP at 85°F, from ref. 1

Then, substituting in Eq. (8),

$$G_{\text{area},f} = \{[(\text{ORD})\,D_G\,D_L\,g_c)] / [(a/e^3)(\mu_L{}^{0.2})]\}^{0.5} \quad (8)$$
$$G_{\text{area},f} = \{(0.15)(0.071)\,(62.18)(32.2) / [(28 / 0.74^3)\,(0.85^{0.2})]\}^{0.5}$$
$$G_{\text{area},f} = 0.56 \text{ lb/s-ft}^2 \text{ (at flooding)}$$

If the fraction of flooding $f = 0.6$, then Eq. (9) becomes

$$G_{\text{area}} = fG_{\text{area},f} = (0.6)(0.56 \text{ lb/s-ft}^2) \quad (9)$$
$$G_{\text{area}} = 0.34 \text{ lb/s-ft}^2$$

Use Eq (10) to find the area of the column:

$$A_{\text{column}} = G/(3600\,G_{\text{area}}) \quad (10)$$
$$A_{\text{column}} = 13{,}200 / [(3600)(0.34)]$$
$$A_{\text{column}} = 10.8 \text{ ft}^2$$

The column diameter is determined as follows:

$$D_{\text{column}} = 1.13\,(A_{\text{column}})^{0.5} = 1.13\,(10.8)^{0.5} \quad (11)$$
$$D_{\text{column}} = 3.7 \text{ using 4 ft}$$

Example 6

A step-by-step procedure to determine column (tower) height and packed depth of a wet scrubber (*see* Fig. 1b) follows.

Solution

A. Using Eq. (13) and Eq. (14) to determine N_{og} (number of gas transfer units) as well as Fig. 3,

HAP concentration in the emission stream (HAP_e) = ______________ ppmv

With the efficiency, RE, of the wet scrubber, HAP_e, and Eq. (14),

$$\text{HAP}_o = \text{HAP}_e\,[1 - (\text{RE}/100)] \quad (14)$$

Then the outlet concentration can be determined:

HAP_o = ______________ ppmv

$$N_{\text{og}} = \ln\{\,[\,(\text{HAP}_e / \text{HAP}_o)\,(1 - (1/\text{AF})) + (1/\text{AF})\,]\,\} / (1-1/\text{AF}) \quad (13)$$

N_{og} = ______________

Using Fig. 3, determine

$\text{HAP}_e / \text{HAP}_o$ = ______________

Assume an adsorption factor of 1.6, then

$1/\text{AF} = 1/1.6 = 0.63.$

Use these values in Fig. 3 to determine N_{og}:

N_{og} = ______________

B. With Eqs. (16), (17), and (15), H_G (height of gas transfer unit), H_L (height of liquid transfer unit), and H_{og} (height of overall gas transfer) are determined. Packing constants needed in Eqs. (16) and (17) are found in Tables 1 and 2.

b = ______________ c = ______________ d = ______________
Y = ______________ s = ______________

Also using Tables 3 and 4, the Schmidt numbers are

Sc_G = ________ Sc_L = ________

Thus, the liquid flow rate is determined from Eq. (18).

$$L'' = L/A_{column} \tag{18}$$

L'' = ______________ lb/h-ft^2

From ref. 1, the solvent viscosity value is found:

μ_L'' = ______________ lb/h-ft^2

Now determine the values for H_G and H_L from Eqs. (16) and (17), respectively:

$$H_G = [b(3600G_{area})^c / (L'')^d] (Sc_G)^{0.5} \tag{16}$$

H_G = ______________ ft

$$H_L = Y (L''/\mu_L'')^s (Sc_L)^{0.5} \tag{17}$$

H_L = ______________ ft

Calculate H_{og} using Eq. (15) and an assumed value for AF of 1.6:

$$H_{og} = H_G + (1/\text{AF}) H_L \tag{15}$$

H_{og} = ______________ ft

C. Equation (12) is now used to determine the height of the column:

$$\text{Ht}_{column} = N_{og} \times H_{og} \tag{12}$$

Ht_{column} = ______________ ft

D. Now use Eq. (19) to determine Ht_{total}

$$\text{Ht}_{total} = \text{Ht}_{column} + 2 + (0.25D_{column}) \tag{19}$$

Ht_{total} = ______________ ft

E. The volume of packing needed to fill the tower is determined from Eq. (20).

$$V_{packing} = (\pi/4)(D_{column})^2 (\text{Ht}_{column}) \tag{20}$$

$V_{packing}$ = ______________ ft^3

Example 7

A wet scrubber (packed tower; Fig. 1b) is proposed to treat Emission Stream 5 in Table 11. With the results from Examples 3 and 5, the column height and packing volume are determined using the procedure explained in Example 6.

Solution

A. Calculation of N_{og} (number of gas transfer units) using Eq. (13) and with an assumed value for AF of 1.6:

$HAP_e = 20{,}000$ ppmv
RE = 98%
$HAP_o = HAP_e[1 - (RE/100)]$ (14)
$HAP_o = 20{,}000\ [1 - (98/100)] = 400$ ppmv

$N_{og} = \ln\{[(HAP_e / HAP_o)(1 - (1/AF)) + (1/AF)]\} / (1 - 1/AF)$ (13)
$N_{og} = \ln\{[(20{,}000 / 400)(1 - (1/1.6)) + (1/1.6)]\} / (1 - 1/1.6)$
$N_{og} = 7.97$

B. Calculation of H_{og} using Eqs. (15)–(18).

$L'' = L/A_{column}$ (18)

Use the following values:

$L = 17{,}410$ lb/h
$A_{column} = 10.8\ ft^2$ from Example 5

Then, substitute these values into Eq. (18).

$L'' = 17{,}410 / 10.8 = 1{,}612$ lb/h-ft^2

Use the following value

$G_{area} = 0.34$ lb/s-ft^2 from Example 5
$3600G_{area} = (3600)(0.34) = 1224$ lb/h-ft^2

With these values for L'' and 3,600 G_{area}, the packing factors are obtained from Tables 1 and 2. The following constants (used for determining height of a gas film transfer units) are obtained from Table 1 for 2 in. Raschig rings:

$b = 3.82, \quad c = 0.41, \quad d = 0.45$

Because 1224 lb/hr-ft^2 is outside the range for 3600 G_{area} of Table 1, one must proceed on the assumption that the above packing factors do apply and that any error introduced into the calculation will be minimal. For 2 in. Raschig rings, the following constants (used for determining height of a liquid film transfer units) are obtained from Table 2:

$Y = 0.0125$ and $s = 0.22$

From Tables 3 and 4, the Schmidt numbers are obtained:

$Sc_G = 0.66$ and $Sc_L = 570$

Also,

$\mu_L'' = 0.85$ centipoises (cP) at 85°F (see Example 5)
$\mu_L'' = (0.85)(2.42) = 2.06$ lb/h-ft (the conversion factor from cP to lb/h-ft is 2.42)

Therefore, H_G (height of gas transfer unit) and H_L (height of liquid transfer unit) are determined from Eqs. (16) and (17), respectively.

$H_G = [b(3600G_{area})^c / (L'')^d](Sc_G)^{0.5}$ (16)
$H_G = [3.82(1224)^{0.41}/(1612)^{0.45}](0.66)^{0.5}$
$H_G = 2.06$ ft

$H_L = Y(L''/\mu_L'')^s (Sc_L)^{0.5}$ (17)
$H_L = 0.0125\,(1612 / 2.06)^{0.22}\,(570)^{0.5}$
$H_L = 1.29$ ft

Use AF = 1.6. Then Eq. (15) can be used to determine H_{og} (height of the overall gas transfer unit):

$H_{og} = H_G + (1/\text{AF})\, H_L$ (15)
$H_{og} = 2.06 + (1/1.6)\ 1.29 = 2.87$, or 2.9

C. Calculation of Ht using Eq. (12).

$\text{Ht}_{column} = N_{og}\ H_{og}$ (12)
$\text{Ht}_{column} = (7.97)(2.9) = 23.1$, or 23 ft

D. Now use Eq. (19) to determine Ht_{total}

$\text{Ht}_{total} = \text{Ht}_{column} + 2 + (0.25D_{column})$ (19)
$\text{Ht}_{total} = 23 + 2 + [(0.25)\ (4)] = 26$ ft

E. The volume of packing needed to fill the tower is determined from Eq. (20).

$V_{packing} = (\pi/4)(D_{column})^2\,(\text{Ht}_{column})$ (20)
$V_{packing} = (0.785)(4)^2\,(23) = 290\ \text{ft}^3$

Example 8

In this example, a step-by-step procedure for determining the pressure drop of the packed bed of the wet scrubber (Fig. 1b) is presented.

Solution

A. Use Eq. (21) to the calculate pressure drop, P_a, select a packing, and determine the constants (g and r) using Table 5.

$g =$ ______________ $r =$ ______________

$$P_a = \left(g \times 10^{-8}\right)\left[10^{\left(rL''/D_L\right)}\right]\left(3600G_{area}\right)^2 / D_G \quad (21)$$

$P_a =$ ______________ pressure drop (lb/ft^2-ft)

B. Use L'', D_L, G_{area}, and D_G from previous exercises.

C. Use Eq. (22) to calculate P_{total}. The total pressure drop through a packed tower (Fig. 1b) or wet scrubber is

$P_{total} = P_a\,\text{Ht}_{column}$ (22)
$P_{total} =$ ______________ lb/ft^2
$P_{total}/5.2 =$ ______________ in. H_2O

Example 9

Emission Stream 5 (Table 11) is again a candidate for pollution control using a packed tower (wet scrubber, Fig. 1b). The total pressure drop through the entire scrubber tower is determined from the results of Examples 1, 3, 5, and 7 and the procedure just explained in Example 8.

Solution

From Table 5, the packing constants $g = 11.13$ and $r = 0.00295$ and

$L'' = 1612$ lb/h-ft^2 (Example 7)
$3600G_{area} = (3600)(0.34) = 1224$ lb/h-ft^2 (Example 7)
$D_G = 0.071$ lb/ft^3 (Example 5)
$D_L = 62.18$ lb/ft^3 at 85°F (Example 5)

Substituting these values into Eq. (21) to determine pressure drop (P_a):

$$P_a = (g \times 10^{-8}\ [10^{(r\ L''/D_L)}]\ (3600G_{area})^2/D_G \tag{21}$$

$$P_a = (11.13 \times 10^{-8})\ [10^{\ (0.00295)(1612\ )\ /\ 62.18}]\ (1224)^2\ /\ 0.071$$

$$P_a = (11.13 \times 10^{-8})\ [10^{\ (0.07648)}]\ (1{,}498{,}176)\ /\ 0.071$$

$$P_a = (11.13 \times 10^{-8})\ [1.1926]\ (1{,}498{,}176)\ /\ 0.071$$

$$P_a = 2.78 \text{ lb/ft}^2\text{-ft}$$

The total pressure drop through a packed tower or wet scrubber can then be calculated:

$\text{Ht}_{column} = 23$ ft (from Example 7)

$$P_{total} = P_a \text{Ht}_{column} \tag{22}$$

$$P_{total} = (2.8)(23) = 64.4 \text{ lb/ft}^2$$

$$P_{total}/5.2 = 12.4 \text{ in. } H_2O$$

Example 10

In this exercise, the initial (capital) costs and operating costs of a wet scrubber are estimated. These costs include initial cost of purchase as well as direct and indirect annual costs (or total annual costs). Recovery credit is assumed to be negligible.

Solution

A. Total Cost of Absorption (Wet Scrubber) Systems

The cost of the absorption system can be found in Figure 7. The cost of packing required to fill the tower is estimated using Table 2 with the volume of packing determined in the previous discussion. Likewise, secondary costs for related equipment were also previously determined. Summation of these costs will determine the total capital outlay needed for this wet scrubber.

1. Absorber tower cost = $______________
2. Packing cost = $______________
3. Auxiliary equipment cost = $______________
4. Equipment cost (EC) = 1 + 2 + 3 = $______________

The purchased equipment cost, PEC, is obtained from the equipment cost determined above and the factors found in Table 6.

PEC = EC + instrumentation + taxes + freight charges
PEC = $______________

Total capital cost (TCC) is estimated using the above calculated PEC and the factors found in Table 6.

TCC = 2.20PEC + SP + building cost
TCC = $______________
where SP is the site preparation cost ($).

B. Direct Annual Cost of a Wet Scrubber System (Absorber)

The total annual costs of a scrubber system include direct costs of power, solvent (absorbent), labor, and so form. Additionally, indirect costs such as taxes, administrative costs, insurance, and so on. must also be considered to fully appreciate the total capital demanded to operate a scrubber system for a full year.

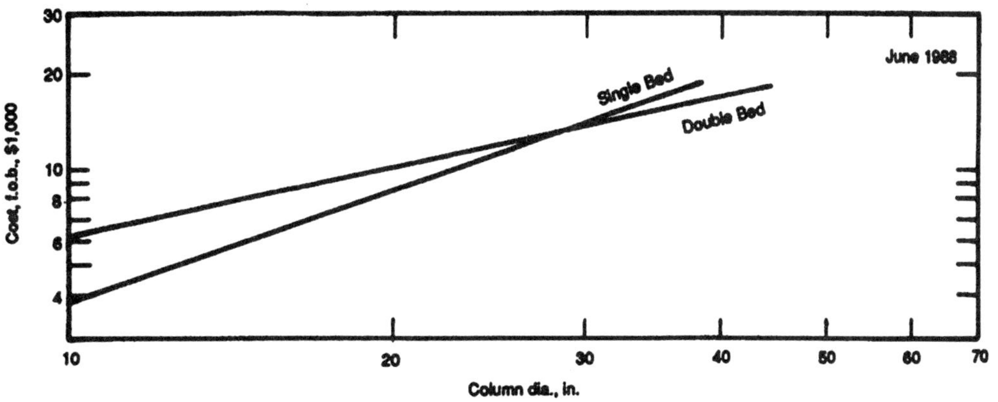

Fig. 7. Costs of absorber towers (from US EPA).

1. Direct Annual Costs (such as electricity, chemicals, labor, etc. expenses)
   a. Power (electricity) cost: Use Eq. (23) to estimate the fan power requirements, F_p:

   $$F_p = 1.81 \times 10^{-4}(Q_{e,a})(P_{\text{total}})(\text{HRS}) \tag{23}$$

   F_p = ________________kWh/yr

   and

   $$Q_{e,a} = Q_e\,(T_e + 460) / 537 \tag{24}$$

   $Q_{e,a}$ = ________________ acfm

   where Q_e is the actual emission stream flow rate (scfm) and T_e is the emission stream temperature (°F).

   It assumed that 2 wk out of the year the factory utilizing the absorber is shut down for inventory and retooling, which provides an annual operating period of 50 wk/yr. Additionally, it is assumed the factory uses the scrubbing system 3–8 h shift, 5 d a week.

   HRS = annual operating hours (h/yr)

   HRS = (3 shift/d) (8 h/shift) (5 d/wk) (50 wk/yr) = 6,000 h/yr

   b. The annual electric cost (AEC) can be determined assuming that UEC is equal to $0.059/kWh and using Eq. (25):

   $$\text{AEC} = (\text{UEC})(F_p) = \$0.059(F_p) \tag{25}$$

   AEC = $________________

2. Solvent cost
   a. Equation (26) estimates the annual solvent requirement (ASR):

   $$\text{ASR} = 60(L_{\text{gal}})\text{HRS} \tag{26}$$

   ASR = ________________ gal/yr

   b. The annual solvent cost (ASC) is obtained from multiplying the yearly solvent requirement and the unit solvent cost (USC) found in Table 7. Because water is commonly used as solvent, the USC cost for water is assumed to equal $0.20 per 1,000 gal.

   $$\text{ASC} = (\text{USC})(\text{ASR}) \tag{27}$$

   ASC = ($0.20/1000 gal)(ASR)

   ASC = $________________

3. Operating Labor Cost: It assumes that 2 wk out of the year the factory utilizing the absorber is shut down for inventory and retooling, which provides an annual operating period of 50 wk/yr.
   Operating labor = (0.5 h/shift)(3 shifts/d)(5 d/wk)(50 wk/yr)($12.96 /h)
   Operating labor cost = $____________

   Supervisory cost = 0.15 (Operating labor)
   Supervisory cost = $____________

4. Maintenance Costs

   Maintenance labor cost = [(0.5 h/shift)(3 shifts/d)(5 d/wk)(50 wk/yr)]($14.96/h labor cost)
   Maintenance labor cost = $____________

   Maintenance materials cost = 1.0 (Maintenance labor)
   Maintenance materials cost = $____________

5. Total Annual Direct Costs = 1b + 2b + 3 + 4 = $____________

C. Indirect Annual Costs of Absorbers
These expenses are obtained from the factors given in Table 7, and the summation of these expenses provides the total indirect annual costs.

Overhead	=	$
Property tax	=	$
Insurance	=	$
Administration	=	$
Capital recovery	=	$
Total Indirect Annual Cost	=	$

D. Total Annual Costs = Total Direct Annual Costs + Total Indirect Annual Costs
Total Annual Costs = $____________

Example 11

The cost of a wet scrubber system (*see* Fig. 1b) is estimated in this example. The scrubber upon which this estimate is based is the scrubber designed to treat Emission Stream 5 (Table 11) in Examples 1, 3, 5, 7, and 9. The methodology of the determination is similar to the procedure detailed in Example 10.

Solution

A. Purchase Equipment and Total Capital Costs
The diameter of the column is 4 ft, from Example 5. Extrapolating from Figure 7 (*see* Example 10), the cost of a single bed tower is approx $25,000. The volume of packing, $V_{packing}$ needed to fill the tower is 290 ft^3. Assume that 2-in. ceramic porcelain Raschig rings are used to pack the tower. The cost of this packing is given in Table 12 as $12.75 per cubic foot. The total packing needed for the project therefore will cost $3700. A figure of $10,000 is assumed for the expense of auxiliary equipment. The equipment cost (EC) is the summation of these costs:

Cost of a single bed tower	=	$25,000
Cost of this packing	=	$3,700
Cost of auxiliary equipment	=	$10,000
Equipment cost (EC)	=	$38,700

Table 12
Cost of Packing Material (8)

Packing Type and Material	Cost per ft^3			
Packing Diameter, in.	1	1.5	2	3
<u>Flexisaddles</u>				
Porcelain	19.5	17.75	16.75	16.00
Polypropylene	20.00	18.00	—	—
Stoneware	19.50	17.75	16.75	16.00
<u>Raschig Rings</u>				
Porcelain	16.00	14.50	12.75	13.25
<u>Pall Rings</u>				
Carbon Steel	35.00	30.00	—	—
304L Stainless Steel	73.50	69.80	66.00	—
316L Stainless Steel	103.00	101.00	97.00	—

The factor in Table 6 may now be used to determine the following costs:

Instrumentation cost	= 0.10 ($38,700)	=	$3,870
Taxes	= 0.03 ($38,700)	=	$1,160
Freight cost	= 0.05 ($38,700)	=	$1,940
	Total		$6,970

The purchased equipment cost, PEC, is obtained from the summation of the above costs.

PEC = EC + instrumentation + taxes + freight charges
PEC = $38,700 + ($3,870 + $1,160 + $1,940) = $38,700 + $6,970 = $45,670

The total capital cost (TCC) is estimated using the above calculated PEC (Table 13) and the cost determined using factors found in Table 6.

TCC = 2.20PEC + SP + building cost
TCC = 2.20 ($45,670) + SP + building cost
TCC = $100,474 + SP + building cost ~ $100,000 + SP + building cost

Because costs for site preparation and building are unique for each installation, the TCC for this example is assumed to be equal to $100,000 (Table 13).

B. Direct Annual Costs (DAC)

1. Electric Power Cost
Equation (23) provides the fan power requirement, F_p (kWh/yr):

$$F_p = 1.81 \times 10^{-4}(Q_{e,a})(P_{\text{total}})(\text{HRS}) \tag{23}$$

where

$$Q_{e,a} = Q_e\,(T_e + 460)\,/\,537 \tag{24}$$

T_e = 85°F from Example 1
HRS = (8 h/wk) (3shifts/d)(5 d/wk)(50 wk/yr) = 6000 h/yr
$Q_{e,a}$ = 3000 (85 + 460)/537 = 3,045 acfm

Then substitute into Eq.(23):

$$F_p = 1.81 \times 10^{-4}\,(3.045)(12)(6{,}000)$$

Table 13
Example Case–Capital Costs for Tower Absorbers

Cost Item	Factor	Cost
Direct Costs, DC		
Purchased equipment cost		
Absorber (tower & packing)		
+ auxiliary equipment	As estimated, EC	$38,700
Instrumentation	0.10 EC	3,870
Sales tax	0.03 EC	1,160
Freight	0.05 EC	1,940
Purchased Equipment Cost, PEC	1.18 EC	$45,670
Direct Installation Costs		
Foundation and supports	0.12 PEC	$5,480
Erection and handling	0.40 PEC	18,300
Electrical	0.01 PEC	457
Piping	0.30 PEC	13,700
Insulation	0.01 PEC	457
Painting	0.01 PEC	457
Direct Installation Cost	0.85 PEC	$38,900
Site preparation	As required, SP	
Building	As required, Bldg.	
Total Direct Costs, DC	1.85 PEC + SP + Bldg.	$84,500 + SP + Bldg.
Indirect Costs, IC		
Engineering	0.10 PEC	$4,570
Construction	0.10 PEC	4,570
Contractor fee	0.10 PEC	4,570
Start-up	0.01 PEC	457
Performance test	0.01 PEC	457
Contingencies	0.03 PEC	1,370
Total Indirect Cost	0.35 PEC	$16,000
Total Capital Cost	2.20 PEC + SP + Bldg.	$100.000 + SP + Bldg.

$F_p = 3.97 \times 10^4$ kWh/yr

Then, the annual electrical cost (AEC) is determined from Eq. (25):

$$AEC = (UEC)(F_p) = \$0.059(F_p) \quad (25)$$

$$AEC = \$0.059(3.97 \times 10^4) = \$2{,}340/yr$$

2. The annual solvent requirement (ASR) is found using Eq. (26):

$$ASR = 60(L_{gal})\ HRS \quad (26)$$

$L_{gal} = 35$ gal/min (from Example 3)

$$ASR = 60(35)(6000) = 1.26 \times 10^7 \text{ gal/yr}$$

The cost of solvent is found by multiplication of the cost factor found in Table 7. As of January 1990, the solvent cost of water, on average, was $0.20 per 1,000 gal in the United States (Note: 1 United States gal = 3.785 L). Use Eq. (27) to calculate the annual solvent cost:

$$ASC = (USC)(ASR) \quad (27)$$
$$ASC = (0.20/1000)(1.26 \times 10^7) = \$2{,}520$$

3. Annual operating labor costs are estimated by assuming that for every 8-h shift the scrubbing system is operated, 0.5 h are required to maintain the wet scrubber. With labor cost at $12.96/h (*see* Table 7),

   Annual operating labor hours = (0.5 h/shift)(3 shifts/d)(5 d/wk) (50 wk/yr)
   Annual operating labor hours = 375 h/yr
   Annual operating labor cost = (375 h/yr)($12.96 /h) = $4,860/yr

   Supervisor labor cost is estimated at 15% of operator labor cost (*see* Table 7):

   Annual supervisor labor cost = (15%)(annual operator labor cost)
   Annual supervisor labor cost = (0.15)($4,860) = $729/yr

4. The annual maintenance labor cost is estimated by assuming that for every 8-h shift the scrubbing system is operated, 0.5 h are required to maintain the wet scrubber. The maintenance labor cost is based on rate of $14.26 per hour. This cost for this example is determined as follows:

   Annual maintenance labor hours = (0.5 h/shift)(3 shifts/d)(5 day/wk)(50 wk/yr)
   Annual maintenance labor hours = 375 h/yr
   Annual maintenance labor cost = (375 h/yr) ($14.26/h) = $5,350

5. Annual Maintenance Materials Cost
   The maintenance material cost is estimated by assuming it is equivalent to 100% maintenance labor cost. This cost for this example is determined as follows:

   Annual maintenance materials cost = (100%)(annual maintenance labor cost)
   Annual maintenance materials cost = (1.0)($5,350) = $5,350/yr

6. The total direct annual costs is the summation of direct costs:

AEC	= $2,340
ASC	= $2,520
Annual operating labor cost	= $4,860
Annual supervisor labor cost	= $729
Annual maintenance cost	= $5,350
Annual maintenance materials cost	= $5,350
Total Annual Direct Cost	= $21,100

C. Indirect Annual Costs
Table 7 lists indirect cost factors. These factors are utilized to determine indirect annual costs (ICA) for this example as follows:

Overhead	= 0.60(annual operating labor and maintenance costs)
Overhead	= 0.60($4,860 + $729 + $5,350 + 5,350) = $9,770
Property tax	= 0.01(TCC)
Property tax	= 0.01($100,000) = $1,000
Insurance	= 0.01(TCC)
Insurance	= 0.01($100,000) = $1,000
Administration	= 0.02(TCC)
Administration	= 0.02($100,000) = $2,000

Capital recovery = 0.1628(TCC)
Capital recovery = 0.1628($100,000) = $16, 280 or $16,300

D. Total Indirect Annual Cost

Total indirect annual cost = Overhead + property tax + insurance + administration + capital recovery
Total indirect annual cost = $9,770 + $1,000 + $1,000 + $2,000 + $16,300
Total indirect annual cost = $30,100

E. The total annual cost is determined by summation of Annual Direct and Annual Indirect costs

Total annual costs = Total direct annual costs + Total indirect annual costs
Total annual costs = $21,100 + $30,100 = $51,200

Example 12

What advantages or disadvantages does dry scrubbing versus wet scrubbing technologies have when these two absorption methods are compared?

Solution

As previously discussed, incineration processes are often found to have some type of dry scrubbing system treating the exhaust being formed in the process. Halogenated compounds, if present, will produce acid gases as the result of combustion. The scrubber system (wet or dry) is used to limit the release of such gases into the atmosphere. Both wet and dry systems use absorption to collect the acid gases present. The dry process will operate at a lower pressure loss than the wet scrubber system. This has the consequence of reducing power costs if a dry system is chosen versus a wet scrubbing system. Also, if using a dry scrubber, the gas exiting the stack will be warm. Additionally, waste product produced in a dry scrubbing system is collected as a solid.

If a wet scrubber is used, the gas exiting the stack is often reheated. Therefore, in addition to having higher operating costs (as the result of a higher pressure drop compared to a dry scrubber), the wet scrubber also may have additional energy costs from this need to warm the exit gas. Additionally, as the wet scrubber uses slurry as the absorbent, waste product is also collected in slurry form. This slurry will have a greater total volume than the dry end product collected in the dry scrubber, and the cost of final disposal may be higher as a result.

At this point of discussion, the advantage/disadvantage comparison indicates that a dry scrubber will be a more economical air pollution control choice than a wet scrubber. However, a wet scrubber does not need a downstream collection device, whereas a dry scrubber does. Also, if SO_2 must be removed from the gas stream, removal efficiency >90% is most often only achieved economically with a wet scrubber. Also, as dry scrubbers must be overdosed with absorbent more so than for wet scrubbers, solvent costs will be higher in dry versus wet scrubbing. (Explained another way, a certain percentage of absorbent in a dry scrubber is not utilized. Such excess solvent cannot be economically separated from the final waste product for reuse. As such, unused solvent in a dry scrubber is a cost that yields no return.)

Example 13

A proposed solution to an air pollution control project is a Venturi scrubber system. A blank data sheet is given to collect the data necessary for the environmental review process to go forward.

Solution

A data collection exercise for the purpose of preparing a permit application suitable for submission to the appropriate authorities must accommodate the following: pretreatment considerations, construction materials, pressure drop estimation, and actual Venturi size estimation. The following is a data sheet suitable for this exercise.

Calculation Sheet for Venturi Scrubbers

A. HAP Characteristics—needed data

1. Flow rate ($Q_{e,a}$) = ______________ acfm
2. Temperature (T_e) = ______________ °F
3. Moisture content (M_e) = ______________ %
4. Required collection (removal) efficiency (CE) = ______________ %
5. Mean particle diameter (D_p) = ______________ μm
6. Particulate content = ______________ grain/scf
7. HAP content = ______________ % of total mass

B. Permit Review Data to be presented by applicant

1. Reported pressure drop (across Venturi) (P_v) = ______________ in. H_2O
2. Pertinent performance curve for Venturi scrubber (from supplier)
3. Reported collection (removal) efficiency (CE) = ______________ %

C. Determination of Pretreatment Requirements
As previously discussed, for the Venturi principle to be applied to an air pollution control need, the air being treated should be at 50 –100°F above its saturation (dew) temperature. If the air to be treated does not meet this condition, then pretreatment of the air will be required. Such pretreatment of the air will change two important design parameters:

1. Maximum flow rate at actual conditions ($Q_{e,a}$) = ______________ acfm
2. Temperature (T_e) = ______________ °F

D. Projected Venturi Pressure Drop
The suppliers of the Venturi scrubber system supply the operating data curves (*see* Fig. 5). These curves can be used to estimate the pressure drop, P_v, for the proposed Venturi scrubber, at a given removal efficiency.

P_v = ______________ in. H_2O

Also previously noted, if this pressure drop exceeds 80 in. of H_2O, alternative control technology needs to be considered, as the Venturi scrubber will most likely not achieve the desired removal efficiency.

E. Proposed Material of Construction
Selection of material used to actually fabricate a Venturi scrubber is normally recommended the system supplier. A useful first estimate of the type of material required can also be made by consulting Table 14.

Material of construction = ______________

F. Proposed Venturi Scrubber Sizing
Performance curves supplied by system suppliers may be derived for saturated emission stream flow rate ($Q_{e,s}$). If so, $Q_{e,s}$ may be determined:

$$Q_{e,s} = [Q_{e,a} (T_{e,s} + 460) / (T_e + 460)] + Q_w \quad (28)$$

Table 14
Construction Materials for Typical Venturi Scrubber Applications (9, 15, 18)

Application	Construction Material
Boilers	
Pulverized coal	316L stainless steel
Stoker coal	316L stainless steel
Bark	Carbon steel
Combination	316L stainless steel
Recovery	Carbon steel or 316L stainless steel
Incinerators	
Sewage sludge	316L stainless steel
Liquid waste	High nickel alloy
Solid waste	
Municipal	316L stainless steel
Pathological	316L stainless steel
Hospital	High nickel alloy
Kilns	
Lime	Carbon steel or stainless steel
Soda ash	Carbon steel or stainless steel
Potassium chloride	Carbon steel or stainless steel
Coal Processing	
Dryers	304 stainless steel or 316L stainless steel
Crushers	Carbon steel
Dryers	
General spray dryer	Carbon steel or stainless steel
Food spray dryer	Food-grade stainless steel
Fluid bed dryer	Carbon steel or stainless steel
Mining	
Crushers	Carbon steel
Screens	Carbon steel
Transfer points	Carbon steel
Iron and Steel	
Cupolas	304-316L stainless steel
Arc furnaces	316L stainless steel
BOFs	Carbon steel (ceramic lined)
Sand systems	Carbon steel
Coke ovens	Carbon steel
Blast furnaces	Carbon steel (ceramic lined)
Open hearths	Carbon steel (ceramic lined)
Nonferrous Metals	
Zinc smelters	Stainless steel or high nickel
Copper and brass smelters	Stainless steel or high nickel
Sinter operations	Stainless steel or high nickel
Aluminum reduction	High nickel
Phosphorus	
Phosphoric acid	
Wet process	316L stainless steel
Furnace grade	316L stainless steel

Continued

Table 14 *(Continued)*

Application	Construction Material
Asphalt	
Batch plants – dryer	Stainless steel
Transfer points	Carbon steel
Glass	
Container	Stainless steel
Plate	Stainless steel
Borosilicate	Stainless steel
Cement	
Wet process kiln	Carbon steel or stainless steel
Transfer points	Carbon steel

where $Q_{e,s}$ is the saturated emission stream flow rate (acfm), $T_{e,s}$ is the temperature of the saturation emission stream (°F), T_e is the temperature of the emission stream at inlet air (°F), $Q_{e,a}$ is the actual emission flow rate from Eq. (24) (acfm), and Q_w is the volume of water added (ft^3/min or cfm).

$T_{e,s}$ is estimated using the psychrometric chart shown in Fig. 4 with values for $L_{w,a}$ and T_e. The inlet lb of H_2O per lb of dry air ($L_{w,a}$) is determined by converting M_e from percent volume to the lb of H_2O per lb of dry air as follows:

$L_{w,a} = (M_e / 100)(18/29) =$ ______________ lb water/lb dry air

The adiabatic saturation line is determine on the psychrometric chart by determining the intersection of the humidity ($L_{w,a}$) and the inlet temperature (T_e). This adiabatic saturation line is followed to the left until it intersects the 100% relative saturation line. At this intersection, the temperature of the saturated emission ($T_{e,s}$) is read from the ordinate and the saturated emission ($L_{w,s}$) is read from the abscissa.

$T_{e,s} =$ ______________ °F

$$Q_w = Q_{e,\text{ad}}(D_e)(L_{w,s} - L_{w,a})(1/D_w) \qquad (29)$$

where $Q_{e,\text{ad}} = (1 - L_{w,a})\,Q_{e,a}$ (acfm) (29a)

D_e is the density of the polluted air stream (lb/ft^3), $L_{w,s} = 0.10$ saturated lb water/lb dry air (from Fig. 4), $L_{w,a} = 0.031$ inlet lb water/lb dry air (from Fig. 4), and D_w is the density of water vapor (lb/ft^3).

Using the ideal gas law, an approximate density of any gas encountered in an air pollution control project can be made:

$$D = (PM) / (RT) \qquad (30)$$

where $D = D_e$ is the density of emission (lb/ft^3), P is the pressure of the emission stream (1 atm), M is the molecular weight of the specific pollutant gas (lb/lb-mole), R is the gas constant (0.7302 atm-ft^3/lb-mole °R), and $T = T_{e,s}$ is the temperature of the gas (°R).

The density of the emission stream is calculated from Eq. (30):

$D_e =$ ______________ ft^3/lb

The density of water vapor is determined from Eq. (30):

$D_w =$ ______________ ft^3/lb

where $D = D_w$ is the density of emission (lb/ft^3), P is the pressure of the emission stream (1 atm), M is the molecular weight of the specific pollutant gas (18 lb/lb-mol), R is the gas constant (0.7302 atm-ft^3/lb-mol °R), and $T = T_{e,s}$ is the temperature of the gas (°R).

Substituting into Eq. (29) yields

$Q_w =$ ______________ cfm

The saturated emission stream flow rate is determined by substituting into Eq. (29):

$Q_{e,s} =$ ______________ acfm

G. Permit Application for Submission
The results of this exercise, with supporting data, are summarized in the following table. Note P_v (estimated pressure drop) versus the observed pressure drop for this Venturi scrubber. When estimated and reported results differ significantly, such discrepancies may be the result of the following:

1. Use of an incorrect performance curve
2. Disagreement between required and observed removal efficiencies

This will necessitate a discussion of system details (design and operational procedures) with the applicant. If the estimated and operational pressure losses are in agreement, one may assume that both design and operation of the Venturi scrubber are satisfactory based on the assumptions used in this handbook.

	Calculated value	Observed value
Particle mean diameter, D_p		
Collection efficiency, CE		
Pressure drop across venture, P_v		

Example 14

This example uses an instance in which a municipal incinerator is conducting an investigation evaluating the usefulness of the Venturi scrubber (*see* Fig. 1c) to solve an air pollution control issue. The Venturi is one of several possible methods that could be used to treat this air emission stream (*see* Table 15). Using the Calculation Sheet for Venturi Scrubbers from Example 13, this study will proceed according to the following:

A. Gather the pertinent pollutant characterization data.
B. Estimate the Venturi scrubber pressure drop.
C. Decide the needed fabrication material.
D. Determine saturated gas flow rate for Venturi scrubber sizing purposes.
E. Consult with the appropriate regulatory authorities (federal, state, and local) how the agency will evaluate the incinerator's Venturi scrubber permit application.

Solution

A. HAP Characteristics—needed data:
Because a Venturi scrubber is one of the selected control techniques for the theoretical municipal emission stream, the pertinent data for these procedures are taken from the HAP Emission Stream Data Form (Table 15).

Table 15
Effluent Characteristics for a Municipal Incinerator Emission Stream

HAP EMISSION STREAM DATA FORM*

Company: Incineration, Inc	Plant contact: Mr. Phil Brothers
Location (Street): 123 Main Street	Telephone No.: (999) 555-5024
(City): Somewhere	Agency Contact: Mr. Ben Hold
(State, Zip):	No. of Emission Streams Under Review: 1

A. Emission Stream Number/Plant Identification: #1 / Incineration

		(a)	(b)	(c)
B.	HAP Emission Source	municipal incinerator		
C.	Source Classification	process point		
D.	Emission Stream HAPs	cadmium		
E.	HAP Class and Form	inorganic particulate		
F.	HAP Content (1,2,3)**	10%		
G.	HAP Vapor Pressure (1,2)			
H.	HAP Solubility (1,2)			
I.	HAP Adsorptive Prop. (1,2)			
J.	HAP Molecular Weight (1,2)			

K. Moisture Content (1,2,3): 5% vol	P. Organic Content (1)***:
L. Temperature (1,2,3): 400° F	Q. Heat/O_2 Content (1):
M. Flow Rate (1,2,3): 110,000 acfm	R. Particulate Content (3): 3.2 gr/acf, flyash
N. Pressure (1,2): atmospheric	S. Particle Mean Diam. (3): 1.0 um
O. Halogen/Metals (1,2): none / none	T. Drift Velocity/SO_3 (3): 0.31 ft/sec/200 ppmv

U. Applicable Regulation(s):
V. Required Control Level: assume 99.9% removal
W. Selected Control Methods: fabric filter, ESP, venturi scrubber

* The data presented are for an emission stream (single or combined streams) prior to entry into the selected control method(s). Use extra forms if additional space is necessary (e.g., more than three HAPs) and note this need.

** The numbers in parentheses denote what data should be supplied depending on the data on lines C and E:
1 = organic vapor process emission
2 = inorganic vapor process emission
3 = particulate process emission

*** Organic emission stream combustibles less HAP combustibles shown on lines D and F.

1. Flow rate ($Q_{e,a}$) = 110,000 acfm
2. Temperature (T_e) = 400°F
3. Moisture content (M_e) = 5%
4. Required collection (removal) efficiency (CE) = 99.9%
5. Mean particle diameter (D_p) = 1.0 μm
6. Particulate content = 3.2 grains/scf fly ash
7. HAP content = 10% of total mass of cadmium

B. Permit Review Data to be presented by applicant:
Given a required collection (removal) efficiency of 99.9% and the mean particle diameter in the emission stream from the incinerator is estimated as 1.0 μm, from Fig. 5. Note that the value of P_v is outside the range of data presented in Table 8. The value of 47 in. H_2O is less than 50 in. H_2O, therefore, it may still be safely used in this exercise. The Venturi scrubber should perform properly.

1. Reported pressure drop (across Venturi) (P_v) = 47 in. H_2O (*see* Fig. 5)
2. Pertinent performance curve for Venturi scrubber (from supplier)
3. Reported collection (removal) efficiency (CE) = 99.9%

C. Determination of Pretreatment Requirements
As previously discussed, for the Venturi principle to be applied to an air pollution control need, the air being treated should be at 50–100°F above its saturation (dew) temperature. If the air to be treated does not meet this condition, then pretreatment of

the air will be required. Such pretreatment of the air will affect two important design parameters:

1. Maximum flow rate at actual conditions ($Q_{e,a}$) = 100,000 acfm
2. Temperature (T_e) = 400°F

D. Projected Venturi Pressure Drop
The suppliers of the Venturi scrubber system supplies the operating data curves (*see* Fig. 5). These curves can be used to estimate the pressure drop, P_v, for the proposed Venturi scrubber, at a given removal efficiency.

$$P_v = 47 \text{ in. } H_2O$$

Also previously noted, if this pressure drop exceeds 80 in. H_2O, alternative control technology needs to be considered, as the Venturi scrubber will most likely not achieve the desired removal efficiency.

E. Proposed Material of Construction
Selection of material used to actually fabricate a Venturi scrubber is normally recommended by the system supplier. A useful first estimate of the type of material required can also be made by consulting Table 14.

Material of construction = 316L stainless steel

F. Proposed Venturi Scrubber Sizing
Performance curves supplied by system suppliers may be derived for saturated emission stream flow rate ($Q_{e,s}$). If so, $Q_{e,s}$ may be determined:

$$Q_{e,s} = [Q_{e,a}\,(T_{e,s} + 460) / (T_e + 460)] + Q_w \tag{28}$$

where $Q_{e,s}$ is the saturated emission stream flow rate (acfm), $T_{e,s}$ is the temperature of the saturation emission stream (°F), T_e is the temperature of the emission stream at inlet air (°F), $Q_{e,a}$ is the actual emission flow rate from Eq. (24) (acfm), and Q_w is the volume of water added (ft^3/min or cfm).

$T_{e,s}$ is estimated to be 127°F using the psychrometric chart shown in Fig. 4 with values for $L_{w,a}$ and T_e. The inlet lb of H_2O per lb of dry air ($L_{w,a}$) is determined by converting M_e (now known to be 5%) from percent volume to the lb of H_2O per lb of dry air as follows.

$$L_{w,a} = (M_e / 100)\,(18/29) = (5/100)\,(18/29) = 0.031 \text{ lb } H_2O / \text{lb dry air (Fig. 4)}$$

The adiabatic saturation line is determined on the psychrometric chart by determining the intersection of the humidity ($L_{w,a} = 0.031$) and the inlet emission stream temperature ($T_e = 400$°F). This adiabatic saturation line is followed to the left until it intersects the 100% relative saturation line. At this intersection, the temperature of the saturated emission ($T_{e,s}$) is read from the ordinate and the saturated emission $L_{w,s}$ is read from the abscissa.

$$T_{e,s} = 127°\text{F}$$

$$Q_w = Q_{e,\text{ad}}\,(D_e)\,(L_{w,s} - L_{w,a})\,(1/D_w) \tag{29}$$

where

$$Q_{e,\text{ad}} = (1 - L_{w,a})\,Q_{e,a} = (1 - 0.031)\,(110{,}000) = 106{,}590 \text{ acfm} \tag{29a}$$

D_e is the density of the polluted air stream (lb/ft^3), $L_{w,s}$ = 0.10 saturated lb water/lb dry air (from Fig. 4), $L_{w,a}$ = 0.031 inlet lb water/lb dry air (from Fig. 4), and D_w is the density of water vapor, (lb/ft^3).

Using the ideal gas law, an approximate density of any gas encountered in an air pollution control project can be made:

$$D = (PM) / (RT) \tag{30}$$

where $D = D_e$ is the density of emission (lb/ft^3), P is the pressure of the emission stream (1 atm), M is the molecular weight of the specific pollutant gas (29 lb/lb-mol), R is the gas constant (0.7302 atm-ft^3/lb-mol °R), and $T = T_{e,s}$ is the temperature of the gas (°R).

The density of the emission stream is calculated from Eq. (30):

$$D_e = (1)\ (29) / [(0.7302)\ (127 + 460)] = 0.0676 \text{ ft}^3/\text{lb}$$

The density of water vapor is determined from Eq. (30):

$$D_w = (1)\ (18) / [(0.7302)\ (127 + 460)] = 0.042 \text{ ft}^3/\text{lb}$$

where $D = D_w$ is the density of emission (lb/ft^3), P is the pressure of the emission stream (1 atm), M is the molecular weight of the specific pollutant gas (18 lb/lb-mol), R is the gas constant (0.7302 atm-ft^3/lb-mol °R), and $T = T_{e,s}$ is the temperature of the gas (°R).

Substituting into Eqs. (29a) and (29) yields

$$Q_w = (106{,}590)(0.0676)(0.10 - 0.031)(1/0.042) = 11{,}838 \text{ cfm or } 11{,}800 \text{ cfm} \tag{29a}$$

The saturated emission stream flow rate is determined by substituting into Eq. (29):

$$Q_{e,s} = [(110{,}000)\ (127 + 460) / (400 + 460)\ ] + 11{,}800 = 86{,}800 \text{ acfm} \tag{29}$$

The calculated saturated gas flow rate will then be used for sizing a commercially available venture scrubber.

G. Permit Application for Submission

The controlling regulatory agency (most likely, but not necessarily limited to, the state) will use the following (or similar) table evaluate the merits of the permit application. The values in the table are calculated from the example used here for discussion purposes:

Calculated versus observed (reported) performance data

	Calculated value	Observed value
Particle mean diameter, D_p	1.0 μm	—
Collection efficiency, CE	99.9%	—
Pressure drop across Venturi, P_v	47 in. H_2O	—

The regulatory agency can be expected to conduct an independent investigation, as per Example 14. If such are independent review yields results not significantly different than the results submitted by the applicant (Example 14, in this example), approval of the permit application for the Venturi scrubber can be expected by the agency.

General industry experience indicates that one of the critical parameters that the agency will scrutinize is actual and estimated pressure drops for the Venturi scrubber. As pre-

Table 16
Venturi Scrubber Equipment Costs

Flow rate (acfm)	Venturi scrubber cost ($)
$10{,}000 \le Q_{e,a} < 50{,}000$	$VSC = \$7250 + 0.585(Q_{e,a})$
$50{,}000 \le Q_{e,a} \le 150{,}000$	$VSC = \$11.10(Q_{e,a})^{0.7513}$

Note: Carbon steel construction; includes cost of instrumentation.
Source: refs. 26 and 27.

viously discussed, if good agreement is not found, this indicates the use of the wrong performance chart to size the scrubber, improper operation of the scrubber, or both. The net result will be delay in permit approval, if not outright rejection of the permit by the regulating agency.

Example 15

A step-by-step estimation of capital and annual costs of a Venturi scrubber is presented.

Solution

A. Purchased Equipment and Total Capital Costs
As with other scrubber systems discussed in this handbook, the capital cost of a Venturi scrubber system is the total of the initial purchase price of the system and the direct and indirect costs of installation. Initial purchase cost (PC) includes shipping and taxes in addition to agreed sale price of the system. The equipment cost must also take into account the cost of any auxiliary equipment cost (Aex) purchased as part of the Venturi scrubber project. The PC may be estimated from Table 16 and the EC was discussed previously.

1. The Venturi scrubber cost (VSC) can be estimated from Table 16:
   VSC = $______________

   If the material construction for scrubber requires 304L stainless steel (SS), 2.3 is the cost multiplier; if 316 SS is used, the multiplier is 3.2. The upgrade VSC is

   VSC = $______________ × 2.3 or 3.2 (if applicable)

2. Equipment cost is computed as follows:

   EC = VSC + Aex = $______________

3. Purchased equipment cost

   PEC = 1.08 × EC = $______________

4. Total capital cost (TCC): The factors given in Table 9 are now used to estimate the TCC:

   TCC = 1.91 × PEC + SP + building = $______________

B. Direct Annual Costs—Venturi Scrubber
The direct annual costs include electric power cost, water, operating labor, and maintenance costs.

Table 17
Annual Cost Factors for Venturi Scrubbers

Cost item	Factor
Direct Anmual Costs, DAC	
Utilities	
Electricity	\$0.059/kWh
Water	\$0.20/10^3 gal
Operating Labor	
Operator labor	\$12.96/h
Supervisory labor	15% of operator labor
Maintenance	
Labor	\$14.26/h
Materials	100% of maintenance labor
Wastewater treatment	Variable; consult source for specific information
Indirect Annual Costs, IAC	
Overhead	0.60 (operating labor + maintenance)
Administrative	2% of TCC
Insurance	1% of TCC
Property tax	1% of TCC
Capital recovery	0.1628 (TCC)

Note: The capital recovery cost is estimated as $i\,(1 + i)^n / [i\,(1 + i)^n - 1]$, where i is the interest 10% and n is the equipment life (10 yr).

Source: ref. 12.

1. Power (electricity) cost: Use Eq. (23) to estimate the fan power requirements, F_p:

$$F_p = 1.81 \times 10^{-4} (Q_{e,a})(P_{\text{total}})(\text{HRS}) \tag{23}$$

$F_p =$ ____________ kWh/yr

and

$$Q_{e,a} = Q_e(T_e + 460) / 537 \tag{24}$$

$Q_{e,a} =$ ____________ acfm

where Q_e is the actual emission stream flow rate (scfm) and T_e is the emission stream temperature (°F).

It assumed that 2 wk out of the year the factory utilizing the Venturi scrubber is shut down for inventory and retooling, which provides an annual operating period of 50 wk/yr. Additionally, it is assumed the factory uses the scrubbing system 3 to 8 h shift, 5 d a week.

HRS = Annual operating = (3 shifts/d)(8 h/shift)(5 d/wk) (50 wk/yr) = 6,000 h/yr

The annual electric cost (AEC) with UEC equal to \$0.059 kWh is determined using Eq. (25):

$$\text{AEC} = (\text{UEC})(F_p) = \$0.059(F_p) \tag{25}$$

AEC = \$ ____________

2. The cost of scrubbing liquor (most likely water) is given by

   $$\text{WR} = 0.6(Q_{e,a})\ \text{HRS} \tag{31}$$

   where WR is the water consumption (gal/yr).

   The annual water cost, which equals ASC in Eq. (27), is obtained from the multiplication of the yearly water requirement and the unit water cost, which equals USC in Eq (27) found in Table 17. The USC cost water is assumed to equal $0.20 per 1,000 gal:

   $$\text{ASC} = (\text{USC})(\text{ASR}) \tag{27}$$

   ASC = ($0.20/1000 gal)(ASR)
   ASC = $____________

3. The annual operating labor costs are estimated by assuming that for every 8-h shift the Venturi scrubbing system is operated, 2 h are required to operator the wet scrubber. With labor cost at $12.96/h (*see* Table 7),

   Annual operating labor hours = (2.0 h/shift) (no. shifts/day) (no. days/wk) (no. wk/yr)
   Annual operating labor hours = HRS = ____________ h
   Annual operating labor cost = ($12.96 / h) HRS = $____________

   The supervisor labor cost is estimated at 15% of operating labor cost (*see* Table 7).

   Annual supervisor labor cost = (0.15)(annual operating labor cost)
   Annual supervisor labor cost = $____________

4. The annual maintenance labor cost is estimated by assuming that for every 8-h shift the scrubbing system is operated 1.0 h is required to maintain the wet scrubber. The maintenance labor cost is based on the rate of $14.96 per hour (*see* Table 17). This cost is determined as follows:

   Annual maintenance labor hours = (1.0 h/shift) (no. shifts/day)(no. days/wk) (no. wk/yr)
   Annual maintenance labor hours = HRS = ____________ h/yr
   Annual maintenance cost = ($14.26 / h) HRS = $____________

5. The annual maintenance materials cost is estimated by assuming that it is equivalent to 100% maintenance labor cost. This cost is determined as follows:

   Annual maintenance materials cost = (100 %)(annual maintenance labor cost)
   Annual maintenance materials cost = (1.0)(annual maintenance labor cost) = $____________

C. Indirect Annual Costs—Venturi Scrubber
Table 17 lists indirect cost factors. These factors and value for TCC are utilized to determine indirect annual costs (ICA) as follows:

Overhead = 0.60(annual operating labor & maintenance costs)
Overhead = $____________

Property Tax = 0.01(TCC)
Property Tax = $____________

Insurance = 0.01(TCC)
Insurance = $____________

Administration = 0.02(TCC)
Administration = $____________

Capital Recovery = 0.1628(TCC)
Capital Recovery = $____________

D. The total annual cost is determined by summing the annual direct and annual indirect costs:

Total annual costs = Total direct annual cost + Total indirect annual costs
Total annual costs = Total DAC + Total IAC
Total annual costs = $____________

E. Cost Update

Please refer to the book entitled *Advanced Air and Noise Pollution Control* (also published by Humana Press, Totowa, NJ). Note Chemical Engineering Indices and New Record Cost Indices in refs. 29 and 30.

Example 16

In this example, the total costs of the Venturi scrubber (*see* Fig. 1c) sized in Example 14 are developed. Recall that this scrubber was deemed to be the correct pollution control technology to treat the pollutants present in the emission stream of a municipal incinerator (*see* Table 15). These costs will include the purchase, direct and indirect annual, and capital costs for the scrubber system. As per Example 15, a step-by-step procedure is explained as follows

Solution

A. Purchase and Total Capital Costs
The cost of the Venturi scrubber is determined using Table 16, which yields the figure of $66,000 for a Venturi scrubber treating 110,000 acfm ($Q_{e,a}$). The purchase cost (PC) may be estimated from Table 16 and the equipment cost (EC) was discussed previously.

The Venturi scrubber cost (VSC) can be estimated from Table 16:

$$\text{VSC} = \$11.10(Q_{e,a})^{0.7513} = \$11.10\,(110{,}000)^{0.7513}$$
$$\text{VSC} = \$68{,}064 \sim \$68{,}000$$

The material construction for scrubber is 316 SS and the multiplier is 3.2. The upgrade VSC is

$$\text{VSC} = 3.2(\$68{,}000) = \$217{,}600$$

Auxiliary equipment (Aex) is assumed to cost $5400. Note that the cost of instrumentation for the Venturi scrubber is included in this estimation.

EC	= VSC + Aex	
EC	= $217,600 + $5,400	= $223,000
Sales tax	= 0.03($223,000)	= $6,390
Freight	= 0.05($223,000)	= $11,200
Purchased equipment cost, PEC		= $241,000

The PEC = $241,000 (*see* Table 18 for detailed calculations)

Table 18
Example Case Capital Costs for Venturi Scrubbers

Cost item	Factor	Cost($)
Direct Costs, DC		
Purchased Equipment Costs		
Venturi scrubber and auxiliary equipment	As estimated, EC	$223,000
Instrumentation	Included with EC	—
Sales tax	0.03 EC	6,690
Freight	0.05 EC	11.200
Purchased Equipment Cost, PEC	1.08 ECs	$241,000
Direct Installation Costs		
Foundation and supports	0.06 PEC	$14,500
Erection and handling	0.40 PEC	96,400
Electrical	0.01 PEC	2,400
Piping	0.05 PEC	12,000
Insulation	0.03 PEC	7,200
Painting	0.01 PEC	2,400
	0.56 PEC	$134,000
Site Preparation	As required, SP	
Building	As required, Bldg.	
Total Direct Cost, DC	1.56 PEC + SP + Bldg.	$375,000
Indirect Costs, IC		
Engineering	0.10 PEC	$24,000
Construction	0.10 PEC	24,000
Contractor fee	0.10 PEC	24,000
Start-up	0.01 PEC	2,400
Performance test	0.01 PEC	2,400
Contingency	0.03 PEC	7,200
Total Indirect Cost, IC	0.35 PEC	$84,000
Total Capital Cost TCC = DC + IC	1.91 PEC + SP + Bldg.	$460,310 + SP + Bldg.

Table 9 is used to determine the direct installation costs (1.56PEC), indirect costs (0.35PEC), and total capital costs (TCC) of the Venturi scrubber system. The factors given in Table 9 are now used to estimate the TCC:

$$TCC = (1.56PEC + 0.35PEC) + SP + \text{Building} \quad \text{(Tables 9 and 18)}$$
$$TCC = 1.91PEC + SP + \text{Building}$$
$$TCC = 1.91(\$241{,}000) + SP + \text{Building}$$
$$TCC = \$460{,}310 + SP + \text{Building}$$

Because the site preparation (SP) cost and building cost are specific to each installation, these costs are not included in this example.

B. The direct annual costs include electric power cost, water, operating labor, and maintenance costs (Table 17).
   1. Power (electricity) Cost: Use Eq. (23) to estimate the fan power requirements, F_p.
      a. It assumed that 2 wk out of the year the factory utilizing the Venturi scrubber is shut down for inventory and retooling, which provides annual operating period

of 50 wk/yr. Additionally, it is assumed that the factory uses the scrubbing system 3 to 8 hr shift, 5 d a week.

$P_{total} = P_v = 47$ in. H_2O (Example 14)

HRS = annual operating = (3 shifts/d) (8 h/shift) (5 d/wk) (50 wk/yr) = 6,000 h/yr

$F_p = 1.81 \times 10^{-4} (Q_{e,a})(P_{total})(\text{HRS})$ (23)

$F_p = 1.81 \times 10^{-4} (110{,}000)(47)(6{,}000)$

$F_p = 5.61 \times 10^6$ kWh/yr

b. The annual electric cost (AEC) with UEC equal to \$0.059 kWh is determined using Eq. (25):

$\text{AEC} = (\text{UEC})(F_p) = \$0.059(F_p)$ (25)

$\text{AEC} = \$0.059\ (5.61 \times 10^6) = \$331{,}000/\text{yr}$

2. The cost of scrubbing liquor (most likely water) is given by

$\text{WR} = 0.6(Q_{e,a})\ \text{HRS}$ (31)

$\text{WR} = 0.6(110{,}000)(6{,}000) = 3.96 \times 10^8$ gal/yr

The annual water cost [equals ASC in Eq. (27)] is obtained by multiplying the yearly water requirement by the unit water cost [equals USC in Eq. (27)] found in Table 17. The USC cost water is assumed to equal \$0.20 per 1,000 gal.

ASC = (USC) (ASR) (27)

Annual water cost = (\$0.20 /1000 gal) (3.96×10^8)

Annual water cost = \$79,200/yr

3. Annual operating labor costs are estimated by assuming that for every 8-h shift the Venturi scrubbing system is operated, 2 h are required to operator the wet scrubber. With labor cost at \$12.96/h (*see* Table 7)

Annual operating labor hours = (2 h/shift) (3 shifts/d) (5 d/wk) (50 wk/yr)

Annual operating labor hours = HRS = 1,500 h/yr

Annual operating labor cost = (\$12.96/h) HRS = (\$12.96/h) (1,500) = \$19,400/yr

Supervisor labor cost is estimated at 15% of operator labor cost (*see* Table 7).

Annual supervisor labor cost = (0.15) (annual operating labor cost)

Annual supervisor labor cost = (0.15)(\$19,400) = \$2910/yr

4. The Annual maintenance labor cost is estimated by assuming that for every 8-h shift the scrubbing system is operated, 1.0 h is required to maintain the wet scrubber. The maintenance labor cost is based on a rate of \$14.26 per hour (*see* Table 17). This cost is determined as follows:

Annual maintenance labor hours = (1 h/shift) (3 shifts/d) (5 d/wk)(50 wk/yr)

Annual maintenance labor hours = HRS = 750 h/yr

Annual maintenance labor cost = (\$14.26 /h) HRS = (\$14.26 /h)(750)

Annual maintenance labor cost = \$10,695 or \$10,700/yr

5. The annual maintenance materials cost is estimated by assuming it is equivalent to 100% maintenance labor cost. This cost is determined as follows:

Annual maintenance materials cost = (100%) (annual maintenance labor cost)

Annual maintenance materials cost = (1.0)(\$10,700) = \$10,700/yr

Note: The Venturi scrubber will generate wastewater during normal operation. The cost of treating this water is not included in this discussion.

6. The total annual direct cost is determine from the summation of direct annual costs:

   Total annual direct cost = \$331,000 + \$79,200 + \$19,400 + \$2,910 + \$10,700 + \$10,700
   Total annual direct cost = \$453,910 or \$454,000/yr

C. Indirect Annual Costs—Venturi Scrubber
The TCC from Table 18 is \$459,000 for equipment only (without site preparation and building). Although the actual TCC = \$459,000 + SP + building, it is assumed that the annual costs of SP and building will be estimated separately together with all other sites and buildings. Accordingly, the annual costs of the subject Venturi scrubber are estimated assuming TCC = \$459,000 for equipment only. Table 17 lists indirect cost factors. These factors and the value for TCC are utilized to determine indirect annual costs (IAC) as follows:

Overhead	=	0.60(annual operating labor and maintenance costs)
Overhead	=	0.60(\$19,400 + \$2,910 + \$10,700 + \$10,700)
Overhead	=	\$ 26,226–\$26,200/yr
Property Tax	=	0.01(TCC) = 0.01(\$459,000)
Property Tax	=	\$4,590/yr
Insurance	=	0.01 (TCC) = 0.01(\$459,000)
Insurance	=	\$4,590/yr
Administration	=	0.02(TCC) = 0.02(\$459,000)
Administration	=	\$9180/yr
Capital Recovery	=	0.1628(TCC) = 0.1628(\$459,000)
Capital Recovery	=	\$74,725/yr

The total indirect annual cost is the summation of the cost list above.

Total indirect annual costs = \$26,200 + \$4,590 + \$4,590 + \$9,180 + \$74,725
Total indirect annual costs = \$119,285/yr

D. Total Annual Cost is determined by summation of Annual Direct and Annual Indirect costs

Total annual costs = Total direct annual cost + Total indirect annual costs
Total annual costs = Total DAC + Total IAC
Total annual costs = \$454,000 + \$119,285
Total annual costs = \$573,285/yr

E. Cost Update—see Example 15, which describes the cost indices found in refs. 28, 31, and 32.

Example 17

In this example, the definition and classification of air pollutants is discussed. Several terms commonly encountered when discussing air pollutants need to be defined:

- Volatile organic compounds (VOCs)
- Semivolatile organic compounds (SVOCs)
- Volatile inorganic compounds (VICs)

- Semivolatile inorganic compounds (SVICs)
- Particulate Matter (PM)

Such compounds and substances when present in an air emission stream are termed hazardous air pollutants (HAPs) and, as such, are subject to regulatory scrutiny. The proper classifications of HAPs help in understanding the needs of a given air pollution project.

Solution

A. Organic compounds that have a vapor pressure of greater than 1 mm Hg at 25°C are defined to be volatile. As such, these compounds constitute the family of "volatile organic compounds" or VOCs. Commonly encountered VOCs are the following:

1. All monochlorinated solvents and several other chlorinated solvents such as trichloroethylene, trichloroethane, and tetrachloroethane.
2. The simple aromatic solvents such as benzene, xylene, toluene, ethyl benzene, and so forth.
3. Most of the alkane solvents up to decane (C_{10}).

B. Some inorganic compounds also meet the definition of volatility just defined (vapor pressure greater than 1 mm Hg at 25°C. As such, these are the "volatile inorganic compounds" or VICs. They include inorganic gases (e.g., hydrogen sulfide, chlorine, and sulfur dioxide).

C. When an organic compound has a vapor pressure of less than 1 but greater than 10^{-7} mm Hg, such a compound is classified as being in the "semivolatile organic compounds," or SVOCs, family. Such organics commonly encountered are as follows:

1. Most polychlorinated biphenolics, dichlorobenzene, phthalates, nitrogen substituted aromatics such as nitroaniline, and so forth.
2. Most pesticides (e.g., dieldrin, toxaphene, parathion, etc.)
3. Most complex alkanes (e.g., dodecane, octadecane, etc).
4. Most of the polynuclear aromatics (naphthalene, phenanthrene, benz(a)anthracene, etc.)

D. Likewise, at the same vapor pressure as given in Part C, inorganic compounds are defined to be "semivolatile inorganic compounds" or SVICs. Elemental mercury is a semivolatile inorganic.

E. A compound or other substance is defined as being "nonvolatile" if it has a vapor pressure $< 10^{-7}$ mm Hg at 25°C. This is also another way of defining a solid. Therefore, almost all particulate matter (PM) is nonvolatile. Examples of compounds that are found in PM, which result in such PM being considered HAP, are as follows:

1. The large polynuclear (also polycyclic) aromatics such as chrysene.
2. Heavy metals (e.g., lead, chromium, etc.).
3. Other inorganics (e.g., asbestos, arsenic, and cyanides).

F. It is now appropriate to classify hazardous, undesirable, or otherwise unwanted air pollutants:

1. Aromatic hydrocarbons: benzene, toluene, xylenes, ethylbenzene, and so forth.
2. Aliphatic hydrocarbons: hexane, heptane, and so forth.
3. Halogenated hydrocarbons: methylene chloride, chloroform, carbon tetrachloride, 1,1-dichloroethane, trichloroethylene, 1,1,1-trichloroethane, tetrachloroethylene, chlorobenzene, and so forth.
4. Ketones and aldehydes: acetone, formaldehyde, methyl ethyl ketone, and so forth.

5. Oxygenated hydrocarbons: methanol, phenols, ethylene glycol, and so forth.
6. Inorganic gases: hydrogen sulfide, hydrogen chloride, sulfur dioxide, nitrogen oxide, nitrogen dioxide, and so forth.
7. Metals: mercury, lead, cadmium, arsenic, zinc, and so forth.
8. Polynuclear aromatics: naphthalene, benzo(a)pyrene, anthracene, chrysene, polychlorinated biphenyls (PCBs), and so forth.
9. Pesticides, herbicides: chlordane, lindane, parathion, and so forth.
10. Other (miscellaneous): asbestos, cyanides, radionuclides, and so forth.

The appendix A lists compounds classified as hazardous by the US Environmental Protection Agency (US EPA) when present in an air emission stream (33). Original equipment suppliers (OEMs) of commercial absorber (scrubber) systems in North America are available refs. 31 and 34–36.

Example 18

Removal of sulfur dioxide from an air emission stream by wet scrubbing is presented in this example in order to assess the suitability of various packings and materials. Contacting efficiencies and pressure drop of various packings were studied under identical controlled conditions in a packed tower wet scrubber shown in Fig. 1b.

Q-PAC, 3.5 in. Tri-Packs, 2K Tellerettes, and 50-mm Pall Rings were tested in a countercurrent packed scrubber for removal of SO_2 from an air emission stream. The SO_2 system has long been used by environmental engineers for comparison of packings because it allows for precise, reproducible measurement of operating parameters and mass transfer rates not affected by changes in the weather. The efficiency of mass transfer depends on the ability of the packing to create more gas–liquid contacting surface, so the results of this test are a good predictor of the relative performance of the tested packings in an acid gas or similar scrubber.

The test apparatus (36,37) consists of a vertical countercurrent scrubber with a cross-sectional area of 6.0 ft^2 packed with the media being tested to a depth of 3.0 ft. The scrubber is equipped with a variable-speed fan and pump drives allowing an engineer to adjust both the gas flow and the liquid loading of the scrubber. The air was spiked with SO_2 fed from a cylinder under its own vapor pressure. The injection point was 15 duct diameters upstream from the scrubber inlet to ensure adequate mixing. The regulator on the SO_2 cylinder was adjusted manually to give an inlet concentration in the range of 80–120 ppmv (parts per million by volume) at each airflow rate. Inlet and outlet SO_2 concentrations were measured simultaneously using Interscan electrochemical analyzers.

The air emission stream was scrubbed with 2% sodium bicarbonate liquor. An automated chemical feed system added caustic to maintain a constant pH of 9.15 ± 0.05 throughout the test. The airstream and liquid flow rates were used in the ranges typically encountered in a wet scrubber operation. The gas loading was varied from 500 to 3000 lb/h-ft^2 corresponding to superficial gas velocities of 110–670 fpm. The liquid loading ranged from 5 to 8 gpm/ft^2.

The test results are summarized in Table 19 and Figs. 8–14. Gas–liquid contacting efficiency is quantified in terms of the height of transfer unit, or HTU. (This is the depth of packing required to reduce the SO_2 concentration to approximately 37% of its initial value.) Discuss the following:

1. The suitability of packing materials evaluated under this optimization project for SO_2 removal.
2. Chemical reactions involved in SO_2 scrubbing.

Table 19
Scrubbing Sulfur Dioxide Using Three Different Packing Materials

			Q-PAC		3.5 in. Tri-Packs		2K Tellerettes	
V_G (ft/min)	Recirc. (gpm)	L (gpm/ft^2)	$\Delta P/z$ (in. H_2O/ft)	HTU (ft)	$\Delta P/z$ (in. H_2O/ft)	HTU (ft)	$\Delta P/z$ (in. H_2O/ft)	HTU (ft)
900	30	5.0	0.401	1.10				
800	30	5.0	0.307	1.15				
700	30	5.0	0.226	1.23				
600	30	5.0	0.157	1.9	0.317	1.80	0.364	1.33
500	30	5.0	0.101	1.24	0.215	1.79	0.235	1.28
400	30	5.0	0.058	1.17	0.134	1.71	0.145	1.20
300	30	5.0	0.029	1.05	0.075	1.57	0.078	1.06
200	30	5.0	0.014	0.91	0.036	1.28	0.035	0.86
900	40	6.7	0.452	1.05				
800	40	6.7	0.334	1.10				
700	40	6.7	0.241	1.16				
600	40	6.7	0.164	1.23	0.340	1.59	0.385	1.21
500	40	6.7	0.103	1.18	0.227	1.66	0.248	1.17
400	40	6.7	0.061	1.12	0.139	1.55	0.153	1.10
300	40	6.7	0.028	1.00	0.081	1.40	0.084	0.96
200	40	6.7	0.014	0.88	0.038	1.07	0.038	0.81
900	50	8.3	0.497	1.00				
800	50	8.3	0.357	1.03				
700	50	8.3	0.259	1.11				
600	50	8.3	0.168	1.16	0.356	1.55	0.398	1.07
500	50	8.3	0.109	1.12	0.237	1.42	0.257	1.03
400	50	8.3	0.063	1.03	0.149	1.39	0.156	0.99
300	50	8.3	0.030	0.93	0.083	1.17	0.089	0.86
200	50	8.3	0.014	0.80	0.038	0.96	0.040	0.71
900	60	10.0	0.522	0.95				
800	60	10.0	0.383	0.97				
700	60	10.0	0.274	1.03				
600	60	10.0	0.174	1.09	0.370	1.27	0.429	0.97
500	60	10.0	0.114	1.04	0.246	1.26	0.272	0.94
400	60	10.0	0.066	0.97	0.156	1.18	0.164	0.87
300	60	10.0	0.032	0.87	0.087	1.02	0.091	0.77
200	60	10.0	0.014	0.75	0.042	0.80	0.042	0.62
900	70	11.7	0.557	0.92				
800	70	11.7	0.409	0.93				
700	70	11.7	0.285	0.96				
600	70	11.7	0.180	1.03	0.386	1.17	0.463	0.87
500	70	11.7	0.118	0.97	0.251	1.16	0.282	0.87
400	70	11.7	0.069	0.90	0.159	1.02	0.171	0.80
300	70	11.7	0.033	0.82	0.093	0.87	0.094	0.70
200	70	11.7	0.016	0.70	0.046	0.72	0.046	0.55

Continued

Table 19 *(Continued)*

			Q-PAC		3.5 in. Tri-Packs		2K Tellerettes	
V_G (ft/min)	Recirc. (gpm)	L (gpm/ft^2)	$\Delta P/z$ (in. H_2O/ft)	HTU (ft)	$\Delta P/z$ (in. H_2O/ft)	HTU (ft)	$\Delta P/z$ (in. H_2O/ft)	HTU (ft)
900	80	13.3	0.609	0.87				
800	80	13.3	0.438	0.89				
700	80	13.3	0.299	0.91				
600	80	13.3	0.186	0.98	0.414	1.03	0.504	0.84
500	80	13.3	0.122	0.91	0.262	1.02	0.299	0.82
400	80	13.3	0.072	0.86	0.166	0.93	0.176	0.77
300	80	13.3	0.037	0.77	0.098	0.76	0.097	0.65
200	80	13.3	0.017	0.65	0.049	0.64	0.047	0.49

Source: ref. 36.

3. Important applications of a suitable, optimized packing material for wet scrubbing

Solution

1. Evaluation of packing materials for scrubbing removal of sulfur dioxide is presented in Table 19 and Fig. 13. The data in Table 19 and Fig. 13 show that Q-PAC is slightly more efficient than 2K Tellerettes at less than half the pressure drop per foot. Compared with 3.5 in. Tri-Packs (*see* Table 18 and Fig. 12), Q-PAC is approximately 40% more efficient with about half the pressure drop.

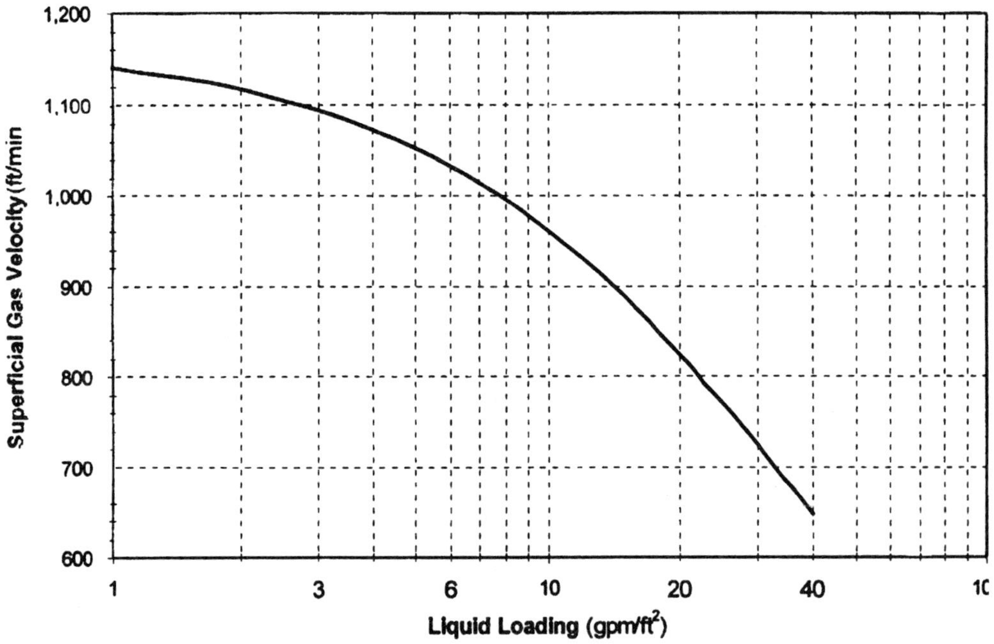

Fig. 8. Flooding curve of a selected packing material. (Courtesy of Lantec Products Inc.)

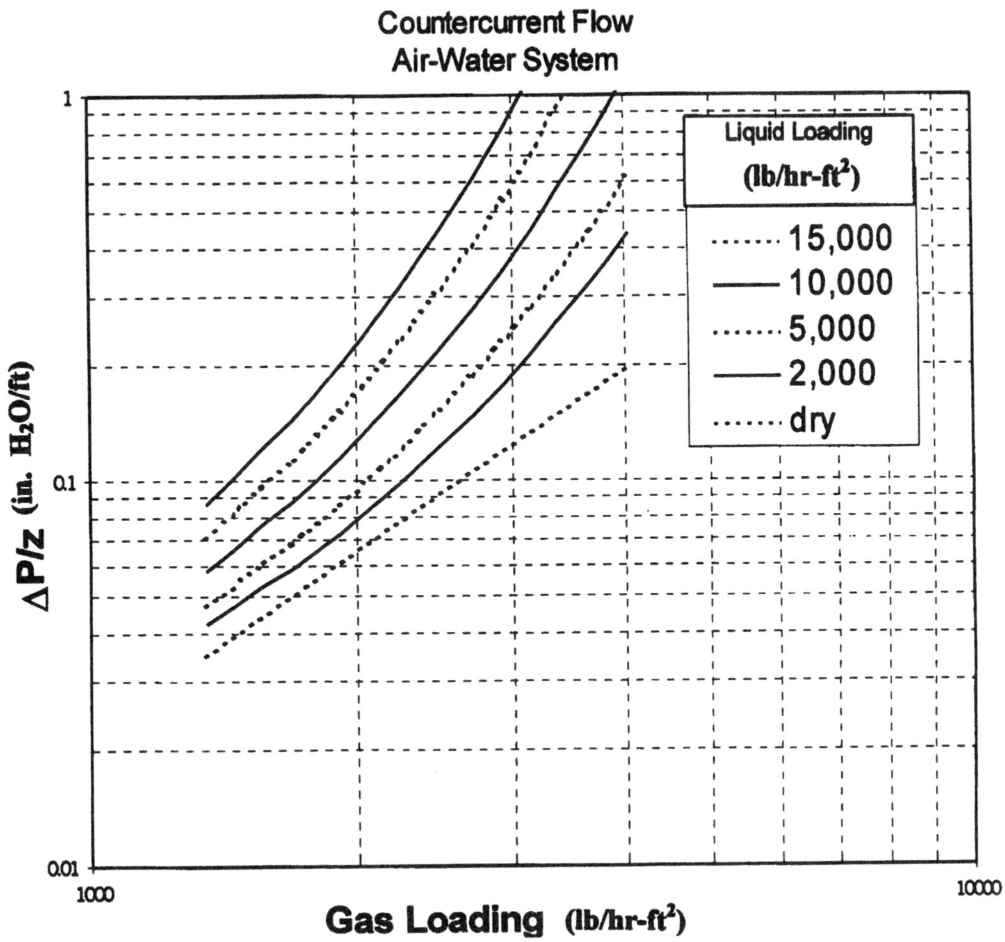

Fig. 9. Pressure drop across a selected packing material—US customary system. (Courtesy of Lantec Products Inc.)

The lower pressure drop of Q-PAC (Table 19, Figs. 12–14) made it possible to continue scrubbing tests all the way up to 900 fpm without exceeding the fan's capacity. At higher velocities, the liquid holdup on the packing increases and the more turbulent airflow helps break the water into smaller droplets, resulting in increased gas–liquid contacting surface. As traditional chemical engineering texts describe (38), maximizing the gas–liquid surface contact is critical to maximizing mass transfer efficiency. However, packings have historically been designed to spread the liquid into a thin film to maximize contact with the passing gas phase, the unique design of Q-PAC (rounded surfaces and many slender needles) forces the liquid into droplets to maximize the surface available to the gas phase for mass transfer. Additionally, note that when a liquid is spread into a thin film over a packing, only that liquid surface facing the gas flow is available for mass transfer. The liquid film facing the packing support cannot participate in mass transfer. As a result, when using Q-PAC, the HTU value actually begins to decrease as the gas velocity increases beyond 600 fpm. (The same behavior of HTU is observed with conventional packings, but the rapid increase in pressure

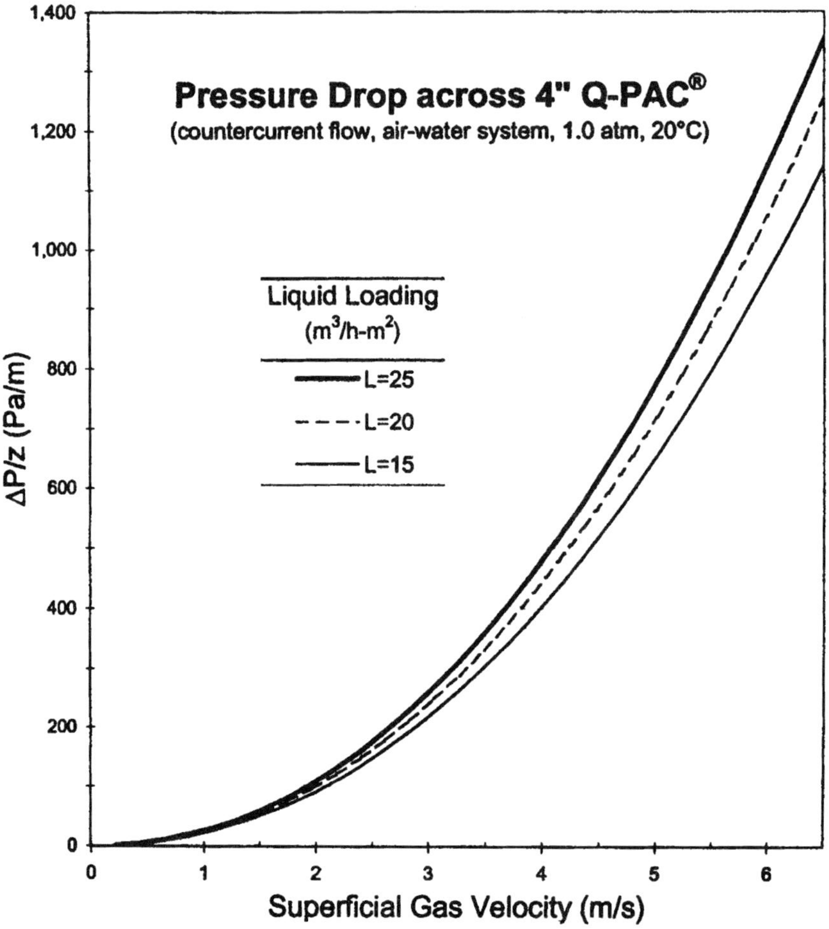

Fig. 10. Pressure drop across a selected packing material—metric system. (Courtesy of Lantec Products Inc.)

drop makes it impractical to operate a scrubber so packed at much over 500 fpm because of increased power consumption as well as because of flooding concerns.)

As a result of the gas–liquid contacting, the efficiency of Q-PAC is better than that of conventional random plastic packings, because in addition to providing a large surface to spread the liquid, Q-PAC also forces the liquid to form droplets with greatly extended surfaces that enhance mass transfer. Q-PAC also provides for substantially higher gas handling capacity in a scrubber tower.

2. Chemical reactions involved in sulfur dioxide scrubbing (Fig. 1b). The inlet SO_2 concentration was controlled in the range of 80–120 ppm at each flow rate. Both inlet and outlet SO_2 concentrations were measured simultaneously using an analyzer. The air emission stream was scrubbed using a buffered solution of 2% sodium bicarbonate and caustic. Over 99.9% of the SO_2 was removed from the air emission stream. The following chemical reactions occur in a packed tower scrubber:

$$SO_2 + NaHCO_3 \rightarrow NaHSO_3 + CO_2$$

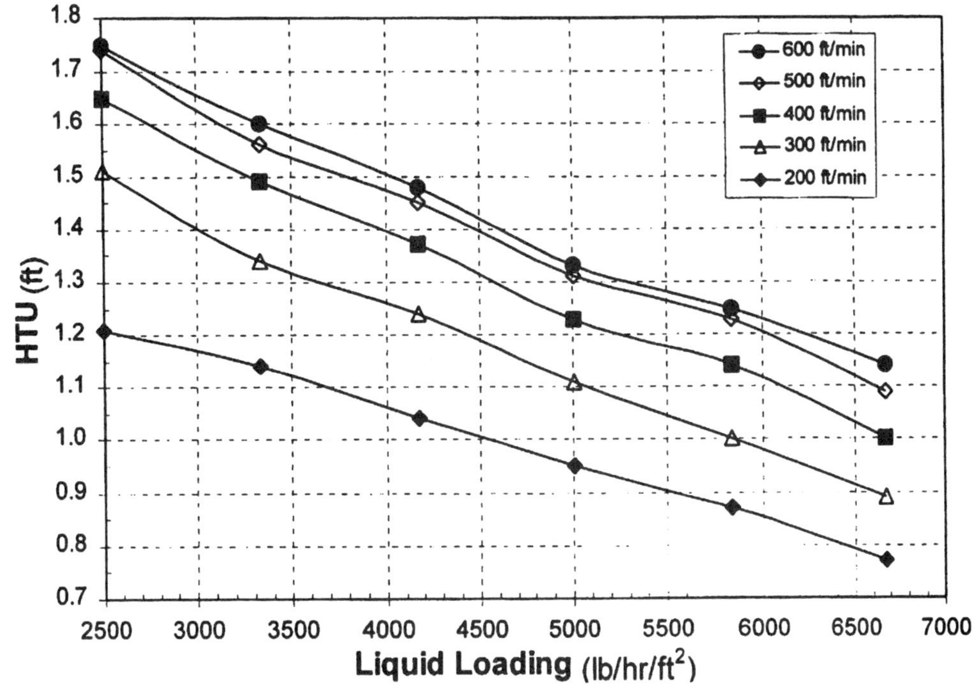

Fig. 11. Concurrent absorption of sulfur dioxide using a selected packing material and scrubbing solution. (Courtesy of Lantec Products Inc.)

$$CO_2 + NaOH \rightarrow NaHCO_3$$

$$NaHSO_3 + NaOH \rightarrow Na_2SO_3 + H_2O$$

3. Important applications of a suitable, optimized packing material such as Q-PAC for wet scrubbing are presented next. The high capacity of an optimized packing material can be utilized in different ways. When designing new equipment, the cross-sectional area of a scrubber can be reduced in order to reduce fabrication costs of the vessel and capital expense of the recirculation pump. Additionally, even the cost of packing required to fill the scrubber is reduced because less is needed to pack the scrubber. This is a consequence of the fact that as the diameter of tower size is reduced, the volume of the tower is reduced geometrically. The fan size need not be increased nor the operating costs of the scrubber system increased. Additional added benefits to reduced tower size is a smaller footprint for the scrubber system as well as reduced noise during normal operations of the system.

   As an alternative, wet scrubbers can be sized for conventional gas velocities (375–475 fpm) but packed with an optimized packing material, such as Q-PAC, in order to reduce the pressure drop for resultant reduced fan power consumption and, hence, lower power costs for the lifetime of the system.

   Retrofitting an existing scrubber with an optimized packing material makes it possible to increase the air being treated in the scrubber without changing the fan. In this

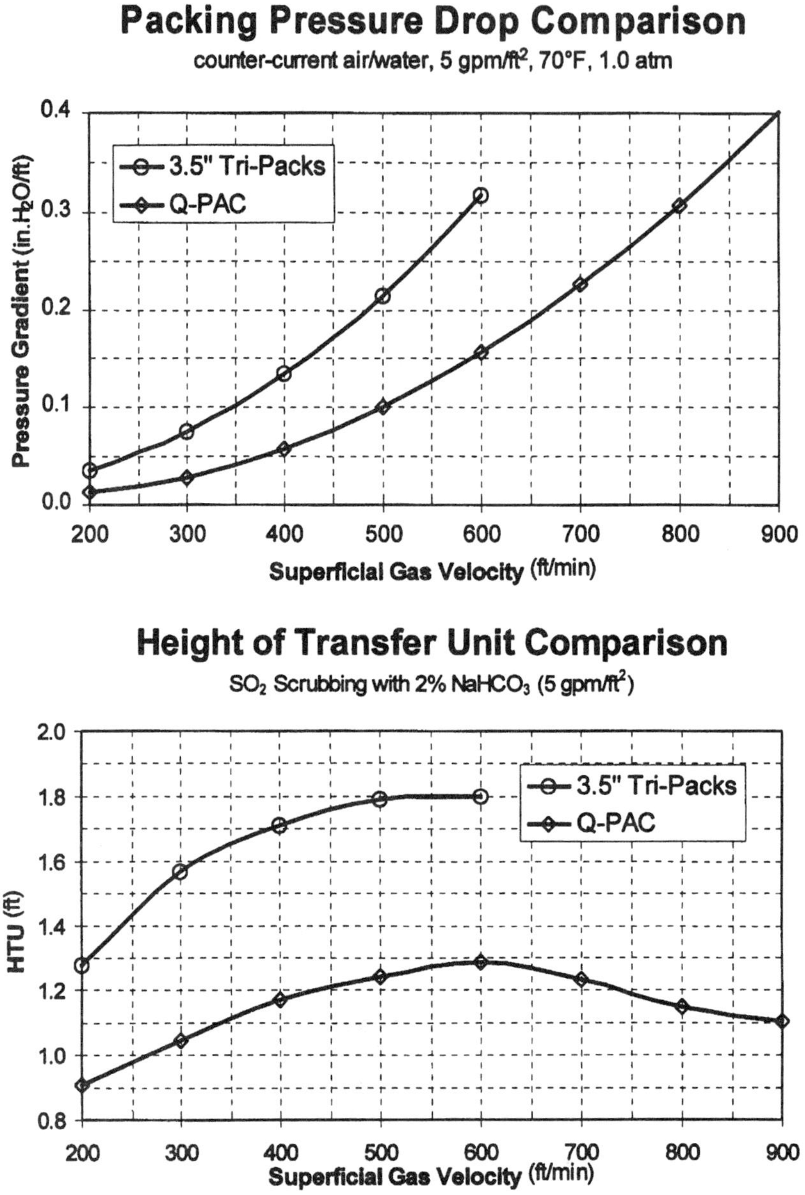

Fig. 12. Sulfur dioxide scrubbing using Q-PAC and 3.5 in. Tri-Packs. (Courtesy of Lantec Products Inc.)

way, a simple repack of a tower may avoid a costly rebuild project. An optimized packing material should also be considered for wet scrubbers in which media fouling is a problem. The uniform spacing of the all rounded plastic elements, in addition to the void space of >97% for a packing, such as Q-PAC, minimizes the tendency of solids to accumulate on the packing surface. Hard water or high particulate loadings will eventually foul any packing, but with the optimized packing material, a scrubber prone to plugging can be operated longer before shutdown is required to clean the packing.

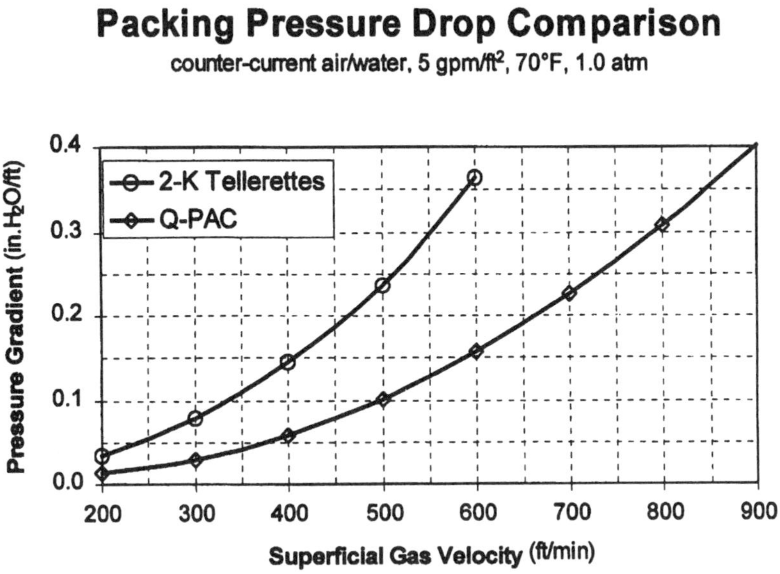

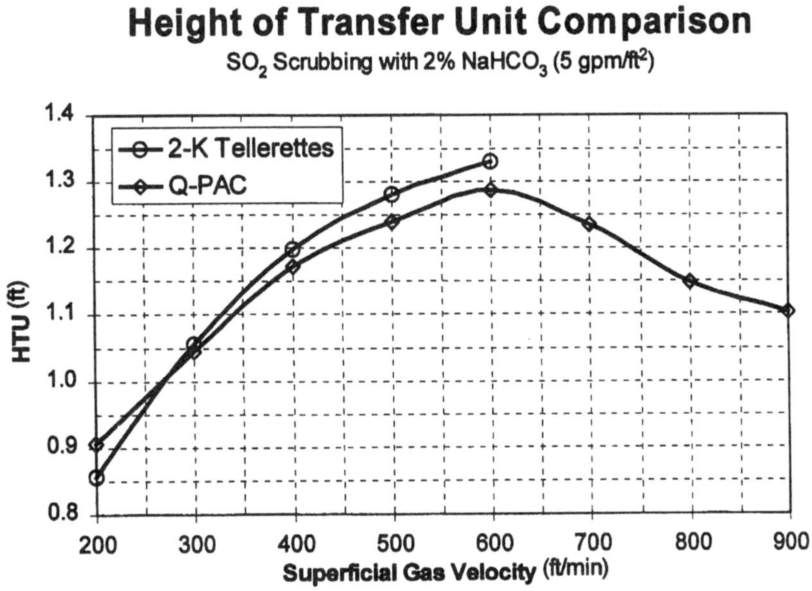

Fig. 13. Sulfur dioxide scrubbing using Q-PAC and 2-K Tellerettes. (Courtesy of Lantec Products Inc.)

Example 19

Hydrogen sulfide is the most common source of odor complaints resulting from normal operations of a typical waste-treatment facility. Removal of hydrogen sulfide from a municipal air emission stream using a packed tower wet scrubber is presented in this example. (Note: Refer to Example 25 for removal of hydrogen sulfide from an industrial air emission stream using a nontraditional wet scrubber.)

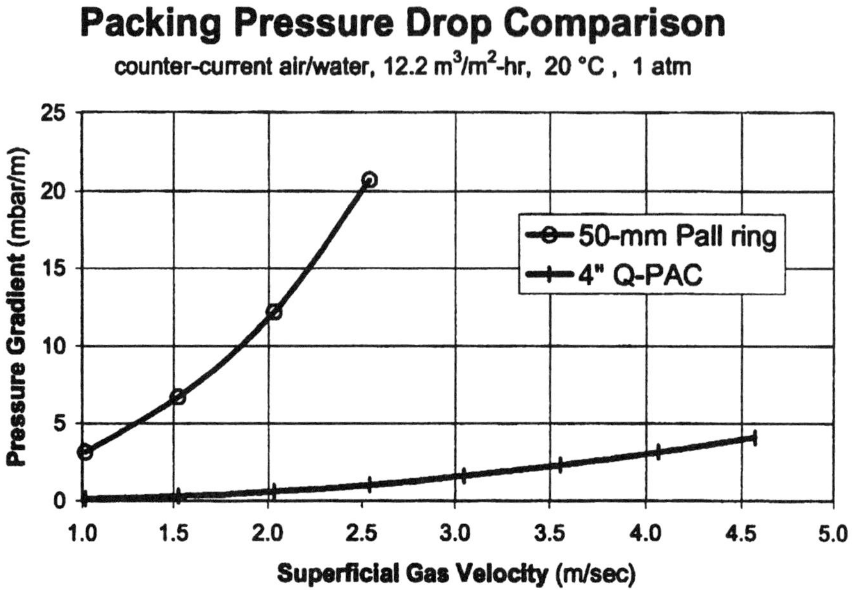

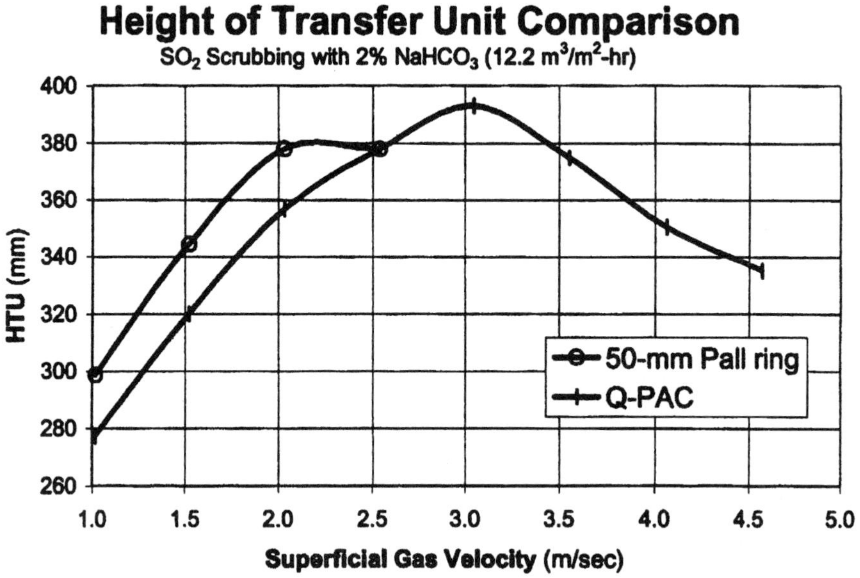

Fig. 14. Sulfur dioxide scrubbing using Q-PAC and 50-mm Pall Ring. (Courtesy of Lantec Products Inc.)

Discuss the following for hydrogen sulfide scrubbing using a packed tower wet scrubber (Fig. 1b):

1. The feasible wet scrubber design, tower media (packing), and scrubbing liquor for a single-stage scrubbing process.
2. The chemical reactions involved in scrubbing hydrogen sulfide (single stage) and the wet scrubber performance.

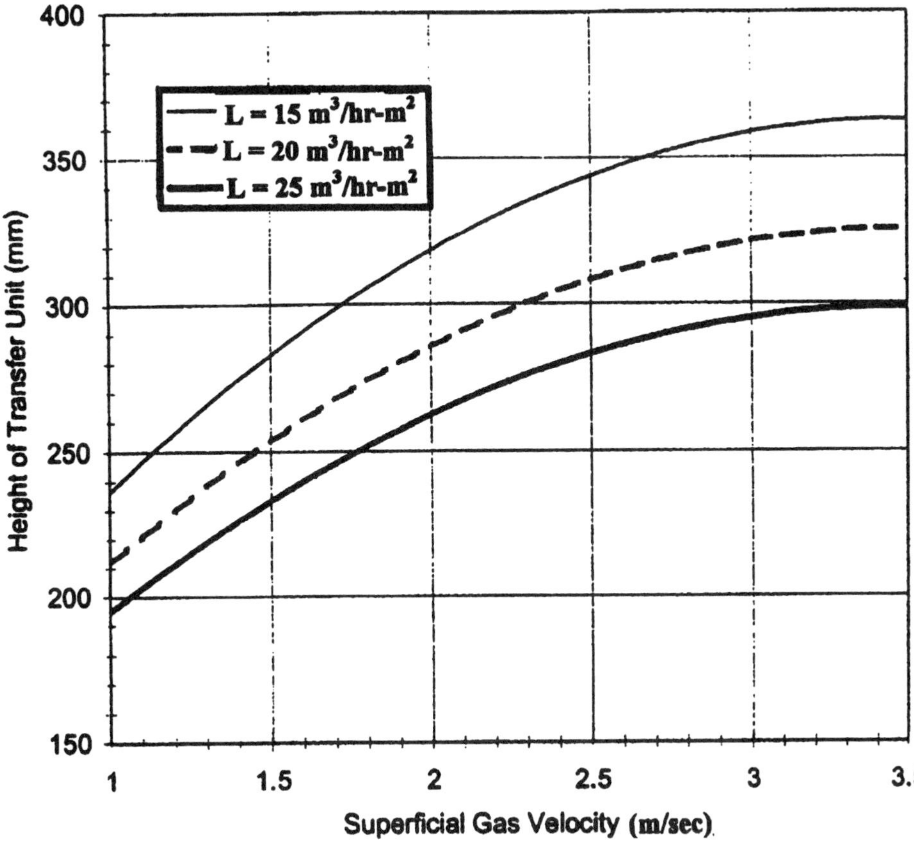

Fig. 15. Height of transfer unit for scrubbing sulfur dioxide. (Courtesy of Lantec Products Inc.)

3. The situations when two-stage or three-stage scrubbing process may be used for hydrogen sulfide removal.
4. Other alternative technologies for odor control at wastewater-treatment plants (WWTPs).

Technical information is presented in Fig. 15 and Tables 20 and 21, which are actual operating data generated from a countercurrent wet scrubbing tower. With a suitable scrubbing solution and an optimized wet scrubber, hydrogen sulfide air emissions in the typical range of 100 ppm or less at municipal sewage-treatment plants are easily controlled. Table 20 presents minimum vessel diameters needed to treat various airflows to remove 99.9% of the H_2S present. The depth of packing is 10 ft; the pressure drop across the packed bed is 2 in. of H_2O. The packing used is Q-PAC from Lantec Products. Table 21 and Fig. 15 give additional operational parameters for such a scrubbing system.

Solution

1. The feasible wet scrubber design, tower media (packing), and scrubbing liquor for a single-stage scrubbing process are presented in the following discussion. One feasible air pollution control system for removal of malodorous hydrogen sulfide commonly emitted by municipal sewage-treatment plants can be a single-stage

Table 20
Hydrogen Sulfide Absorption Using Packed Tower Wet Scrubbers with Various Diameters

Airflow rate (acfm)	Tower diameter (ft)	Gas velocity (ft/min)	Removal (%)	Liquid flux (gpm/ft^2)	HTU (ft)
90,000	12	796	99.9+	6.0	1.18
75,000	11	790	99.9+	6.0	1.19
60,000	10	764	99.9+	6.0	1.20
50,000	9	786	99.9+	6.0	1.19
40,000	8	796	99.9+	6.0	1.18
30,000	7	780	99.9+	6.0	1.19
22,000	6	778	99.9+	6.0	1.19
15,000	5	764	99.9+	6.0	1.20
10,000	4	796	99.9+	6.0	1.18

Note: Design considerations: H_2S removal efficiencies: >99.9%; temperature: 80°F at sea level; height of packed bed: 10 ft.
Source: ref. 36.

Table 21
Scrubbing H_2S for 99.9% Removal

V_G (ft/min)	L (gpm/ft^2)	Pressure drop (in. WC/ft)	HTU (ft)	Bed height (ft)
800	8	0.379	0.954	8.00
800	6	0.341	1.071	9.00
800	4	0.306	1.242	10.00
700	8	0.257	0.981	8.00
700	6	0.234	1.089	9.00
700	4	0.213	1.278	10.00
600	8	0.171	0.972	8.00
600	6	0.158	1.080	9.00
600	4	0.146	1.269	10.00
500	8	0.113	0.936	7.50
500	6	0.106	1.044	8.50
500	4	0.099	1.215	10.00
400	8	0.073	0.864	7.00
400	6	0.069	0.963	8.00
400	4	0.066	1.125	9.50
300	8	0.046	0.774	6.50
300	6	0.044	0.864	7.00
300	4	0.042	1.008	8.00
200	8	0.027	0.648	5.00
200	6	0.026	0.720	6.00
200	4	0.026	0.837	7.00

Note: Scrubbing solution: 0.1% caustic and 0.3% hypochlorite; pH 9.0–9.5; ORP 500–600 mV; temperature 80°F.
Source: ref. 36.

packed tower wet scrubber (Fig. 1b) using a suitable packing material (such as Q-PAC from Lantec Products, or an equivalent packing from another manufacturer) and a typical scrubbing liquor (such as 0.1% caustic and 0.3% sodium hypochlorite), with pH control to 9.0–9.5 and ORP control to 550–600 mV at 80°F and atmospheric pressure.

2. Described here are the chemical reactions involved in scrubbing hydrogen sulfide (single stage) and the wet scrubber performance. The wet scrubber described in part 1 has been proven capable of a removal efficiency of 99.9% of hydrogen sulfide from a contaminated airstream at various airflow rates, superficial gas velocities, liquid flux rates, tower diameters, and HTU values (Tables 20 and 21, Fig. 15). Different scrubbing liquors can be used in hydrogen sulfide control. It is important to realize that whatever scrubbing liquor is chosen, the chemistry of a hydrogen sulfide scrubber is essentially two step. First, the hydrogen sulfide becomes soluble in the presence of caustic and is then oxidized by an oxidizing agent such as hydrogen peroxide, chlorine, or potassium permanganate. The following reactions are for a single-stage scrubbing system using 0.1% caustic and 0.3% sodium hypochlorite to control hydrogen sulfide emissions:

$$H_2S + 2\ NaOH \rightarrow Na_2S + 2\ H_2O$$

$$NaOCl + H_2O \rightarrow HOCl + NaOH$$

$$3\ HOCl + Na_2S \rightarrow Na_2SO_3 + 3\ HCl$$

$$HOCl + Na_2SO_3 \rightarrow Na_2SO_4 + HCl$$

$$HCl + NaOH \rightarrow NaCl + H_2O$$

A single-stage scrubbing system as described in this example will therefore always need to be overdosed with oxidizing agent. Additionally, sodium hypochlorite decomposes slowly in storage, which represents additional long-term costs to a municipality or other industry controlling hydrogen sulfide with a wet scrubber system.

Sodium hypochlorite may also be consumed if other VOCs or SVOCs are present in the airborne emissions from a sewage- or water-treatment plant. The presence of such compounds can be difficult to predict, as these compounds will occur because of the materials to be processed, time of year, and other factors. If such compounds are present in the scrubber system, then the discharge Na_2S will need to be treated and disposed of by chemical precipitation. For instance,

$$3\ Na_2S + 2\ FeCl_3 \rightarrow Fe_2S_3 + 6\ NaCl$$

where the sodium sulfide is the soluble pollutant, ferric chloride is the soluble precipitation agent and the ferric sulfide (fool's gold) is the insoluble precipitate produced in the aforementioned chemical reaction. Several other precipitation reactions using different chemicals for control of Na_2S have been reported by Wang et al. (39).

Although other oxidizing agents may be used, as previously mentioned, sodium hypochlorite remains the predominant choice of oxidizing chemical in H_2S odor control scrubbing systems in North America. This is so because it is far less expensive and less dangerous than hydrogen peroxide, it is more active than potassium permanganate (which will also stain purple everything it touches), and it does not have to be stored in pressurized containers as does chlorine gas.

3. The situations when two-stage or three-stage scrubbing process may be used for hydrogen sulfide removal are described in the following discussion. If ammonia is produced in normal operations of a WWTP, then a two-stage scrubbing system is required. The first stage will remove the ammonia with dilute acid scrubbing liquor and the second stage uses a caustic/oxidizing step for removal of hydrogen sulfide.

   Additionally, as this text is being prepared, increasing concern with other malodorous reduced sulfur compounds is being noted throughout North America. Historically, waste-treatment plants as well as other odor producing industries (such as rendering) have been located in remote areas to minimize odor complaints from neighbors. As urban growth has accelerated, such plants often find their location surrounded by new development, where previously only open fields or forests had been neighbors (40).

   As a result, more concern must be given to control of malodorous-reduced sulfur compounds such as methyl disulfide (MDS), dimethyl disulfide (DMDS), and mercaptans that previously were allowed to simply disperse into the atmosphere. Although commonly present in very low concentrations, typically ppb rather than ppm as with hydrogen sulfide, these other reduced sulfur compounds have extremely low odor thresholds.

   When present in an airstream, a three-stage odor scrubbing system may be called for: (a) first stage to remove NH_3 with dilute acid scrubbing; (b) second stage to remove DMS, DMDS, and mercaptans with oxidative scrubbing at neutral pH; and (c) third stage "traditional" oxidative scrubbing at high pH to control H_2S (41–43). This type of three-stage odor control scrubbing is common in Europe, where urban congestion around industrial facilities has long been a problem.

   When hydrogen sulfide concentrations are approx 100 ppm and higher, a two-stage scrubber system to control hydrogen sulfide will be justified based on chemical costs. In such a system, approx 80% of the H_2S present is solubolized in a first-stage caustic-only scrub, then the remainder of the H_2S present is oxidized in a second caustic/oxidation scrubber such as the scrubber described in this example. The advantage of this two-stage system is that all the oxidizing chemical (the most expensive chemical consumed) will be utilized as the blowdown from the second stage is directed to the sump of the first stage.

   Additionally, when using sodium hypochlorite as the oxidizing agent, a competitive chemical reaction is present in the scrubber:

$$H_2S + NaOCl \rightarrow S + NaCl + H_2O$$

   Although this reaction accounts for only about 1% of the chemistry in a caustic/hypochlorite scrubbing system, at higher concentrations of H_2S, the elemental sulfur formed can form deposits on the tower packing. An open packing with all rounded surfaces and high void space, such as Q-PAC from Lantec Products, may minimize fouling problems in a scrubber. Nevertheless, this reaction will consume additional chemicals.

4. Other alternative technologies for odor control at wastewater-treatment plants (WWTPs) are presented in the following discussion. There are many alternative technologies for removing hydrogen sulfide and other malodorous substances from

a contaminated airstream (32,36–38,40–48). The most commonly used odor control processes include (1) wet scrubbers, (2) regenerative thermal oxidizers, and (3) bioscrubbers.

The single- and multiple-stage wet scrubbers have been discussed in this example, although different types of wet scrubber (other than packed towers), other packing products (in addition to Q-PAC), and other scrubbing liquors (per previous discussion) may also be used. Example 25 introduces a totally different type wet scrubber (47) for hydrogen sulfide removal.

Regenerative thermal oxidizers (RTOs) are introduced in another chapter of this handbook. Chemical porcelain heat recovery media (manufactured by Lantec Products) have revolutionized the design of RTO units. Thermal oxidation was once thought to be practical for odor and VOC control only when airflow rates were large (>25,000 scfm). Lantec's new Multi-Layered Media (MLM) has reduced the size and fabrication costs of a RTO to the point where a unit as small as 800 scfm is now practical. Because of the low-pressure-drop characteristics of MLM, electric power consumption and hence operating cost has been reduced for a RTO unit threefold to fivefold. All Operating and manufacturing costs have been reduced as well, because of the non-plugging characteristics of MLM heat recovery media.

Bioscrubbers are also called biofilters or biofiltration units, which are discussed in another chapter of this handbook series. Conventional wet scrubbers such as this example will predictably consume large quantities of chemicals in WWTPs. Per previous discussions, chemical costs often dictate consideration of additional stage scrubbing. Odor-causing compounds such as hydrogen sulfide, mercaptans, DMS, and DMDS often require chemicals for treatment as well as an additional scrubber stage. The bioscrubber (37,44,45), on the other hand, utilizes a dense biofilm to control these malodorous sulfur compounds. A biological substrate has recently been demonstrated (41) as an effective (99.9% + removal) odor control method with greatly reduced chemical costs.

Example 20

Removal of carbon dioxide from an air emission stream by wet scrubbing is presented in this example. Please note the following:

1. The environmental engineering significance of carbon dioxide removal.
2. A feasible wet scrubbing system for removing carbon dioxide from an air emission stream.
3. The chemical reactions involved in scrubbing carbon dioxide using a scrubbing liquor of 25% sodium carbonate and 4% caustic (concentrations by weight percentage).

Solution

1. The environmental engineering significance of reducing carbon dioxide emissions has been noted previously by Wang and Lee (49). Greenhouse gases such as carbon dioxide, methane, and so forth caused global warming over the last 50 yr. Average temperatures across the world could climb between 1.4°C and 5.8°C over the coming century. Carbon dioxide emissions from industry and automobiles are the major causes of global warming. According to the United Nations Environment Programme Report released in February 2001, the long-term effects could cost the world about 304 billion U.S. dollars a year down the road. This is the result of the

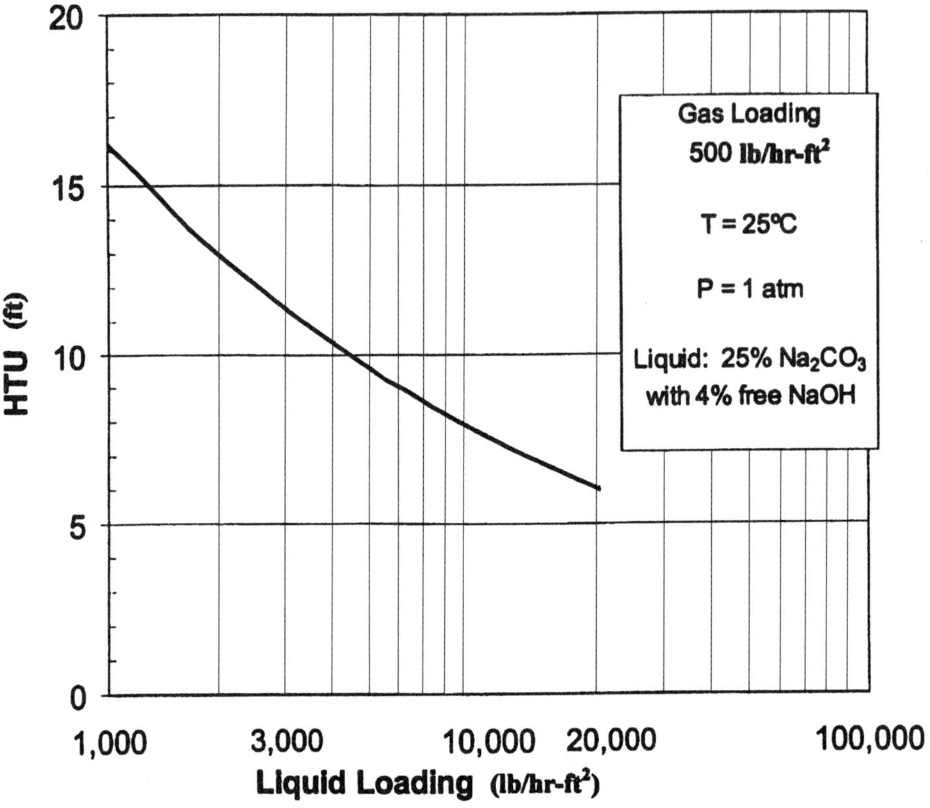

Fig. 16. Carbon Dioxide Absorption. (Courtesy of Lantec Products Inc.)

following projected losses: (1) human life loss and property damages as a result of more frequent tropical cyclones; (2) land loss as a result of rising sea levels; (3) damages to fishing stocks, agriculture, and water supplies; and (4) disappearance of many endangered species. Technologically, carbon dioxide is a gas that can easily be removed from the industrial stacks by a scrubbing process using any alkaline substances.

2. The following discussion presents a feasible wet scrubbing system for removing carbon dioxide from an air emission stream. Several wet scrubbing (absorption) processes are possible for carbon dioxide removal from an air emission stream. One plausible scrubbing solution is presented in Fig. 16.

   This single-stage scrubbing system will remove 99.9% of the carbon dioxide present in the airstream at these given conditions. A scrubber characteristics diagram is shown in Fig. 16. The packed tower wet scrubber, per Fig.1b, has the following characteristics:

   Gas loading = 500 lb/h-ft^2
   Temperature = 25°C
   Pressure = 1 atm
   Scrubbing liquor of 25% sodium carbonate and 4% caustic
   Packing media used Q-PAC packing in polypropylene from Lantec Products

Packing height = 24 in.
Gas loading = 500 lb/h-ft^2
Liquid temperature = 75°F

3. The chemical reactions involved in scrubbing carbon dioxide are dependent on the actual scrubbing solution chosen. In this example using a scrubbing liquor of 25 % sodium carbonate and 4 % caustic, the chemical reactions are

$$CO_2 + NaOH \rightarrow NaHCO_3$$

$$CO_2 + H_2O + Na_2CO_3 \rightarrow 2\ NaHCO_3$$

Example 21

Because the greenhouse effect of carbon dioxide gas is well established and human activities are the largest source of CO_2 entering the atmosphere, the burden rests upon humanity to solve the economic conundrum currently limiting efforts to bring CO_2 emission under control. The wet scrubber system introduced in Example 20 or a similar wet scrubber system is quite capable of removal of CO_2 from an air emission stream. However, the economic and political considerations currently are such that even in developed nations, such as the United States and the European Union, governments are not willing to force their domestic industries to institute such well-proven methods of CO_2 emission control. Discuss the following:

1. Technical limitations to removal of carbon dioxide from an air emission stream.
2. The economic and political solutions and driving forces for carbon dioxide control.
3. The possible combined technical and economic solutions to carbon dioxide emission reduction.

Solution

1. The technical limitations for removal of carbon dioxide from an air emission stream are presented in the following discussion. From a strictly engineering viewpoint, CO_2 could be easily removed from the air emission of any industrial facility (such as a coal-fired power plant or other single-source site of CO_2) using an alkali (such as sodium carbonate/caustic) and a wet scrubber packed with highly efficient mass transfer media (such as Q-PAC or similar).

   There is no technical limitation to removing CO_2 from an air emission stream. The only limitation to such a scrubber would be that absorption of CO_2 beyond 360 ppmv, the ambient level of CO_2 in the atmosphere, would obviously not be an effective use of resources.

2. Following is a discussion on the economic and political solutions and driving forces for carbon dioxide control: Economic solutions for CO_2 emission reduction and control are very difficult to find. The wet scrubbing technology for CO_2 emission control (shown in Example 20) is widely rejected as being too costly for industry and society, in general, to accept. Although the societal benefits of CO_2 emission reduction and control are no longer widely debated, the realities of the marketplace, global competition, and the parochial individual interests of various nations have all combined to prevent forward movement and application of proven methods of keeping CO_2 from entering the atmosphere. When products are sold in the world marketplace that are of equal or similar quality, the lowest-cost product will eventually dominate the marketplace. Higher-cost competition will be driven to extinction as a result.

As an example, the US government will not force CO_2 emission limitations on its coal-burning electrical generating facilities because this would force its industries, such as steel, automotives, and chemicals, to accept much higher electrical power costs. Such costs would need to be recovered by the various industries affected through higher pricing, and as just mentioned, this will not be allowed by the global forces driving the current world economy. Only if all industries in all nations are required to implement CO_2 reduction technologies will the competitive disadvantage of single nations or groups of nations placing CO_2 emission requirements on their respective industries be negated.

At present the international community is attempting to educate world leaders in the hope that a political solution, even if temporary, may be found for this problem (63). Also, as awareness of the harmful effects of CO_2 emissions grows, governments will become more likely to commit resources to development of alternative technologies to limit CO_2 emission.

3. The possible combined technical and economic solutions to carbon dioxide emission reduction are presented as follows. A plausible alternative technology is collection of CO_2 emission streams for reuse. Research for utilization and reduction of CO_2 emissions has been conducted by Wang and colleagues (48,49). Wang and Lee (49) have reported that collection of carbon dioxide emissions at tanneries, dairies, water-treatment plants, and municipal wastewater-treatment plants for in-plant reuse as chemicals will be technically and economically feasible.

   About 20% of organic pollutants in a tannery's wastewater are dissolved proteins, which can be recovered using the tannery's own stack gas (containing mainly carbon dioxide). Similarly, 78% of dissolved proteins in a dairy factory can be recovered by bubbling its stack gas (containing mainly carbon dioxide) through its waste stream using a new type of wet scrubber (see Example 25). The recovered proteins from both tanneries and dairies can be reused as animal feeds. In water-softening plants for treating hard-water removal using a chemical precipitation process, the stack gas can be reused as a precipitation agent for hardness removal. In municipal wastewater-treatment plants, the stack gas containing carbon dioxide can be reused as both a neutralization agent and a warming agent. Because a large volume of carbon dioxide gases can be immediately reused as chemicals in various in-plant applications, the plants producing carbon dioxide gas actually may save chemical costs, produce valuable byproducts, conserve heat energy, and reduce the global warming problem (48,49).

Example 22

Discuss the following:

1. The similarities and dissimilarities between wet scrubbing process and gas stripping process.
2. The possibility of a combined wet scrubbing and gas stripping process.

Solution

1. The following presents a discussion on the similarities and dissimilarities between the wet scrubbing process and the gas stripping process. Wet scrubbing and gas stripping are both mass transfer unit operations. Only the direction of the movement of a given pollutant species is different. In a wet scrubber, such as in the previous example of

odor control (Example 19), the offending species is H_2S (or some other malodorous gas) and it is controlled by being absorbed into a passing liquid phase. Mass transfer from gas to liquid defines a scrubbing situation.

On the other hand, in certain situations the desire is to remove a given species from a liquid into a passing gas phase. This is the definition of air stripping. Using the example of H_2S gas, when H_2S is dissolved in groundwater in small amounts, the offending odor of H_2S may prevent the use of an otherwise potable water source. Therefore, a stripping tower is one plausible technique to use to remove the H_2S from the water. Other possible solutions are aeration or tray tower technologies (47,50–55). The actual choice of removal technology will depend on the space available for the equipment used to treat the water. An aeration basin will require a large available area and will lose significant amounts of water to evaporation, as well as have high power (and hence operating) costs. A tray tower will be less costly to fabricate than a stripper system. However, if the water flow being treated is large, the large pressure loss in a tray tower and subsequent cost of operation will make a stripper tower the logical choice to treat the water.

Briefly, in mass transfer, a species must leave one phase and enter another phase. This movement of a molecule from one phase to another is treated extensively in standard academic texts by McCabe et al. (38). The two-film theory presented by McCabe et al. (38) is widely accepted as the model to explain how mass transfer occurs in both a scrubber and a stripper tower.

A simple, graphical explanation of the two-film theory of mass transfer is presented by Heumann (56). The concentrations of the species being scrubbed/stripped at the film interface will be less than the bulk concentrations of the species in the bulk phases as the specie transfers from one phase to the other. The difference in concentration between the bulk phases, and actually between the two-film interface, is the driving force to mass transfer. If the concentrations of the species at the film interface are equal to the bulk concentrations of the same species in the bulk phases, no mass transfer will occur.

In actual practice, the specie being treated in the system will have limited solubility in one of the two phases. In a scrubbing situation, the specie being scrubbed must cross the barrier of the gas film in order to pass into the liquid film. This resistance of passage of the molecule out of the gas film is the limiting factor to mass transfer in a scrubber system. So, with exceptions noted below, scrubbing is said to be gas film controlled.

The exceptions referred to are CO_2, NO_x, phosgene, or similar scrubbing situations. Although these gases have high solubility in water and one would think that as such gas film resistance would limit their mass transfer in a scrubber, in reality these and similar compounds are liquid film controlled in a scrubber system. This is so because, although readily absorbed into water, the subsequent chemical reactions of these compounds in water are relatively slow, therefore, the liquid film resistance is the controlling factor when scrubbing these compounds from an air emission stream. In a stripping situation, the specie of concern is moving in the other direction, out of the liquid film into the gas film. Thus, in a stripping situation, the limiting factor to mass transfer is the ability of the molecule in question to break out of the liquid film to enter the gas phase. Thus, with very few exceptions, stripping is said to be liquid film controlled.

It is important to know the detailed relationship between the scrubbing process and the stripping process. The reader is referred to another chapter, "Gas Stripping," of this handbook for a more detailed explanation of the stripping process than that given here. This chapter places emphasis on scrubbing process design and applications. Nevertheless, the reader should understand both the similarities and dissimilarities of the two processes.

For instance, if a packed tower reactor (Fig. 1b) or another reactor (Fig. 1a,c,d) is available, an environmental engineer may wish to use the same reactor both as a scrubbing process and a stripping process. In each instance, the scrubber or stripper will have two separate streams: (1) gas stream and (2) liquid stream.

It is a scrubbing process if (1) the gas stream is the target contaminated air emission stream from which one or more airborne pollutants (such as SO_2, H_2S, HAPs, VOCs, SVOCs, PM, heavy metals) will be removed by the reactor and (2) the liquid stream is the scrubbing solution (such as water with or without chemicals depending on the airborne pollutant(s) that need to be removed).

It is a stripping process if (1) the gas stream is the scrubbing agent (such as air with or without gaseous chemicals depending on the waterborne pollutants to be removed) and (2) the liquid stream contains the targeted pollutant (such as ammonia, chlorine, VOCs) that will be removed by the reactor. Normal instances of use of stripping towers is potable groundwater remediation, other contaminated groundwater treatment, or some other water pollution control need.

This discussion of the difference between scrubbing processes and stripping processes is more than an academic exercise. The optimum performance of a scrubber or a stripper tower most often depends on the correct selection of packing media with which to fill the tower. A given packing may perform better in promoting mass transfer in a scrubbing (gas film controls) process as opposed to promoting mass transfer in a stripping (liquid film controls) process. The opposite is true as well: A packing media may be better suited to enhancement of mass transfer in a stripping process and be less effective (less efficient, larger HTU value) in a scrubbing process.

2. The possibility of a combined wet scrubbing and gas stripping process is presented in the following discussion. A combined wet scrubbing and stripping process has been attempted by Wang and colleagues (48,49) for groundwater decontamination and reuse. The contaminated groundwater contains high concentrations of total hardness and volatile organic compounds (VOCs). An industrial plant near the contaminated site is discharging an air emission stream containing high concentration of carbon dioxide and is in need of additional industrial water supply.

   It has been demonstrated by Wang et al.(48,49) in a small pilot-plant study that a combined wet scrubbing and stripping process system using the aeration or tray tower technology is technically feasible for achieving (1) reduction of CO_2 from the air emission stream by scrubbing (i.e., groundwater is the scrubbing solution) and (2) reduction of VOCs by simultaneous stripping (i.e., the carbon dioxide gas is the stripping gas). Thus, the air emission stream and the groundwater stream treat each other. After treatment, the former is free from CO_2, whereas the latter is free from VOCs.

   The treated groundwater that is free from both VOCs and hardness may be recycled for the in-plant application as the industrial water supply. The treated air emission stream free from CO_2 is discharged into the ambient air.

The hardness in the groundwater contains mainly calcium bicarbonate, magnesium bicarbonate, magnesium sulfate, and calcium sulfate, which are to be removed. CO_2 in the flue gas is reused as a chemical for hardness removal from the groundwater. Lime (calcium hydroxide or calcium oxide) and soda ash (sodium carbonate) are additional chemicals required for groundwater treatment as well as carbon dioxide gas stripping. The following are chemical reactions for the combined flue gas (air emission stream) and groundwater treatment in the combined wet scrubbing and stripping process system:

$$\text{Contaminated flue gas} \rightarrow \text{air} + CO_2$$

$$\text{Contaminated groundwater} \rightarrow H_2O + \text{VOCs} + Ca(HCO_3)_2 + Mg(HCO_3)_2$$

$$CO_2 + Ca(OH)_2 \rightarrow CaCO_3 \text{ (precipitate)} + H_2O$$

$$Ca(HCO_3)_2 + Ca(OH)_2 \rightarrow 2CaCO_3 \text{ (precipitate)} + 2H_2O$$

$$Mg(HCO_3)_2 + Ca(OH)_2 \rightarrow CaCO_3 \text{ (precipitate)} + MgCO_3 + 2H_2O$$

$$MgCO_3 + Ca(OH)_2 \rightarrow Mg(OH)_2 \text{ (precipitate)} + CaCO_3 \text{ (precipitate)}$$

$$MgSO_4 + Ca(OH)_2 \rightarrow Mg(OH)_2 \text{ (precipitate)} + CaSO_4$$

$$CaSO_4 + Na_2CO_3 \rightarrow CaCO_3 \text{ (precipitate)} + Na_2SO_4$$

$$\text{Air effluent} \rightarrow \text{air} + \text{VOCs (to be removed by gas phase GAC)}$$

$$\text{Purified air} \rightarrow \text{ambient environment}$$

$$\text{Purified groundwater } (H_2O) \rightarrow \text{industrial water supply}$$

The precipitates produced from the above chemical reactions occurred in the combined wet scrubbing and stripping process and must be further removed by one of the following water–solid separation processes (49,57), before the purified groundwater can be reused as an industrial water supply: (1) dissolved air flotation and filtration, (2) sedimentation and filtration, or (3) ultrafiltration or microfiltration.

The air effluent from the combined wet scrubbing and stripping process will contain air and stripped VOCs. Before the air effluent can be discharged into ambient environment, it must be further purified by gas-phase granular activated carbon (GAC) or an equivalent air pollution process.

More research on simultaneous air and water pollution by a combined wet scrubbing and stripping process system should be conducted aiming at water recycle, greenhouse gas reduction, and resource recovery (i.e., CO_2 is a useful chemical for pH control, hardness precipitation, protein precipitation).

Example 23

What is the sound engineering solution to a described process situation? A chemical company in southern Louisiana manufactures 100,000 tons per day of herbicide and was faced with a potentially costly dilemma. The plant needed to treat the plant's output but required a Cl_2 stripper with a capacity of 75 gpm to do so. This represented a 50% increase in the capacity of the existing Cl_2 stripper or the need to build (1) a second stripper, (2) a new stripper, or (3) find a packing that would allow for the 50% increase in capacity in the existing stripper. If possible, solution 3 is the most economical choice. This means that the plant would need to find a packing that had a substantially lower pressure drop compared to the current packing (to allow for the increased throughput) and the new packing would also have to have an increased transfer efficiency to be able to meet the effluent specifications at the higher flow rate.

In addition to the capacity issue, the stripper (or packed tower, shown in Fig.1b), having a diameter of 18 in. and packing height of 28.5 ft, performs the function of stripping elemental chlorine (Cl_2) from hydrochloric (muriatic) acid (HCl). The existing packing (media) in the stripper tower is 2-in. Pall rings. Originally developed in the 1920s, Pall rings have traditionally been used for scrubbing applications. However, as needs dictated over the years, the Pall rings (and other similar packings) have found their way into stripping process applications.

Solution

For the plant in Louisiana, in this process situation the target contaminated liquid stream is hydrochloric acid from which elemental chlorine must be stripped. The gas stream is simply the air driven by an air blower (fan). The new packing material, in addition to the requirements stated above, also needed to have adequate acid and chlorine resistances to ensure a service lifetime of longer than 10 yr.

When approached by the Louisiana plant with this problem, the environmental engineer in charge considered several possible packing materials. Previous discussion has indicated that Q-PAC (supplied by Lantec Products) can be used to optimize a scrubber process system. However, in this instance, being a stripping process system, a different packing was found to be the solution to the needs of the Louisiana chemical plant.

The packing material recommended by the environmental engineer was #2 NUPAC in polyethylene. Although slightly more expensive than polypropylene, polyethylene offers better resistance to oxidative attack than polyethylene. This packing material also offers both improved mass transfer properties as well as reduced pressure drop compared to Pall rings.

In February 1999, the 28.5-ft bed of the packed tower was packed with #2 NUPAC. The performance of the tower after repack was excellent, so plant personnel were relieved that no new capital project would be required. Stripping of elemental chlorine remained at 99% efficiency in the hydrochloric acid liquid system at the increased flow of 75 gpm in the existing air stripping system. The upgraded stripping process system is summarized as follows:

Reactor design = packed tower (Fig. 1b)
Packing material = #2 NUPAC (Lantec Products)
Tower diameter = 18 in.
Packing height = 28.5 ft.
Target pollutant liquid stream = hydrochloric acid containing elemental chlorine
Liquid design flow = 75 gpm
Liquid maximum flow = 85 gpm
Gas stream = clean air
Service life of packing = 10+ yr
Design stripping efficiency = 99+%
Flow pattern = liquid flows downward, air flows upward in a countercurrent flow pattern

Example 24

Discuss the past problem and the recent developments in packing materials for scrubbers and stripper absorption systems.

Solution

1. The past problem is present as follows (36). Packing materials (hereafter referred to as packing) have been used to enhance gas–liquid contact in chemical engineering scrubber and stripping unit operation systems (hereafter referred to as tower) as standard engineering practice for several generations. Many environmental engineers originally graduated as chemical engineers from their respective university or college and modern environmental engineering degree programs require several courses be taken in chemical engineering before graduation. Therefore, the principle that gas–liquid contact must be maximized for optimum absorber tower performance is universally understood and accepted.

   The packing in an absorption tower is placed there to optimize contact between the two phases present (liquid and gas) so that a target pollutant in one phase will transfer into the other phase. As has been previously described in this handbook, the actual direction of this movement defines a scrubbing (from the gas into the liquid) process or a stripping (from the liquid into the gas) process. The historic solution to maximizing gas–liquid contact has been to design the packing with more and more complicated shapes (38). Famous packing materials (mass transfer media), such as Saddles, Pall rings, Tellerettes, and Tri-Packs, were patented in 1908, 1925, 1964, and 1973–1978, respectively. When packed in the absorption tower, such media tend to spread the liquid into a thin film over the surface of the packing to maximize the liquid–gas contact (38).

   A standard measure by which competitive packing products have been historically compared is as specific surface of media in square feet per cubic foot of the media (ft^2/ft^3). With this parameter, the environmental or other design engineer could assess the area of available packing surface upon which the liquid in the tower could form a film. A higher specific surface of a media product is equated to larger film surfaces. This, in turn, meant that when comparing media products, the media with the largest specific surface most likely promoted the most efficient mass transfer in an absorption tower.

   Therefore, suppliers of packings responded by inventing products of with increasingly complicated designs, as well as smaller sizes per individual piece, to increase the specific surface area of their products. The problem with these early packings is that more surface and more pieces of media per cubic foot increase costs. This is the result of increased raw material needed to produce a smaller packing. Operating costs of an absorption system also increase as a result of an increased pressure drop when a smaller packing is chosen. Therefore, capital costs increase as towers are sized larger to minimize pressure drop.

2. The engineering solution is described here (36). Not satisfied with the need to trade increased costs for improved performance in absorber tower performance, starting in 1987 innovative engineers of many packing manufacturers began introducing new, high-technology packings. Lantec Products alone patented LANPAC, NUPAC, and Q-PAC in 1988, 1992, and 1996, respectively. The authors choose one of the best, Q-PAC, for illustrating how an advanced packing has been conceptually developed, introduced, tested, manufactured, and eventually patented for commercial applications.

   The introduction of the latest mass transfer media has revolutionized tower designs. Previously discussed in this handbook were examples of how towers are significantly smaller when designed with modern media than is possible with any other early commercial media products.

The success of Q-PAC is a result of the insight of Dr. K. C. Lang of Lantec Products. His realization was that additional opportunity to force liquid–gas contact existed that had been ignored in previous packing designs. In addition to having the liquid spread into a thin film on the solid surface of the packing, if the packing design could be such that the liquid was forced to pass through the tower as a shower of droplets, each and every droplet would offer surface for gas–liquid contact through which mass transfer would occur.

Prior to this innovation, the primary means of creating liquid surfaces was to spread the liquid over the media, as previously discussed. However, also as previously discussed, this additional liquid surface was obtained at a price: (1) higher media costs as the consequence of smaller media size that requires more raw material and more pieces per cubic foot; (2) increased operating costs as the consequence of smaller media size causing pressure drop increases; and (3) increased capital costs as the result of the need to design larger installations to minimize pressure drop.

In addition to using a specific surface as a comparison parameter, packing suppliers have provided a parameter called void fraction (or free volume) to describe a given packing. This parameter is expressed as percent (%) of free space. Although useful, void fraction is nevertheless always subjective and therefore susceptible to manipulation. This is so because in addition to the free volume of the packing, the numbers presented to industry also include the percentage of free space within an absorber tower that results from the "random dump" of the media into the tower. This tower free volume will obviously depend on the tower diameter, the overall packing depth, the type of supports used within the tower, and numerous other variables. A general industry standard has been to use an estimate of 39% tower free volume, which is used to determine the free volume or void fraction published for a given media product. This is, as stated, only a general standard; therefore, individual suppliers are free to choose their own standard as well as to keep such choice proprietary.

Industry would be better served if a quantitative parameter free of any possible manipulation were available for use to evaluate packings. Therefore, it is suggested here that the absolute void volume (AVV) be introduced and used as the standard parameter for the free space of a packing. The absolute void volume is independent of any subjective interpretation as the result of its definition:

$$\text{AVV} = \{1 - (W_{\text{media}}/W_{\text{water}})(\text{SG}_{\text{water}}/\text{SG}_{\text{media}})\}\ (100\ \%)$$

where AVV is the absolute void volume (dimensionless), W_{media} is the weight of the media (lb/ft^3), W_{water} is the weight of the water (62.4 lb/ft^3), SG_{media} is the specific gravity of the plastic or other material used to produce the media, and SG_{water} is the specific gravity of water (= 1). In the case of Q-PAC, W_{media}, W_{water}, SG_{media}, and SG_{media} are 2.1 lb/ft^3, 62.4 lb/ft^3, 0.91 (polypropylene), and 1 (for water), respectively. The AVV of Q-PAC is calculated to be 96.3%, whereas the AVV of all other commercial packings using the same plastic material (polypropylene) will be below 95%. As a result of this definition of AVV, it is now possible to evaluate, independent of any subjective manipulation, the void volume of a single piece of packing or 1000 pieces of packings, where the AVV parameter is absolute. Using this new parameter, an environmental engineer will be able to compare various commercial packing products.

The design of Q-PAC using rounded surfaces and needles to support droplet formation was arrived at through extensive trial and error. A liquid stream is forced into a shower of droplets by the media; in turn, the media's mass transfer efficiency increases. It is interesting to note that the new media, in addition to reduced capital costs (the result of smaller tower diameter) and lower operating costs (a direct result of lower operating costs due to reduced pressure drop) that have been previously discussed in this handbook, offer additional savings to industry.

Regardless of the design of a given packing, the cost of that packing will be fixed based on the amount of material (plastic resin, metal, ceramic, etc.) that is required to produce the packing. It is important to note that the amount of plastic needed to produce a cubic foot of Q-PAC as well as the number of pieces of Q-PAC needed to fill a cubic foot is far less than any other early contemporary packings. This is very significant when it is realized that the cost of plastic resin represents approx 40% of the final cost of a packing when using polyethylene to mold the packing. If a more expensive plastic resin such as Teflon must be used (because of chemical- or temperature-resistance considerations), the cost of resin can escalate to 95% of the final cost of the packing.

Also, as plastic media are produced by injection molding, the number of pieces per cubic foot will directly impact the final cost of a packing. The greater the number of pieces needed for a cubic foot, the more costly to mold and, hence, the more expensive a given packing will be.

It should be noted that in order to reduce the cost of injection molding ($/ft^3) a greater number of pieces need to be molded in a single cycle of the injection-molding machine. However, to accomplish this, a multipiece mold is required for the injection-molding process. The fabrication cost of this type of mold increases geometrically as the number of pieces the mold is capable of producing is increased. Therefore, although the number of pieces being produced in a single cycle of the molder can be increased, the savings thus realized in reduced labor costs are quickly consumed by increased capital expenditure and amortization of the mold.

Modern mass transfer media should maximize mass transfer in scrubber towers (gas-film-limited systems, per previous discussion). The use of modern mass transfer media provides several advantages:

- Smaller tower diameters: reduced capital and fabrication costs, smaller system footprint!
- Lower pressure drop: smaller blower motor, lower electrical energy costs, less noise!
- Smaller chemical recirculation pumps: less costly!
- Smaller mist eliminators: less costly!
- Less total packing volume: reduced capital and fabrication costs, smaller system footprint!
- Greater mass transfer media: lower cost packing ($/ft^3)!
- Increase fouling and plugging resistant: reduced maintenance costs!
- Increase capacity of existing towers

Some commercial packings (such as LANPAC), on the other hand, maximize mass transfer in liquid film-limited systems (per previous discussion), which is commonly encountered in stripping situations. Other commercial packings (such as NUPAC) are highly efficient media that have found a niche in keeping tower heights to the absolute minimum, such as with are indoor tower.

Each mass transfer problem is unique and deserves individual attention in order that the most cost-effective and productive solution is obtained for the given set of circumstances. A responsible environmental engineer should always conduct an extensive literature study and a pilot-plant study to evaluate and select the most suitable mass transfer media for the specific scrubbing/stripping applications of his/her clients.

Example 25

Traditional scrubbing/stripping systems involve distribution of small liquid/slurry droplets or thin films into the bulk of a flowing airstream (32). Innovative scrubbing/stripping systems, on the other hand, involve distribution of small air bubbles into a bulk of a flowing water stream (47–55). Provide a discussion on the following:

1. The flow patterns, advantages, and disadvantages of the innovative scrubbing/stripping systems in comparison with comparable traditional scrubbing/stripping systems.
2. A typical case history of an innovative wet scrubbing system for hydrogen sulfide reduction from an air emission stream.

Solution

1. The flow patterns, advantages, and disadvantages of the innovative scrubbing/stripping systems in comparison with comparable traditional scrubbing/stripping systems are discussed here. It has been known that wet scrubbing and gas stripping are both mass transfer unit operations. Only the direction of the movement of a given pollutant species is different.

   Either the wet scrubbing or gas stripping process will have two streams: a gas steam and a liquid stream. When the two streams meet in a scrubbing/stripping reactor, the mass transfer occurs. Because the scrubbing reactor and stripping reactor are similar to each other, only the wet scrubbers are discussed.

   In a traditional wet scrubber, for instance, such as in the previous example of H_2S reduction (Example 19), the offending specie is H_2S (or some other malodorous gas) present in an air emission stream or gas phase. The liquid phase is the scrubbing solution, which is distributed into the wet scrubber as small liquid droplets or thin films. The traditional wet scrubber is controlled by distributing the liquid phase (i.e., scrubbing solution containing the scrubbing chemicals) as liquid droplets or thin films into a passing bulk gas phase (i.e., air emission stream containing the target pollutant, H_2S). The flow pattern in the traditional wet scrubber can be either counterflow or cross-flow. Mass transfer from gas to liquid defines a scrubbing situation.

   In an innovative wet scrubber, such as in a new case history for H_2S reduction to be presented in the second portion of this example, the offending specie is still H_2S (or some other malodorous gas) present in an air emission stream or gas phase. The liquid phase is still the scrubbing solution, but it is distributed into the innovative wet scrubber as a flowing bulk liquid. The innovative wet scrubber is controlled by distributing the gas phase (i.e., air emission stream containing the target pollutant, H_2S) as small gas bubbles into a passing bulk liquid phase (i.e., scrubbing solution containing the scrubbing chemicals). The flow pattern in the innovative wet scrubber can also be either counterflow or cross-flow. Mass transfer from gas to liquid also defines this scrubbing situation.

   The mass transfer efficiency of a traditional scrubbing/striping process can be enhanced by packing materials. In comparison with a comparable traditional scrub-

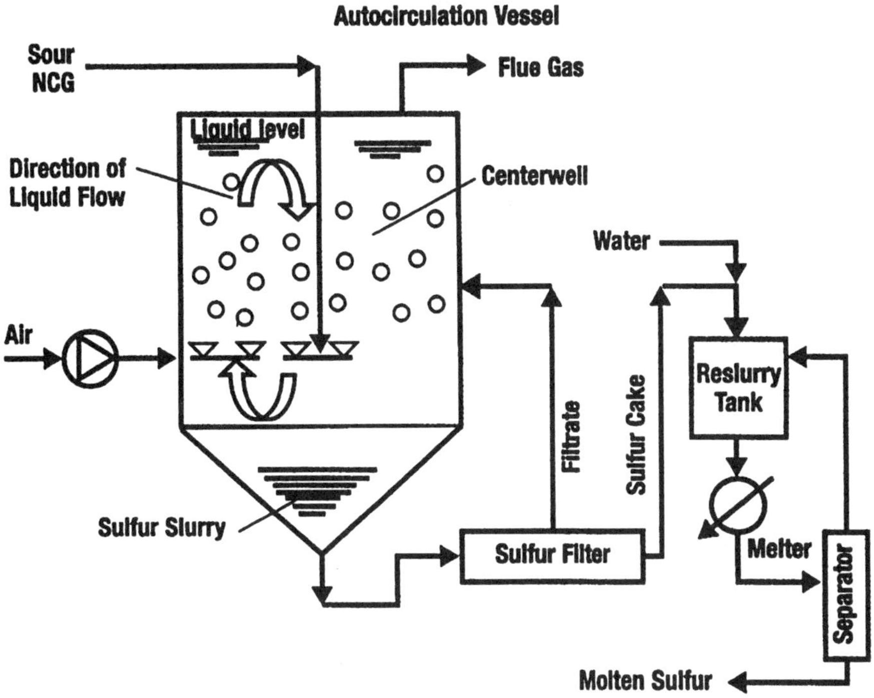

Fig. 17. Innovative wet scrubbing system for the removal of hydrogen sulfide from a geothermal power plant's emission stream.

bing/stripping process system treating the same flow rate of same polluting stream, the innovative scrubbing/stripping process system will have lower mass transfer efficiency and lower capital costs.

Both the traditional scrubbing/stripping process systems and the innovative scrubbing/stripping process systems have their proper places in modern pollution control. The innovative design of scrubbers and strippers are attractive when the liquid stream contains a high concentration of suspended solids either before or after scrubbing/stripping.

2. A typical case history of an innovative wet scrubbing system for hydrogen sulfide reduction from an air emission stream is presented in the discussion as follows. Geothermal power plants are environmentally attractive because they employ a renewable energy source; however, a geothermal stream contains varying amounts of noncondensable gases (NCGs), such as carbon dioxide and hydrogen sulfide, which, if not disposed of correctly, may cause environmental, health, and safety problems. In addition, if the carbon dioxide is to be further processed to produce beverage-grade carbon dioxide, the hydrogen sulfide must be removed to extremely low levels. In some locations, H_2S may be present in high enough quantities to represent an economical raw material for recovering elemental sulfur, which can then be sold as a product for further processing into sulfuric acid or fertilizers (48).

When the amount of H_2S in the air emission stream is above 140 kg/d (as H_2S), an innovative wet scrubbing system, also known as the liquid redox system, can be generally employed to treat the NCG because it can achieve very high H_2O removal efficiencies (99+%) and because it has very high turndown capabilities. The liquid redox system is considered by some (58,59) to be the best available control technology for geothermal power plants. The process employs a nontoxic, chelated iron catalyst that accelerates the oxidation reaction between H_2S and oxygen to form elemental sulfur. The oxidation process is

$$H_2S + 0.5\ O_2 \rightarrow S \text{ (elemental sulfur)} + H_2O$$

As implied by its generic name, liquid redox, all of the reactions in the process occur in the liquid phase in spite of the fact that the above reaction is a vapor-phase reaction. In the process, the NCG is contacted in a wet scrubber (shown in Fig. 17) with the aqueous, chelated iron solution where the H_2S is absorbed and ionizes into sulfide and hydrogen ions. This process is presented as follows:

$$H_2S \text{ (vapor)} + H_2O \rightarrow 2\ H^+ + S^{2-}$$

The dissolved sulfide ions then react with chelated ferric ions to form elemental sulfur:

$$S^{2-} + 2\ Fe^{3+} \rightarrow S \text{ (elemental sulfur)} + 2\ Fe^{2+}$$

The solution is then contacted with air in an oxidizer, where oxygen (in air bubbles) is absorbed into the solution and converts the ferrous ions back to the active ferric state for reuse as follows:

$$0.5\ O_2 \text{ (vapor)} + H_2O + 2\ Fe^{2+} \rightarrow 2\ Fe^{3+} + 2OH^-$$

Combining the above three reactions yields the following reaction,

$$H_2S + 0.5\ O_2 \rightarrow S \text{ (elemental sulfur)} + H_2O$$

As illustrated in Fig. 17, the air emission stream (containing H_2S) enters the wet scrubber's absorption section, where it is contacted with the scrubbing solution (LO-CAT solution) and where the H_2S is converted to elemental sulfur. The partially reduced solution then circulates to the oxidation section where it is contacted with air, which reoxidizes the iron. The exhaust air from the oxidation section and the sweet NCG from the absorption section are exhausted to the atmosphere.

In the conical portion of the vessel (*see* Fig. 17), the sulfur will settle into a slurry of approx 15 % (by weight). A small stream is withdrawn from the cone and sent to a vacuum belt filter, where the sulfur is further concentrated to approx 65% (by weight) sulfur. Some units stop at this stage and sell the sulfur cake as a fertilizer. If molten sulfur is required, the cake is reslurried and melted as shown in Fig. 17.

Although an innovative wet scrubbing system (i.e., the liquid redox system) has slightly high capital cost, it is very inexpensive to operate. Operating costs usually range between \$0.20/kg of H_2S to \$0.25/kg of H_2S (48).

Example 26

Wet scrubbing using lime/limestone is one of the feasible processes for flue gas desulfurization (32). Discuss the following:

1. The process description, performance, and future of the wet scrubbing flue gas desulfurization process.
2. The chemical reactions of the wet scrubbing flue gas desulfurization process.

Solution

1. The discussion on wet scrubbing flue gas desulfurization process is presented here. Flue gas desulfurization (FGD) is a process by which sulfur is removed from the combustion exhaust gas. Wet scrubbing FGD using lime/limestone is the most commonly used method of removing sulfur oxides resulting from the combustion of fossil fuels. It is also the method that is best suited to control SO_x emissions from copper smelters. SO_x is a symbol meaning oxides of sulfur (e.g., SO_2 and SO_3).

   The FGD processes result in SO_x removal by inducing exhaust gases to react with a chemical absorbent as they move through a long vertical or horizontal chamber. The absorbent is dissolved or suspended in water, forming a solution or slurry that can be sprayed or otherwise forced into contact with the escaping gases. The chamber is known as a wet scrubber, and the process is often referred to as wet scrubbing FGD. More than 60 different FGD processes have been developed, but only a few have received widespread use. Of the systems currently in operation, over 90% use lime or limestone as the chemical absorbent. In a lime slurry system, the sulfur dioxide reacts with lime to form calcium sulfite and water. For cases where limestone is used instead of lime, the sulfur dioxide reacts with limestone to form calcium sulfite, water, and carbon dioxide gas.

   Wet scrubbing FGD typically removes 90+% of the sulfur dioxide in a flue gas stream. A few problems have arisen in the operation of the lime or limestone wet scrubbing FGD systems, and US EPA's Industrial Environmental Research Laboratory in Research Triangle Park, North Carolina, has been successful in developing solutions. Current efforts are directed toward using the limestone more efficiently, removing more SO_2 from the exhaust gases, improving equipment reliability, and altering the composition of the waste sludge so that it can be more easily disposed of in landfills.

   Calcium sulfite that is formed during the scrubbing process presents another important problem. This substance settles and filters poorly, and it can be removed from the scrubber slurry only in a semiliquid or pastelike form that must be stored in lined ponds. The US government has developed a method to solve this engineering problem through a process called forced oxidation.

   Forced oxidation is a defined as a process in which sulfite-containing compounds are further oxidized to sulfate compounds by aeration with air or pure oxygen to promote dewatering, ease of handling, and/or stability in the waste product. Forced oxidation requires air to be blown into the tank that holds the used scrubber slurry, composed primarily of calcium sulfite and water. The air oxidizes the calcium sulfite to calcium sulfate.

   The calcium sulfate formed by this reaction grows to a larger crystal size than does calcium sulfite. As a result, the calcium sulfate can easily be filtered to a much drier and more stable material, which can be disposed of as landfill. In some areas, the material may be useful for cement or wallboard manufacture or as a fertilizer additive.

   Another problem associated with limestone wet scrubbers is the clogging of process equipment as a result of calcium sulfate scale. Forced oxidation can help control scale by removing calcium sulfite from the slurry and by providing an abundance of pure gypsum (calcium sulfate) to rapidly dissipate the supersaturation normally present. The process also requires less fresh water for scrubber operation, which is scarce in many western US locations. Current experiments at the US Research Triangle Park are directed toward testing various forced oxidation designs to find the best oxidation system using the least energy.

2. The chemical reactions of the wet scrubbing FGD process are discussed and presented next. As stated previously, in a lime slurry system, the sulfur dioxide and sulfur trioxide react with lime (CaO) to form calcium sulfite and water, based on the following reaction:

$$SO_x + CaO + H_2O \rightarrow CaSO_3 + H_2O$$

When limestone ($CaCO_3$) is used instead of lime, it results in a similar chemical reaction, but also yields carbon dioxide:

$$SO_x + CaCO_3 + H_2O \rightarrow CaSO_3 + H_2O + CO_2$$

In the forced oxidation reaction, the oxygen in air oxidizes the calcium sulfite $CaSO_3$ to calcium sulfate $CaSO_4$ as present in the following reaction:

$$CaSO_3 + H_2O + 0.5\ O_2 \rightarrow CaSO_4 + H_2O$$

where $SO_x = SO_2 + SO_3$, CaO = lime, calcium oxide, $CaSO_3$ = calcium sulfite, $CaCO_3$ = limestone and, $CaSO_4$ = calcium sulfate.

Example 27

Dry scrubbing is a feasible process FGD (32,66,67). Discuss the following:

1. The process description, performance, and future of the dry scrubbing FGD process.
2. The chemical reactions of the dry scrubbing FGD process.

Solution

1. The following presents the process description, performance, and future of the dry scrubbing desulfurization process. Dry scrubbing is a modification of the wet scrubbing FGD technology. As in other FGD systems, the exhaust gases combine with a fine slurry mist of lime or sodium carbonate. This system, however, takes advantage of the heat of the exhaust gases to dry the reacted slurry into particles of calcium sulfite and sodium sulfite.

   The particles generated by this dry scrubbing process are then collected along with other particles from coal combustion in a baghouse collector. This collector uses fabric bags that function similarly to those in a vacuum cleaner, which collect particles while permitting cleaned gases to escape.

   Dry scrubbing typically removes 70% of the sulfur dioxide in a waste gas stream. It is 15–30% less expensive to install and operate than a conventional wet scrubbing system. However, because dry scrubbing is less efficient than wet scrubbing, the technology has been limited to use with low-sulfur coal.

   Plans for future research include evaluating the performance and reliability of a full-scale utility boiler equipped with a spraydryer SO_2 control system. Improvements could make these dry scrubbing systems acceptable for general use by the early 2000s.

2. The chemical reactions of the dry scrubbing desulfurization process are as follows:

$$SO_2 + CaO \rightarrow CaSO_3$$

$$SO_2 + Na_2CO_3 \rightarrow Na_2SO_3 + CO_2$$

where CaO = lime, calcium oxide, $CaSO_3$ = calcium sulfite, Na_2CO_3 = sodium carbonate, and Na_2SO_3 = sodium sulfite.

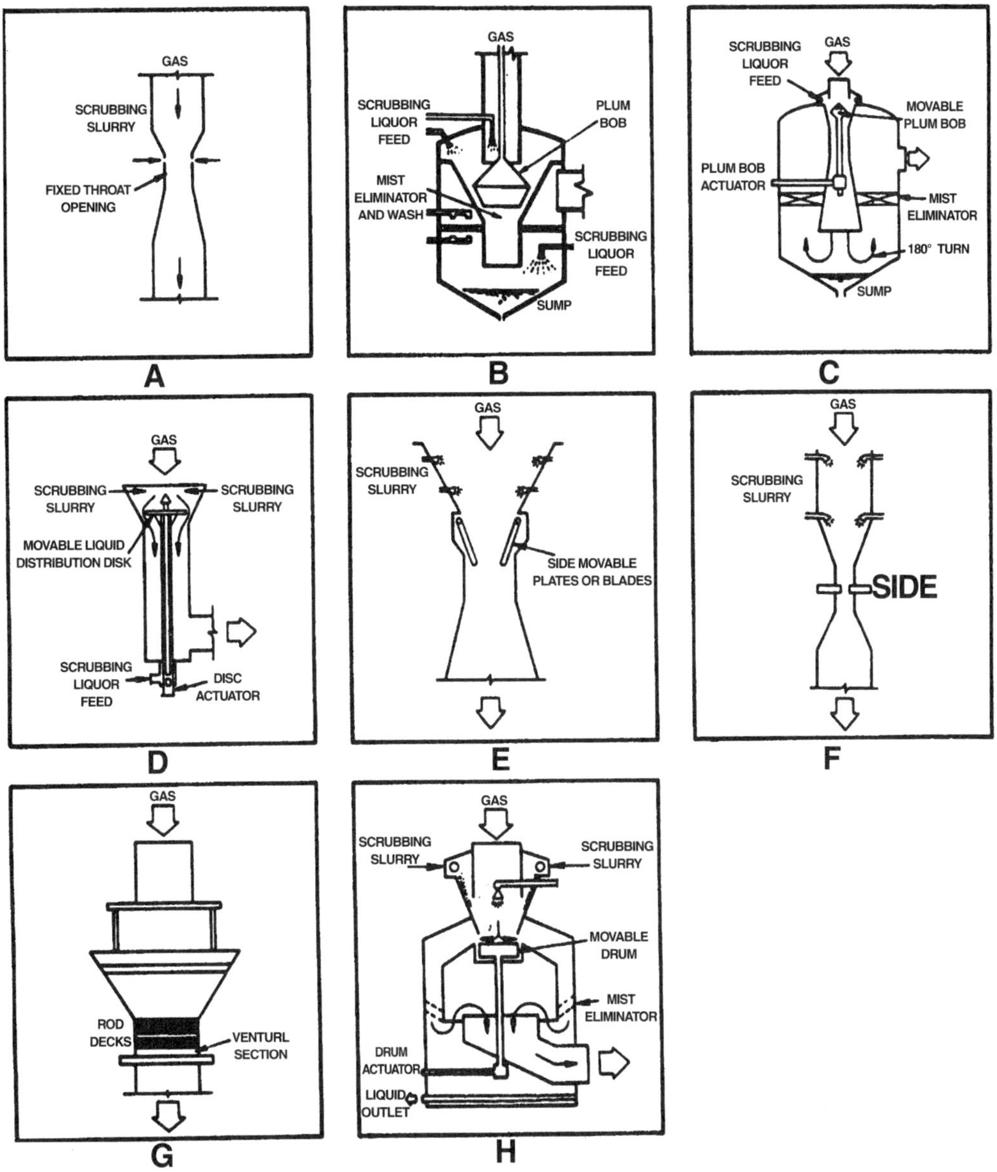

Fig. 18. Venturi tower configurationswer plant's emission stream: (a) Fixed Throat; (b) Variable-Throat top–entry plumb bob; (c) Variable-Throat bottom-entry plumb bob; (d) Variable-Throat top-entry liquid distribution disk; (e) Variable -Throat side veriable-plates or blades; (f) Variable-Throat side-movable blocks; (g) Variable -Throat vertically adjusted fod decks; (h) Variable-Throat vertically adjusted drum. (From US EPA.)

Example 28

The lime/limestone FGD process has been previously discussed in this chapter. This example is a discussion of various scrubber (absorption towers) designs that may be considered by an environmental engineer to use to treat the emissions from the FGD process. The possible choices are as follows (32,66,67):

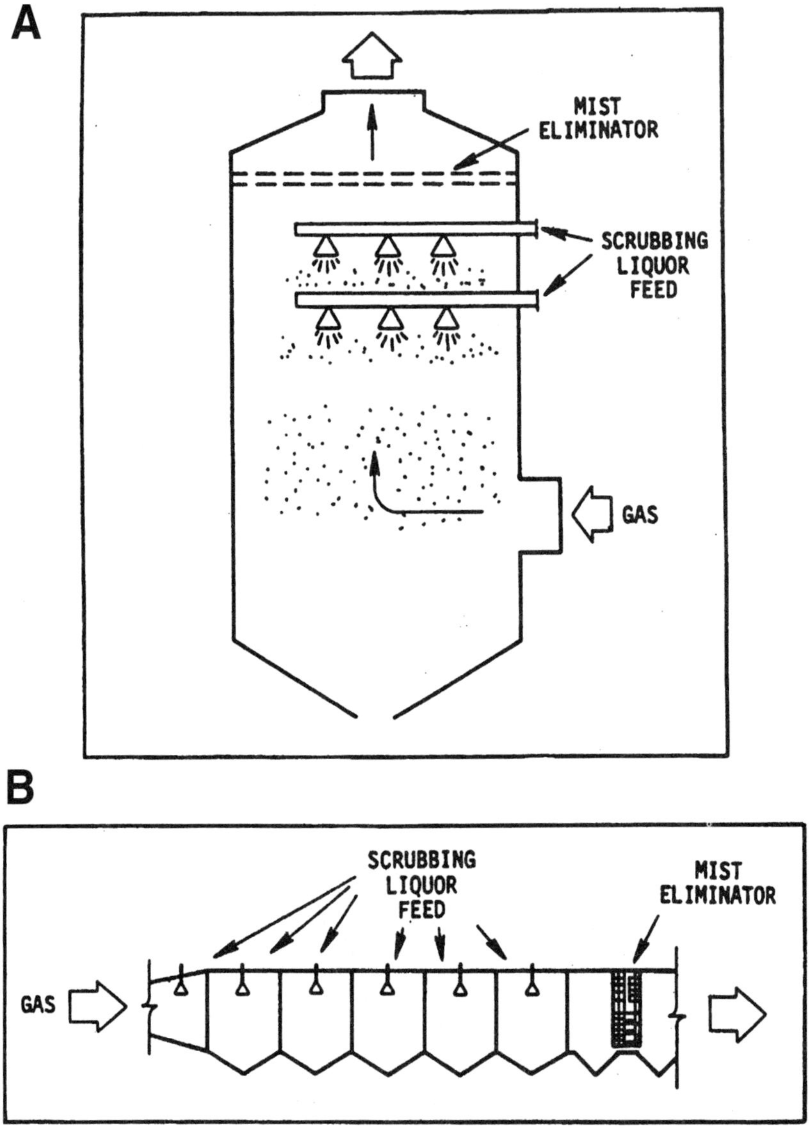

Fig. 19. Spray tower types: (a) open countercurrent; (b) open cross-current. (From US EPA.)

1. Venturi scrubber
2. Spray tower
3. Tray tower
4. Packed (wet) scrubber
5. A combination tower utilizing two or more of the above choices

Solution

1. First-generation Venturi scrubbers are presented in Fig. 18. Such systems typically used a fixed-throat design for the Venturi. As such, the throat opening used to form

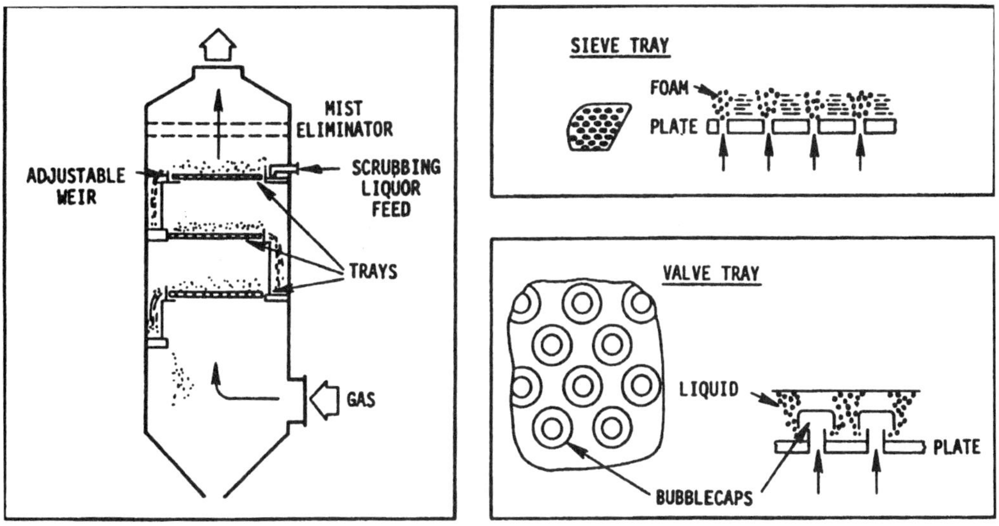

Fig. 20. Tray tower and tray types. (From US EPA.)

the Venturi remains fixed in these designs. Industry preference to use a single Venturi scrubber to process variable gas flow rates led to designs of Venturi scrubbers with adjustable throat openings. Examples of variable-throat Venturi scrubbers are presented in Fig. 18b–h.

Venturi scrubbers typically have a pressure drop of 10 to as high as 30 in. of water. As such, the Venturi scrubber is normally classified as a high-power-consumption unit operation. In addition to having a high energy demand, the choice of the Venturi scrubber is also limited by polluted gas–absorbent (the slurry) contact time within the tower.

2. Typical spray towers are presented in Fig. 19. In the spray tower, the absorbent (slurry) is injected into the polluted gas stream being treated through atomizing nozzles. The slurry is forced into a mist of fine microdroplets by the action of the nozzles. Droplet formation is also supported by the velocity of the gas being treated within the tower. The resultant extremely high surface area of the many droplets provides. for excellent contact between gas and liquid surfaces. In normal operations, slurry droplets are formed with diameters of 50–4000 mm.

3. In addition to promoting excellent gas–liquid contact, a spray tower accomplishes this with minimal pressure loss. This is a result of the fact that the spay tower has no internal components that will impede the upward flow of air as the slurry droplets pass downward through the tower countercurrent to the gas flow, as seen in Fig. 19a. This simple design allows for spray towers to operate with pressure losses in the range of 1–4 in. of water.

Spray towers are also sometimes designed using a crosscurrent flow scheme as presented in Fig. 19b. This design may be chosen over the countercurrent design as the result of height restrictions or other concerns regarding a vertical tower. As the result of the lower height, the slurry pump size will be reduced somewhat. A cross-flow tower will always require increased spatial area than a vertical tower. The need to have

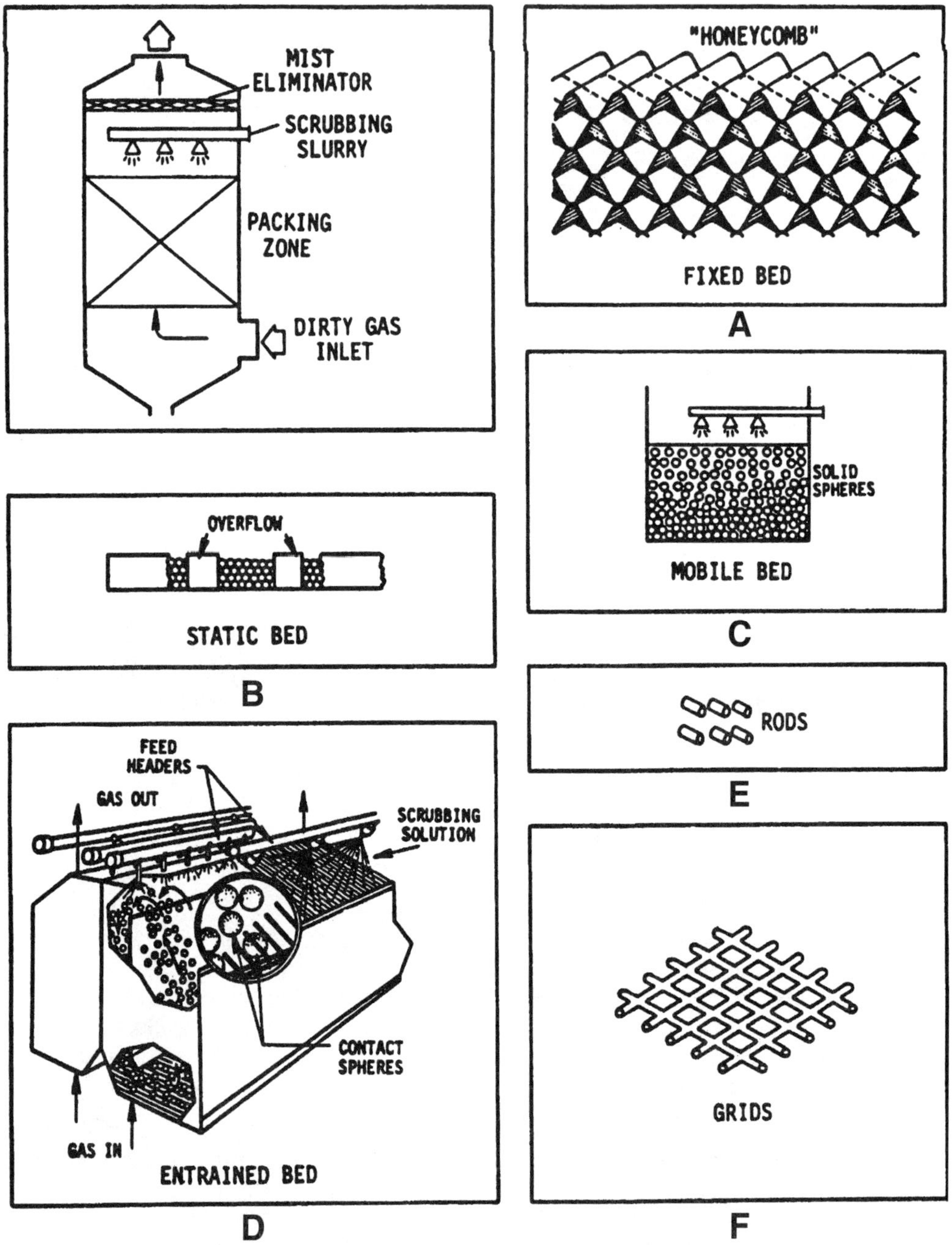

Fig. 21. Packed tower and packing types. (From US EPA.)

a larger tower fabricated will result in increased capital expense for a cross-flow tower versus a countercurrent flow spray tower.

4. A tray tower will always utilize the classic countercurrent flow scheme. As the name implies, the tray tower has one or more internal trays. These trays have openings to

a certain open area per the tower's design. As polluted gas being treated enters the tower, the gas passes upward through the tower. The slurry liquid (the absorbent) is introduced into the top of the tower and flows downward. Therefore, the slurry flows across each tray and the liquid finds its way to the tray openings. As a result, the two opposing flows are forced to interact, with resultant gas–liquid surface contacts that allow for the pollutant present in the gas to absorb into the liquid, as seen in Fig. 20.

Tray towers known as sieve towers utilize a design gas velocity sufficient to force the gas to form bubbles as the gas passes through the tray openings. Figure 20 illustrates this method of forcing gas–liquid contacts. An alternate design for tray towers is the valve tray tower. These towers use a "bubble cap" on each tray opening. Each bubble cap is also surrounded with a cage intended to constrain the flow of liquid (*see* Fig. 20). As the polluted gas flows upward through the tray openings, these caps keep the downward flowing slurry in an agitated condition. This forces the gas to exit each cap at near Venturi scrubber velocity. Tray towers typically operate at a pressure drop below that of a Venturi scrubber but well above the pressure drop of a spray tower. A typical pressure loss for a tray tower is about 20 in. of water. The power consumption of such towers is therefore significant.

5. Most packed towers are of vertical design so as to utilize countercurrent flow between the gas and liquid (*see* Fig. 21). Inside the tower is a packed bed. The packing that comprises the packed bed is in the tower to force increased gas–liquid contact to improve absorption efficiency. Packings of a wide variety of shapes, sizes, and material of construction are available. Additionally, packings can form several structures. A fixed structure such as the honeycomb packing seen in Fig. 21a is possible. Also, a random yet fixed structure such as the glass spheres presented in Fig. 21b may be used. The packing may also be mobile, as illustrated in Fig. 21c. The glass spheres become fluidized with sufficient gas velocity through the tower. In normal operation of the fluidized-bed scrubber, the packing actually passes out of the tower, where it is normally collected for reuse. Finally, rods, decks, vanes, or some other fixed structure may be used inside the tower as in Fig. 21e,f. As such, in this last choice, there is actually no "packing" *per se*; the rods are used to force gas–liquid contact.

   When properly designed, packed towers do not need high power. Packed towers are normally designed for pressure losses that overlap or are slightly higher than with tray towers. A packed (wet scrubber) will normally operate with pressure drop in the range of 2–8 in. of water.

A combination tower, as implied by the name, is the use of two or possibly more of these absorption techniques in a single tower. As such, the combination tower will allow for targeted pollutant removal (absorption) and/or operational flexibilities not possible with a tower that utilizes only one absorption technique. In the combination tower, discrete chemical and physical conditions are possible in different sections of the tower. Thus, one unit installation may be used to accomplish multiple goals. A combination tower will obviously be larger than a single absorption technique tower. Therefore, initial capital costs will be greater for a combination tower versus the single absorption technique tower. However, the costs of the single tower may be favorable when compared to the costs of two individual absorption towers. Combination towers that have been successfully used in industry are spray/Venturi and spray/packed tower combinations.

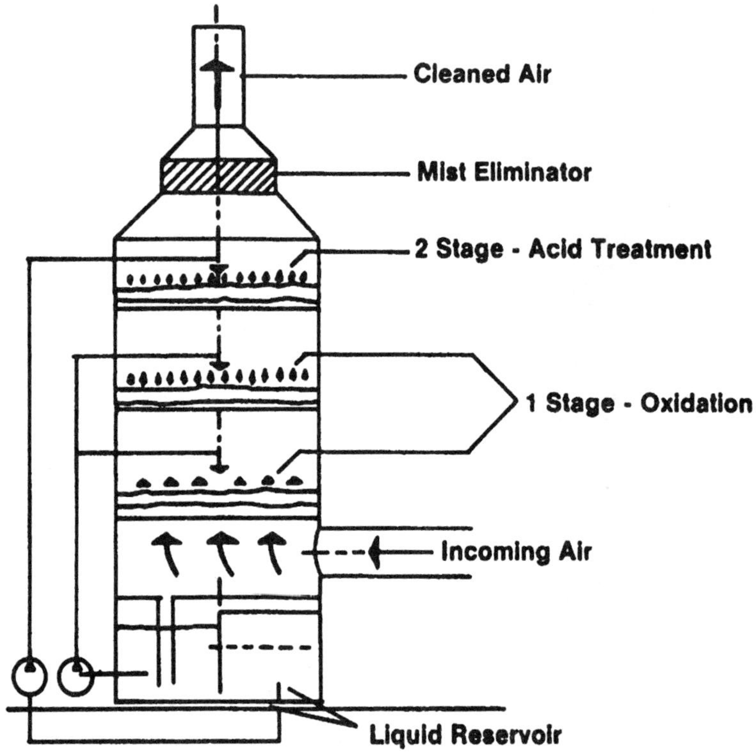

Fig. 22. Two-stage chemical scrubber. (From US EPA.)

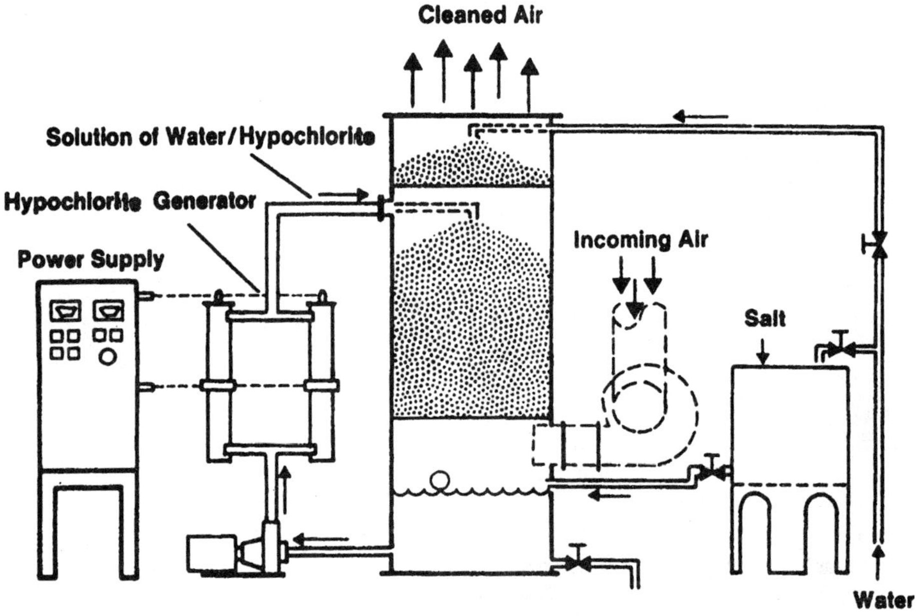

Fig. 23. Chemical scrubber with hypochlorite generator. (From US EPA.)

Example 29

Environmental engineers who design geothermal power plants (47), reverse-osmosis water plants (60), wastewater-treatment facilities (61), septage receiving facilities (62), or sanitary landfill sites (63) must address H_2S/odor control during the design process rather than a retrofit measure in response to pressure from nearby residents. H_2S/odor problems of waste-treatment or waste handling facilities can be solved by proper siting and application of exist technologies, including chemical scrubber, filters, combustion, biological processes, and so forth. As an environmental engineer in charge of a design project, please provide discussions on the following:

1. Siting considerations
2. Commerically available chemical scrubbers (62)

Solution

1. Siting considerations that should be considered by an environmental engineer are in the following discussion. It is very important to identify the main source of H_2S/odor at the facility and treat only the odorous gases. A simple approach to isolating the odorous gases would be to enclose the component of the facility generating the odors. The gases would be confined in a housing structure and thereby isolated from non-odorous air. This would reduce the volume of air to be treated and thus the overall cost. Designer must understand the dangers (toxic and explosive potential) of the closed spaces to operating personnel.

   During the site-selection process, consideration should be given to the impact that offensive odors may have on nearby residents. Zoning ordinances and land development patterns must be reviewed. An isolated area, if residentially zoned, may develop in the near future and result in pressure being applied to retrofit a facility without odor control. Care should be taken to locate the facility in a well-ventilated area (e.g., an open space on a hilltop) and downwind from existing or projected population centers. Provisions for adding odor control systems in the future should be considered.

2. The following presents a discussion on commercial chemical scrubbers. Sodium hypochlorite has been used successfully as an oxidizing agent in commercial chemical scrubbers to control odor at many waste-handling or waste-treatment facilities. Single-stage, two-stage, or three-stage scrubbers have been used. In Fig. 22, a two-stage scrubber is shown. The first stage is alkaline oxidation (NaOH+NaOCl), and the second stage is an acidic wash using H_2SO_4. Automatic dosage systems are a necessity in preventing accidents when using the concentrated chemicals required for this system.

   Another type of chemical scrubber used at treatment plants that receive septage (shown in Fig. 23) generates sodium hypochorite by electrolysis of salt (NaCl). Because this scrubber produces hypochlorite (concentration less than 2%) and no acidic step is involved, there is less need for special care concerning the delivery, handling, and dosing of dangerous chemical.

   The results from total odor strength measurements of different chemical scrubbers show odor reduction efficiencies between 95% and 98%. The air has been characterized as being "free from sewage odors, but it smells like chemicals." It seems that a chemical scrubber always gives this "scrubber odor." However, if the scrubber is incorrectly operated, this "scrubber odor" changes to a chlorine odor.

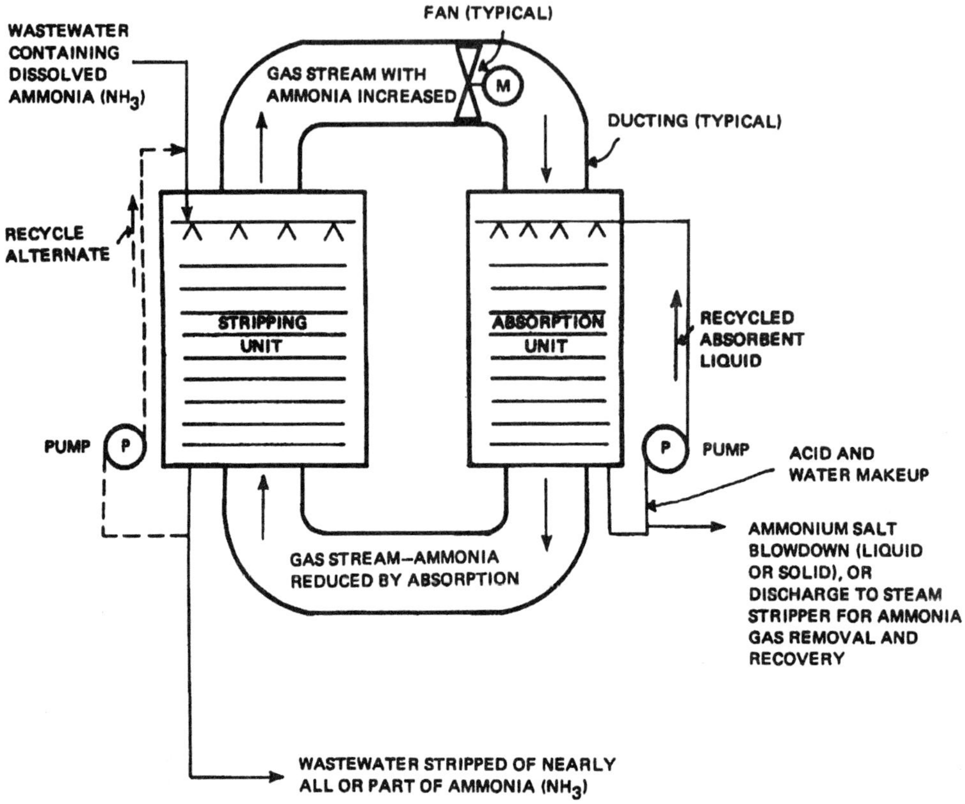

Fig. 24. A combined gas stripping and absorption process system for simultaneous ammonia removal and recovery. (From US EPA.)

The cost for operating the chemical scrubber can be divided into chemical cost and energy cost. Energy will always contribute most to the total cost of operation. For the two-stage scrubber (*see* Fig. 22), the energy cost will be approximately two-thirds of the total operational cost. Although some simpler types are available, chemical scrubbers are generally applicable only at large treatment plants, where biological methods of control are not feasible.

Example 30

A combined gas stripping and scrubbing (absorption) process system was introduced and discussed previously in Example 22. Example 30 introduces another new type of combined gas stripping and scrubbing (absorption) process system for ammonia removal and recovery. The new process system, shown in Fig. 24, has been developed by CH2M/HILL Consulting Engineers, and is highly recommended by the US EPA (64,67). Please review the theory, principles, and applications of stripping and scrubbing (absorption) and examine the process system shown in Fig. 24 carefully. Please then discuss the following:

1. This new process system shown in Fig. 24, including its applications, technology status, historical background, operation, and possible process modifications.

2. The difference between the process system introduced in Example 22 and the process system introduced in this example (*see* Fig. 24).

Solution

1. This special combined gas stripping and scrubbing (absorption) process system shown in Fig. 24 is for simultaneous ammonia removal and recovery and is a significant advance in the state of the art of nitrogen management. The new process overcomes most of the limitations of a conventional gas stripping process and has the advantage of recovering ammonia as a byproduct.

   It appears that the improved process (Fig. 24) includes an ammonia-stripping unit and an ammonia-absorption unit (or scrubbing unit). Both units are essentially sealed from the outside air but are connected by appropriate ducting. The stripping gas, which initially is air, is maintained in a closed cycle. The stripping unit operates essentially in the same manner that is now being or has been used in a number of conventional gas stripping systems, except that this system recycles the gas stream rather than using single-pass outside air.

   It can been seen from Fig. 24 that most of the ammonia discharged to the gas stream from the stripping unit is removed in the absorptio n unit. Because ammonia is an alkaline substance, the absorbing liquid should be maintained at a low pH to convert absorbed ammonia gas to soluble ammonium ion. This technique may effectively trap the ammonia and also may have the effect of maintaining the full driving force for absorbing the ammonia, because ammonia gas does not build up in the absorbent liquid.

   The absorption unit can be a slat tower, packed tower, or sprays similar to the stripping unit, but will usually be smaller owing to kinetics of the absorption process.

   The absorbent liquid initially should be water with acid added to obtain a low pH, usually below 7.0. In the simplest case, as ammonia gas is dissolved in the absorbent and converted to ammonium ions, acid should be added to maintain the desired pH. If sulfuric acid is added, for example, an ammonium sulfate salt solution is formed. This salt solution may continue to build up in concentration and the ammonia may be finally discharged from the absorption device as a liquid or solid (precipitate) blowdown of the absorbent. With current shortages of ammonia-based fertilizers, a salable byproduct may result. This is the advantage of this new process system.

   Other methods of removing the ammonia from the absorbent may also be applicable, depending on the acid used and the desired byproduct. Ammonia gas or aqua ammonia could be produced, for example, by steam stripping the absorbent. In this case, acid makeup would be unnecessary.

   It is believed that the usual scaling problem associated with ammonia-stripping towers will be eliminated by the improved process system (Fig. 24) because the carbon dioxide that normally reacts with the calcium and hydroxide ions in the water to form the calcium carbonate scale will be eliminated from the stripping air during the first few passes. The freezing problem can be eliminated owing to the exclusion of nearly all outside air. The treatment system may normally operate at the temperature of the wastewater.

2. The process system introduced in Example 22 involves only one process unit in which the emission stream treats contaminated groundwater, and the groundwater serves as a scrubbing liquid for purifying the emission stream at the same time. From an air

pollution control perspective (for treatment of the contaminated air emission stream), it is a scrubbing (absorption) process, whereas from a water pollution control perspective (for treatment of the contaminated groundwater), it is a gas stripping process. The process system introduced here employs two separate but almost identical process units connected to each other. A gas-stripping process unit removes ammonia from a contaminated wastewater and a scrubbing (absorption) process unit recovers the stripped ammonia gas from the emitted gas stream for reuse.

NOMENCLATURE

a	Empirical packing constant (dimensionless)
A_{bed}	Bed area (ft^2)
acfh	Actual cubic feet/hour
acfm	Actual cubic feet/minute
A_{column}	Column (tower) cross-sectional area (ft^2)
AF	Absorption factor
am^3/h	Actual cubic meters/hour
ABS	Flooding correlation absicca
AEC	Annual electricity cost ($)
Aex	Auxiliary equipment cost ($)
ASC	Annual cost of solvent ($/yr)
ASR	Annual solvent required (gal/yr)
AVV	Absolute void volume (dimensionless)
b	Empirical packing constant (dimensionless)
c	Empirical packing constant (dimensionless)
Ca	Size of existing system
Cb	Size of system being considered
CE	Collection efficiency (%)
d	Empirical packing constant (dimensionless)
D	Density (lb/ft^3)
DAC	Direct annual cost ($)
D_{column}	Column (tower) diameter (ft)
D_e	Density of polluted airstream (lb/ft^3)
D_G	Density of gas stream (lb/ft^3)
D_{HAP}	Density of HAP (lb/ft^3)
D_L	Density of liquid (lb/ft^3)
D_p	Particle diameter (μm)
D_w	Density of water vapor (lb/ft^3)
e	Empirical packing constant (dimensionless)
EC	Equipment cost ($)
f	Fraction of flooding
F_p	Fan power requirement (kWh/yr)
g	Empirical packing constant (dimensionless)
G	Gas flow rate (lb/h)
g_c	Gravitational constant (32.2 ft/s^2)
G_{area}	Gas stream flow rate based on tower cross-sectional area (lb/ft^2-s)

$G_{area,f}$	Gas stream flow rate based on tower cross-sectional area at flooding point (lb/ft^2-s)
G_{mol}	Gas flow rate (lb-mol/h)
HAP	Hazardous air pollutant
HAP_e	HAP emission stream concentration (ppmv)
$HAP_{e,m}$	Moles of HAP in inlet stream (mol/min)
HAP_o	HAP outlet concentration (ppmv)
$HAP_{o,m}$	Moles of HAP in outlet stream (mol/min)
H_G	Height of gas transfer unit (ft)
H_L	Height of liquid transfer unit (ft)
H_{og}	Height of an overall gas transfer unit based on overall gas film coefficients (ft)
hp	Horsepower
HRS	System operating hours (h/yr)
Ht_{column}	Column (packed tower) height (ft)
Ht_{total}	Total column height (ft)
Ia	Known cost of existing system
Ib	Cost of system being sized
IAC	Indirect annual cost
L	Solvent flow rate (lb/h)
L''	Liquid flow rate per cross-sectional area of column (lb/h-ft^2)
L_{gal}	Solvent flow rate (gal/min)
L_{mol}	Liquid (absorbent) flow rate (lb-mol/h)
$L_{w,s}$	Saturated lb water/lb dry air (from psychrometric chart)
$L_{w,a}$	Inlet lb water/lb dry air (from psychrometric chart)
m	Empirical parameter or slope of the equilibrium curve
M	Molecular weight of specific pollutant gas (lb/lb-mol)
M_e	Moisture content (%)
MW	Molecular weight of the scrubbing liquor (lb/lb-mol)
MW_e	Molecular weight of the emission stream (lb/lb-mol)
MW_{HAP}	Molecular weight of HAP (lb/lb-mol)
$MW_{solvent}$	Molecular weight of solvent (lb/lb-mol)
n	Sizing exponent
N_{og}	Number of gas transfer units (based on overall gas film coefficients) (dimensionless)
ORD	Flooding correlation ordinate
P	System pressure drop (in. H_2O or atm)
P_a	Pressure drop (lb/ft^2-ft)
PM	Particulate matter
P_{total}	System pressure drop (in. H_2O)
P_v	Venturi scrubber pressure drop (in. H_2O)
P_e	Pressure of emission stream (mm Hg)
PEC	Purchased equipment cost ($)
Q_e	Emission stream flow rate (scfm)
Q_d	Dilution air required (scfm)

$Q_{e,a}$ Actual emission stream flow rate (acfm)
$Q_{e,ad}$ Actual flow rate of dry air (acfm)
$Q_{e,s}$ Saturated emission stream flow rate (acfm)
Q_w Volume of water added (ft^3/min)
r Empirical packing constants (dimensionless)
R Gas constant (0.7302 atm-ft^3/lb-mol°R)
RE Removal efficiency (%)
$RE_{reported}$ reported removal efficiency (%)
s Empirical packing constant (dimensionless)
Sc_G Schmidt number for gas stream
Sc_L Schmidt number for liquid stream
S_g Specific gravity of fluid
SG_{media} Specific gravity of the plastic or other material used to produce the media
SG_{water} Specific gravity of water (=1)
T Temperature of gas (°R)
TAC Total annual cost ($)
TCC Total capital cost ($)
T_e Temperature of the emission stream at inlet air (°F)
$T_{e,s}$ Temperature of the saturation emission stream (°F)
UEC Unit electricity cost ($/kWh)
USC Unit cost of the solvent ($/gal)
$V_{packing}$ Volume of packing (ft^3)
VSC Venturi scrubber cost ($)
W_{media} Weight of media (lb/ft^3)
WR Water consumption (gal/yr)
W_{water} Weight of water (62.4 lb/ft^3)
Y Empirical packing constant (dimensionless)
μ_L Viscosity of scrubbing liquor (cP [1 when using water])
$\mu_{L''}$ Liquid viscosity (lb/ft-h)
μ_m Micrometer (1×10^{-3}m)

REFERENCES

1. R. H. Perry, and, C. H. Chilton (eds.), *Chemical Engineer's Handbook*, 6th ed., McGraw-Hill, New York, 1980.
2. R. E. Treybal, *Mass Transfer Operations*, 3rd ed., McGraw-Hill, New York, 1980.
3. US EPA, *Wet Scrubber System Study, Vol. 1: Scrubber Handbook*, EPA-R2-72-118a (NTIS PB 213016), US Environmental Protection Agency, Washington, DC, 1972.
4. US EPA, *Organic Chemical Manufacturing, Vol. 5: Adsorption, Condensation and Absorption Devices*, EPA-450/3-80-027 (NTIS PB 81-220543), US Environmental Protection Agency, Research Triangle Park, NC, 1980.
5. W. M. Vatavuk, and R. B. Neveril, *Chem. Eng. NY*, Oct. 4 (1962).
6. A. Kohl, and F. Riesenfield *Gas Purification*, 2nd ed., Gulf Publishing, Houston, TX, 1974.
7. R. S. Hall, W. M. Vatavuk, and J. Matley, *Chem. Eng. NY*, Nov. 21 (1988).
8. Michael Sink of Pacific Environmental Services, Inc. to Koch Engineering, Inc. and Glitsch, Inc., Costs of tower packings, *Telecommunications*, Research Triangle Park, NC, Jan., 1990.

9. US EPA, *Control Technologies for Hazardous Air Pollutants*, EPA 625/6-91/014, US Environmental Protection Agency, Washington, DC, 1991.
10. R. E. Robinson, *Chemical Engineering Reference Manual*, 4th ed., Professional Publications, Belmont, CA, 1987.
11. A. J. Buonicore, and L. Theodore, *Industrial Control Equipment for Gaseous Pollutants*, CRC, Cleveland, OH, 1975, Vol. 1.
12. US EPA, *OAQPS Control Cost Manual*, 4th ed., EPA 450/3-90-006 (NTIS PB 90-169954), US Environmental Protection Agency, Research Triangle Park, NC, 1960.
13. M. Fogiel, (ed.), *Modern Pollution Control Technology*, Air Pollution Control, Research and Education Association, New York, 1978, Vol. 1.
14. R. F. Shringle, *Random Packings and Packed Towers*, Gulf Publishing, Houston, TX, 1987.
15. US EPA, *Control of Air Emissions from Superfund Sites*, EPA 625/R-92/012, US Environmental Protection Agency, Washington, DC, 1992.
16. US EPA, *Organic Air Emissions from Waste Management Facilities*, EPA 625/R92/003, US Environmental Protection Agency, Washington, DC, 1992.
17. E. R. Attwicker, Wet scrubbing. *Handbook of Environmental Engineering*, (L. K. Wang, and N. C. Pereira, eds.), Humana Press, Totowa, NJ, 1979, Vol. 1. pp. 145–198.
18. P. N. Cheremisinoff, and R. A. Young (eds.) *Air Pollution Control and Design Handbook*, Marcel Dekker, New York, 1977, Part 2.
19. B. G. Liptak, (ed.), *Environmental Engineers Handbook, Vol. II, Air Pollution*, Chilton, Radnor, PA, 1974.
20. US EPA, *The Cost Digest, Cost Summaries of Selected Environmental Control Technologies*, EPA 600/8-84-010 (NTIS PB 85-155695), US Environmental Protection Agency, Washington, DC, 1984.
21. US EPA, *Wet Scrubber Performance Model*, EPA 600/2-77-172 (NTIS PB 2715125), US Environmental Protection Agency, Washington, DC, 1977.
22. H. E. Hesketh, *Air Pollution Control: Traditional and Hazardous Pollutants,* Technomic, Lancaster, PA, 1991.
23. US EPA, *TI-59 Programmable Calculator Programs for Opacity, Venturi Scrubbers and Electrostatic Precipitators*, EPA 600/8-80-024 (NTIS PB 80-193147), US Environmental Protection Agency, Washington, DC, 1980.
24. US EPA, *Handbook: Guidance on Setting Permit Conditions and Reporting Trial Burn Results*, EPA 625/6-86-019, US Environmental Protection Agency, Washington, DC, 1989.
25. US EPA, *Control Technologies for Fugitive VOC Emissions from Chemical Process Facilities*, EPA 625/R-93/005, US Environmental Protection Agency, Cincinnati, OH, 1994.
26. M. Sink and M. Borenstein, Costs of venturi scrubbers, *Telecommunications* between Pacific Environmental Services, Inc. and Air Pollution Inc., Jan. 5, 1990.
27. M. Sink and C. O'Conner, *Telecommunication and Fax* between Pacific Environmental Services, Inc. and American Air Filter, Inc., Jan. 5, 1990.
28. J. R. Donnelly, *Proceedings of the 12th National Conference on Hazardous Materials Control*, 1991, HMCRI. *Silver Spring, MD.*
29. Anon, *ENR Cost Indices*, Engineering News Record, McGraw-Hill, New York (2002).
30. Anon, *Equipment Indices*, Chemical Engineering, McGraw-Hill, New York (2002).
31. Tri-Mer Corp., Environ. Protect. **12(11)**, 56 (2001).
32. L. K. Wang, N. C. Pereira, and Y. T. Hung (eds.), *Advanced Air and Noise Pollution Control.*, Humana Press, Totowa, NJ, 2004.
33. US EPA, *Tutorial Manual for Controlling Air Toxics*, EPA 600/8-88/092, US Environmental Protection Agency, Cincinnati, OH, 1988.
34. Anon, Annual 2001/2002 Buyer's guide—scrubbers, *Water Wastes Digest* **40(6)**, 93–94 (2001).
35. Anon, 2003 Buyer's Guide—absorption equipment and scrubbers, *Environ. Protect.* **14(2)**, 120, 163–164 (2003).

36. L. K. Wang, and J. Eldridge, Telecommunications between Zorex Corporation, Newtonville, NY and Lantec Products Inc., Canton, MA, 2004.
37. O. Reynoso, *Acid Gas Scrubber Packing Test*, Lantec Products, Agoura Hills, CA, 2003; available at www.lantecp.com.
38. W. L. McCabe, J. C. Smith and H. Harriot, *Unit Operations of Chemical Engineering*, 5th ed., McGraw-Hill New York, 1993.
39. L. K. Wang, L. Kurylko, and M. H. S. Wang, US patent 5354458, 1994.
40. K. Rutledge, D. H. Langley, A. Hogge, et al., Water Environment Conference, 2001.
41. R. P. G. Bowker, and B. M. Blades, *Odors and VOC Emissions*, Water Environment Federations, 2000.
42. A. Ellis, *Environ. Protect.* **14(1)**, 10–11 (2003).
43. D. Kiang, J. Yoloye, J. Clark, et al., *Water Environ. Technol.* **14(4)**, 39–44 (2002).
44. M. Wu, *Trickling Biofilters for Hydrogen Sulfide Odor Control*, Lantec Products, Agoura Hills, CA, 2003; available at www.lantecp.com, Feb.
45. J. Devinny and M. Webster, *Biofiltration for Air Pollution Control*, Lewis Boca Raton, FL, 1999, p. 74.
46. M. Lutz and G. Farmer, *Water Environ. Feder. Oper. Forum* **16(7)**, 10–17 (1999).
47. G. L. Nagl, *Environ. Technol.*, **9(7)**, 18–23 (1999).
48. L. K. Wang, *Hazardous Waste Management: A United States Perspective*, Eolss Publishers, London, 2002; available at www.eolss.co.uk.
49. L. K. Wang, and S. L. Lee, 2001 Annual Conference of Chinese American and Professional Society (CAAPS), 2001.
50. L. K. Wang, J. V. Krouzek, and U. Kounitson, *Case Studies of Cleaner Production Site Remediation*, Training Manual No. DTT-5-4-95, United Nations Industrial Development Organization (UNIDO), Vienna, Austria 1995.
51. L. K. Wang, L. Kurylko, and M. H. S. Wang, US patent 5240600, 1993.
52. L. K. Wang, L. Kurylko, and O. Hrycyk, US patents 5122165 and 5122166, 1992.
53. L. K. Wang, L. Kurylko, and O. Hrycyk, US patent 5399267, 1995.
54. L. K. Wang, L. Kurylko, and O. Hrycyk, US patent 5552051, 1996.
55. L. K. Wang, M. H. S. Wang, and P. Wang, *Management of Hazardous Substances at Industrial Sites*, UNIDO Registry No. DTT-4-4-95, United Nations Industrial Development Organization, Vienna, Austria 1995.
56. W. L. Heumann, *Industrial Air Pollution Control Systems*, McGraw-Hill, New York, 1997, pp. 514–519.
57. L. K. Wang, *Treatment of Groundwater by Dissolved Air Flotation Systems*, NTIS PB85-167229/AS, National Technical Information Service, Springfield, VA, 1984.
58. S. A. Bedall, C. A. Hammond, L. H. Kirby, et al., *H_2S Abatement in Geothermal Power Plants*, Geothermal Resources Council, 1996.
59. T. R. Mason, GRC Bulle. 233–235 (1996).
60. L. K. Wang, and N. Kopko, *City of Cape Coral Reverse Osmosis Water Treatment Facility*, PB97-139547, National Technical Information Service, Springfield, VA, 1997.
61. L. K. Wang, *Emissions and Control of Offensive Odor in Wastewater Treatment Plants*, PB88-168042/AS, National Technical Information Service, Springfield, VA, 1985.
62. US EPA, *Septage Treatment and Disposal*, EPA-625/6-84/009, US Environmental Protection Agency, Washington, DC, 1984.
63. L. K. Wang, *Identification and Control of Odor at Ferro Brothers Sanitary Landfill Site*, P902FB-3-89-1, Zorex Corp. Newtonville, NY, 1989.
64. US EPA and Zorex Corp. Tech. Communications between the US Environmental Protection Agency, Washington, DC, and Zorex Corp. Newtonville, NY, 2004.
65. John Zink Co., *Technical Bulletins HSS 0003A and HSS 0004A and Technical Paper 7802A*, tulsa, OK, 1988.

66. US EPA, *Flue Gas Desulfurization Inspection and Performance Evaluation*, EPA-625/I-88/019, US Environmental Protection Agency, Washington, DC, 1985.
67. US EPA, Air Pollution Control Technology Series Training Tool: Wet Scrubbers. www.epa.gov. US Environmetal Protection Agency, Washington, DC, 2004.
68. US EPA, Air Quality Planning and Standards. www.epa.gov. US Environmental Protection Agency, Washington, DC, 2004.

Appendix
Listing of Compounds Currently Considered Hazardous[1]

CAS Number	HAP Name	CAS Number	HAP Name
83-32-9	Acenaphthene	106-99-0	Butadiene,1,3-
208-96-8	Acenaphthylene	106-97-8	Butane
75-07-0	Acetaldehyde	109-79-5	Butanethiol
64-19-7	Acetic Acid	78-93-3	Butanone,2-
108-24-7	Acetic Anhydride	1338-23-4	Butanoneperoxide,2-
67-64-1	Acetone	106-98-9	Butene,1-
75-05-8	Acetonitrile	140-32-2	Butyl Acrylate
74-86-2	Acetylene	71-36-3	Butyl Alcohol
50-78-2	Acetylsalicylic Acid	85-68-7	Butyl Benzyl Phthalate
107-02-8	Acrolein	123-86-4	Butylacetate,n-
79-06-1	Acrylamide	105-46-4	Butylacetate,sec-
79-10-7	Acrylic Acid	540-88-5	Butylacetate,tert-
107-13-1	Acrylonitrile	141-32-2	Butylacrylate,n-
15972-60-8	Alachlor	78-92-2	Butylalcohol,sec-
107-18-6	Allyl Alcohol	75-65-0	Butylalcohol,t-
7429-90-5	Aluminum	109-73-9	Butylamine,n-
133-90-4	Amiben	128-37-0	Butylated Hydroxytoluene
92-67-1	Aminobiphenyl,4-	2426-08-6	Butylglycidylether,n-
504-29-0	Aminopyridine,2-	138-22-7	Butyllactate,n-
7664-41-7	Ammonia	123-72-8	Butyraldehyde
12124-97-9	Ammonium Bromide	107-92-6	Butyric Acid
12125-02-9	Ammonium Chloride-Fume	7440-43-9	Cadmium
7783-20-2	Ammonium Sulfate	10108-64-2	Cadmium Chloride
628-63-7	Amylacetate,n-	1306-19-0	Cadmium Oxide
62-53-3	Aniline	2223-93-0	Cadmium Stearate
120-12-7	Anthracene	7440-70-2	Calcium
7440-36-0	Antimony	1305-62-0	Calcium Hydroxide
1327-33-9	Antimony Oxide	1305-78-8	Calcium Oxide
11097-69-1	Aroclor 1254	76-22-2	Camphor, Synthetic
7440-38-2	Arsenic and Compounds (as As)	105-60-2	Caprolactam
1303-28-2	Arsenic Pentoxide	133-06-2	Captan
1327-53-3	Arsenic Trioxide	63-25-2	Carbaryl
7784-42-1	Arsine	1563-66-2	Carbofuran
1332-21-4	Asbestos	7440-44-0	Carbon
8052-42-4	Asphalt (Petroleum) Fumes	1333-86-4	Carbon Black
1912-24-9	Atrazine	124-38-9	Carbon Dioxide
7440-39-3	Barium	75-15-0	Carbon Disulfide
10294-40-3	Barium Chromate	630-08-0	Carbon Monoxide
114-26-1	Baygon	56-23-5	Carbon Tetrachloride
56-55-3	Benz(a)Anthracene	353-50-4	Carbonyl Fluoride
98-87-3	Benzal Chloride	120-80-9	Catechol
100-52-7	Benzaldehyde	7782-50-5	Chlorine
71-43-2	Benzene	10049-04-4	Chlorine Dioxide
92-87-5	Benzidine	79-11-8	Chloroacetic Acid
205-99-2	Benzo(b)Fluoranthene	108-90-70	Chlorobenzene
191-24-2	Benzo(ghi)Perylene	75-45-6	Chlorodifluoromethane
207-08-9	Benzo(k)Fluoranthene	53449-21-9	Chlorodiphenyl
50-32-8	Benzo(a)Pyrene	75-00-3	Chloroethane
65-85-0	Benzoic Acid	67-66-3	Chloroform
120-51-4	Benzoic Acid, Benzyl Ester	107-30-2	Chloromethyl Methyl Ether, bis
98-07-7	Benzotrichloride	100-00-5	Chloronitrobenzene,
98-88-4	Benzoyl Chloride	4-76-15-3	Chloropentafluoroethane
94-36-0	Benzoyl Peroxide	3691-35-8	Chlorophacinone
12-05-58	Benzyl Benzoate	75-29-6	Chloropropane,2-
100-44-7	Benzyl Chloride	107-05-1	Chloropropene,3-
7440-41-7	Beryllium	2921-88-2	Chloropyrifos
92-52-4	Biphenyl	123-09-1	Chlorothioanisole
80-05-7	Bisphenol A	108-41-8	Chlorotoluene,M-
7440-42-8	Boron	95-49-8	Chlorotoluene,O-
1303-86-2	Boron Oxide	13907-45-4	Chromate, (CrO_4^{2-})
10294-33-4	Boron Tribromide	68131-98-6	Chrome Tanned Cowhide
7637-07-2	Boron Trifluoride	7738-94-5	Chromic Acid
7726-95-6	Bromine	13548-38-4	Chromic Nitrate
74-97-5	Bromochloromethane	7440-47-3	Chromium

[1]U.S. EPA, Shareef, G.S., M.T. Johnston, E.P. Epner, D. Ocamb, and C. Berry. *Tutorial Manual for CAT (Controlling Air Toxics) Version 1.0* EPA/600/8-88/092, August 1988.

Continued

Appendix (cont)

CAS Number	HAP Name	CAS Number	HAP Name
18540-29-9	Chromium (VI) Compounds (as Cr)	26761-40-0	Diisodecyl Phthalate
1333-82-0	Chromium Oxide	109-87-5	Dimethoxymethane
14977-61-8	Chromyl Chloride	127-19-5	Dimethyl Acetamide
7788-96-7	Chromyl Fluoride	115-10-6	Dimethyl Ether
218-01-9	Chrysene	131-11-3	Dimethyl Phthalate
8001-58-9	Coal Tar	77-78-1	Dimethyl Sulfate
8007-45-2	Coal Tar Pitch Volatiles	75-18-3	Dimethyl Sulfide
7440-48-4	Cobalt	124-40-3	Dimethylamine
1317-42-6	Cobalt Sulfide	60-11-7	Dimethylaminoazobenzene,4-
7440-50-8	Copper	1300-73-8	Dimethylaminobenzene,4-
1317-38-0	Copper Oxide (CuO)	121-69-7	Dimethylaniline,n,n-
13071-79-9	Counter 15 G	68-12-2	Dimethylformamide,n,n-
1319-77-3	Cresol (All Isomers)	123-91-1	Dioxane,1,4-
108-39-4	Cresol,m-	101-84-8	Diphenyl Oxide
95-48-7	Cresol,o-	101-68-8	Diphenylmethane Diisocyanate,4,4'-
106-44-5	Cresol,p-	1314-56-3	Diphosphorus Pentoxide
8021-39-4	Creosote	110-98-5	Dipropylene Glycol
14464-46-1	Cristobalite (SiO_2)	34590-94-8	Dipropylene Glycol Methyl Ether
123-73-9	Crotonaldehyde	64742-47-8	Distillates (Petroleum)
98-82-8	Cumene	106-89-8	Epichlorohydrin
101-14-4	Curene	75-08-1	Ethanethion
420-04-2	Cyanamide	64-17-5	Ethanol
590-28-3	Cyanic Acid, Potassium Salt	110-80-5	Ethoxyethanol,2-
917-61-3	Cyanic Acid, Sodium Salt	111-15-9	Ethoxyethylacetate,2-
57-12-5	Cyanide	141-78-6	Ethyl Acetate
143-33-9	Cyanides	140-88-5	Ethyl Acrylate (Inhibited)
506-68-3	Cyanogen Bromide	541-85-5	Ethyl Amyl Ketone
506-77-4	Cyanogen Chloride	100-41-4	Ethyl Benzene
110-82-7	Cyclohexane	60-29-7	Ethyl Ether
108-93-0	Cyclohexanol	759-94-9	Ethyldipropylcarbamsthioat, s-
108-94-1	Cyclohexanone	109-94-4	Ethyl Formate
110-83-8	Cyclohexene	78-10-4	Ethyl Silicate
542-92-7	Cyclopentadiene	74-85-1	Ethylene
112-31-2	Decanal	106-93-4	Ethylene Dibromide
124-18-5	Decane	107-06-2	Ethylene Dichloride
2238-07-5	di-2,3-Epoxypropyl Ether	107-21-1	Ethylene Glycol
84-74-2	di-n-Butyl Phthalate	110-49-6	Ethylene Glycol Methyl Ether Acetate
117-84-0	di-n-Octyl Phthalate	111-76-2	Ethylene Glycol Monobutyl Ether
123-42-2	Diacetone Alcohol	75-21-8	Ethylene Oxide
39393-37-8	Dialkyl Phthalates	107-15-3	Ethylenediamine
124-09-4	Diaminohexane,1,6-	151-56-4	Ethyleneimine
95-80-7	Diaminotoluene,2,4-	117-81-7	Ethylhexylphthalate,Bis,2-
333-41-5	Diazinon	16219-75-3	Ethylidene-2-Norbornene
53-70-3	Dibenz(a,h)Anthracene	12604-58-9	Ferrovanadium Dust
19287-45-7	Diborane	206-44-0	Fluoranthene
96-12-8	Dibromochloropropane,1,2-,3-	86-73-7	Fluorene
95-50-1	Dichlorobenzene,1,2-	53-96-3	Fluorenylacetamide,n-,2-
91-94-1	Dichlorobenzidine,3,3'-	16984-48-8	Fluorides
110-56-5	Dichlorobutane,1,4-	7782-41-4	Fluorine
75-71-8	Dichlorodifluoromethane	75-69-4	Fluorotrichloromethane
75-34-3	Dichloroethane,1,1-	50-00-0	Formaldehyde
75-35-4	Dichloroethylene,1,1-	75-12-7	Formamide
540-59-0	Dichloroethylene,1,2-,Cis-,Trans-	64-18-6	Formic Acid
156-60-5	Dichloroethylene,1,2-,Trans-	110-00-9	Furan
75-43-4	Dichloromonofluoromethane	98-01-1	Furfural
594-72-9	Dichloronitroethane,1,1,1-	98-00-0	Furfuryl Alcohol
120-83-2	Dichlorophenol,2,4-	110-17-8	Fumaric Acid
94-75-7	Dichlorophenoxyacetic-acid,2,4-	8006-61-9	Gasoline
78-87-5	Dichloropropane,1,2-	111-30-8	Glutaraldehyde
542-75-6	Dichloropropene,1,3-	56-81-5	Glycerol
77-73-6	Dicyclopentadiene	556-52-5	Glycidol
111-42-2	Diethanolamine	111-71-7	Heptanal
96-22-0	Diethyl Ketone	142-82-5	Heptane
84-66-2	Diethyl Phthalate	87-68-3	Hexachloro-1,3- Butadiene
109-89-7	Diethylamine	118-74-1	Hexachlorobenzene
100-37-8	Diethylaminoethanol	77-47-4	Hexachlorocyclopentadiene
111-46-6	Diethylene Glycol	34465-46-8	Hexachlorodibenzodioxin,1,2,3,6,7,8-
111-40-0	Diethylenetriamine	67-72-1	Hexachloroethane
108-83-8	Diisobutyl Ketone	684-16-2	Hexafluoroacetone

Continued

Appendix (cont)

CAS Number	HAP Name	CAS Number	HAP Name
66-25-1	Hexanal	563-80-4	Methyl Isopropyl Ketone
110-54-3	Hexane,n-	74-94-1	Methyl Mercaptan
591-78-6	Hexanone,2-	80-62-6	Methyl Methacrylate
142-92-7	Hexylacetate,sec-	110-43-0	Methyl n-Amyl Ketone
107-41-5	Hexylene Glycol	98-83-9	Methyl Styrene
302-01-2	Hydrazine	78-94-4	Methyl Vinyl Ketone
122-66-7	Hydrazobenzene	137-05-3	Methyl-2-Cyanoacrylate
7647-01-1	Hydrochloric Acid	100-61-8	Methylaniline,n-
10035-10-6	Hydrogen Bromide	583-60-8	Methylcyclohexanone,o-
7647-01-0	Hydrogen Chloride	75-09-2	Methylene Chloride
74-90-8	Hydrogen Cyanide	101-77-9	Methylenedianiline,4,4'-
7664-39-3	Hydrogen Fluoride	108-11-2	Methylisobutylcarbin
7722-84-1	Hydrogen Peroxide(30%)	108-10-1	Methylpentanone,4-,2-
7783-07-5	Hydrogen Selenide	872-50-4	Methylpyrrolidone,n-,2-
7783-06-4	Hydogen Sulfide	12001-26-2	Mica (VAN8CI9CI)
123-31-9	Hydroquinone	7439-98-7	Molybdenum
193-39-5	Indeno(1,2,3-c,d)Pyrene	108-90-7	Monochlorobenzene
7440-74-6	Indium	141-43-5	Monoethanolamine
7553-56-2	Iodine	75-04-7	Monoethylamine
74-88-4	Iodomethane	74-89-5	Monomethylamine
15438-31-0	Iron	110-91-8	Morpholine
1309-37-1	Iron Oxide Fume	107-87-9	n-Methyl Propyl Ketone
123-92-2	Isoamyl Acetate	684-93-5	n-Nitroso-n-Methylurea
110-19-0	Isobutyl Acetate	62-75-9	n-Nitrosodimethylamine
78-83-1	Isobutyl Alcohol	8030-30-6	Naphtha
78-84-2	Isobutyraldehyde	91-20-3	Naphthalene
26952-21-6	Isooctyl Alcohol	134-32-7	Naphthylamine,1-
78-59-1	Isophorone	91-59-8	Naphthylamine,2-
78-79-5	Isoprene	7440-02-2	Nickel
67-63-0	Isopropanol	13463-39-3	Nickel Carbonyl
109-59-1	Isopropoxyethanol	1313-99-1	Nickel Oxide
108-21-4	Isopropyl Acetate	7440-02-0	Nickel Powder
108-20-3	Isopropyl Ether	7697-37-2	Nitric Acid
8001-20-6	Kerosene	10102-43-9	Nitric Oxide
463-51-4	Ketene	100-01-6	Nitroaniline,p-
301-04-2	Lead Acetate	98-95-3	Nitrobenzene
18454-12-1	Lead Chromate	92-93-3	Nitrodiphenyl,4-
1309-60-0	Lead Dioxide	79-24-3	Nitroethane
7439-92-1	Lead Powder	10102-44-0	Nitrogen Dioxide
8032-32-4	Ligroine	55-63-0	Nitroglycerine
1310-65-2	Lithium Hydroxide	75-52-5	Nitromethane
7439-95-4	Magnesium	100-02-7	Nitrophenol,p-
1309-48-4	Magnesium Oxide	108-03-2	Nitropropane,1-
1309-48-8	Magnesium Oxide Fume	79-46-9	Nitropropane,2-
121-75-5	Malathion	99-99-0	Nitrotoluene,p-
108-31-6	Maleic Anhydride	124-19-6	Nonanal
7439-96-5	Manganese	111-84-2	Nonane,n-
104-14-4	Mboca	3268-87-9	Octachlorodibenzo-p-Dioxin
7725-93-1	Mercaptomethylthiazolylmethylketone,2	124-13-0	Octanal
7439-97-6	Mercury	111-65-9	Octane
141-79-7	Mesityl Oxide	8012-95-1	Oil Mist, Mineral
79-41-4	Methacrylic Acid	1317-71-1	Olivine
74-93-1	Methanethiol	144-62-7	Oxalic Acid (Anhydrous)
67-56-1	Methanol	7783-41-7	Oxygen Difluoride
16752-77-5	Methomyl	10028-15-6	Ozone
72-43-5	Methoxychlor	106-51-4	p-Quinone
109-86-4	Methoxyethanol,2-	8002-74-2	Paraffin Wax Fume
79-20-9	Methyl Acetate	87-86-5	Pentachlorophenol
74-99-7	Methyl Acetylene	504-60-9	Pentadiene Isomer
96-33-3	Methyl Acrylate	115-77-5	Pentaerythritol
74-83-9	Methyl Bromide	109-66-0	Pentane
78-78-4	Methyl Butane	79-21-0	Peracetic Acid (40% Solution)
74-87-3	Methyl Chloride	594-42-3	Perchloromethyl Mercaptan
107-30-2	Methyl Chloromethyl Ether	7616-94-6	Perchloryl Fluoride
8022-00-2	Methyl Demeton	64741-88-4	Petro Distill (Heavy)
624-92-0	Methyl Disulfide	8002-05-9	Petroleum Distillates
107-31-3	Methyl Formate	85-01-8	Phenanthrene
110-12-3	Methyl Isoamyl Ketone	108-95-2	Phenol
624-83-9	Methyl Isocyanate	122-60-1	Phenyl Glycidyl Ether

Continued

Appendix (cont)

CAS Number	HAP Name	CAS Number	HAP Name
122-39-4	Phenylbenzenamine,n-	13494-80-9	Tellerium and Compounds (as Te)
106-50-3	Phenylenediamine,p-	26140-60-3	Terphenyl
298-02-2	Phorate	75-65-1	t-Butyl Alcohol
75-44-5	Phosgene	634-66-2	Tetrachlorobenzene,1,2,3,4-
7803-51-2	Phosphine	1746-01-6	Tetrachlorodibenzo-p-Dioxin,2,3,7,8-
7664-38-2	Phosphoric Acid	76-12-0	Tetrachlorodifluoroethane,1,1,2,2-,1,2-
7723-14-0	Phosphorous (Yellow)	79-34-5	Tetrachloroethane,1,1,2,2-
10025-87-3	Phosphorous Oxychloride	127-18-4	Tetrachloroethylene
7719-12-2	Phosphorous Trichloride	3689-24-5	Tetraethyl Dithiopyrophosphate
85-44-9	Phthalic Anhydride	78-00-2	Tetraethyl Lead
88-89-1	Picric Acid	1320-37-2	Tetrafluorodichloroethane
80-56-8	Pinene,a-	109-99-9	Tetrahydrofuran
7440-06-4	Platinum	27813-21-4	Tetrahydropthalimide
1336-36-3	Polychlorinated Biphenyls	7722-88-5	Tetrasodium Pyrophosphate
25322-69-4	Polypropylene Glycol	7440-28-0	Thallium, Soluble Compounds (as Tl)
9002-86-2	Polyvinyl Chloride Latex	463-71-8	Thiophosgene
7440-09-7	Potassium	110-02-1	Thiophene
7789-00-6	Potassium Chromate	62-56-6	Thiourea
151-50-8	Potassium Cyanide	137-26-8	Thiram
7778-50-9	Potassium Dichromate	7440-31-5	Tin (as Sn)
1310-58-3	Potassium Hydroxide	13463-67-7	Titanium Dioxide
75-28-5	Propane, 2-Methyl-	108-88-3	Toluene
1120-71-4	Propane Sultone	26471-62-5	Toluene Diisocyanate
57-55-6	Propanediol,1,2-	584-84-9	Toluene,2,4,Diisocyanate
57-57-8	Propiolacetone,b-	95-53-4	Toluidine,o-
123-38-6	Propionaldehyde	126-73-8	Tributyl Phosphate
79-09-4	Propionic Acid	87-61-6	Trichlorobenzene,1,2,3-
71-23-8	Propyl Alcohol	120-82-1	Trichlorobenzene,1,2,4-
109-60-4	Propylacetate,n-	71-55-6	Trichloroethane,1,1,1-
107-10-8	Propylamine	79-00-5	Trichloroethane,1,1,2-
115-07-1	Propylene	79-01-6	Trichloroethylene
6423-43-4	Propylene Glycol Dinitrate	96-18-4	Trichloropropane,1,2,3-
107-98-2	Propylene Glycol Monomethyl Ether	76-13-1	Trichlorotrifluoromethane,1,1,2-
75-56-9	Propylene Oxide	121-44-8	Triethylamine
503-30-0	Propylene Oxide,1,3-	75-63-8	Trifluormonobromomethane
129-00-0	Pyrene	1582-09-8	Trifluralin
121-29-9	Pyrethrin	552-30-7	Trimellitic Anhydride
8003-34-7	Pyrethrum	75-50-3	Trimethylamine
110-86-1	Pyridine	25551-13-7	Trimethylbenzene
14808-60-7	Quartz (Silica Dust)	58784-13-7	Trimethylphenyl n-Methylcarbamate
91-22-5	Quinoline	110-88-3	Trioxane,1,3,5-
108-46-3	Resorcinol	115-86-6	Triphenyl Phosphate
10049-07-7	Rhodium Chloride	7440-33-7	Tungsten and Compounds (as W)
7782-49-2	Selenium Compounds (as Se)	8006-64-2	Turpentine
7803-62-5	Silane	7440-61-1	Uranium
7631-86-9	Silica	51-79-6	Urethane
60676-86-0	Silica Vitreous	7440-62-2	Vanadium
7440-21-3	Silicon	1314-62-1	Vanadium Pentoxide
409-21-2	Silicon Carbide	62-73-7	Vapona
7440-22-4	Silver	108-05-4	Vinyl Acetate
7631-90-5	Sodium Bisulfate	75-01-4	Vinyl Chloride
10588-01-9	Sodium Dichromate	75-02-5	Vinyl Fluoride
7681-49-4	Sodium Fluoride	25013-15-4	Vinyl Toluene
1310-73-2	Sodium Hydroxide	75-05-4	Vinylidene Chloride
7681-57-4	Sodium Metabisulfate	1330-20-7	Xylene
131-52-2	Sodium Pentachloro-	108-38-3	Xylene,m-
10102-18-8	Sodium Selenite	95-47-6	Xylene,o-
1302-67-6	Spinel	106-42-3	Xylene,p-
8052-41-3	Stoddard Solvent	7440-65-5	Yttrium
7789-06-2	Strontium Chromate	7440-66-6	Zinc
100-42-5	Styrene	7699-45-8	Zinc Bromide
7446-09-5	Sulfur Dioxide	7646-85-7	Zinc Chloride, Fume
2551-62-4	Sulfur Hexafluoride	13530-65-9	Zinc Chromate (as Cr)
10025-67-9	Sulfur Monochloride	1314-13-2	Zinc Oxide (Fume)
7446-11-9	Sulfur Trioxide	1314-84-7	Zinc Phosphide
7664-93-9	Sulfuric Acid	13597-46-1	Zinc Selenite
14807-96-6	Talc	557-05-1	Zinc Stearate
7727-43-7	Tbariumsulfate, Total Dust	440-67-2	Zirconium Compounds (as Zr)

6
Condensation

Lawrence K. Wang, Clint Williford, and Wei-Yin Chen

CONTENTS

1. INTRODUCTION

1.1. Process Description

Condensation is a separation process in which one or more volatile components of a vapor mixture are separated from the remaining vapors through saturation followed by a phase change (*see* Fig. 1). The phase change from gas to liquid can be accomplished in two ways: (1) the system pressure may be increased at a given temperature or (2) the system temperature may be reduced at a given pressure. Condensation occurs when the vapor-phase partial pressure of a volatile component exceeds that of the component in the liquid phase (or the vapor pressure for a pure liquid phase). Condensers are the unit operations primarily used to remove volatile organic compounds (VOCs) from gas streams prior to other controls such as incinerators or absorbers, but can sometimes be used alone to reduce emissions from high-VOC-concentration gas streams.

Figure 1 illustrates a simple process flow diagram for condensation. A typical condensation system consists of the condenser, refrigeration system, storage tanks, and pumps. Figure 2 further details an entire condensation and recovery process: (1) VOC off-gas is compressed as it passes through a blower; (2) the exiting hot gas flows to an aftercooler commonly constructed of copper tubes with external aluminum fins. Air is passed over the fins to maximize the cooling effect. Some condensation occurs in the aftercooler; (3) the gas stream cools further in an air-to-air heat exchanger; (4) the condenser cools the gas to below the condensing temperature in an air-to-refrigerant heat

From: *Handbook of Environmental Engineering, Volume 1: Air Pollution Control Engineering*
Edited by: L. K. Wang, N. C. Pereira, and Y.-T. Hung © Humana Press Inc., Totowa, NJ

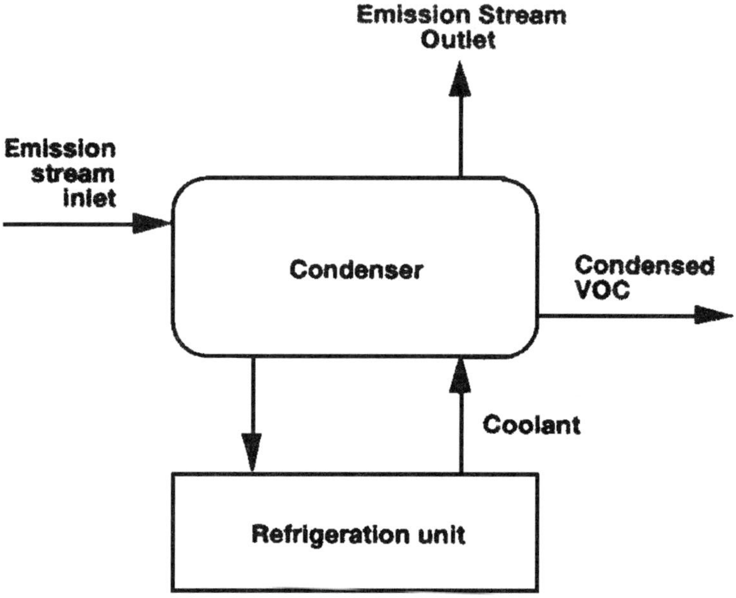

Fig. 1. Process flow diagram for condensation. (From US EPA.)

exchanger; (5) the cold gas then passes to a centrifugal separator where the liquid is removed to a collecting vessel. Not all condensing systems will require the aftercooler and heat exchanger. Final polishing typically requires further treatment (e.g., use of a carbon adsorption unit) before the stream can be vented to the atmosphere (1–14).

1.2. Types of Condensing Systems

Condensing systems usually contain either a contact condenser or surface condenser.

1.2.1. Contact Condensing Systems

Contact condensing systems cool the gas stream by spraying ambient or chilled liquid directly into the gas stream. Typically, use of a packed column maximizes surface area and contact time. Some contact condensers use simple spray chambers with baffles, whereas others have high-velocity jets designed to produce a vacuum. The direct mixing of the coolant and contaminant necessitates separation or extraction before coolant reuse. This separation process may lead to a disposal problem or secondary emissions. Contact condensers usually remove more air contaminants as a result of greater condensate dilution (14–16).

1.2.2. Surface Condensing Systems

In surface condensing systems (or surface condensers), the coolant does not mix with the gas stream, but flows on one side of a tube or plate. The condensing vapor contacts the other side, forms a film on the cooled surface, and drains into a collection vessel for storage, reuse, or disposal. Condensation can occur in the tubes (tube side) or on the shell (shell side) outside of the tubes. Condensers are usually of the shell and tube or plate/fin type, the most common being the former with the coolant flowing on the inside of the tubes countercurrent to the gas stream. Condensation occurs on the outside of the tubes

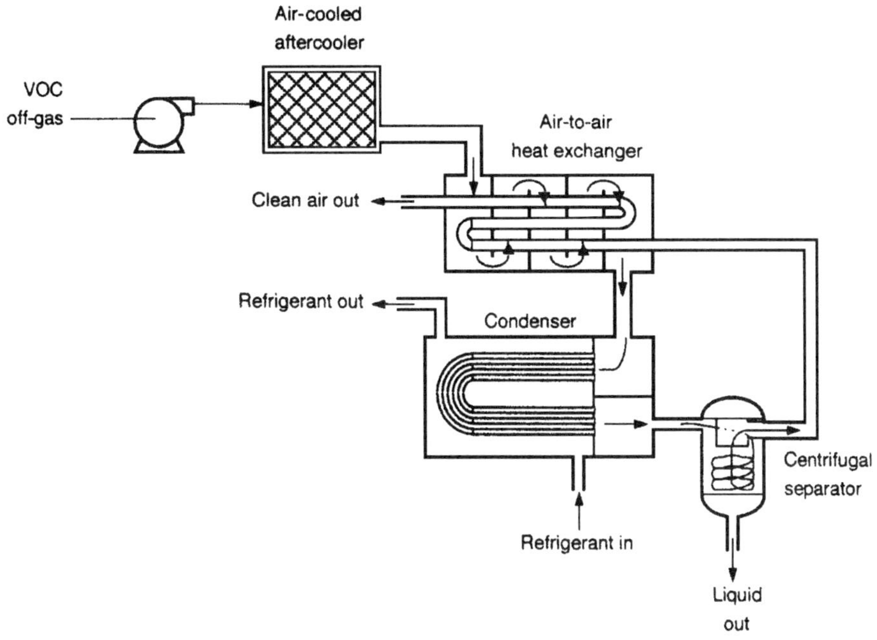

Fig. 2. Schematic diagram of a vapor condensation system. (From US EPA.)

in this arrangement. The condenser tubes usually run horizontally, but may run vertically. Surface condensers require less water and produce 10–20 times less condensate than contact condensers. Surface condensers are more likely to produce a salable product. However, these types of condenser have a greater amount of maintenance, because of the required auxiliary equipment (14–16).

1.3. Range of Effectiveness

Condensation can remove 50–95% of condensable VOCs. Removal efficiency depends on characteristics of the vapor stream, the concentration of emission stream components, and the condenser operating parameters. The removal efficiency depends on the nature and concentration of emission stream components. High-boiling (low-volatility) compounds condense more efficiently than low-boiling ones. Thus, the design condensation temperature and coolant selection depend on vapor pressure and temperature data. Practical limits for coolant selection are presented in Table 1. Figure 3 shows that removal efficiency rises with lower condenser temperatures and follows a higher curve for the less volatile (higher-boiling-point) xylene. Removal efficiency increases with contaminant boiling point, for a given inlet concentration, and condensing temperature.

2. PRETREATMENT, POSTTREATMENT, AND ENGINEERING CONSIDERATIONS

2.1. Pretreatment of Emission Stream

Water vapor in the emission stream may form ice on condenser tubes carrying coolants such as chilled water or brine solutions, decreasing the heat transfer efficiency

Table 1
Coolant Selection

Required condensation temperature, ${}^aT_{con}$ (°F)	Coolant	^{b}Coolant inlet temperature, $T_{cool,i}$ (°F)
${}^cT_{con}$: 60–80	Water	T_{con} − 15
60 > T_{con} > 45	Chilled water	T_{con} − 15
45 > T_{con} ≥ −30	Brine solutions (e.g., calcium chloride, Ethylene glycol)	T_{con} − 15
−30 > T_{con} ≥ −90	Chlorofluorocarbons (e.g., Freon-12)	T_{con} − 15

[a]Also emission stream outlet temperature.
[b]Assume the approach as 15°F.
[c]Summer limit.
Source: ref. 4.

and the condenser's removal efficiency. Dehumidification eliminates icing by using a heat exchanger to cool the vapor to about 35°F, prior to the condenser. Even with dehumidification, water vapor can remain a problem for subzero condensation systems, requiring provisions such as dual condensers, in which heated air melts ice from the off-line condenser (12).

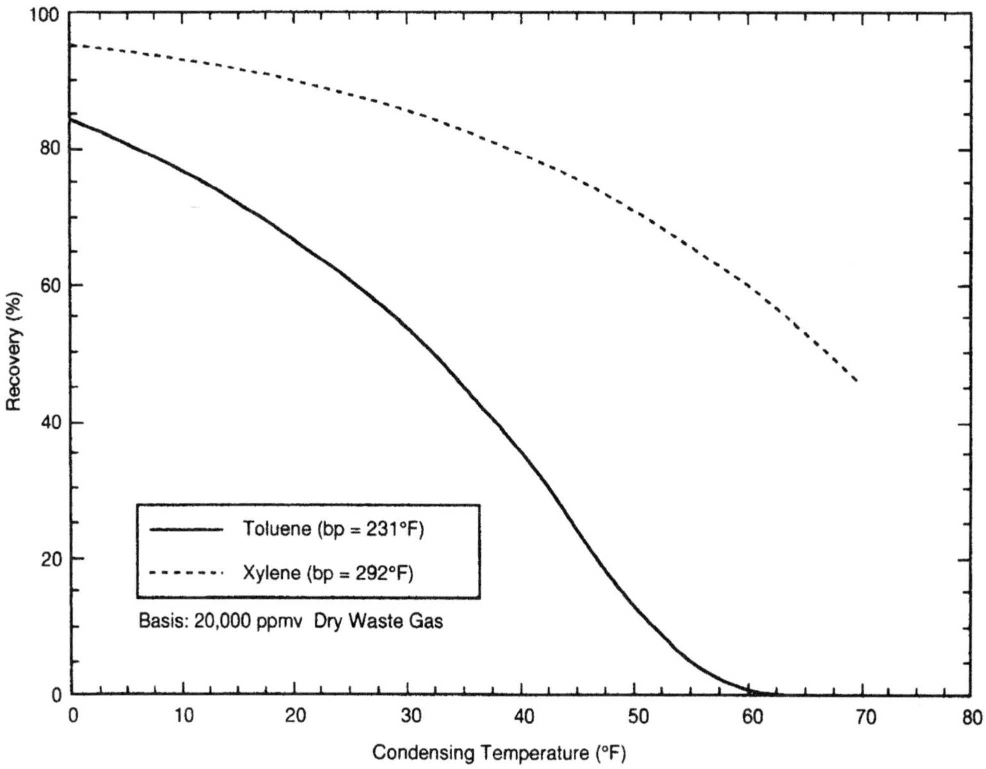

Fig. 3. Example of condenser performance. (From US EPA.)

2.2. *Prevention of VOC Emission from Condensers*

In most control applications, the emission stream contains large quantities of noncondensible gases and small quantities of condensible compounds. Design and operation must limit emissions of VOCs from discharged condensate (i.e., secondary emissions). Subcooling of the condensate may be required. Uncondensable air contaminants in the gas stream must be either dissolved in the condensate or vented to other control equipment. Gas streams at Superfund sites usually contain a variety of contaminants, and the recovered stream may fail purity specifications and be unsalable. Such streams must be disposed of by incineration or other methods. Another consideration is the moisture content of the gas stream. Any water condensing with the organic vapors dilutes the solvent stream. Finally, condenser off-gas not meeting emission standards will require further treatment, usually with activated carbon. Disposal problems and high power costs are some of the disadvantages associated with condensation.

2.3. *Proper Maintenance*

Proper maintenance of a condenser system is essential to maintaining performance. Scale buildup over time fouls condenser systems. This significantly increases fluid pressure drop or decreases heat transfer, resulting in higher fluid outlet temperature and decreased efficiency. Adequate control of hazardous air pollutants (HAPs) requires continuous monitoring of the emission stream outlet temperature. Cleaning must be performed without delay because scale buildup becomes much harder to remove over time (1,11,16).

2.4. *Condenser System Design Variables*

The required condensation temperature represents the key design variable for condenser systems. As stated previously, a condenser's removal efficiency greatly depends on the concentration and nature of emission stream components. For example, compounds with high boiling points (i.e., low volatility) condense more readily compared to those with low boiling points. Assume, as a conservative starting point, that condensation will be considered as a HAP emission control technique for VOCs with boiling points above 100°F. Therefore, the concentration and nature of an emission stream are also important design variables.

The temperature necessary to achieve a given removal efficiency (or outlet concentration) depends on the vapor pressure of the HAP in question at the vapor–liquid equilibrium. The removal efficiency for a given HAP can be determined from data on its vapor pressure–temperature relationship. Vapor pressure–temperature data for typical VOCs appear graphically in Cox charts (*see* Fig. 4). Coolant selection depends on the required condensation temperature. All aforementioned parameters must be considered for a proper condenser system design. See Table 1 for a summary of practical limits for coolant selection.

3. ENGINEERING DESIGN

3.1. *General Design Information*

This section describes a shell-and-tube heat exchanger with the hot fluid (emission stream) on the shell side and the cold fluid (coolant) on the tube side. Condensate forms

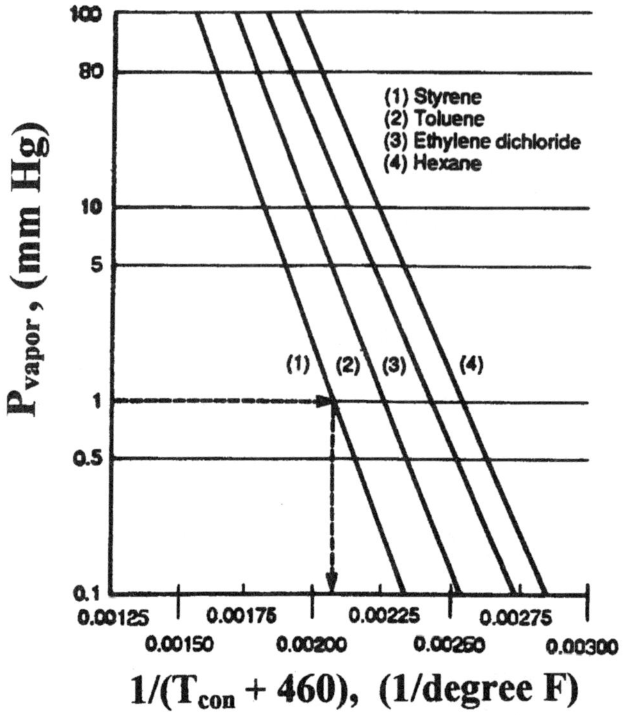

Fig. 4. Vapor pressure–temperature relationship. (From US EPA.)

on the shell side of the tubes. Depending on the application, the tube and shell side fluids may be reversed. The emission stream is assumed to consist of a two-component mixture: one condensible component (VOC, possible HAP), and one noncondensible component (air). Such systems typically exhibit nonisothermal condensation. However, the calculations here assume isothermality, which usually does not introduce large errors.

3.2. *Estimating Condensation Temperature*

Sizing the condenser involves steps to determine the surface area of the condenser. The following design procedure applies to a condenser system with a shell and tube heat exchanger, with condensate forming on the shell side. The waste gas stream is assumed to be the VOC–air mixture noted earlier. Calculations for cases involving mixtures of HAPs and supersaturated streams are quite complex and will not be treated here because they are beyond the scope of this chapter. References 5 and 6 contain information on these streams. For this case, estimation of the condensation temperature assumes the gas (air) stream to be saturated with a VOC component. For a given removal efficiency, the partial pressure (in mm Hg) for the contaminant in the exiting stream, P_{partial}, can be calculated:

$$P_{\text{partial}} = 760\left\{\frac{(1-0.01\text{RE})}{1-\left(\text{RE}\times10^{-8}\,\text{HAP}_e\right)}\right\}\text{HAP}_e\times10^{-6} \tag{1}$$

where RE is the removal efficiency (%), HAP_e is the contaminant concentration in entering gas stream (ppmv), and $P_{partial}$ is the partial pressure (mm Hg) of the HAP in the exit stream assuming the pressure in the condenser is constant and at atmospheric.

For this air–VOC system at equilibrium, the partial pressure of the HAP equals its vapor pressure at that temperature. Determining this temperature permits specification of the condensation temperature (T_{con}). This calculation requires vapor pressure–temperature data for the specific HAP (*see* Fig. 4), which can be obtained from refs. 3 and 7. Equation (1) gives the partial pressure as a function of the desired removal efficiency for the range likely to be encountered. Importantly, a high removal efficiency (and thus low partial pressure) might require an unrealistically low condensation temperature (T_{con}). In this case, a lower removal efficiency must be accepted or a different control technique adopted. Information on coolants necessary for a given condensation temperature (T_{con}) appear in Table 1. At this step, the coolant can be selected from Table 1 based on the calculated T_{con}.

3.3. *Condenser Heat Load*

Condenser heat load is the quantity of heat extracted from the emission stream to achieve specified removal. It is determined from an energy balance, combining the heat of condensation and sensible heat change of the HAP, and the sensible heat change in the emission stream. This calculation neglects enthalpy changes associated with noncondensible vapors (i.e., air), which is typically a very small value. The calculation steps are as follows:

1a. Calculate moles of HAP in the inlet emission stream (basis: 1 min):

$$HAP_{e,m} = (Q_e/392)HAP_e \times 10^{-6} \tag{2}$$

The factor 392 is the volume (ft^3) occupied by 1 lb-mol of an ideal gas at standard conditions (77°F and 1 atm).

1b. Calculate moles of HAP remaining in the outlet emission stream (basis: 1 min):

$$HAP_{o,m} = (Q_e/392)\left[1-(HAP_e \times 10^{-6})\right]\left[P_{vapor}/(P_e - P_{vapor})\right] \tag{3}$$

where P_{vapor} is equal to $P_{partial}$ given in Eq. (1).

1c. Calculate moles of HAP condensed (basis: 1 min):

$$HAP_{con} = HAP_{e,m} - HAP_{o,m} \tag{4}$$

2a. Determine the HAP's heat of vaporization (ΔH): Typically, the heat of vaporization will vary with temperature. Using vapor pressure–temperature data as shown in Fig. 4, ΔH can be estimated by linear regression for the vapor pressure and temperature range of interest (see ref. 3 for details). Compare the estimated ΔH with that of the permit application and ensure that they are in the same units. If these values differ significantly, contact the permit applicant to determine the reason for the difference.

2b. Calculate the enthalpy change associated with the condensed HAP (basis: 1 min):

$$H_{con} = HAP_{con}\left[\Delta H + C_{p_{HAP}}(T_e - T_{con})\right] \tag{5}$$

Table 2
Design Equations for Condensing Systems

$H_{\text{con}} = \text{HAP}_{\text{con}}[\Delta H + C_{p_{\text{HAP}}}(T_e - T_{\text{con}})]$
$H_{\text{uncon}} = \text{HAP}_{o,m}\, C_{p_{\text{HAP}}}(T_e - T_{\text{con}})$
$H_{\text{noncon}} = [(Q_e/392) - \text{HAP}_{e,m}]\, C_{p_{\text{air}}}(T_e - T_{\text{con}})$
$\text{HAP}_{\text{con}} = \text{HAP}_{e,m} - \text{HAP}_{o,m}$
$\text{HAP}_{o,m} = (Q_e/392)[1 - (\text{HAP}_e \text{ x } 10^{-6})][P_{\text{vapor}}/(P_e - P_{\text{vapor}})]$
$\text{HAP}_{e,m} = (Q_e/392)\text{HAP}_e \times 10^{-6}$

Note: $C_{p_{\text{HAP}}}$ = average specific heat of compound (Btu/lb-mol °F)
HAP_e = entering concentration of HAP (ppmv)
$\text{HAP}_{e,m}$ = molar flow of HAP inlet (lb-mol/min)
$\text{HAP}_{o,m}$ = molar flow of HAP outlet (lb-mol/min)
ΔH = heat of evaporation (Btu/lb-mol)
P_e = system pressure (mm Hg)
P_{vapor} = P_{partial}
Q_e = maximum flow rate (scfm at 77°F and 1 atm)
T_{con} = condensing temperature (°F)
T_e = entering emission stream temperature (°F)
Source: US EPA (1991).

where $C_{p_{\text{HAP}}}$ is the average specific heat of the HAP for the temperature interval T_{con}–T_e (Btu/lb-mol °F). (See the Appendix.)

2c. Calculate the enthalpy change associated with the noncondensible vapors (i.e., air) (basis: 1 min):

$$H_{\text{noncon}} = \left[(Q_e/392) - \text{HAP}_{e,m}\right] C_{p_{\text{air}}}\left(T_e - T_{\text{con}}\right) \quad (6)$$

where $C_{p_{\text{air}}}$ is the average specific heat of air for the temperature interval T_{con}–T_e (Btu/lb-mol °F). (See the Appendix.)

3a. Calculate the condenser heat load (Btu/h) by combining Eqs. (5) and (6):

$$H_{\text{load}} = 1.1 \times 60\left(H_{\text{con}} + H_{\text{noncon}}\right) \quad (7)$$

where H_{load} is the condenser heat load (Btu/h), H_{con} is the enthalpy of condensed HAP, and H_{noncon} is the enthalpy of the noncondensible vapors. The factor 1.1 is included as a safety factor.

Table 2 summarizes design equations for condensing systems.

3.4. Condenser Size

Condenser systems are typically sized based on the total heat load and the overall heat transfer coefficients of the gas stream and the coolant. An accurate estimate of individual coefficients can be made using physical/chemical property data for the gas stream, the coolant, and the specific shell-and-tube system to be used. Because the calculation of individual heat transfer coefficients lies beyond the scope of this manual, a conservative estimate is made for the overall heat transfer coefficient. This yields a conservatively large surface area estimate. (For the calculation of individual heat transfer coefficients, consult refs. 1–3.) The calculation procedure here assumes countercurrent flow, commonly found in industrial applications. However, some applications employ cocurrent flow or use fixed heat exchangers. The following procedure is valid for cocurrent flow, but requires an adjustment to the logarithmic mean temperature difference (1–3).

To size countercurrent condensers, use the following equation to determine the required heat transfer area:

$$A_{\text{con}} = H_{\text{load}} / U \, \Delta T_{\text{LM}} \tag{8}$$

where A_{con} is the condenser (heat exchanger) surface area (ft^2), U is the overall heat transfer coefficient (Btu/h-ft^2 °F), ΔT_{LM} is the logarithmic mean temperature difference (°F);

$$\Delta T_{\text{LM}} = \frac{\left(T_e - T_{\text{cool},o}\right) - \left(T_{\text{con}} - T_{\text{cool},i}\right)}{\ln\left[\left(T_e - T_{\text{cool},o}\right) / \left(T_{\text{con}} - T_{\text{cool},i}\right)\right]} \tag{9a}$$

where T_e is the emission stream temperature (°F), $T_{\text{cool},o}$ is the coolant outlet temperature (°F), T_{con} is the condensation temperature (°F), and $T_{\text{cool},i}$ is the coolant inlet temperature (°F). For cocurrent flow, this equation becomes

$$\Delta T_{\text{LM}} = \frac{\left(T_e - T_{\text{cool},i}\right) - \left(T_{\text{con}} - T_{\text{cool},o}\right)}{\ln\left[\left(T_e - T_{\text{cool},i}\right) / \left(T_{\text{con}} - T_{\text{cool},o}\right)\right]} \tag{9b}$$

Assume that the approach temperature at the condenser exit is 15°F. In other words,

$$T_{\text{cool},i} = (T_{\text{con}} - 15) \tag{9c}$$

Also, the temperature rise of the coolant fluid is specified as 25°F; that is,

$$T_{\text{cool},o} = (T_{\text{cool},i} + 25) \tag{9d}$$

where $T_{\text{cool},o}$ is the coolant exit temperature.

In estimating A_{con}, the overall heat transfer coefficient can be conservatively assumed as 20 Btu/h-ft^2 °F. The actual value will depend on the specific system under consideration. This calculation is based on refs. 2 and 6, which report guidelines on typical overall heat transfer coefficients for condensing vapor–liquid media.

3.5. Coolant Selection and Coolant Flow Rate

The next step is to select the coolant based on the condensation temperature required. Use Table 1 to specify the type of coolant. For additional information on coolants and other properties, see refs. 3 and 7.

The heat extracted from the emission stream is transferred to the coolant. From the energy balance, the flow rate of the coolant can be calculated as follows:

$$Q_{\text{coolant}} = H_{\text{load}} / \left[C_{p\text{coolant}}\left(T_{\text{cool},o} - T_{\text{cool},i}\right)\right] \tag{10}$$

where $Q_{coolant}$ is the coolant flow rate (lb/h) and $C_{p_{coolant}}$ is the average specific heat of the coolant over the temperature interval $T_{cool,i}$ to $T_{cool,o}$ (Btu/lb-°F). Specific heat data for coolants are available in refs. 3 and 7.

3.6. Refrigeration Capacity

A refrigeration unit is assumed to supply the coolant at the required temperature to the condenser. For costing purposes, the required refrigeration capacity is expressed in terms of refrigeration tons as follows:

$$\text{Ref} = H_{\text{load}}/12{,}000 \tag{11}$$

where Ref is the refrigeration capacity (tons).

3.7. Recovered Product

To calculate costs, the quantity of the recovered product that can be sold and/or recycled to the process must be determined. Use the following equation:

$$Q_{\text{rec}} = 60 \times \text{HAP}_{\text{con}} \times \text{MW}_{\text{HAP}} \tag{12}$$

where Q_{rec} is the quantity of the product recovered (lb/h) and HAP_{con} is the HAP condensed (lb-mol), based on 1 min of operation.

4. MANAGEMENT

4.1. Permit Review and Application

In a permit evaluation, use Table 1 to check the consistency of the condensation temperature (T_{con}) and the type of coolant selected. Also, ensure that the coolant inlet temperature is based on a reasonable approach temperature (a conservative value of 15°F is used in the table). If the reported values are appropriate, proceed with the calculations. The permit reviewer may then follow the calculation procedure outlined next. Otherwise, the applicant's design is considered unacceptable, unless supporting documentation indicates that the design is feasible.

Compare all results from the calculations and the values supplied by the permit applicant using Table 3. The calculated values in the table are based on the example case. If the calculated values of T_{con}, coolant type, A_{con}, $Q_{coolant}$, Ref, and Q_{rec} are different from the reported values of these variables, the differences may be the result of the assumptions involved in the calculations. The reviewer may then wish to discuss the details of the proposed design with the permit applicant. If the calculated values agree with the reported values, the design and operation of the proposed condenser system may be considered appropriate, based on the assumptions made in this chapter.

4.2. Capital and Annual Costs of Condensers

4.2.1. Capital Costs for Condensers

The capital costs of a condenser system consist of purchased equipment costs (equipment costs and auxiliary equipment) and direct and indirect installation costs. Table 4 provides factors for these costs. References 4 and 8 serve as sources for equipment costs

Table 3
Comparison of Calculated Values and Values Supplied by the Permit Application for Condensation

	Calculated value[a] (example case)	Reported value
Continuous monitoring of exit stream temperature	Yes	—
Condensation temperature, T_{con}	20°F	—
Coolant type	Brine solution	—
Coolant flow rate, $Q_{coolant}$	14,700 lb/h	—
Condenser surface area, A_{con}	370 ft^2	—
Refrigeration capacity, Ref	20 tons	—
Recovered product, Q_{rec}	373 lb/h	—

[a]Based on emission stream 6.

Table 4
Capital Cost Factors for Condensers

Cost item	Factor
Direct Costs	
Purchased Equipment Costs	
Condenser and auxiliary equipment	As estimated, EC
Instrumentation[a]	0.10 EC
Sales tax	0.03 EC
Freight	0.05 EC
Purchased Equipment Cost (PEC)	1.08 EC
Direct Installation Costs	
Foundation and supports	0.08 PEC
Erection and handling	0.14 PEC
Electrical	0.08 PEC
Piping	0.02 PEC
Insulation	0.10 PEC
Painting	0.01 PEC
Direct Installation Cost	0.43 PEC
Site preparation	As required, SP
Buildings	As required, Bldg.
Total Direct Costs (DC)	1.43 PEC + SP +Bldg.
Indirect Installation Costs	
Engineering	0.10 PEC
Construction	0.05 PEC
Contractor fee	0.10 PEC
Start-up	0.02 PEC
Performance test	0.01 PEC
Contingencies	0.03 PEC
Total Indirect Cost (IC)	1.31 PEC
Total Capital Costs[b]	1.74 PEC + SP + Bldg.

[a]Typically included with the condenser cost.
[b]Does not include cost of refrigeration system.
Source: Data from refs. 4 and 9.

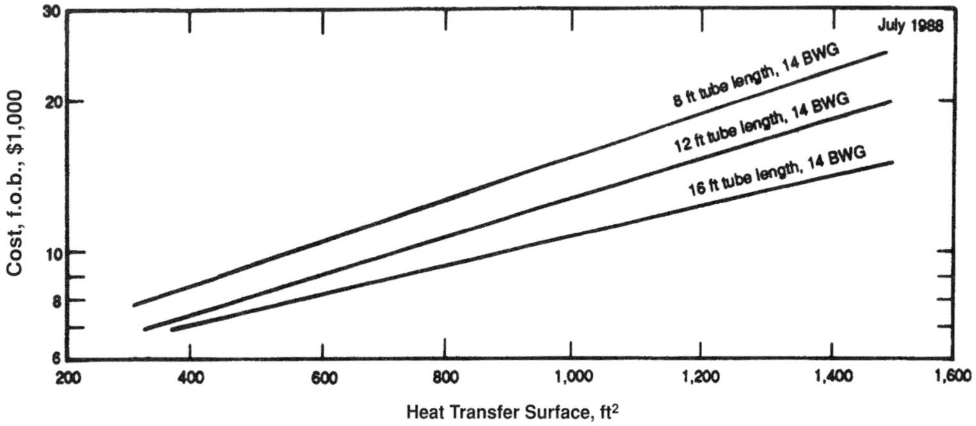

Fig. 5. Costs for fixed-tubesheet condensers. (From US EPA.)

for cold-water condenser systems. Equipment costs for fixed-tubesheet and floating-head heat exchangers are given in Figs. 5 and 6 for heat transfer surface areas (A_{con}) from 300 to 1500 ft^2. The equipment costs are in July 1988 dollars. The cost of auxiliary equipment includes ductwork, dampers, fan, and stack costs, which can be obtained from another chapter of this handbook series specifically dealing with cost estimations (24).

For condenser systems requiring a coolant based on Table 1, Table 5 can be used to estimate the total capital cost (RTCC) of a refrigerant system, as a function of refrigeration capacity (Ref) and condensation temperature (T_{con}). This cost must be added to the condenser capital cost (TCC) obtained from Fig. 6 or Fig. 6 and Table 4. Although refrigerated units are often sold as packaged systems, splitting the cost of the basic condenser system and refrigerant system in this manner allows for more flexibility in estimating the cost of a given system. A refrigerant system may not be necessary for

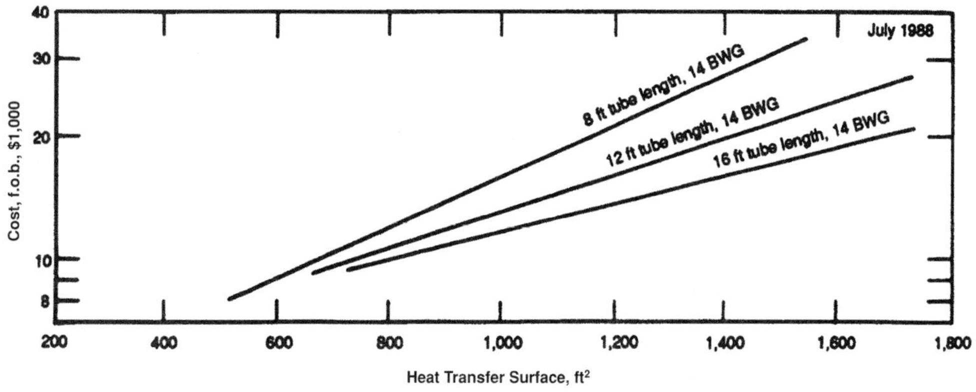

Fig. 6. Costs for floating-head condensers. (From US EPA.)

Table 5
Capital Costs for Refrigerant Systems

Required condensation temperature (°F)	Refrigerant system capital cost, RTCC ($)
≥40°F	1,989.5(Ref) + 10,671
≥20°F	4,977(Ref) + 7,615
≥0°F	7,8876.8(Ref) + 9,959
≥–20°F	6,145.4(Ref) + 26,722
≥–45°F	10,652(Ref) + 13,485
≥–85°F	12,489(Ref) + 28,993

Note: See Eq. (11) for a definition of Ref. A refrigerant system may be required for condensation temperatures between 40–60°F, although this will be dependent on the cooling water available. If cooling water of a sufficiently low temperature is available, a refrigerant system is not required.
Source: ref. 13.

condensation temperatures above 40°F, depending on the cooling water available. The costs given in Table 5 are in Spring 1990 dollars and were obtained from ref. 13.

4.2.2. Annual Costs for Condensers

The annual costs for a condenser system consist of direct and indirect annual costs, minus recovery credits. Table 6 provides appropriate factors for estimating annual costs.

Table 6
Annual Cost Factors for Condenser Systems

Cost item	Factor
Direct Annual Cost (DAC)	
Utilities	
Electricity	$0.059/kWh
Refrigerant	0
Operating Labor	
Operator labor	$12.96/h
Supervisor	15% of operator labor
Maintenance	
Maintenance labor	$14.26/h
Materials	100% of maintenance labor
Indirect Annual Costs (IAC)	
Overhead	0.60(operating labor and maintenance)
Administrative	2% of TCC
Property tax	1% of TCC
Insurance	1% of TCC
Capital recovery[a]	0.1628(TCC)
Recovery Credits	As applicable

[a]Capital recovery factor is estimated as $i(1+i)^n/[(1+i)^n-1]$, where i is the interest rate (10%) and n is the equipment life (10 yr).
Source: Data from refs. 4 and 9.

4.2.2.1. Direct Annual Costs

Direct annual costs consist of utilities (electricity, refrigerant) and operating labor and maintenance costs. The electricity cost is a function of the fan power requirement. Equation (13) can be used to obtain this requirement, assuming a fan-motor efficiency of 65 % and a fluid specific gravity of 1.0:

$$F_p = 1.81 \times 10^{-4}(Q_{e,a})(P)(\text{HRS}) \tag{13}$$

where F_p is the fan power requirement (kWh/yr), $Q_{e,a}$ is the emission stream flow rate (acfm), P is the system pressure drop (in. H_2O [default = 5 in. H_2O]), and HRS is the system operating hours per year (h/yr).

To obtain $Q_{e,a}$ from Q_e, use the formula

$$Q_{e,a} = Q_e\left(T_e + 460\right)/537 \tag{14}$$

The cost of refrigerant replacement varies with the condenser system, but is typically very low. Therefore, assume that refrigerant replacement costs are zero unless specific information is available. The operator labor is estimated as 0.5 h per 8-h shift, with the wage rate given in Table 6. Supervisory costs are assumed to be 15 % of operator labor cost. Maintenance labor is estimated as 0.5 h per 8-h shift, with the maintenance wage ratc provided in Table 6. Material costs are assumed to be 100 % of maintenance labor costs.

4.2.2.2. Indirect Annual Costs

These costs consist of overhead, property tax, insurance, administrative, and capital recovery costs. Table 6 provides the appropriate cost factors.

4.2.2.3. Recovery Credits

A condenser system may have significant recovery credits. The amount of recovered HAP can be estimated using Eq. (12). Multiplying this amount by the value of the recovered product gives the recovery credit.

5. ENVIRONMENTAL APPLICATIONS

Air strippers (*see* Fig. 7) are frequently used to treat aqueous wastes and contaminated groundwater (14–18). Units may consist of a spray tower, packed column, or a simple aerated tank. They commonly remove parts per million or lower levels of volatiles from dilute aqueous wastes. Many air strippers with lower emissions simply vent directly to the atmosphere. Those with higher organic concentrations or those located in zones of regulatory (air pollutant) noncompliance are followed by a control device (shown in Fig. 7). The control device can be a gas-phase carbon adsorption unit, an incineration unit, or others. Condensers alone placed directly after air strippers generally prove ineffective, because of low vapor-phase concentrations and high volumetric flow rates. For high-concentration emission streams, however, condensation efficiently removes and recovers VOCs from the emission streams prior to other final polishing control technologies, such as carbon adsorption. There are situations in which condensation can be used alone, in some applications, to control emissions at high VOC concentrations (i.e., greater than 5000 ppmv). This type of VOC control is not suitable for low-boiling-point organics (i.e., very low condensation temperatures [< 32°F]) or high concentrations of inert of noncondensable gases (air, nitrogen, or methane).

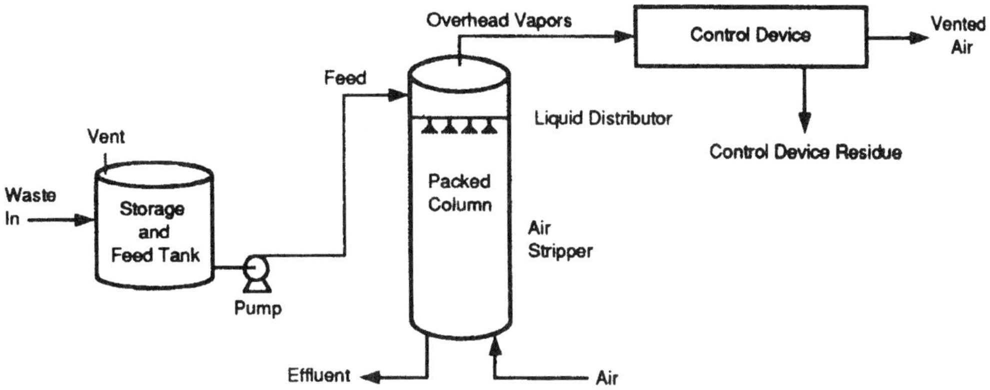

Fig. 7. Schematic of an air stripping system. (From US EPA.)

When VOC concentrations are too low for direct condensation, gas-phase carbon adsorption can provide proper initial treatment. Activated carbon will adsorb the VOC until the carbon particles are saturated, reaching capacity. The carbon then undergoes regeneration (heating) to desorb the VOC vapors at a higher concentration, which may be condensed for removal/recovery. In this case, the control device shown in Fig. 7 will be gas-phase carbon adsorption followed by condensation. The condenser in this case is used for recovery of VOC from the adsorber during its desorption stage for adsorbent regeneration (17–18).

A freeze-condensation vacuum system (19) has been developed for both chemical and environmental engineering applications. The freeze-condenser operates with heat transfer surfaces below the freezing point of the vapors. VOC or steam solidify on the heat transfer surfaces through condensation followed by freezing, by direct deposition. Placing a freeze-condenser upstream of an ejector system traps unwanted vapors before they enter the ejector system. The ejector system becomes less expensive and requires less utility consumption. Furthermore, the production of waste and vent streams is lessened, reducing environmental impacts.

Condensers are sized and their costs estimated by environmental engineers. Once an air pollution control system involving the use of one or more condensers is properly designed, individual condensers can be purchases commercially (20).

An effective training program can be provided to a condenser operator, in turn, to operate the condenser efficiently (22–25). According to Buecker (21), a monitoring program can improve operations and yield substantial energy savings for a condenser.

6. DESIGN EXAMPLES

Example 1

Perform the following design steps for a condensing system to remove VOCs from an air emission stream. Stream characteristics appear in Table 7.

1. Gather important air emission stream characteristics data.
2. Confirm the required VOC removal efficiency (RE).

Table 7
Effluent Characteristics for Emission Stream 6

HAP Emission Stream Data Form[a]

Company	Glaze Chemical Company	Plant Contact	Mr. John Leake
Location (Street)	87 Octane Drive	Telephone No.	(999) 555-5024
(City)	Somewhere	Agency Contact	Mr.Efrem Johnson
(State, Zip)	______	No. of Emission Streams Under Review	7

A. Emission Steam Number/Plant Identification		#6/Styrene Recovery Condenser Unit	
B. HAP Emission Source	(a) condensor vent	(b) ______	(c) ______
C. Source Classification	(a) process point	(b) ______	(c) ______
D. Emission Stream HAPs	(a) styrene	(b) ______	(c) ______
E. HAP Class and Form	(a) organic vapor	(b) ______	(c) ______
F. HAP Content (1,2,3)[b]	(a) 13,000 ppmv	(b) ______	(c) ______
G. HAP Vapor Pressure (1,2)	(a) provided	(b) ______	(c) ______
H. HAP Solubility (1,2)	(a) insoluble in water	(b) ______	(c) ______
I. HAP Adsorptive Prop. (1,2)	(a) not given	(b) ______	(c) ______
J. HAP Molecular Weight (1,2)	(a) 104 lb/lb-mole	(b) ______	(c) ______
K. Moisture Content (1,2,3)	Negligible	P. Organic Content (1)[c]	none
L. Temperature (1,2,3)	90ºF	Q. Heat/O2 Content (1)	61.5 Btu/scf/20.7 vol
M. Flow Rate (1,2,3)	2,000 scfm (max)	R. Particulate Content (3)	
N. Pressure (1,2)	atmospheric	S. Particle Mean Diam. (3)	
O. Halogen/Metals (1,2)	none/none	T. Drift Velocity/SO3 (3)	
U. Applicable Regulation(s)	______		
V. Required Control Level	assume 90% removal		
W. Selected Control Methods	absorption, condensation		

[a]The data presented are for an emission stream (single or combined streams) prior to entry into the selected control method(s). Use extra forms, if additional space is necessary (e.g., more than three HAPs) and note this need.

[b]The number in parentheses denote what data should be supplied depending on the data on lines C and E:
1 = organic vapor process emission
2 = inorganic vapor process emission
3 = particulate process emission

[c]Organic emission stream combustibles less HAP combustibles shown on lines D and F.

3. Determine the partial pressure of the HAP in the condenser effluent, assuming the pressure in the condenser is constant and at atmospheric.
4. Determine the condensation temperature, T_{con}.
5. Select an appropriate coolant.

Solution

1. These stream characteristics are taken from Emission Stream 6 in Table 7.

   Maximum flow rate, Q_e = 2000 scfm
   Temperature, T_e = 90°F
   HAP = styrene
   HAP concentration, HAP_e = 13,000 ppmv (corresponding to saturation conditions)
   Moisture content, M_e = negligible
   Pressure, P_e = 760 mm Hg

2. Based on the control requirements for the emission stream,

   Required removal efficiency, RE = 90 %

3. Using Eq. (1) and Fig. 4,

$$\text{HAP}_e = 13,000 \text{ ppmv (styrene)}$$
$$\text{RE} = 90\%$$

$$P_{\text{partial}} = 760\left\{\frac{(1-0.01\,\text{RE})}{\left[1-\left(\text{RE}\times 10^{-8}\text{HAP}_e\right)\right]}\right\}\text{HAP}_e \times 10^{-6} \quad (1)$$

$$P_{\text{partial}} = 760\left\{\left[1-(0.01\times 90)\right]/\left[1-\left(90\times 10^{-8}\times 13,000\right)\right]\right\}13,000\times 10^{-6}$$
$$P_{\text{partial}} = 1.0 \text{ mm Hg}$$

4. For styrene, the value of $1/(T_{con}+460)$ corresponding to 1.0 mm Hg in Fig. 4 is about 0.00208. Solve for T_{con} = 20°F. Based on T_{con} = 20°F, the appropriate coolant is a brine solution. Assume that the brine solution is a 29% (wt) calcium chloride solution, which can be cooled down to –45°F (see ref. 3).

Example 2

The air emission stream documented in Table 7 is to be treated by a condenser. Determine the following condenser design parameters:

1a. The moles of HAP in the inlet emission stream, $HAP_{e,m}$
1b. The moles of HAP in the outlet emission stream, $HAP_{o,m}$
1c. The moles of HAP condensed, HAP_{con}
2a. The HAP's heat of vaporization, ΔH
2b. The enthalpy change associated with the condensed HAP, H_{con}
2c. The enthalpy change associated with the noncondensible vapors (i.e., air), H_{noncon}
3. The condenser heat load, H_{load}

The following technical data are known:

Q_e = 2000 scfm
T_e = 90°F
HAP = styrene
HAP_e = 13,000 ppmv

M_e = negligible
P_e = 760 mm Hg
RE = 90%
$P_{partial}$ = 1 mm Hg
T_{con} = 20°F

Solution

Use Eqs. (2)–(7):

1a. Q_e = 2000 scfm

$$\text{HAP}_{e,m} = (Q_e/392)\text{HAP}_e \times 10^{-6} \quad (2)$$

$$\text{HAP}_{e,m} = (2000/392)(13,000) \times 10^{-6}$$

$$\text{HAP}_{e,m} = 0.06633 \text{ lb-mol/min}$$

1b. P_{vapor} = 1.0 mm Hg

P_e = 760 mm Hg

$$\text{HAP}_{o,m} = (Q_e/392)\left[1-(\text{HAP}_e \times 10^{-6})\right]\left[P_{\text{vapor}}/(P_e - P_{\text{vapor}})\right] \quad (3)$$

$$\text{HAP}_{o,m} = (2000/392)\left[1-13,000\times10^{-6}\right]\left[1.0/(760-1.0)\right]$$

$$\text{HAP}_{o,m} = 0.00663 \text{ lb-mol/min}$$

1c.

$$\text{HAP}_{\text{con}} = \text{HAP}_{e,m} - \text{HAP}_{o,m} \quad (4)$$

$$\text{HAP}_{\text{con}} = 0.0663 - 0.00663$$

$$= 0.0597 \text{ lb-mol/min}$$

2a. ΔH = 17,445 Btu/lb-mol (see ref. 3)

2b. MW_{HAP} = 104.2 lb/lb-mol

$C_{p\text{HAP}}$ = 24 Btu/lb-mol °F(extrapolated from data in ref. 7)

$$H_{\text{con}} = \text{HAP}_{\text{con}}\left[\Delta H + C_{p\text{HAP}}(T_e - T_{\text{con}})\right] \quad (5)$$

$$H_{\text{con}} = 0.0597[17,445 + 24(90-20)]$$

$$H_{\text{con}} = 1140 \text{ Btu/min}$$

2c. $C_{p\text{air}}$ = 7.05 Btu/lb-mol °F(see ref. 3 or Appendix for details)

$$\text{HAP}_{\text{noncon}} = \left[(Q_e/392) - \text{HAP}_{e,m}\right]C_{p\text{air}}(T_e - T_{\text{con}}) \quad (6)$$

$$\text{HAP}_{\text{noncon}} = \left[(2000/392) - 0.0663\right]7.05\times(90-20)$$

$$\text{HAP}_{\text{noncon}} = 2,480 \text{ Btu/min}$$

3.

$$H_{\text{load}} = 1.1\times60\ (H_{\text{con}} + H_{\text{noncon}}) \quad (7)$$

$$H_{\text{load}} = 1.1\times60\ (1140 + 2480)$$

$$H_{\text{load}} = 239,000 \text{ Btu/h}$$

Example 3

An air emission stream with its characteristic data shown in Table 7 is to be treated by a condenser. Determine the following:

1. The logarithmic mean temperature difference (ΔT_{LM})
2. The condenser (heat exchanger) surface area (A_{con})
3. The coolant flow rate ($Q_{coolant}$)
4. The refrigeration capacity (Ref)
5. The quantity of recovered product (Q_{rec})

The following data are given:

T_e = 90°F
T_{con} = 20°F
H_{load} = 239,000 Btu/h
Condenser = countercurrent flow
U = overall heat transfer coefficient
U = 20 Btu/h-ft^2 °F (assumed)
$HAP_{e,m}$ = 0.06633 lb-mol/min
$HAP_{o,m}$ = 0.006633 lb-mol/min

Solution

1. Determine ΔT_{LM} using Equation (9a) for counter-current flow.

$$T_e = 90°F$$
$$T_{con} = 20°F$$
$$T_{cool,i} = 20 - 15 = 5°F \tag{9c}$$
$$T_{cool,o} = T_{cool,i} + 25 = 30°F \tag{9d}$$
$$\Delta T_{LM} = \frac{(T_e - T_{cool,o}) - (T_{con} - T_{cool,i})}{\ln[(T_e - T_{cool,o})/(T_{con} - T_{cool,i})]} \tag{9a}$$
$$\Delta T_{LM} = \frac{(90 - 30) - (20 - 5)}{\ln[(90 - 30)/(20 - 5)]}$$
$$\Delta T_{LM} = 32°F$$

2. Determine A_{con} using Eq. (8).

$$\Delta T_{LM} = 32°F$$
$$H_{load} = 239,000 \text{ Btu/h}$$
$$U = 20 \text{ Btu/h-ft}^2 \text{ °F (assumed)}$$
$$A_{con} = H_{load}/U\Delta T_{LM} \tag{8}$$
$$A_{con} = 239,000/(20 \times 32)$$
$$A_{con} = 370 \text{ ft}^2$$

3. Determine $Q_{coolant}$ using Eq. (10).

$$H_{load} = 239,000 \text{ Btu/h}$$
$$T_{cool,i} = 20 - 15 = 5°F \tag{9c}$$
$$T_{cool,o} = T_{cool,i} + 25 = 30°F \tag{9d}$$
$$C_{p_{coolant}} = 0.65 \text{ Btu/lb °F (ref. 3)}$$
$$Q_{coolant} = H_{load}/\left[C_{p_{coolant}}(T_{cool,o} - T_{cool,i})\right] \tag{10}$$
$$Q_{coolant} = 239,000/[0.65(30 - 5)]$$
$$Q_{coolant} = 14,700 \text{ lb/h}$$

4. Determine Ref using Eq (11).

$$H_{load} = 239,000 \text{ Btu/h}$$
$$\text{Ref} = H_{load}/12,000 \quad (11)$$
$$\text{Ref} = 239,000/12,000$$
$$\text{Ref} = 20 \text{ tons}$$

5. Determine Q_{rec} using Eq. (12) and Table 2.

$$\text{HAP}_{con} = \text{HAP}_{e,m} - \text{HAP}_{o,m}$$
$$\text{HAP}_{con} = 0.06633 - 0.00663$$
$$\text{HAP}_{con} = 0.0597 \text{ lb-mol/min, from Eq. (4) and Example 2}$$
$$\text{MW}_{HAP} = 104.2 \text{ lb/lb-mol}$$
$$Q_{rec} = 60 \times \text{HAP}_{con} \times \text{MW}_{HAP} \quad (12)$$
$$Q_{rec} = 60 \times 0.0597 \times 104.2$$
$$Q_{rec} = 373 \text{ lb/h}$$

After a condensation system is properly designed and sized, many main components of the condensation system can be purchased from the manufacturers (20). One of the condenser manufacturers or suppliers is Swenson Process Equipment, Inc. in Harvey, IL (USA).

Example 4

Given the following information about a condenser system, determine the fan power requirement, F_p:

1. Emission stream flow rate ($Q_{e,a}$) = 2050 acfm
2. Condenser system pressure drop (P) = 5 in. H_2O
3. Condenser system operating hours per year (HRS) = 6000 h/yr

Solution

Determine F_p using Eq. (13):

$$F_p = 1.81 \times 10^{-4} (Q_{e,a})(P)(\text{HRS}) \quad (13)$$
$$F_p = 1.81 \times 10^{-4} (2050)(5)(6000)$$
$$F_p = 11,100 \text{ kWh/h}$$

NOMENCLATURE

A_{con}	Condenser (heat exchanger) surface area (ft^2)
$C_{p_{air}}$	Average specific heat of air for the temperature interval ($T_{con} - T_o$) (Btu/lb-mol °F)
$C_{p_{coolant}}$	Average specific heat of the coolant over the temperature interval $T_{cool,i}$ to $T_{cool,o}$ (Btu/lb °F)
$C_{p_{HAP}}$	Average specific heat of the HAP for the temperature interval ($T_{con} - T_o$) (Btu/lb-mol °F)
F_p	Fan power requirement (kWh/yr)
H_{con}	Enthalpy change associated with the condensed HAP (Btu/h)
H_{load}	Condenser heat load (Btu/h)
H_{noncon}	Enthalpy change associated with the noncondensible vapors (Btu/h)

HAP	Hazardous air pollutant
HAP_e	Contaminant concentration in entering gas stream (ppmv)
HAP_{con}	Moles of HAP condensed (mol)
$HAP_{e,m}$	Moles of HAP in the inlet emission stream (mol)
$HAP_{o,m}$	Moles of HAP remaining in the outlet emission stream (mol)
ΔH	Heat of vaporization
HRS	System operating hours per year (h/yr)
M_e	Moisture content
MW_{HAP}	Molecular weight of an HAP (lb/lb-mol)
P	System pressure drop (in. H_2O)
P_e	Pressure of emission stream (mm Hg)
$P_{partial}$	Partial pressure (mm Hg) of the HAP in the exit stream assuming the pressure in the condenser is constant and at atmospheric
P_{vapor}	Vapor pressure = $P_{partial}$
$Q_{coolant}$	Coolant flow rate (lb/h)
Q_e	Maximum flow rate (scfm at 77°F and 1 atm)
$Q_{e,a}$	Emission stream flow rate (acfm)
Q_{rec}	Quantity of the product recovered (lb/h)
RE	Removal efficiency (%)
Ref	Refrigeration capacity (tons)
RTCC	Total capital cost of refrigeration system ($)
T_{con}	Condensation temperature (°F)
$T_{cool,i}$	Coolant inlet temperature (°F)
$T_{cool,o}$	Coolant outlet temperature (°F)
T_e	Entering emission stream temperature (°F)
ΔT_{LM}	Logarithmic mean temperature difference (°F)
TCC	Total capital cost of condenser ($)
U	Overall heat transfer coefficient (Btu/h-ft^2 °F)
VOC	Volatile organic compound

REFERENCES

1. W. L. McCabe and J. C. Smith, *Unit Operations of Chemical Engineering,* 3rd ed., McGraw-Hill, New York, 1976.
2. R. N. Robinson, *Chemical Engineering Reference Manual,* 4th ed., Professional Publications, Belmont, CA, 1987.
3. R. H. Perry and C. H. Chilton, *Chemical Engineers Handbook,* 6th ed., McGraw-Hill, New York, 1980.
4. US EPA, *Handbook: Control Technologies for Hazardous Air Pollutants,* EPA 625/6-86-014, (NTIS PB91-228809), US Environmental Protection Agency, Cincinnati, OH, 1984.
5. D. O. Kem, *Process Heat Transfer,* McGraw-Hill Koga Kusha Co., Tokyo, 1950.
6. E. E. Ludwig, *Applied Process Design for Chemical and Petrochemical Plants, Volume III.* Gulf Publishing, Houston, TX, 1965.
7. J. A. Dean, *Lange's Handbook of Chemistry,* 12th ed., McGraw-Hill, New York, 1979.
8. R. S. Hall, W. M. Vatavuk, and J. Malley, *Chem. Eng.,* 1988.
9. US EPA, *OAQPS Control Cost Manual,* 4th ed., EPA 450/3-90-006 (NTIS PB90169954), US Environmental Protection Agency, Research Triangle Park, NC, 1990.
10. C. L. Yaws, H. M. Ni, and P. Y. Chiang, *Chem. Eng.* **95(7)** (1988).

11. D. L. Fijas, *Chem. Eng.* **96(12)** (1989).
12. J. S. Foffester and J. G. LeBlanc, *Chem. Eng.* **95(8)** (1988).
13. R. Waldrop (Edwards Engineering), personal correspondence with M. Sink (PES), August 20, 1990.
14. US EPA, *Control of Air Emissions from Superfund Sites,* EPA/625/R-92/012, US Environmental Protection Agency, Washington, DC, 1992.
15. US EPA, *Organic Emissions from Waste Management Facilities,* EPA/625/R-92/003, US Environmental Protection Agency, Washington, DC, 1992.
16. US EPA, *Control Technologies for Hazardous Air Pollutants,* EPA/625/6-91/014, US Environmental Protection Agency, Washington, DC, 1991.
17. L. K. Wang, L. Kurylko, and O. Hrycyk, US patents 5122165 and 5122166, 1992.
18. L. K. Wang and L. Kurylko, US patent 5399267, 1996.
19. J. R. Lines, *Chem. Eng. Process.* **97(9)**, 46–51 (2001).
20. Anon. Buyers' guide, *Chem. Eng.* **109(9)** (2002).
21. B. Buecker, *Chem. Eng. Prog.* **99(2)**, 40–43 (2003).
22. L. K. Wang, J. V. Krouzek, and U. Kounitson, *Case Studies of Cleaner Production and Site Remediation.* United Nations Industrial Development Organization, Vienna, 1995 (Manual No. DTT-5-4-95).
23. US EPA, *Control Techniques for Fugitive VOC Emissions from Chemical Process Facilities,* EPA/625/R-93/005, US Environmental Protection Agency, Cincinnati, OH, 1994.
24. L. K. Wang, N. C. Pereira and Y. T. Hung (eds.). Advanced Air and Noise Pollution Control, Humana Press, Totowa, NJ, 2004
25. US EPA, Air Pollution Control Technology Series Training Tool: Condensation. www.epa.gov. US Environmental Protection Agency, Washington, DC, 2004.

APPENDIX:

Average Specific Heats of Vapors[a]

Temperature (°F)	Average specific heat, C_p (Btu/scf °F)[b,c]									
	Air	H_2O	O_2	N_2	CO	CO_2	H_2	CH_4	C_2H_4	C_2H_6
77	0.0180	0.0207	0.0181	0.0180	0.0180	0.0230	0.0178	0.0221	0.0270	0.0326
212	0.0180	0.0209	0.0183	0.0180	0.0180	0.0239	0.0179	0.0232	0.0293	0.0356
392	0.0181	0.0211	0.0186	0.0181	0.0181	0.0251	0.0180	0.0249	0.0324	0.0395
572	0.0183	0.0212	0.0188	0.0182	0.0183	0.0261	0.0180	0.0266	0.0353	0.0432
752	0.0185	0.0217	0.0191	0.0183	0.0184	0.0270	0.0180	0.0283	0.0379	0.0468
932	0.0187	0.0221	0.0194	0.0185	0.0186	0.0278	0.0181	0.0301	0.0403	0.0501
1112	0.0189	0.0224	0.0197	0.0187	0.0188	0.0286	0.0181	0.0317	0.0425	0.0532
1292	0.0191	0.0228	0.0199	0.0189	0.0190	0.0292	0.0182	0.0333	0.0445	0.0560
1472	0.0192	0.0232	0.0201	0.0190	0.0192	0.0298	0.0182	0.0348	0.0464	0.0587
1652	0.0194	0.0235	0.0203	0.0192	0.0194	0.0303	0.0183	0.0363	0.0481	0.0612
1832	0.0196	0.0239	0.0205	0.0194	0.0196	0.0308	0.0184	0.0376	0.0497	0.0635
2012	0.0198	0.0243	0.0207	0.0196	0.0198	0.0313	0.0185	0.0389	0.0512	0.0656
2192	0.0199	0.0246	0.0208	0.0197	0.0199	0.0317	0.0186	0.0400	0.0525	0.0676

[a]Average for the temperature interval 77°F and the specified temperature.

[b]Based on 70°F and 1 atm.

[c]To convert to Btu/lb-°F basis, multiply by 392 and divide by the molecular weight of the compound. To convert to Btu/lb-mol-°F, multiply by 392.

Source: ref. 3, Section 4.8.6.

7
Flare Process

Lawrence K. Wang, Clint Williford, and Wei-Yin Chen

CONTENTS

1. INTRODUCTION AND PROCESS DESCRIPTION

"Flares" are open flames used for disposing of waste gases during normal operations and emergencies (1–8). Flares are an open combustion process in which surrounding air supplies oxygen to the flame. They are operated either at ground level (usually with enclosed multiple burner heads) or at elevated positions. Elevated flares use steam injection to improve combustion by increasing mixing or turbulence and pulling in additional combustion air. Properly operated flares can achieve destruction efficiencies of at least 98%. Figure 1 is a schematic of the components of a flare system (9–11). Flares are typically used when the heating value of the waste gases cannot be recovered economically because of intermittent or uncertain flow or when the value of the recovered product is low. In some cases, flares are operated in conjunction with baseload gas recovery systems (e.g., condensers). Flares handle process upset and emergency gas releases that the baseload system is not designed to recover.

Several types of flare exist. The most common are the steam assisted, air assisted, and pressure head flares. Typical flare operations can be classified as "smokeless," "nonsmokeless," and "fired" or "endothermic." For smokeless operation, flares use outside momentum sources (usually steam or air) to provide efficient gas–air mixing and turbulence for complete combustion. Smokeless flaring is required for the destruction of organics heavier than methane. Nonsmokeless operation is used for organic or other vapor streams that burn readily and do not produce smoke. Fired or endothermic flaring requires additional energy in order to ensure complete oxidation of the waste streams, such as for sulfur tail gas and ammonia waste streams. The US Environmental Protection Agency

From: *Handbook of Environmental Engineering, Volume 1: Air Pollution Control Engineering*
Edited by: L. K. Wang, N. C. Pereira, and Y.-T. Hung © Humana Press Inc., Totowa, NJ

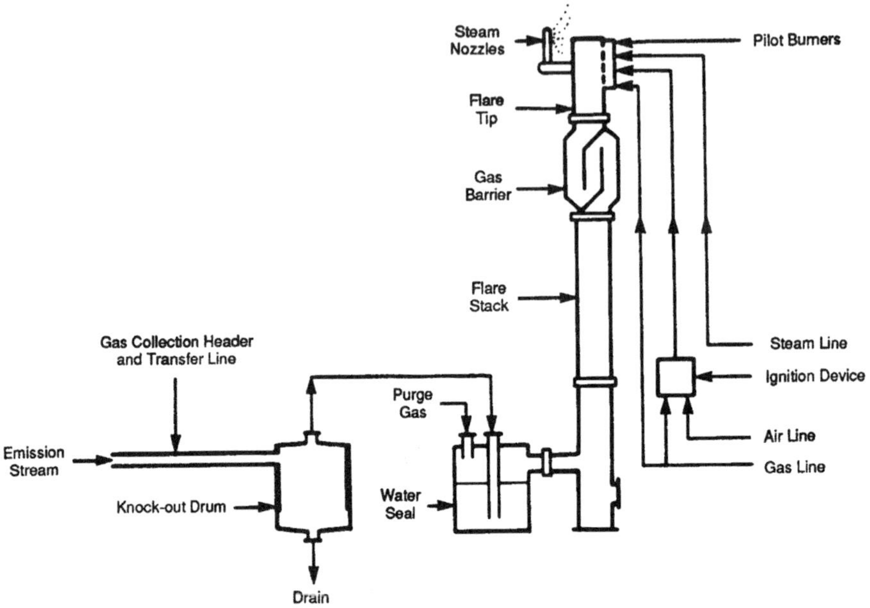

Fig. 1. Typical steam-assisted flare.

(EPA) has developed regulations for the design and operation of flares that include tip exit velocities for different types of flare and different gas stream heating values.

In general, flare performance depends on flare gas exit velocity, emission stream heating value, combustion zone residence time, waste gas–oxygen mixing, and flame temperature. This discussion focuses on steam-assisted smokeless flares, the most frequently used form. Figure 1 shows a typical steam-assisted flare system. First, process off-gases enter the flare through the collection header. Passing the off-gases through a knockout drum may be necessary to remove water or organic droplets. Water droplets can extinguish the flame, and organic droplets can result in burning particles (1–8).

Once the off-gases enter the flare stack, flame flashback can occur if the emission stream flow rate is too low. Flashback may be prevented by passing the gas through a gas barrier, a water seal, or a stack seal. Purge gas is another option. At the flare tip, the emission stream is ignited by pilot burners. If conditions in the flame zone are optimum (oxygen availability, adequate residence time, etc.), the volatile organic compounds (VOCs) in the emission stream may be completely burned (near 100% efficiency). In some cases, it may be necessary to add supplementary fuel (natural gas) to the emission stream to achieve destruction efficiencies of 98% and greater if the net heating value of the emission stream is less than 300 Btu/scf (1,2).

Typically, existing flare systems are used to destroy hazardous air pollutants (HAPs) in emission streams. The following sections describe how to evaluate whether an existing flare system is likely to achieve 98% destruction efficiency under expected flow conditions (e.g., continuous, start-up, shutdown). The discussion will be based on the recent regulatory requirements of 98% destruction efficiency for flares. The calculation procedure will be illustrated for emission stream 3 described in Table 1 using a steam-assisted flare system. Note that flares often serve more than one process unit and the total flow rate to the flare needs to be determined before the following calculation procedure can be applied. A number of flare sizing software packages have been developed. One example is the Pegasus algorithm, described elsewhere (8). Related hazardous-waste-treatment technologies and HAP emission control technologies can be found from two United Nations Industrial Development Organization (UNIDO) technical reports (12,13).

2. PRETREATMENT AND ENGINEERING CONSIDERATIONS

2.1. *Supplementary Fuel Requirements*

Based on studies conducted by the EPA, relief gases having heating values less than 300 Btu/scf are not ensured of achieving 98% destruction efficiency when they are flared in steam-assisted or air-assisted flares. Therefore, the first step in the evaluation procedure is to check the heat content of the emission stream and determine if additional fuel is needed (1–3).

In a permit review case, if the heating value of the emission stream is less than 300 Btu/scf and no supplementary fuel has been added, the application is considered unacceptable. The reviewer may then wish to follow the following calculations. If the reported value for the emission stream heat content is above 300 Btu/scf, the reviewer should skip to Section 2.3.

If the emission stream heating value is less than the 300 Btu/scf required to achieve a destruction level of 98%, it is assumed that natural gas will be added to the emission stream to bring its heat content to 300 Btu/scf. Calculate the required natural gas requirements using

$$Q_f = \left[(300 - h_e)Q_e\right]/582 \quad (1)$$

where Q_e is the emission stream flow rate (scfm), Q_f is the natural gas flow rate (scfm), h_e is the emission stream heating content or value (Btu/scf), and 582 = 882–300; 882 is the lower heating content or value of natural gas (Btu/scf). If the emission stream heating value is greater than or equal to 300 Btu/scf, then $Q_f = 0$.

2.2. *Flare Gas Flow Rate and Heat Content*

The flare gas flow rate is determined from the flow rates of the emission stream and natural gas using

$$Q_{flg} = Q_e + Q_f \quad (2)$$

where Q_{flg} is the flare gas flow rate (scfm). Note that if $Q_f = 0$, then $Q_{flg} = Q_e$.

The heating value of the flare gas (h_{flg}) is dependent on whether supplementary fuel is added to the emission stream. When h_e is greater than or equal to 300 Btu/scf, then $h_{flg} = h_e$. If h_e is less than 300 Btu/scf, supplementary fuel is added to increase h_e to 300 Btu/scf, and h_{flg} = 300 Btu/scf.

Table 1
Effluent Characteristics for Emission Stream 3

HAP EMISSION STREAM DATA FORM*

Company	Glaze Chemical Company	Plant contact	Mr. John Leake
Location (Street)	87 Octane Drive	Telephone No.	(999) 555–5024
(City)	Somewhere	Agency contact	Mr. Efrem Johnson
(State, Zip)	______	No. of Emission Streams Under Review	7

A. Emission Stream Number/Plant Identification		#3/Acetaldehyde Manufacturing Absorber Vent	
B. HAP Emission Source	(a) absorber vent	(b)______	(c)______
C. Source Classification	(a) process plant	(b)	(c)
D. Emission Stream HAPs	(a) methylene chloride	(b)______	(c)______
E. HAP Class and Form	(a) organic vapor	(b)	(c)
F. HAP Content (1,2,3)**	(a) 44,000 ppmv.	(b)______	(c)______
G. HAP Vapor Pressure (1,2)	(a) 436 mmHg of 77°F	(b)	(c)
H. HAP Solubility (1,2)	(a) insoluble in water	(b)______	(c)______
I. HAP Adsorptive Prop. (1,2)	(a) not given	(b)	(c)
J. HAP Molecular Weight (1,2)	(a) 85 lb/lb-mole	(b)______	(c)______
K. Moisture Content (1,2,3)	none	P. Organic Content (1)***	17.8% vol CH_4
L. Temperature (1,2,3)	100°F	Q. Heat/O_2 Content (1)	180 Btu/scf/ none
M. Flow Rate (1,2,3)	30,000 scfm expected	R. Particulate Content (3)	
N. Pressure (1,2)	atmospheric	S. Particle Mean Diam. (3)	______
O. Halogen/Metals (1,2)	none/none	T. Drift Velocity/SO_3 (3)	
U. Applicable Regulation(s)	______		
V. Required Control Level	assume 98% removal		
W. Selected Control Methods	flare, boiler, process heater		

*The data presented are for an emission stream (single or combined streams) prior to entry into the selected control method(s). Use extra forms, if additional space is necessary (e.g., more than three HAPs) and note this need.

**The numbers in parentheses denote what data should be supplied depending on the data on lines C and E:

1 = organic vapor process emission
2 = inorganic vapor process emission
3 = particulate process emission

***Organic emission stream combustibles less HAP combustibles shown on lines D and F.

Table 2
Flare Gas Exit Velocities for 98% Destruction Efficiency

Flare gas heat constant[a] h_{flg}(Btu/scf)	Maximum exit velocity U_{max} (ft/s)
<300	[b]
$300 \leq h_{flg} < 1000$	$3.28[10^{(0.00118h_{flg} + 0.908)}]$
>1000	400

[a]If no supplementary fuel is used :$h_{flg} = h_e$.
[b]Based on studies by the US EPA, waste gases having heating values less than 300 Btu/scf are not assured of achieving 98% destruction efficiency when they are flared in steam-assisted flares.

2.3. *Flare Gas Exit Velocity and Destruction Efficiency*

Table 2 presents maximum flare gas exit velocities (U_{max}) necessary to achieve at least 98% destruction efficiency in a steam-assisted flare system. These values are based on studies conducted by EPA. Flare gas exit velocities are expressed as a function of flare gas heat content. The maximum allowable exit velocity can be determined using the values in Table 2 (1,2,9). The information available on flare destruction efficiency as a function of exit velocity does not allow for a precise determination of this value. All that can be ascertained is whether the destruction efficiency is greater than or less than 98%, based on the exit velocity.

If a flare is controlling an intermittent process stream (or streams), a continuous monitoring system should be employed to ensure that the pilot light has a flame. If a flare is controlling a continuous process stream, continuous monitoring of either the flare flame or the pilot light is acceptable.

From the emission stream data (expected flow rate, temperature) and information on flare diameter, the flare gas exit velocity (U_{flg}) may be calculated and compared with U_{max}. An engineer may use Eq. (3) to calculate U_{flg} (6):

$$U_{flg} = \frac{\left(5.766 \times 10^{-3}\right)\left(Q_{flg}\right)\left(T_{flg} + 460\right)}{D_{tip}^{\ 2}} \tag{3}$$

where U_{flg} is the exit velocity of flare gas (ft/s), Q_{flg} is the flare gas flow rate (scfm), and D_{tip} is the flare tip diameter (in.).

If U_{flg} is less than U_{max}, then the 98% destruction level can be achieved. However, if U_{flg} exceeds U_{max}, this destruction efficiency level may not be achieved. This indicates that the existing flare diameter is too small for the emission stream under consideration and may lead to reduced efficiency. Note, at very low flare gas exit velocities, flame instability may occur, affecting destruction efficiency. In this text, the minimum flare gas exit velocity for a stable flame is assumed as 0.03 ft/s. Thus, if U_{max} is below 0.03 ft/s, the desired destruction efficiency may not be achieved. In summary, U_{flg} should fall in the range of 0.03 ft/s and U_{max} for a 98% destruction efficiency level.

In a permit review case, if U_{flg} exceeds U_{max}, then the application is not acceptable. If U_{flg} is below U_{max} and exceeds 0.03 ft/s, then the proposed design is considered acceptable, and the reviewer may proceed with the design or analysis calculations.

2.4. Steam Requirements

Steam requirements for steam-assisted flare operation depend on the composition of the flare gas and the flare tip design. Typical design values range from 0.15 to 0.50 lb steam/lb flare gas. In this handbook, the amount of steam required for 98% destruction efficiency is assumed as 0.4 lb steam/lb flare gas. The following equation is used to determine steam requirements (5):

$$Q_s = 1.03 \times 10^{-3} \times Q_{\text{flg}} \times \text{MW}_{\text{flg}} \tag{4}$$

where Q_s is the steam requirement (lb/min), MW_{flg} is the molecular weight of the flare gas (lb/lb-mole)

$$\text{MW}_{\text{flg}} = [(Q_f)(16.7) + (Q_e)(\text{MW}_e)]/Q_{\text{flg}} \tag{5}$$

MW_e is the molecular weight of the emission stream (lb/lb-mole).

3. ENGINEERING DESIGN

3.1. Design of the Flame Angle

The flare tip diameter, D_{tip}, should be rounded up to the next largest commercially available size (14–16). The minimum diamcter is 1 in. with larger diameters available in 2-in. increments between 2 and 24 in., and 6-in. increments betwecn 24 and 60 in.

The flame angle, θ, is calculated using

$$\theta = \tan^{-1}[1.47V_w/(550(\Delta P/55)^{1/2}] \tag{6a}$$

where θ is the flame angle (deg), V_w is the wind velocity (assumed to equal 60 mph), and ΔP is the pressure drop (in. H_2O) = 55 $(U_{\text{flg}}/550)^2$, where U_{flg} is obtained from Section 2.3. This reduces to

$$\theta = \tan^{-1}(88.2/U_{\text{flg}}) \tag{6b}$$

3.2. Design of Flare Height

The flare height is calculated using

$$H = (0.012185)(Q_{\text{flg}} \times h_{\text{flg}})^{1/2} - (6.05 \times 10^{-3})(D_{\text{tip}})(U_{\text{flg}})(\cos\theta) \tag{7}$$

where H is the flare height (ft), Q_{flg} is the flare gas flow rate (scfm), h_{flg} is the flare gas heat content (Btu/scf), D_{tip} is the flare tip diameter (in.), and U_{flg} is the exit velocity of flare gas (ft/s).

3.3. Power Requirements of a Fan

The electricity cost results mainly from a fan needed to move the gas through the flare. Equation (8) can be used to estimate the power requirements for a fan. This equation assumes a fan-motor efficiency of 65% and a fluid specific gravity of 1.0:

$$F_p = 1.81 \times 10^{-4}(Q_{\text{flg},a})(P)(\text{HRS}) \tag{8}$$

where F_p is the power requirement for the fan (kWh/h), $Q_{\text{flg},a}$ is the actual flare gas flow rate (scfm), P is the system pressure drop (in. H_2O [typically 16 in. of H_2O]), and HRS is the annual operating hours (h/yr).

Table 3
Comparison of Calculated Values and Values Supplied by the Permit Application for Flares

System parameters	Calculated value (example Case)[a]	Reported value
Appropriate continuous monitoring system	Yes	—
Emission stream heating value, h_e	180 Btu/scf	—
Supplementary fuel flow rate, Q_f	6200 scfm	—
Flare gas exit velocity, U_{flg}	40 ft/s	—
Flare gas flow rate, Q_{flg}	36200 scfm	—
Steam flow rate, Q_s	1140 lb/min	—

[a]Based on emission stream 3.

4. MANAGEMENT

4.1. *Data Required for Permit Application*

The data necessary to perform the calculations consist of HAP emission stream characteristics previously compiled on the HAP Emission Stream Data Form (Table 1), flare dimensions, and the required HAP control as determined by the applicable regulations.

In the case of a permit review, the data outlined below should be supplied by the applicant. The calculations in this section would then be used to check the applicant's values. Flare system variables at standard conditions (77°F, 1 atm) should include the following:

Flare tip diameter, D_{tip} (in.)
Expected emission stream flow rate, Q_e (scfm)
Emission stream heat content, h_e (Btu/scf)
Temperature of emission stream, T_e (°F)
Mean molecular weight of the emission stream, MW_e (lb/lb-mol)
Steam flow rate, Q_s (lb/min)
Flare gas exit velocity, U_{flg} (ft/s)
Supplementary fuel flow rate, Q_f (scfm)
Supplementary fuel heat content, h_f (Btu/scf)
Temperature of flare gas, T_{flg} (°F)
Flare gas flow rate, Q_{flg} (scfm)
Flare gas heat content, h_{flg} (Btu/scf)

4.2. *Evaluation of Permit Application*

Compare the results from the calculated and reported values using Table 3. If the calculated values of Q_f, U_{flg}, Q_{flg}, and Q_s are different from the reported values for these variables, the differences may be the result of assumptions (e.g., steam to flare gas ratios) involved in the calculations. In such a case, the reviewer may wish to discuss the details of the proposed system with the permit applicant.

If the calculated values agree with the reported values, then the operation of the proposed flare system may be considered appropriate based on the assumptions made in this handbook. Selecting thermal treatment equipment to destroy organic vapors is a challenge. The large number of treatment operations and the myriad of possible vent stream conditions create a very large set of choices to be evaluated, each of which has benefits and disadvantages. Martin et al. (17) state that it is difficult to effectively capture and treat acid gas products from a flare, because a flare is designed to exhaust directly to atmosphere. An enclosed flare is basically an open flare installed at the bottom of a refractory-lined stack. Moretti and Mukhopadhyay (18) compare the flare process with other VOC control technologies, such as catalytic oxidation, condensation, adsorption, absorption, biofiltration, membrane separation, ultraviolet (UV) oxidation, and heaters.

4.3. Cost Estimation

4.3.1. General Information

In many cases, existing flares are used, and it is not necessary to obtain flare purchase costs. For cases where cost information is necessary, this section can be used to estimate capital and annual flare costs for budget and planning purposes. In addition, the capital costs can be obtained from the flare manufacturers (14–16).

4.3.2. Capital Cost of Flares

The capital cost for a flare consists of purchased equipment costs and direct and indirect installation costs. The purchased equipment cost is the sum of the equipment costs (flare + auxiliary equipment), instrumentation costs, freight, and taxes. Factors for these costs are presented in Table. 4. The cost of auxiliary equipment for a flare can be obtained from another chapter of this handbook series (24). This includes the cost of ductwork, dampers, and fans.

The equipment cost of a flare is a function of the flare tip diameter (D_{tip}), height (H), and the cost of auxiliary equipment. The procedure used to obtain the flare height, H, is taken from ref. 6, whereas the flare cost equations were obtained from the Emissions Standards Division of the Office of Air Quality Planning and Standards, US EPA, Research Triangle Park, NC. The flare cost is dependent on the type of flare stack used. Typical configurations include the self-supporting configuration (used between 30 and 100 ft), guy towers (used for up to 300 ft), and derrick towers (used for heights above 200 ft).

Flare equipment costs in March 1990 dollars are presented in Eqs. (9)–(11) as a function of flare height and diameter. Equation (9) is used for self-supporting flares, Eq. (10) is used for guy support flares, and Eq. (11) is used for derrick support flares.

$$\mathrm{FC} = [78 + 9.14\,(D_{tip}) + 0.749\,(H)]^2 \tag{9}$$

where FC is the flare cost for self-support,

$$\mathrm{FC} = [103 + 8.68\,(D_{tip}) + 0.470\,(H)]^2 \tag{10}$$

where FC is the flare cost for guy support, and

$$\mathrm{FC} = [76.4 + 2.72\,(D_{tip}) + 1.64\,(H)]^2 \tag{11}$$

where FC is the flare cost for derrick support.

Table 4
Capital Cost Factors for Flares

Cost item	Factor
Direct Costs	
Purchased Equipment Costs	
Flare (FC) and auxiliary equipment, EC	As estimated, EC
Instrumentation	0.10 EC
Sales tax	0.03 EC
Freight	0.05 EC
Purchased equipment cost, (PEC)	PEC = 1.18 EC
Direct Installation Costs	
Foundation and supports	0.12 PEC
Handling and erection	0.40 PEC
Electrical	0.01 PEC
Piping	0.01 PEC
Insulation for ductwork	0.01 PEC
Painting	0.01 PEC
Direct installation cost	0.56 PEC
Site preparation	As required, SP
Buildings	As required, Bldg.
Total direct costs (DC)	1.56 PEC + SP + Bldg.
Indirect Costs (Installation)	
Engineering	0.10 PEC
Construction and field expenses	0.10 PEC
Contractor fee	0.10 PEC
Start-up	0.01 PEC
Performance test	0.01 PEC
Contingencies	0.03 PEC
Total indirect cost (IC)	0.35 PEC
Total capital costs = DC + IC	1.74 PEC + SP + Bldg.

Note: Obtained from Emissions Standards Division, QAQPS (Office of Air Quality Planning and Standards), US EPA, Research Triangle Park, NC.

For all three cases cost includes the flare stack and support, burner tip, pilots, utility piping, 100 ft of vent stream piping, utility metering, utility control, water seals, gas seals, platforms, and ladders. The costs are based on carbon steel construction except for the upper 4 ft and the burner tip, which is constructed of 316L stainless steel.

Once FC has been obtained, Table 4 is used to obtain the flare capital costs. The flare equipment cost (EC) is obtained by adding FC to any auxiliary equipment, and the purchased equipment cost (PEC) is obtained using the factors given in Table 4.

The total capital cost (TCC) of a flare is the sum of the purchased equipment cost and the direct and indirect installation cost factors. These factors are given in Table 4 as a percentage of the purchased equipment cost.

Table 5
Example Case Capital Costs

Cost item	Factor	Cost ($)
Direct Costs		
Purchased Equipment Costs		
Flare (FC) and auxiliary equipment, EC	As required	$396,000
Instrumentation	0.10 EC	39,600
Sales tax	0.03 EC	11,900
Freight	0.05 EC	19,800
Purchased equipment cost, (PEC)	PEC = 1.18 EC	$467,000
Direct Installation Costs		
Foundation and supports	0.12 PEC	$56,000
Handling and erection	0.40 PEC	187,000
Electrical	0.01 PEC	4,670
Piping	0.01 PEC	4,670
Insulation for ductwork	0.01 PEC	4,670
Painting	0.01 PEC	4,670
Direct installation cost	0.56 PEC	$262,000
Site preparation		As required, SP
Buildings		As required, Bldg.
Total direct costs (DC)	1.56 PEC+SP +Bldg	$467,000+$262,000 +SP+Bldg.
Indirect Costs (Installation)		
Engineering	0.10 PEC	$46,700
Construction and field expenses	0.10 PEC	46,700
Contractor fee	0.10 PEC	46,700
Start-up	0.01 PEC	4,670
Performance test	0.01 PEC	4,670
Contingencies	0.03 PEC	14,000
Total indirect cost (IC)	0.35 PEC	$163,000
Total capital costs = DC+IC	1.91 PEC+SP+Bldg.	$892,000+SP+Bldg.

4.3.3. Flare Annual Costs

The total annual cost (TAC) of a flare is the sum of the direct and indirect annual costs, which are discussed in more detail here. Table 5 contains the appropriate factors necessary to estimate the TAC.

The direct annual cost (DAC) includes the cost of fuel, electricity, pilot gas, steam, operating and supervisory labor, and maintenance labor and materials. Fuel usage (in scfm) is calculated in Section 2.1. Once this value (Q_f) is calculated, multiply it by 60 to obtain the fuel usage (in scfh), and multiply this by the annual operating hours to obtain the annual fuel usage. Then, multiply the annual fuel usage by the cost of fuel provided in Table 6 to obtain the annual fuel usage. Then, multiply the annual fuel usage by the cost of fuel provided in Table 6 to obtain annual fuel costs. The steam requirement

Table 6
Annual Cost Factors for Flares

Cost item	Factor
Direct cost[a] (DAC)	
Utilities	
Fuel[b] (natural gas)	$3.30/10^3 ft^3
Electricity	$0.059/k Wh
Steam	$6.00/10^3 lb steam
Operating labor	
Operator labor	$12.96/h
Supervisor	15% of operator labor
Maintenance	
Maintenance labor	$14.26/h
Materials	100% of maintenance labor
Indirect annual cost (IAC)	
Overhead	0.60 (operating labor and maintenance costs)
Administrative	2% of TCC
Property tax	1% of TCC
Insurance	1% of TCC
Capital recovery[c]	0.1315 (TCC)

[a]1988 $.
[b]This cost may vary. When possible, obtain a value more appropriate for the situation.
[c]The capital recovery factor is calculated as $i(1+i)^n/[(1+i)^n-1]$, where i is the interest rate (10%) and n is the equipment life (15 yr).
Note: Data from refs. 6 and 7.

for the flare is calculated in Section 2.4. This value (Q_s) is multiplied by 60 to obtain the steam requirement on an hourly basis. This is multiplied by the annual operating hours and by the cost of steam provided in Table 6 to obtain annual steam costs. Operating labor requirements are estimated as 0.5 h per 8-h shift. The operator labor wage rate is provided in Table 6. Supervisory costs are estimated as 15% of operator labor costs. Maintenance labor requirements are estimated as 0.5 h per 8-h shift, with a slightly higher labor rate (*see* Table 6) to reflect increased skill levels. Maintenance materials are estimated as 100% of maintenance labor.

4.3.4. Calculation of Present and Future Costs

If equipment costs must be indexed (adjusted) to the current year, the Chemical Engineering (CE) Equipment Cost Index can be used (19). Monthly indices for 5 yr are provided in another chapter of this handbook as typical examples. The following equation can be used for converting the past cost to the future cost, or vice versa:

$$\text{Cost}_b = \text{Cost}_a\,(\text{Index}_b)/(\text{Index}_a) \tag{12}$$

where $Cost_a$ is the cost in the month-year *a* ($), $Cost_b$ is the cost in the month-year *b* ($), $Index_a$ is the CE Equipment Cost Index in the month-year *a*, and $Index_b$ is the CE Equipment Cost Index in the month-year *b*.

It should be noted that although the CE Equipment Cost Indices (19) are recommended here for $Index_a$, and $Index_b$, the ENR Cost Indices (20–22), the US EPA Cost Indices (23), or the US Army Cost Indices (24) can also be adopted for updating the costs. Wang et al. (21) have shown how mathematical models can be developed for various cost indices, which, in turn, can be used for forecasting future costs.

5. DESIGN EXAMPLES

Example 1

Emission stream 3 (*see* Table 1) is to be properly treated. Assume the HAP control requirement for emission stream 3 is 98% reduction. In this case, the inlet HAP concentration falls outside the operating range of thermal incineration, catalytic incineration, carbon adsorption, absorption, and condensation; therefore, none of the control devices is applicable. Note that dilution air could be used to decrease the HAP concentration. Alternatively, this stream may warrant consideration as a fuel gas stream (25,26). However, for example purposes, assume that this stream is to be flared. Flares can be used to control emission streams with high heat contents; hence, flaring can be considered an option. Assume a steam-assisted elevated flare system as shown in Fig. 1, with a flare tip diameter of 54 in. The molecular weight of the emission stream is 33.5 lb/lb-mol. The minimum flare exit velocity for a stable flame is 0.03 ft/s. Determine the following:

1. Required destruction efficiency, DE
2. Supplementary fuel requirements in terms of natural gas flow rate, Q (scfm)
3. Flare gas flow rate, Q_{flg} (scfm)
4. Flare gas heat content (Btu/scf)
5. Maximum flare gas exit velocity, U_{max} (ft/s)
6. Flare gas exit velocity, U_{flg} (ft/s)
7. Steam requirements (lb/min)

Solution

1. Determine the required destruction efficiency using the air emission stream characteristics data (from Table 1):

   Expected emission stream flow rate, Q_e = 30,000 scfm
   Emission stream temperature, T_e = 100°F
   Heat content, h_e = 180 Btu/scf
   Mean molecular weight of emission stream, MW_e = 33.5 lb/lb-mol
   Flare tip diameter, D_{tip} = 54 in.

   Based on the control requirements for the emission stream, destruction efficiency (DE) = 98%.

2. Determine the supplemental fuel requirements, using Eq. (1).
   Because h_e is less than 300 Btu/scf, supplementary fuel is needed:

   h_e = 180 Btu/scf
   Q_e = 30,000 scfm
   $Q_f = [(300 - h_e)Q_e]/582$

$$Q_f = [(300 - 180)30{,}000]/582$$
$$Q_f = 6200 \text{ scfm of natural gas flow rate}$$

3. Determine the flare gas flow rate, using Eq. (2):

$$Q_e = 30{,}000 \text{ scfm}$$
$$Q_f = 6200 \text{ scfm}$$
$$Q_{flg} = Q_e + Q_f = 30{,}000 + 6200 \text{ scfm}$$
$$Q_{flg} = 36{,}200 \text{ scfm}$$

4. Determine the flare gas heat content. Given $h_e < 300$ Btu/scf, then $h_{flg} = 300$ Btu/scf.
5. Determine the maximum flare gas exit velocity, U_{max}, using Table 2. Given $h_e < 300$ Btu/scf, use the equation in Table 2 to calculate U_{max}. Thus,

$$U_{max} = 3.28\ [10^{(0.0118 h_{flg} + 0.908)}]$$
$$= 3.28\ [10^{(0.00118 \times 300 + 0.908)}]$$
$$U_{max} = 60 \text{ ft/s}$$

6. Determine the flare gas exit velocity, using Eq. (3). Given

$$Q_{flg} = 36{,}200 \text{ scfm}$$
$$T_{flg} = 95°\text{F}$$
$$D_{tip} = 54 \text{ in.}$$

then

$$U_{flg} = \frac{(5.766 \times 10^{-3})(Q_{flg})(T_{flg} + 460)}{D_{tip}^{\ 2}}$$

$$U_{flg} = \frac{(5.766 \times 10^{-3})(36{,}200)(95 + 460)}{54^2}$$
$$U_{flg} = 39.7 = 40 \text{ ft/s}$$

Because 0.03 ft/s < U_{flg} = 40 ft/s < U_{max} = 60 ft/s, the required level of 98% DE can be achieved under these conditions.

7. Determine the steam requirement using Eqs. (4) and (5). Given

$$Q_{flg} = 36{,}200 \text{ scfm}$$
$$MW_e = 33.5 \text{ lb/lb-mol}$$
$$MW_{flg} = [(Q_f)(16.7) + (Q_s)(MW_e)]/Q_{flg}$$
$$MW_{flg} = [(6200)(16.7) + (30{,}000)(33.5)]/36{,}200$$
$$= 30.6 \text{ lb/lb-mol}$$

Then,

$$Q_s = 1.03 \times 10^{-3}\ (Q_{flg})(MW_{flg})$$
$$Q_s = 1.03 \times 10^{-3}\ (36{,}200)(30.6)$$
$$Q_s = 1{,}140 \text{ lb/min}$$

Example 2

Assume that the flare gas exit velocity U_{flg} = 40 ft/s (see Example 1); determine the flame angle of a stream-assisted elevated flare system shown in Fig. 1.

Solution

$$\theta = \tan^{-1}(88.2/U_{flg}) = 65.6°$$

Example 3

Determine the flare height assuming the following system data are known for a steam-assisted elevated flare system (*see* Fig. 1):

Flare gas flow rate = 36,200 scfm
Flare gas heat content = 300 Btu/scf
Flare tip diameter = 60 in.
Flare gas exit velocity = 40 ft/s

Solution

Using Eq. (7)

$$H = (0.02185)(Q_{flg} \times h_{flg})^{1/2} - (6.05 \times 10^{-3})(D_{tip})(U_{flg})(\cos\theta)$$
$$H = (0.02185)(36{,}200 \times 300)^{1/2} - (6.05 \times 10^{-3})(60)(40)(\cos 65.6°)$$
$$H = 66 \text{ ft}$$

Example 4

Determine the flare cost (FC), the purchased equipment cost (PEC), the total capital cost (TCC), and annual cost of the steam-assisted elevated flare system illustrated in Fig. 1 and analyzed in Example 3. Assume that the March 1990 site preparation cost (SP) and building cost (Bldg.) are $50,000 and $100,000, respectively.

Solution

1. Determine the flare cost (FC) using Eq. (9) because H is between 30 and 100 ft.

$$H = 66 \text{ ft.}$$
$$FC = [78 + 9.14\,(D_{tip}) + 0.749\,(H)]^2$$
$$FC = [78 + 9.14\,(54) + 0.749\,(66)]^2$$
$$FC = \$386{,}000 \text{ March 1990 cost}$$

2. Determine the purchased equipment cost (PEC). Assume auxiliary equipment costs (i.e., ductwork, dampers, and fans) estimated from another chapter of this handbook series (22) are $10,000. The equipment cost EC is then $386,000 + $10,000 = $396,000. Next, use Table 4 to obtain the purchased equipment cost, PEC, as shown follows.

The equipment cost (EC) = flare cost (FC) + auxiliary equipment cost
= $386,000 + $10,000
= $396,000 March 1990 cost.

The purchased equipment cost (PEC) = EC + 0.1 EC (instrumentation)
+ 0.03 EC (sales taxes)
+ 0.05 EC (freight)
= 1.18 EC
= 1.18 ($396,000)
= $467,000 March 1990 cost

3. Determine the total capital cost (TCC) in March 1990. From Table 4, the following are known:

$$\begin{aligned}\text{Total direct costs (DC)} &= 1.56\ \text{PEC} + \text{SP} + \text{Bldg.}\\ &= 1.56 \times 467{,}000 + 50{,}000 + 100{,}000\\ &= \$878{,}520 \text{ March 1990 cost}\end{aligned}$$

$$\begin{aligned}\text{Total indirect costs (IC)} &= 0.35\ \text{PEC}\\ &= 0.35 \times 467{,}000\\ &= \$163{,}450\end{aligned}$$

$$\begin{aligned}\text{Total capital costs (TCC)} &= \text{DC} + \text{IC}\\ &= 1.91\ \text{PEC} + \text{SP} + \text{Bldg.}\\ &= 1.91 \times 467{,}000 + 50{,}000 + 100{,}000\\ &= \$878{,}520 + \$163{,}450\\ &= \$1{,}041{,}970 \text{ March 1990 costs}\end{aligned}$$

Table 5 also summarizes the calculation procedures for TCC determination.

4. The annual cost is determine and presented in Table 6.

Example 5

The purchased equipment cost (PEC) determined in Example 4 was $396,000 (March 1990 cost data). Please explain how a future or present PEC can be calculated.

Solution

Equation (12) can be used for converting the March 1990 PEC to the present or future PEC as follows:

$$\text{Cost}_b = \text{Cost}_a\ (\text{Index}_b)/(\text{Index}_a)$$

where Cost_a is the \$ cost in the month-year a (March 1990 in this case), Cost_b is the \$ cost in the month-year $_b$ (present or future), Index_a is the CE Equipment Cost Index in the month-year a (March 1990 in this case), and Index_b is the CE Equipment Cost Index in the month-year b (present or future).

Index_a can be found from the March 1990 issue of *Chemical Engineering*, and Index_b can be found from the latest issues of *Chemical Engineering* (19).

With the latest CE Equipment Cost Indices (the past 3 yr, for instance), one can plot a curve and forecast the future CE Equipment Cost Indices for cost estimation.

NOMENCLATURE

Bldg.	Building cost (\$)
Cost_a	Cost in the month-year a (\$)
Cost_b	Cost in the month-year b (\$)
D_{tip}	Diameter of flare tip (in.)
DC	Total direct costs (\$)
DE	Destruction efficiency (%)
EC	Flare equipment costs (\$)
F_p	Fan power requirement (kWh/yr)

FC	Flare cost ($)
H	Flare height (ft)
h_e	Emission stream heating value (Btu/scf)
h_f	Supplementary fuel heating value (Btu/scf)
h_{flg}	Flare gas heat content (Btu/scf)
HRS	System operating hours per year (h/yr)
IC	Total indirect costs ($)
Index_a	CE Equipment Cost Index in the month-year *a*
Index_b	CE Equipment Cost Index in the month-year *b*
MW_e	Molecular weight of emission stream (lb/lb-mol)
MW_{flg}	Molecular weight of flare gas (lb/lb-mol)
ΔP	System pressure drop (in. H_2O)
PEC	Purchased equipment cost ($)
Q_e	Emission stream flow rate (scfm at 77°F and 1 atm)
Q_f	Supplementary fuel flow rate (scfm)
Q_{flg}	Flare gas flow rate (scfm)
Q_s	Steam requirement (lb/min)
SP	Site preparation cost ($)
T_e	Emission stream temperature (°F)
T_{flg}	Temperature of flare gas (°F)
TAC	Total annual cost ($)
TCC	Total capital cost ($)
θ	Flare angle (deg)
U_{flg}	Actual exit velocity of flare gas (ft/s)
U_{max}	Maximum flare gas exit velocity (ft/s)
V_w	Wind velocity (mph)

REFERENCES

1. US Congress, *Fed. Reg.* 50, 14,941–14,945 (1985).
2. US EPA, *Evaluation of the efficiency of Industrial Flares: Test Results.* EPA600/2-84-095 (NTIS PB84199371). US Environmental Protection Agency, Washington, DC, 1984.
3. US EPA, *Parametric Evaluation of VOC/HAP Destruction Via Catalytic Incineration.* EPA-600/2-85-041 (NTIS PB85-1987). US Environmental Protection Agency, Washington, DC, 1985.
4. US EPA, *Organic Chemical Manufacturing. Vol. 4: Combustion Control Devices.* EPA-450/3-80-026 (NTIS PB81-220535). US Environmental Protection Agency, Washington, DC, 1980.
5. US EPA, *Reactor Processes in Synthetic Organic Chemical Manufacturing Industry—Background Information for Proposed Standards.* US Environmental Protection Agency, Research Triangle Park, NC, 1984.
6. US EPA, *Polymer Manufacturing Industry—Background Information for Proposed Standards.* EPA450/3-83-019a (NTIS PB88-114996). US Environmental Protection Agency, Washington, DC, 1985.
7. US EPA, OAQPS (Office of Air Quality Planning and Standards), *Control Cost Manual.* Fourth Edition, EPA-450/3-90-006 (NTIS PB90-69954). US Environmental Protection Agency, Washington, DC, 1990.
8. G. S. Mason and R. Kamar. *Chem. Eng.*, **95(9)** (1988).

9. US EPA, *Control Technologies for Hazardous Air Pollutants*, EPA/625/6-91/014. US Environmental Protection Agency, Washington, DC, 1991.
10. US EPA, *Organic Air Emissions from Waste Management Facilities*. EPA/625/R-92/003. US Environmental Protection Agency, Washington, DC, 1992.
11. US EPA, *Control of Air Emissions from Superfund Sites*. EPA/625/R-92/012. US Environmental Protection Agency, Washington, DC, 1992.
12. L. K. Wang, J. V. Krouzek and U. Kounitson. *Case Studies of Cleaner Production and Site Remediation*. Training Manual No. DTT-5-4-95. Untied Nations Industrial Development Organization (UNIDO), Vienna, 1995. 136 p.
13. L. K. Wang, M. H. S. Wang, and P. Wang. *Management of Hazardous Substances at Industrial Sites*. UNIDO Registry No. DTT-4-4-95. United Nations Industrial Development Organization (UNIDO). Vienna, 1995. 105 p.
14. Anon. Buyer's Guide 2001, *Chem. Eng.* **107(9),** 504 (2000).
15. Anon. 2000–2001 Buyer's guide, *Pollut. Eng.* **32(12)** 44 (2000).
16. Anon. 2000 Resource guide, *Environ. Technol.* **9(6)**, 18 (2000).
17. R. J. Martin, T. J. Myers, P. C. Hinze, et al. *Chem. Eng. Prog.* **99(2)**, 36–39 (2003).
18. E. C. Moretti and N. Mukhopadhyay, *Chem. Eng. Prog.* **89(7),** 20–25 (1993).
19. Anon. *Chemical Engineering Equipment Indices*, McGraw-Hill, New York (2004).
20. Anon. Engineering News Record. *ENR Indices*. McGraw-Hill, New York (2004).
21. J. C. Wang, D. A. Aulenbach, L. K. Wang, and M. H. S. Wang. *Energy Models and Cost Models for Water Pollution Control: Clean Production: Environmental and Economic Perspectives* (K. B. Misra, ed.). Springer-Verlag, Heidelberg, 1996, pp. 685–713.
22. L. K. Wang, N. C. Pereira, and Y. T. Hung (eds.). *Advanced Air and Noise Pollution Control*. Humana Press, Totowa, NJ, 2005.
23. US EPA. Innovative and Alternative Technology Assessment Manual. EPA 430/9-78-009. US Environmental Protection Agency, Washington, DC, 1980.
24. US ACE. Civil Works Construction Cost Index System Manual. ACE/1110-2-1304. US Army Corps of Engineers, Washington, DC, 2003.
25. State of Minnesota. Kummer Sanitary Landfill's Flaring System Destroys NMOC. www.atsdr.cdc.gov. Minnesota Health Department, St. Paul, MN, 2004.
26. State of California. Flare and Thermal Oxidizers. State of California, Santa Barbara County Air Pollution Control District. Rule 359. www.arb.ca.gov, 2004.

8
Thermal Oxidation

Lawrence K. Wang, Wei Lin, and Yung-Tse Hung

CONTENTS

1. INTRODUCTION

1.1. Process Description

Thermal oxidation (thermal incineration) is a widely used air pollution control technique whereby organic vapors are oxidized at high temperatures. Incineration (both thermal oxidation and catalytic oxidation) is considered an ultimate disposal method in that organic compounds in a waste gas stream are converted to carbon dioxide, water, and other inorganic gases rather than collected. In thermal incineration, contaminant-laden waste gas is heated to a high temperature (above 1000°F) at which the organic contaminants are burned with air in the presence of oxygen (*see* Figs. 1 and 2). A major advantage of incineration is that virtually any gaseous organic stream can be incinerated safely and cleanly, given proper design, engineering, installation, operation, and maintenance. Also, high (99% and higher) destruction efficiencies are possible with a wide variety of emission streams.

Depending on the types of heat recovery unit, incinerators are further classified as regenerative and recuperative. A recuperative thermal incinerator uses a shell and tube heat exchanger to transfer the heat generated by incineration to preheat the feed stream. Recuperative incinerators can recover about 70% of the waste heat from the exhaust gases. Regenerative thermal incinerators consist of a flame-based combustion chamber that connects two to three fixed beds containing ceramic or other inert packing. Incoming gas enters one of the beds where it is preheated. The heated gas flows into the combustion chamber, burns, and the hot flue gases flow through the packed beds where

From: *Handbook of Environmental Engineering, Volume 1: Air Pollution Control Engineering*
Edited by: L. K. Wang, N. C. Pereira, and Y.-T. Hung © Humana Press Inc., Totowa, NJ

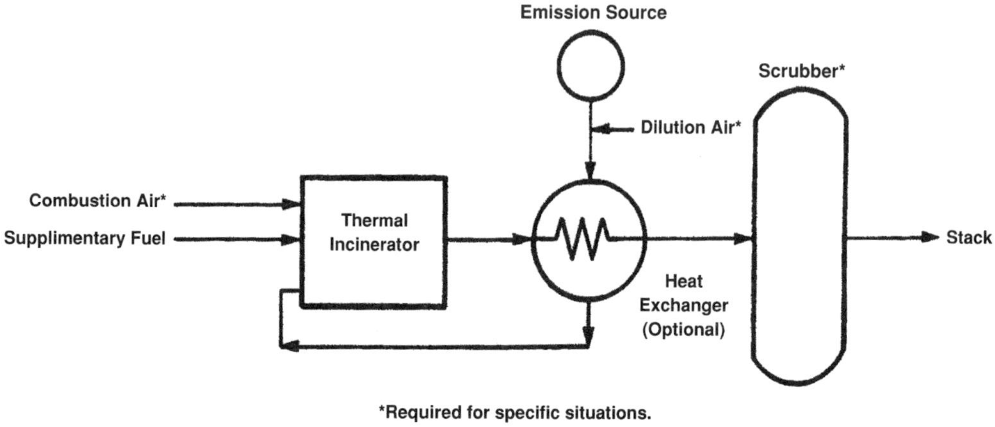

Fig. 1. Schematic of a thermal incinerator. (From US EPA.)

the heat generated during incineration is recovered and stored. The packed beds store the heat energy during one cycle and then release it as the beds preheat the incoming organic-laden gas during the second cycle. Up to 95% of energy in the flue gas can be recovered in this manner (1). The discussion in this chapter focuses on the more common recuperative-type incineration system. A detailed discussion of regenerative thermal incinerators is provided elsewhere (2).

In this chapter, a methodology is provided to quickly estimate thermal incinerator design and cost variables (2–12). The approach taken in this chapter is somewhat less detailed than the approach given in other US Environmental Protection Agency (EPA) references, but it allows for a relatively quick calculation of design and operational parameters. This approach enables the readers to obtain a general indication of design and cost parameters without resorting to more detailed and complex calculations.

When an adequate amount of oxygen is present in the combustion chamber, organic destruction efficiency (DE) of a thermal incinerator is determined by combustion temperature and residence time. Furthermore, at a given combustion temperature and residence time, DE is also affected by the degree of turbulence, or mixing of the emission stream and hot combustion gases, in the incinerator. DE in an incinerator depends on the types of organic pollutants as well. Halogenated organic compounds are more difficult to oxidize than unhalogenated organics; hence, the presence of halogenated compounds in the emission stream requires higher temperatures and longer residence times for complete oxidation. Depending on the goals of emission stream control, thermal incinerators can be designed achieve a wide range of DE. Discussion in this chapter will focus on the processes with organic DE of 98–99% and higher.

The incinerator flue gases are discharged at high temperatures and contain valuable heat energy. Therefore, a strong economic incentive exists for heat recovery. Typical recovery methods include the use of flue gas to preheat the emission stream that is going to be incinerated, to preheat combustion air, and to produce hot water or steam for other heating requirements. In most thermal incinerator applications, the available heat energy in the flue gases is used for preheating the emission stream. Discussion

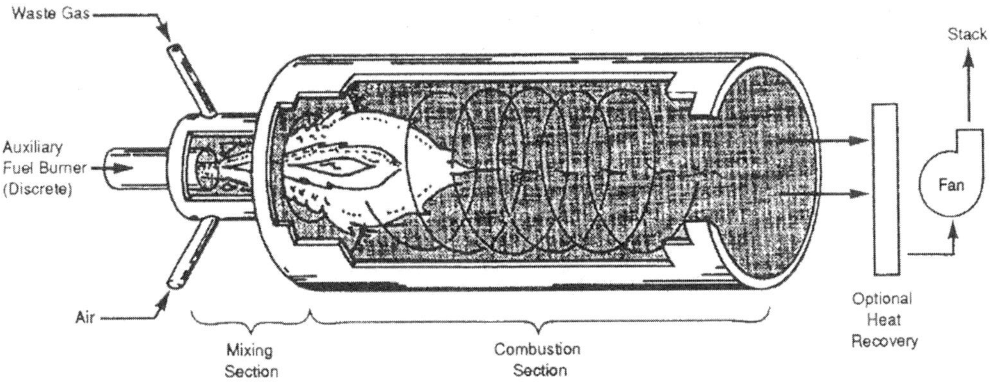

Fig. 2. Thermal incinerator. (From US EPA.)

in this chapter will be based on thermal incineration systems where the emission streams are preheated.

The incineration of emission streams containing organic vapors with halogen or sulfur components may create additional control requirements. For example, if sulfur and/or chlorine are present in the emission stream, the resulting flue gas will contain sulfur dioxide (SO_2) and/or hydrogen chloride (HCl). Depending on the concentrations of these compounds in the flue gas and the applicable regulations, scrubbing may be required to reduce the concentrations of these compounds. The selection and design of scrubbing systems are discussed in another chapter.

In this chapter the calculation procedure will be illustrated using emission stream 1 described in Table 1. Example 1 contains worksheets for design and technical calculations (4).

1.2. Range of Effectiveness

Thermal incineration is a well-established method for controlling volatile organic compound (VOC) emissions in waste gases. The DE for thermal incineration is typically 98% or higher. Factors that affect DE include the three "Ts" (temperature, residence time, and turbulence) as well as the type of contaminant in the waste gas. With a 0.75-s residence time, the suggested thermal incinerator combustion temperatures for waste gases containing nonhalogenated VOCs are 1600°F and 1800°F, respectively, for 98% and 99% VOC DEs. Higher temperatures (about 2000°F) and longer residence times (approx 1 s) are required for achieving DEs of 98% or higher with a halogenated VOC (4,12).

1.3. Applicability to Remediation Technologies

Storage tanks and surface impoundments are the major sources of organic air emissions at hazardous waste treatment, storage, and disposal facilities, and Superfund sites. Based on work performed in the development of the benzene waste National Emission Standards for Hazardous Air Pollutants (NESHAP), wastewater systems are a major source of benzene emissions from wastes that contain benzene (8–11).

Table 1
Effluent Characteristics for Emission Stream 1

Company	Glaze Chemical Company	Plant Contact	Mr. John Leake
Location (Street)	87 Octane Drive	Telephone No.	(999) 555-5024
(City)	Somewhere	Agency Contact	Mr.Efrem Johnson
(State, Zip)		No. of Emission Streams Under Review	7

A. Emission Steam Number/Plant Identification: #1 / #3 Oven Exhaust

	(a)	(b)	(c)
B. HAP Emission Source	paper coating oven		
C. Source Classification	process point		
D. Emission Stream HAPs	toluene		
E. HAP Class and Form	organic vapor		
F. HAP Content (1,2,3)**	960 ppmv		
G. HAP Vapor Pressure (1,2)	28.4 mm Hg at 77°F		
H. HAP Solubility (1,2)	insoluble in water		
I. HAP Adsorptive Prop. (1,2)	provided		
J. HAP Molecular Weight (1,2)	92 lb/lb-mole		

K. Moisture Content (1,2,3)	2% by volume	P. Organic Content (1) ***	100 ppmv CH_4
L. Temperature (1,2,3)	120°F	Q. Heat/O_2 Content (1)	4.1 Btu/scf/20.6 vol. %
M. Flow Rate (1,2,3)	15,000 scfm (max)	R. Particulate Content (3)	
N. Pressure (1,2)	atmospheric	S. Particle Mean Diam. (3)	
O. Halogen/Metals (1,2)	none/none	T. Drift Velocity/SO_3 (3)	
U. Applicable Regulation(s)			
V. Required Control Level			
W. Selected Control Methods	thermal oxidizer		

*The data presented are for an emission stream (single or combined streams) prior to entry into the selected control methods. Use extra form if additional space is necessary (e.g., more than three HAPs) and note this need.

**The number in parentheses denote what data should be supplied depending on the data on lines C and E:

1 = organic vapor process emission
2 = inorganic vapor process emission
3 = particulate process emission

***Organic emission stream combustibles less HAP combustibles shown on lines D and F.

Emissions occur from the surface of open-area sources, and high percentages of the volatiles are lost as emissions in these sources. For enclosed sources, the displacement of vapor containing volatiles from the enclosed air space is the emission mechanism. For both types of source, heating or aeration increases emissions. Emissions also occur from the evaporation of leaks and spills (8–11).

For emission control, open-area sources and containers can be covered or enclosed. Control devices can be installed to collect and remove organics from vented vapors, which is especially important if the sources are heated or aerated. Destruction of organic vapor by incineration is one of the emission control options.

The applicability of thermal incineration depends on the concentration of oxygen and contaminants in the waste gas. The waste gas composition will determine the auxiliary air and fuel requirements. These requirements, in turn, will have a strong influence on whether thermal incineration is an economical approach for controlling air emissions (13–22). Thermal incineration is best suited to applications where the gas stream has a consistent flow rate and concentration.

For most remediation technologies used at Superfund sites, the off-gases that require control are dilute mixtures of VOCs and air. The VOC concentration of these gases tends to be very low, whereas their oxygen content is high. In this case, auxiliary fuel is required but no auxiliary air is needed. However, if the waste has VOC content greater than 25% of its lower explosive limit (LEL), auxiliary air must be used to dilute the emission stream to below 25% of its LEL prior to incineration. The LEL for a flammable vapor is defined as the minimum concentration in air or oxygen at and above which the vapor burns upon contact with an ignition source and the flame spreads through the flammable gas mixture. Emission streams from some soil vapor extraction-based cleanups may contain VOCs greater than 25% of the LEL.

If the remediation activity generates an off-gas that has low oxygen content (below 13–16%), ambient air must be used to raise the oxygen level to ensure the burner flame stability. In the rare case when the waste gas is very rich in VOCs, using it directly as a fuel may be possible.

Information is presented in Table 2 for determining the suitability of a waste gas for incineration and establishing its auxiliary fuel and oxygen requirements. This same information is shown in Fig. 3 in an alternative format.

2. PRETREATMENT AND ENGINEERING CONSIDERATIONS

2.1. Air Dilution

In hazardous air pollutant (HAP) emission streams containing oxygen/air and flammable vapors, the concentration of flammable vapors is generally limited to less than 25% of the LEL. Insurance companies require that if the emission stream is preheated, the VOC concentration must be maintained below 25% of the LEL to minimize the potential for explosion hazards. In some cases, flammable vapor concentrations up to 40–50% of the LEL are permitted if on-line monitoring of VOC concentrations and automatic process control and shutdown are provided. The LELs of some common organic compounds are provided in Chapter 9 of this book.

In general, emission streams from waste-management facilities and Superfund sites are dilute mixtures of VOC and air and typically do not require further dilution. For

Table 2
Categorization of Waste Gas Streams

Waste gas category	Waste gas composition	O_2	VOC	Heat content	Auxiliaries and other requirements
1	Mixture of VOC, air, and inert gas	>16%	<25% LEL	<13 Btu/ft^3	Auxiliary fuel is required. No auxiliary air is required.
2	Mixture of VOC, air, and inert gas	16%	25–50% LEL	13–26 Btu/ft^3	Dilution air is required to lower the heat content to <13 Btu/ft^3. (Alternative to dilution air is installation of LEL monitors.)
3	Mixture of VOC, air, and inert gas	<16%	—	—	Treat this waste stream the same as categories 2,1 and and except augment the portions of the waste gas used for fuel burning with outside air to bring its O_2 content to above 16%
4	Mixture of VOC and inert gas	0—negligible	—	<100 Btu/scf	Oxidize it directly with a sufficient amount of air.
5	Mixture of VOC and inert gas	0—negligible	—	>100 Btu/scf	Premix and use it as a fuel.
6	Mixture of VOC and inert gas	0—negligible	—	Insufficient to raise gas temperature to the combustion temperature	Auxiliary fuel and combustion air for both the waste gas VOC and fuel are required.

Source: Adapted from Katari, et al., 1987 (12).

emission streams with oxygen concentrations less than 20% and heat contents greater than 176 Btu/lb or 13 Btu/scf (in most cases, corresponding to flammable vapor concentrations of approx 25% of LEL), the calculation procedure in this handbook assumes that dilution air is required. Equation (1) can be used to obtain the dilution airflow rate:

$$Q_d = \left[\left(h_e / h_d\right) - 1\right] Q_e \tag{1}$$

where Q_d is the required dilution airflow rate (scfm), h_e is the heat content of the emission stream (Btu/scf), h_d is the desired heat content of the emission stream (≤13 Btu/scf), and, Q_e is the emission stream flow rate (scfm). Note that this dilution will change other emission stream parameters.

2.2. Design Variables

Suggested combustion temperature (T_c) and residence time (t_r) values for thermal incinerators to achieve a given destruction efficiency are presented in Table 3. Two sets

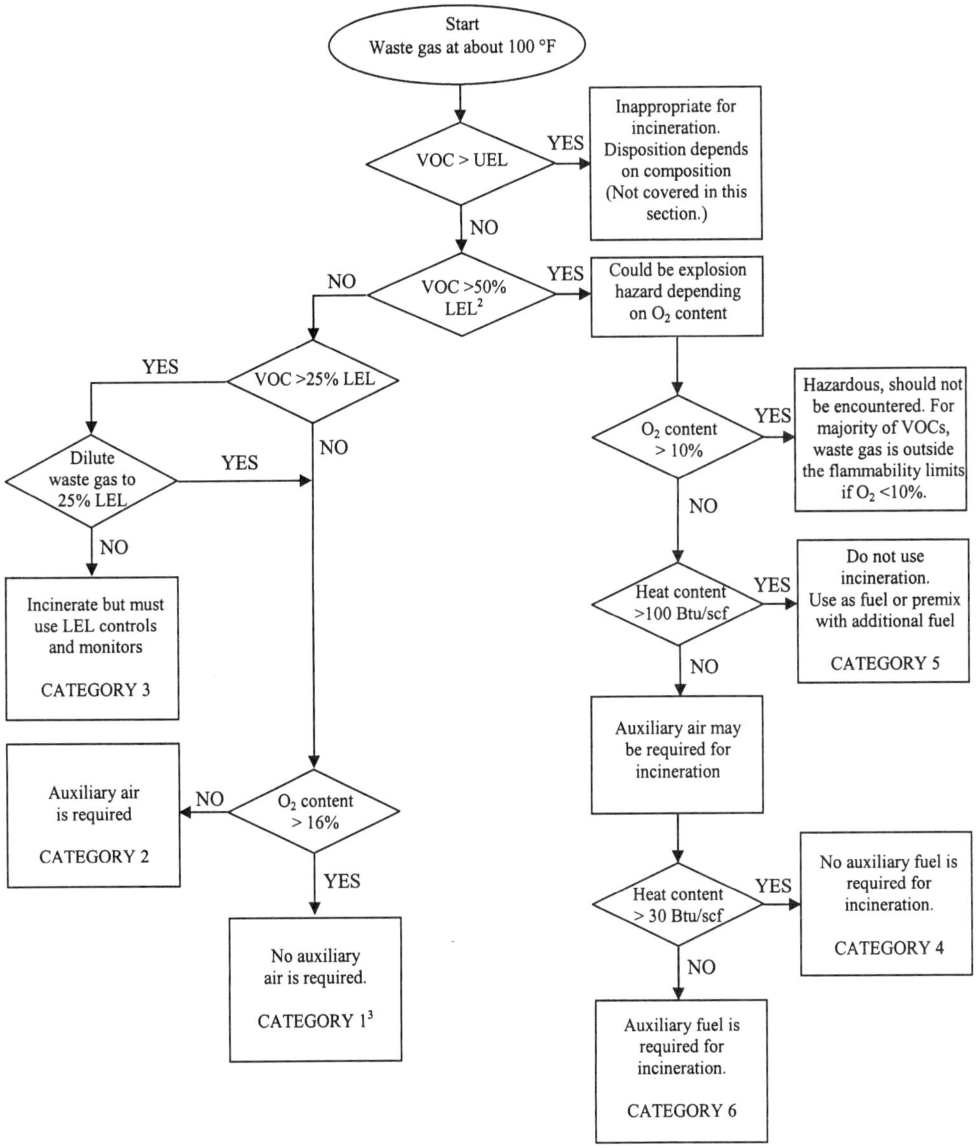

Fig. 3. Flowchart for categorization of a waste gas to determine its suitability for incineration and need for auxiliary fuel and air.

of values are shown in the table: one set for nonhalogenated emission streams and another set for halogenated emission streams. The combustion temperature and residence time values listed are conservative and assume adequate mixing of gases in the incinerator and adequate oxygen in the combustion chamber. The criteria in Table 3 are not the only conditions required for achieving the specified destruction efficiencies. For a given destruction efficiency, HAP emission streams may be incinerated at lower temperatures with longer residence times. However, the values provided in Table 3 reflect temperatures and residence times found in industrial applications. Based on the required

Table 3
Thermal Incinerator System Design Variables

Required destruction efficiency (DE) (%)	Nonhalogenated stream		Halogenated stream	
	Combustion temperature T_c (°F)	Residence time t_r (s)	Combustion temperature T_c (°F)	Residence time t_r (s)
98	1600	0.75	2000	1.0
99	1800	0.75	2200	1.0

Source: ref. 4.

DE, appropriate values for T_c and t_r can be selected from Table 3. For more information on temperature requirements versus destruction efficiency, consult Appendix D of ref. 7.

Because the performance of a thermal incinerator is highly related to the combustion chamber and outlet gas temperature, any thermal incinerator system used to control HAPs should be equipped with a continuous-temperature-monitoring system. Most vendors routinely equip thermal incinerators with such a system (5). However, some older units may not have a continuous-temperature-monitoring system. In this case, a retrofit installation of such a system should be requested.

In addition to temperature and residence time, good mixing of the gas streams is essential for proper operation. Unfortunately, mixing cannot be measured and quantified during design calculations. Typically, mixing is adjusted and improved during the start-up period of an incinerator. It is ultimately the responsibility of the operator to ensure correct operation and maintenance of a thermal incinerator after start-up.

In a system evaluation, if the design values for T_c and t_r are sufficient to achieve the required DE (compare the design values with the values from Table 3), the system design is considered acceptable. If the reported values for T_c and t_r are not sufficient, the design may be corrected by using the values for T_c and t_r from Table 3. (Note: If the DE is less than 98%, obtain information from the literature and incinerator vendors to determine appropriate values for T_c and t_r.)

Table 4 contains theoretical combustion chamber temperatures for 99.99% destruction efficiencies for various compounds with a residence time of 1 s. Note that the theoretical temperatures in Table 4 are considerably lower than those given in Table 3. This difference is because the values in Table 4 are theoretical values for specific compounds, whereas the values given in Table 3 are more general values designed to be applicable to a variety of compounds. Therefore, values in Table 3 are conservatively high. Table 4 is provided to indicate that certain specific applications may not require as high a combustion chamber temperature as those given in Table 3. Because the values given in Table 4 are theoretical, they may not be as applicable in the system design as the values in Table 3.

As a practical matter, a specific temperature to provide a specific destruction efficiency cannot be calculated *a priori*. Typically, incinerator vendors can provide general guidelines for destruction efficiency based on extensive experience. Tables 3 and 4 are presented to show a range of differences between theoretical and general values. In essence, these tables are used as a substitute for design equations relating destruction

Table 4
Theoretical Combustion Temperature Required for 99.99% Destruction Efficiencies

Compound	Combustion temperature (°F)	Residence time (s)
Acrylonitrile	1344	1
Allyl chloride	1276	1
Benzene	1350	1
Chlorobenzene	1407	1
1,2-Dichloroethane	1368	1
Methyl chloride	1596	1
Toluene	1341	1
Vinyl chloride	1369	1

Source: ref. 2.

efficiency to equipment parameters, because design equations are seldom used in hand analysis.

3. SUPPLEMENTARY FUEL REQUIREMENTS

Supplementary fuel is added to the thermal incinerator to attain the desired combustion temperature (T_c). For a given combustion temperature, the amount of heat needed to maintain the combustion temperature in the thermal incinerator is provided by (1) the heat supplied from the combustion of supplementary fuel, (2) the heat generated from the combustion of hydrocarbons in the emission stream, (3) the sensible heat contained in the emission stream as it leaves the emission source, and (4) the sensible heat gained by the emission stream through heat exchange with hot flue gases.

In general, emission streams treated by thermal incineration are dilute mixtures of VOC and air and typically do not require additional combustion air. For purposes of this handbook, it is assumed that the streams treated will have oxygen contents greater than 20% in the waste gas stream, which is typical of the majority of cases encountered. The following simplified equation can be used to calculate supplementary fuel requirements (based on natural gas) for dilute VOC streams:

$$Q_f = \frac{D_e Q_e \left[C_{p_{air}} \left(1.1T_c - T_{he} - 0.1T_r \right) - h_e \right]}{D_f \left[h_f - 1.1C_{p_{air}} \left(T_c - T_r \right) \right]} \qquad (2)$$

where Q_f is the natural gas flow rate (scfm), D_e is the density of the flue gas stream (lb/scf [usually 0.0739 lb/scf]), see Eq. (4), D_f is the density of fuel gas (0.0408 lb/scf for methane), see Eq. (4), Q_e is the emission stream flow rate (scfm), T_c is the combustion temperature (°F), T_{he} is the emission stream temperature after heat recovery (°F), T_r is the reference temperature (77°F), $C_{p_{air}}$ is the mean heat capacity of air between T_c and T_r (Btu/lb-°F) (*see* Table 7), h_e is the heat content of the flue gas (Btu/lb), and h_f is the lower heating value of natural gas (21,600 Btu/lb). T_{he} can be calculated by using the following expression if the value for T_{he} is not specified:

$$T_{he} = (HR/100)T_c + \left[1 - (HR/100)\right]T_e \qquad (3)$$

where HR is the heat recovery in the exchanger (%) and T_e is the temperature of the emission stream (°F). Assume that a value of 70% for HR if no other information is available.

The factor 1.1 in Eq. (2) is to account for an estimated heat loss of 10% in the incinerator. Supplementary heat requirements are typically calculated based on maximum emission stream flow rate, and, hence, will lead to a conservative design.

4. ENGINEERING DESIGN AND OPERATION

4.1. *Flue Gas Flow Rate*

Flue gas is generated as a result of the combustion process. Flue gas flow rate can be calculated using the following equation:

$$Q_{fg} = Q_e + Q_f + Q_d \tag{4}$$

where Q_{fg} is the flue gas flow rate (scfm), Q_e is the emission stream flow rate (scfm), Q_f is the natural gas (fuel) flow rate (scfm), and Q_d is the dilutmon air requirement (scfm).

Because the flow rate auxiliary fuel is usually much lower than the flow rate of emission streams, the flue gas flow rate for dilute waste gases when auxiliary air is not required is approximately equal to the waste gas flow rate. In cases where auxiliary air is required, the flue gas flow rate is roughly equal to the sum of the waste gas flow rate and the auxiliary airflow rate. The flue gas flow rate can be used in many correlations to size the incinerator and estimate equipment costs.

4.2. *Combustion Chamber Volume*

The combustion chamber volume (V_c) can be determined using the actual flue gas flow rate and the desirable residence time (t_r). The actual flue gas flow rate can be calculated using

$$Q_{fg,a} = Q_{fg}\left[\left(T_c + 460\right)/537\right] \tag{5}$$

where $Q_{fg,a}$ is the actual flue gas flow rate (acfm), Q_{fg} is the flue gas flow rate under standard conditions [scfm calculated from Eq. (4)], and T_c is the combustion temperature (°F).

The combustion chamber volume, V_c, is determined from the residence time t_r from Table 3 and $Q_{fg,a}$ obtained from Eq. (5):

$$V_c = \left[\left(Q_{fg,a}/60\right)t_r\right] \times 1.05 \tag{6}$$

The factor of 1.05 is used to account for minor fluctuations in the flow rate and follows industry practice.

4.3. *System Pressure Drop*

The total pressure drop for an incinerator depends on the type of equipment used in the system as well as other design considerations. The total pressure drop across an incinerator system determines the waste gas fan size and horsepower requirements, which, in turn, determine the fan capital cost and electricity consumption (12).

Table 5
Typical Pressure Drops for Thermal Incinerators

Equipment type	Heat recovery (HR)	Pressure drop *P* (in. H_2O)
Thermal incinerator	0	4
Heat exchanger	35	4
Heat exchanger	50	8
Heat exchanger	70	15

Note: The pressure drop is calculated as the sum of the incinerator and heat-exchanger pressure drops.
Source: ref. 2.

An accurate estimate of system pressure drop would require complex calculations. A preliminary estimate can be made using the approximate values listed in Table 5. The system pressure drop is the sum of the pressure drops across the incinerator and the heat exchanger plus the pressure drop through the ductwork.

The pressure drop can then be used to estimate the power requirement for the flue gas fan using the following empirical correlation (12):

$$\text{Power} = 1.17 \times 10^{-4} \times Q_{fg} \times P/\varepsilon \tag{7}$$

where Power is the fan power requirement (kWh), Q_{fg} is the flue gas flow rate (scfm), P is the system pressure drop (inches of water column), and ε is the combined motor fan efficiency (dimensionless [typically 60%]).

5. MANAGEMENT

5.1. Evaluation of Permit Application

Permit evaluators can use Table 6 to compare the results from the calculations and the values supplied by the permit applicant. The values in Table 6 are calculated based on the emission stream 1 example presented in Table 1. The flue gas flow rate (Q_{fg}) is determined from the emission stream flow rate (Q_e), dilution air requirement (Q_d), and supplementary fuel requirement (Q_f). Therefore, any differences between the

Table 6
Comparison of Calculated Values and Values Supplied by the Permit Applicant for Thermal Incineration

	Calculated value (example case)[a]	Reported value
Continuous monitoring of combustion temperature	163 scfm	—
Supplementary fuel flowratc, Q_f	Yes	—
Dilution airflow rate, Q_d	0	—
Flue gas flow rate, Q_{fg}	15,200 scfm	—
Combustion chamber volume, V_c	840 ft^3	—

[a]Based on emission stream 1.

calculated and reported values for Q_{fg} will be dependent on the differences between the calculated and reported values for Q_d and Q_f. If the calculated values for Q_d and Q_f differ from the reported values for these variables, the differences may be the result of the assumptions involved in the calculations. Therefore, further discussions with the permit applicant will be necessary to find out about the details of the design and operation of the proposed thermal incinerator system.

If the calculated values and the reported values are not different, then the design and operation of the proposed thermal incinerator system may be considered appropriate based on the assumptions used in this handbook.

Table 7 presents a HAP Emission Stream Data Form generally required for the permit applications.

5.2. *Operations and Manpower Requirements*

Electricity costs are associated primarily with the fan needed to move the gas through the incinerator. Equation (8) can be used to estimate the power requirements for a fan, assuming a fan motor efficiency of 65%. The fan is assumed to be installed downstream of the incinerator, as shown in Fig. 2.

$$F_p = 1.81 \times 10^{-4}\left(Q_{fg,a}\right)(P)(HRS) \tag{8}$$

where F_p is the power needed for the fan (kWh/yr), $Q_{fg,a}$ is the actual flue gas flow rate (acfm), P is the system pressure drop (in. H_2O [from Table 5]), and *HRS* is the operating hours per year (h/yr).

Operating labor requirements are estimated as 0.5 h per 8-h shift. Supervisory costs are estimated as 15% of operator labor costs. Maintenance labor requirements are estimated as 0.5 h per 8-h shift, with a slightly higher labor rate reflecting increased skill levels. Maintenance materials are estimated as 100% of maintenance labor.

Indirect annual costs include the capital recovery cost, overhead, property taxes, insurance, and administrative charges. The capital recovery cost is based on an estimated 10-yr equipment life, whereas overhead, property taxes, insurance, and administrative costs are percentages of the total capital cost (31).

Most operational problems with thermal incinerators are related to the burner. Typical problems encountered include low burner firing rates, poor fuel atomization (oil-fired units), poor air/fuel ratios, inadequate air supply, and quenching of the burner flame (6). These problems lead to lower DEs for HAPs. Symptoms of these problems include obvious smoke production or a decrease in combustion chamber temperature, as indicated by the continuous monitoring system. If a thermal incinerator system begins to exhibit these symptoms, the facility operator should take immediate action to correct any operational problems. Typical thermal incinerator pressure drops presented in Table 5 can be used as reference for checking the equipment operational conditions.

In the case of permit review for a thermal incinerator, the data outlined below should be supplied by the applicant. The calculations in this section will then be used to check the applicant's values.

Thermal incinerator system variables at standard conditions (77°F, 1 atm) are as follows:

Table 7
HAP Emmission Stream data Form*

Company			Plant Contact		
Location (Street)		______	Telephone No.	______	
(City)			Agency Contact		
(State, Zip)		______	No. of Emission Streams Under Review		7
A. Emission Steam Number/Plant Identification			#1 / #3 Oven		
B. HAP Emission Source	(a)	______	(b) ______	(c)	______
C. Source Classification	(a)		(b)	(c)	
D. Emission Stream HAPs	(a)	______	(b) ______	(c)	______
E. HAP Class and Form	(a)		(b)	(c)	
F. HAP Content (1,2,3)**	(a)	______	(b) ______	(c)	______
G. HAP Vapor Pressure (1,2)	(a)		(b)	(c)	
H. HAP Solubility (1,2)	(a)	______	(b) ______	(c)	______
I. HAP Adsorptive Prop. (1,2)	(a)		(b)	(c)	
J. HAP Molecular Weight (1,2)	(a)	______	(b) ______	(c)	______
K. Moisture Content (1,2,3)			P. Organic Content (1) ***		
L. Temperature (1,2,3)		______	Q. Heat/O_2 Content (1)		______
M. Flow Rate (1,2,3)			R. Particulate Content (3)		
N. Pressure (1,2)		______	S. Particle Mean Diam. (3)		______
O. Halogen/Metals (1,2)			T. Drift Velocity/SO_3 (3)		
U. Applicable Regulation(s)		______			
V. Required Control Level					
W. Selected Control Methods		______			

*The data presented are for an emission stream (single or combined streams) prior to entry into the selected control methods. Use extra form if additional space is necessary (e.g., more than three HAPs) and note this need.

**The number in parentheses denote what data should be supplied depending on the data on lines C and E:

1 = organic vapor process emission
2 = inorganic vapor process emission
3 = particulate process emission

***Organic emission stream combustibles less HAP combustibles shown on lines D and F.

Source: US EPA.

1. Reported destruction efficiency, $DE_{reported}$ (%)
2. Temperature of the emission stream entering the incinerator:
   if no heat recovery, T_e °F
   if a heat exchanger employed, T_{he} (°F)
3. Combustion temperature, T_c (°F)
4. Residence time, t_r (s)
5. Maximum emission stream flow rate, Q_e (scfm)
6. Fuel heating value, h_f (Btu/lb)
7. Combustion chamber volume, V_c (ft^3)
8. Flue gas flow rate, Q_{fg} (scfm)

5.3. Decision for Rebuilding, Purchasing New or Used Incinerators

Technical innovations have changed the landscape of the incinerator market. These advances have dramatically impacted the economics of VOC control and created new options for meeting future clean air regulations. The durability of incinerators themselves contributes to this picture. The physical structure of an incinerator unit may have years of remaining life, but its operating efficiency may no longer meet current needs. What does a plant do when faced with changing regulations or process expansion?

There is no easy answer and multiple options need to be explored. For example, many of the new incinerator innovations can be retrofitted into existing units. This means a plant engineer may be able to improve the VOC destruction, thermal efficiency, and capacity of the existing system while lowering its operating costs and expanding process flexibility.

In addition, the growing popularity of incinerators in industry is creating an increasing supply of used systems for sale. Often these units can be cost-effectively upgraded to increase VOC destruction efficiency or add needed capacity.

Further fueling the changing face of emission control options, the price of new incinerators has dropped significantly in recent years. Two typical examples on how to make decisions on rebuilding, purchasing new, or purchasing used incinerators have been presented in a recent technical article (23). In case the plant manager decides to purchase new process equipment, thermal incinerators (thermal oxidizers) are commercially available (24,25). More technical information on thermal oxidizers is available elsewhere (28–31).

5.4. Environmental Liabilities

The risk management process provides a framework for managing environmental liabilities. It is composed of the identification, quantification, and ultimate treatment of loss exposures. Indelicato (26) has outlined several strategies than can be used singularly or in combination for the risk management to accomplish two objectives: (1) to control losses and (2) to minimize the financial impacts resulting from the loss.

6. DESIGN EXAMPLES

Example 1

Develop a calculation sheet for thermal incineration (4) before performing design and technical calculations.

Solution

The following are important elements of a calculation sheet for thermal incineration:

1. Data requirement for HAP emission stream characteristics:

   1. Maximum flow rate, Q_e = ______________ scfm
   2. Temperature, T_e = ______________°F
   3. Heat content, h_e = ______________ Btu/scf
   4. Oxygen content, O_2 = ______________ %
   5. Halogenated organics: Yes ______________ No ______________
   6. Required destruction efficiency, DE = ______________ %

2. In the case of a permit review, the following data should be supplied by the applicant: Thermal incinerator system variables at standard conditions (77°F, 1 atm):

   1. Reported destruction efficiency, $DE_{reported}$ = ______________ %
   2. Temperature of emission stream entering the incinerator,
      T_e = ______________ °F (if no heat recovery)
      T_{he} = ______________ °F (if a heat exchanger is employed)
   3. Combustion temperature, T_c = ______________ °F
   4. Residence time, t_r = ______________ s
   5. Maximum emission stream flow rate, Q_e = ______________ scfm
   6. Fuel heating value, h_f = ______________ Btu/lb
   7. Combustion chamber volume, V_c = ______________ ft^3
   8. Flue gas flow rate, Q_{fg} = ______________ scfm

It should be noted that (1) if dilution air is added to the emission stream upon exit from the process, the data required are the resulting characteristics after dilution and (2) the oxygen content depends on the oxygen content of the organic compounds (fixed oxygen) and the free oxygen in the emission stream. Because emission streams treated by thermal incineration are generally dilute VOC and air mixtures, the fixed oxygen in the organic compounds can be neglected.

Example 2

Outline a step-by-step procedure for the determination of dilution air requirements, combustion temperature, and residence time of a thermal incinerator.

Solution

1. **Pretreatment of the emissions stream: dilution air requirements.** Typically, dilution will not be required. However, if the emission stream heat content (h_e) is greater than 176 Btu/lb or 13 Btu/scf with an oxygen concentration less than 20%, the dilution airflow rate (Q_d) can be determined using

$$Q_d = \left[(h_e/h_d) - 1\right]Q_e$$

Q_d = ______________ scfm

2. **Design variables, destruction efficiency, and typical operational problems.** Based on the required destruction efficiency (DE), select appropriate values for T_c and t_r from Table 3.

   T_c = ______________ °F
   t_r = ______________ s

For a permit evaluation, if the applicant's values for T_c and t_r are sufficient to achieve the required DE (compare the reported values with the values presented in Table 3), proceed with the calculations. If the applicant's values for T_c and t_r are not sufficient, the applicant's design is unacceptable. The reviewer may then use the values for T_c and t_r from Table 3.

T_c = ______________ °F
t_r = ______________ s

Note: If DE is less than 98%, obtain information from the literature and incinerator vendors to determine appropriate values for T_c and t_r.

Example 3

The data necessary to perform the calculations of this example consist of HAP emission stream characteristics previously compiled on the HAP Emission Stream Data Form (Tables 1 and 7) and the required HAP destruction efficiency as determined by the applicable regulations. The following are the given data for a HAP emission stream:

Maximum flow rate Q_e = 15,000 scfm
Temperature T_e = 120°F
Heat content, h_e = 4.1 Btu/scf
Oxygen content, O_2 = 20.6%
Halogenated organics: Yes ______________ No X______________

Based on the control requirements for the emission stream: DE = 99%.

Determine (1) the dilution air requirements of a thermal incinerator and (2) the combustion temperature and the residence time of this incinerator.

Solution

1. If dilution air is add to the emission stream upon exit from the process, the data that will be used in the calculations are the resulting characteristics after dilution.

Dilution air is required for emission streams with oxygen concentrations less than 20%, and heat contents greater than 176 Btu/lb or 13 Btu/scf, which, in most cases, correspond to flammable vapor concentrations of approx 25% of the lower explosive limit (LEL). To convert Btu/lb to Btu/scf, multiply Btu/lb by the density of the emission stream at standard conditions (typically, 0.0739 lb/ft^3).

Because the oxygen content and heat content of this HAP emission stream are 20.6% and 4.1 Btu/scf, respectively, no dilution air required.

2. The required destruction efficiency is 99% and the HAP emission stream is non-halogenated; therefore,

T_c = 1800°F (Table 3)
t_r = 0.75 s (Table 3)

A continuous monitoring system should ensure operation at 1800°F.

Example 4

Determine the supplementary fuel requirement of a thermal incinerator for treatment of the HAP emission stream stated in Example 1, with the following additional technical data:

1. Density of flue gas stream, D_e = 0.0739 lb/scf
2. Density of fuel (methane or natural gas), D_f = 0.0408 lb/scf
3. Emission stream flow rate, Q_e = 15,000 scfm
4. Mean heat capacity, $C_{p_{air}}$ = 0.269 Btu/lb-°F for the interval 77–1800°F
5. Emission stream temperature, T_{he} = 1296°F
6. Combustion temperature, T_c = 1800°F
7. Reference temperature, T_r = 77°F
8. Lower heating value of natural gas, h_f = 21,600 Btu/lb
9. Heat content of flue gas, h_e = 55.4 Btu/lb

Solution

Because the emission stream is very dilute and has an oxygen content greater than 20%, Eq. (2) is applicable. The natural gas flow rate is then calculated to be

$$Q_f = \frac{(0.0739)(15,000)[0.269(1980 - 1296 - 7.7) - 55.41]}{0.0408[21,600 - 1.1(0.269)(1800 - 77)]}$$

$$Q_f = 163 \text{ scfm}$$

The following should be noted before calculation:

Q_e, h_e	Input data
D_e	0.0739 lb/scf, if no other information available
D_f	0.0408 lb/scf, if no other information available
h_f	Assume a value of 21,600 Btu/lb if no other information available
$C_{p_{air}}$	See Table 8 for values of $C_{p_{air}}$ at various temperatures
T_c	Obtain value from Table 3 or from permit applicant
T_{he}	Use the following equation if the value for T_{he} is not specified:

$$T_{he} = (HR/100)T_c + [1 - (HR/100)]T_e$$

where *HR* is the heat recovery in the heat exchanger (percent); assume a value of 70% for *HR* if no other information available

T_r	77°F, if no other information available

Example 5

Determine (1) the flue gas flow rate under standard conditions, (2) the actual flue gas flow rate, and (3) the combustion chamber volume for a proposed thermal incinerator, under the following design conditions:

1. Maximum HAP emission stream flow Q_e = 15,000 scfm
2. Natural gas flow rate Q_f = 163 scfm
3. Combustion temperature T_c = 1800°F
4. Combustion residence time t_r = 0.75 s

Solution

1. The flue gas flow rate Q_{fg} is

Table 8
Average Specific Heats of Vapors[a]

Temp.	Average specific heat,[a] C_p (Btu/scf-°F)									
(°F)	Air	H_2O	O_2	N_2	CO	CO_2	H_2	CH_4	C_2H_4	C_2H_6
77	0.0180	0.0207	0.0181	0.0180	0.0180	0.0230	0.0178	0.0221	0.0270	0.0326
212	0.0180	0.0209	0.0183	0.0180	0.0180	0.0239	0.0179	0.0232	0.0293	0.0356
392	0.0181	0.0211	0.0186	0.0181	0.0181	0.0251	0.0180	0.0249	0.0324	0.0395
572	0.0183	0.0212	0.0188	0.0182	0.0183	0.0261	0.0180	0.0266	0.0353	0.0432
752	0.0185	0.0217	0.0191	0.0183	0.0184	0.0270	0.0180	0.0283	0.0379	0.0468
932	0.0187	0.0221	0.0194	0.0185	0.0186	0.0278	0.0181	0.0301	0.0403	0.0501
1112	0.0189	0.0224	0.0197	0.0187	0.0188	0.0286	0.0181	0.0317	0.0425	0.0532
1292	0.0191	0.0228	0.0199	0.0189	0.0190	0.0292	0.0182	0.0333	0.0445	0.0560
1472	0.0192	0.0232	0.0201	0.0190	0.0192	0.0298	0.0182	0.0348	0.0464	0.0587
1652	0.0194	0.0235	0.0203	0.0192	0.0194	0.0303	0.0183	0.0363	0.0481	0.0612
1832	0.0196	0.0239	0.0205	0.0194	0.0196	0.0308	0.0184	0.0376	0.0497	0.0635
2012	0.0198	0.0243	0.0207	0.0196	0.0198	0.0313	0.0185	0.0389	0.0512	0.0656
2192	0.0199	0.0246	0.0208	0.0197	0.0199	0.0317	0.0186	0.0400	0.0525	0.0676

Note: Average for the temperature interval 77°F and the specified temperature.

[a]Based on 70°F and 1 atm. To convert to Btu/lb-°F basis, multiply by 392 and divide by the molecular weight of the compound. To convert to Btu/lb-mol°F, multiply by 392.

Source: ref. 4.

$$Q_{fg} = Q_e + Q_f + Q_d$$

$$Q_{fg} = 15{,}000 + 163 + 0$$

$$Q_{fg} = 15{,}200 \text{ scfm (rounded to three significant digits)}$$

2. The actual flue gas flow rate $Q_{fg,a}$ can be determined using Eq. (5).

$$Q_{fg,a} = 15{,}200[(1{,}800 + 460)/537]$$
$$= 64{,}000 \text{ acfm}$$

3. The combustion chamber volume V_c can then be calculated using Eq. (6).

$$V_c = [(64{,}000/60)0.75] \times 1.05$$

$$V_c = 840 \text{ ft}^3$$

The combustion chamber volume for the proposed thermal incinerator should therefore have a value that is approx 840 ft^3.

Example 6

Discuss how the permit application for a thermal incinerator is evaluated.

Solution

Compare the calculated values and reported values using Table 9. The combustion volume (V_c) is calculated from flue gas flow rate (Q_{fg}) and Q_{fg} is determined by emission stream

Table 9
Comparison of Calculated Values and Values Supplied by the Permit Applicant for Thermal Incineration

	Calculated Value	Reported Value
Continuous monitoring of combustion temperature		
Supplementary fuel flow rate, Q_f		
Dilution air flow rate, Q_d		
Flue gas flow rate, Q_{fg}		
Combustion chamber size, V_c		

flow rate (Q_e), supplementary fuel flow rate (Q_f), and dilution air requirement (Q_d). Therefore, if there are differences between the calculated and reported values for V_c and Q_{fg}, these are dependent on the differences between the calculated and reported values for Q_d and Q_f.

If the calculated and reported values are different, the differences may be the result of the assumptions involved in the calculations. Discuss the details for the design and operation of the system with the applicant. Table 6 shows an example.

If the calculated and reported values are not different, then the design and operation of the system can be considered appropriate based on the assumptions employed in this handbook.

NOMENCLATURE

ε	Combined motor fan efficiency (dimensionless [approx 60%])
$C_{p_{air}}$	Mean specific heat of air (Btu/lb-°F)
D_e	Density of emission stream (lb/ft^3)
DE	Destruction efficiency (%)
D_f	Density of fuel gas (lb/ft^3)
Fp	Power requirement (kWh/yr)
h_d	Emission stream desired heat content (Btu/scf)
h_e	Emission stream heat content (Btu/lb or Btu/scf)
h_f	Supplementary fuel heating value (Btu/lb)
HR	Heat recovery in heat exchanger (%)
HRS	Operating hours per year (h/yr)
m	Mass of flue gas (waste gas plus auxiliary air) or flow rate (lb-mol/h)
P	System pressure drop (in H_2O)
Power	Fan power requirement (kWh)
Q_d	Dilution air required (scfm)
Q_e	Emission stream flow rate (scfm)
$Q_{e,a}$	Actual emission stream flow rate (acfm)
Q_f	Supplementary fuel gas flow rate (scfm)
Q_{fg}	Flue gas flow rate (scfm)
$Q_{fg,a}$	Actual flue gas flow rate (acfm)
T_c	Combustion temperature (°F)

T_{he}	Temperature of emission stream exiting heat exchanger (°F)
t_r	Residence time (s)
T_r	Reference temperature (77°F)
V_c	Combustion chamber volume (ft^3)

REFERENCES

1. US EPA, *Survey of Control Technologies for Low Concentration Organic Vapor Gas Streams*. EPA-456/R-95-003. US Environmental Protection Agency, Research Triangle Park, NC, 1995.
2. US EPA, *OAQPS Control Cost Manual*, 4th ed. EPA 450/3-90-006 (NTIS PB90-169954). US Environmental Protection Agency, Washington, DC, 1990.
3. C. Cleveland *Handbook of Chemistry and Physics*, 60th ed., The Chemical Rubber Company, Cleveland, OH, 1980.
4. US EPA, *Handbook: Control Technologies for Hazardous Air Pollutants*. EPA 625/6-91-014. (NTIS PB91-228809). U.S. Environmental Protection Agency, Cincinnati, OH, 1991.
5. PES, Inc., Research Triangle Park, NC. Company data for the evaluation of continuous compliance monitors.
6. C. Nunez, US EPA, AEERL memorandum with attachments to Michael Sink, PES, Research Triangle Park, NC, October 1989.
7. US EPA, *Handbook: Guidance on Setting Permit Conditions and Reporting Trial Burn Results*. EPA 625/6-89-019. US Environmental Protection Agency, Cincinnati, OH, 1989.
8. US EPA, *Control of Air Emissions from Superfund Sites*. EPA/625/R-92/012. US Environmental Protection Agency, Washington, DC 1992.
9. US EPA, *Organic Air Emissions from Waste Management Facilities*. EPA/625/R-92/003. US Environmental Protection Agency, Washington, DC 1992.
10. L. K. Wang, J. V. Krouzek, and U. Kounitson, *Case Studies of Cleaner Production and Site Remediation*. UNIDO Manual No. DTT-5-4-95. United Nations Industrial Development Organization, Vienna, 1995.
11. L. K. Wang, M. H. S. Wang, and P. Wang, *Management of Hazardous Substances at Industrial Sites*. UNIDO Registry No. DTT-4-4-95. United Nations Industrial Development Organization, Vienna, 1995.
12. V. W. Katari, W. Vatavuk, and A. H. Wehe, *JAPCA*. 37 (1) (1987).
13. H. E. Hesketh, *Air Pollution Control*. Technomic, Lancaster, PA, 1991.
14. US EPA, *Soil Vapor Extraction Technology VOC Control Technology Assessment*. EPA/450/4-89/017 (NTIS PB90-216995). US Environmental Protection Agency, Research Triangle Park, NC, 1989.
15. US EPA *Industrial Wastewater Volatile Organic Compound Emissions—Background Information for BACT/LAER Determination*. EPA-450/390-004. US Environmental Protection Agency, Washington, DC, 1990.
16. US EPA, *Alternative Control Technology Document – Organic Waste Process Vents*, EPA—450/3-91-007. US Environmental Protection Agency, Washington, DC, 1990.
17. US EPA, *Hazardous Waste Treatment, Storage, and Disposal Facilities (TSDF)—Air Emission Models*. EPA-450/3-87-026. US Environmental Protection Agency, Washington, DC, 1989.
18. US EPA, *Hazardous Waste Treatment, Storage, and Disposal Facilities (TSDF)—Background Information for Promulgated Organic Emission Standards for Process Vents and Equipment Leaks*. EPA-450/3-89-009. US Environmental Protection Agency, Washington, DC, 1990.

19. US EPA, *Hazardous Waste TSDF—Background Information Document for Proposed RCRA Air Emission Standards.* EPA-450/3-89-23. U.S. Environmental Protection Agency, Washington, DC, 1989.
20. US EPA, *Hazardous Waste TSDF—Technical Guidance Document for RCRA Air Emission Standards for Process Vents and Equipment Leaks.* EPA-450/3-89-21. US Environmental Protection Agency, Washington, DC, 1990.
21. US EPA, *Preliminary Assessment of Hazardous Waste Pretreatment as an Air Pollution Technique.* EPA-600/2-86-028 (NTIS PB86-172095/AS). US Environmental Protection Agency, Washington, DC, 1986.
22. US EPA, *VOC Emission from Petroleum Refinery Wastewater Systems—Background Information for Proposed Standards.* EPA-450/3-85-001a. US Environmental Protection Agency, Washington, DC, 1985.
23. J. Gallo, R. Schwartz, and J. Cash, *Environ. Protect.* **12(3)**, 74–80 (2001).
24. Anon., *Environ. Protect.* **14(2)**, 104–106 (2003).
25. Anon. 2002 Buyer's guide: oxidizers. *Environ. Protect.*, **13(3)**, 148 (2002).
26. G. Indelicato, *Environ. Protect.* **12(3)**, 88–92 (2001).
27. Anon. *Environ. Protect.* **14(2)**, 155–156 (2003).
28. US EPA. *Air Pollution Control Technologies Series Training Tool*: Incineration. www.epa.gov. US Environmental Protection Agency, Washington DC. 2004.
29. State of California. Flare and Thermal Oxidizers. State of California, Santa Barbara County Air Pollution Control District. Rule 359. www.arb.ca.gov. 2004.
30. R. L. Pennington. Options for controlling hazardous air pollutants. *Environmental Technology* **6(6)**, 18–23 (1996).
31. L. K. Wang, N. C. Pereira, and Y. T. Hung (eds.), *Advanced Air and Noise Pollution Control.* Humana Press, Totowa, NJ, 2005.

9
Catalytic Oxidation

Lawrence. K. Wang, Wei Lin, and Yung-Tse Hung

CONTENTS

1. INTRODUCTION

Catalytic oxidation (catalytic incineration) is an oxidation process, shown in Figs. 1 and 2, that converts organic compounds to carbon dioxide and water with the help of a catalyst. A catalyst is a substance that accelerates the rate of a reaction at a given temperature without being appreciably changed during the reaction. In catalytic incinerators, the flame-based incineration concept is modified by adding a catalyst to promote the oxidation reaction, allowing faster reaction and/or reduced reaction temperature. A faster reaction requires a smaller vessel, thus reducing capital costs and low operating temperatures generally and reduced auxiliary fuel requirements, thus reducing operating costs (1).

Catalytic incineration is most suitable for treatment of emission streams containing a low concentration of volatile organic compounds (VOCs). It may allow a more cost-effective operation compared to thermal incineration processes. Catalytic incineration, however, is not as broadly used as thermal incineration because of its greater sensitivity to pollutant characteristics and process conditions (1). Design and operating considerations are therefore critical to applications of catalytic incineration in air pollution control. In this chapter, a methodology is proved to quickly estimate catalytic incinerator design and cost variables (2–14).

1.1. Process Description

Schematics of a catalytic incinerator, also known as catalytic oxidizer and catalytic reactor, are shown in Figs. 1 and 2. In catalytic incineration, a contaminant-laden emission

From: *Handbook of Environmental Engineering, Volume 1: Air Pollution Control Engineering*
Edited by: L. K. Wang, N.C. Pereira, and, Y.-T. Hung © Humana Press Inc., Totowa, NJ

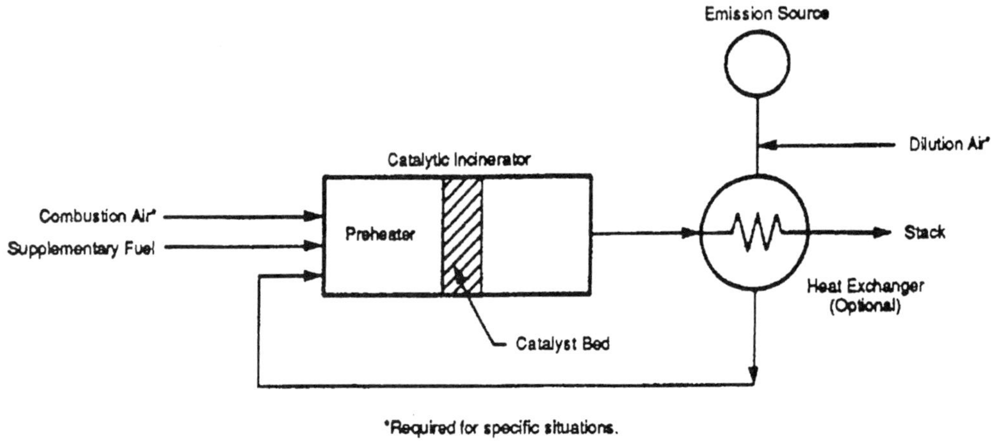

Fig. 1. Schematic of a catalytic incinerator system. (From US EPA.)

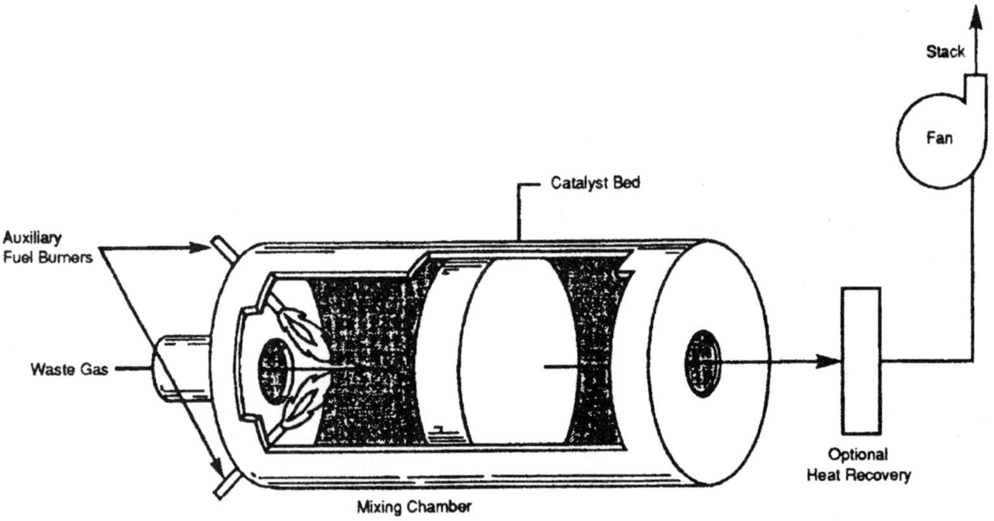

Fig. 2. Catalytic oxidizer. (From US EPA.)

stream is usually first preheated in a primary heat exchanger to recover heat from the exhaust gases. Additional heat is then added to the emission stream in a natural-gas-fired or electric preheater to increase the temperature to 600–900°F. The emission stream is then passed across a catalyst bed where the VOC contaminants react with oxygen to form carbon dioxide and water.

After oxidation of the emission stream, the heat energy in the flue gases leaving the catalyst bed may be recovered in several ways including (1) use of a recuperative heat exchanger to preheat the emission stream and/or combustion air or (2) use of the available energy for process heat requirements (e.g., recycling flue gases to the process, producing hot water or steam). Catalytic incineration systems using regenerative heat exchange are in the developmental stage.

Catalysts typically used for VOC incineration include platinum and palladium; other formulations are also used, including metal oxides for emission streams containing chlorinated compounds. The catalyst bed (or matrix) in the incinerator is generally a metal mesh-mat, ceramic honeycomb, or other ceramic matrix structure designed to maximize catalyst surface area. The catalysts may also be in the form of spheres or pellets.

Recent advances in catalysts have broadened the applicability of catalytic incineration. Catalysts now exist that are relatively tolerant of compounds containing sulfur or chlorine. These new catalysts are often single or mixed metal oxides and are supported by a mechanically strong carrier. A significant amount of effort has been directed toward the oxidation of chlorine-containing VOCs. These compounds are widely used as solvents and degreasers and are often encountered in emission streams. Catalysts such as chrome/alumina, cobalt oxide, and copper oxide/manganese oxide have been demonstrated to control an emission stream containing chlorinated compounds. Platinum-based catalysts are often employed for the control of sulfur-containing VOCs but are sensitive to chlorine poisoning.

Despite catalyst advances, some compounds simply do not lend themselves well to catalytic oxidation. These include compounds containing lead, arsenic, and phosphorus. Unless the concentration of such compounds is sufficiently low or a removal system is employed upstream, catalytic oxidation should not be considered in these cases.

The performance of a catalytic incinerator is affected by several factors including: (1) operating temperature, (2) space velocity (reciprocal of residence time), (3) VOC composition and concentration, (4) catalyst properties, and, as mentioned earlier, (5) presence of poisons/inhibitors in the emission stream. When adequate oxygen is present in the incineration stream, important variables for catalytic incinerator design are the operating temperature at the catalyst bed inlet, the temperature rise across the catalyst bed, and the space velocity. The operating temperature for particular destruction efficiency is dependent on the concentration and composition of the VOC in the emission stream and the type of catalyst used.

Space velocity (SV) is defined as the volumetric flow rate of the combined gas stream (i.e., emission stream plus supplemental fuel plus combustion air) entering the catalyst bed divided by the volume of the catalyst bed. As such, space velocity also depends on the type of catalyst used. At a given space velocity, increasing the operating temperature at the inlet of the catalyst bed increases the destruction efficiency. At a given operating temperature, as space velocity is decreased (i.e., as residence time in the catalyst bed increases), destruction efficiency increases.

The performance of catalytic incinerators is sensitive to pollutant characteristics and process conditions (e.g., flow rate fluctuations). In the following discussion, it is assumed that the emission stream is free from poisons/inhibitors such as phosphorus, lead, bismuth, arsenic, antimony, mercury, iron oxide, tin, zinc, sulfur, and halogens. (Note: Some catalysts can handle emission streams containing halogenated compounds, as discussed above.) It is also assumed that the fluctuations in process conditions (e.g., changes in VOC content) are kept to a minimum.

Temperature control in preheat chamber is important to catalytic incineration (catalytic oxidation) systems. High preheat temperatures accompanied by a temperature increase across the catalyst bed may lead to overheating of the catalyst bed and eventually loss of its activity (6,7).

The following discussion will be based on fixed-bed catalytic incinerator system with recuperative heat exchange (i.e., preheating the emission stream). Throughout this chapter, it is assumed that adequate oxygen (i.e., O_2 content greater than 20%) is present in the emission stream so that combustion air is not required. The calculation procedure will be illustrated using emission stream 2 described in Table 1.

1.2. *Range of Effectiveness*

Catalytic oxidation is a well-established method for controlling VOC emissions in waste gases. The control efficiency (also referred to as destruction efficiency or DE) for catalytic oxidation is typically 90–95%. In some cases, the efficiency can be significantly lower, particularly when the waste stream being controlled contains halogenated VOCs.

Factors that affect the performance of a catalytic oxidation system include the following:

1. Operating temperature
2. Space velocity (the reciprocal of residence time)
3. VOC composition and concentration
4. Catalyst properties
5. Presence of poisons/inhibitors in the waste gas stream
6. Surface area of the catalyst

Poisons/inhibitors that can significantly degrade the catalyst activity include sulfur, chlorine, chloride salts, heavy metals (e.g., lead, arsenic), and particulate matter. The presence of any of these species in the waste gas stream would make catalytic incineration unfavorable.

If halogenated VOCs are present in the influent gas stream, then hydrochloric acid (HCl) may be produced in the catalytic oxidizer. HCl emissions are regulated and off-gas controls for HCl and other acid gases may be required.

Catalytic incineration can achieve overall hazardous air pollutant (HAP) destruction efficiencies of about 95% with SV in the range of 30,000–40,000 h^{-1} using precious metal catalysts, or 10,000–15,000 h^{-1} using base metal catalysts. However, greater catalyst volume and/or higher temperatures required for higher destruction efficiencies (i.e., 99%) my make catalytic incineration uneconomical. In this chapter, discussions on catalytic incineration design and operation will be based on HAP destruction efficiencies of 95%.

The influence of temperature and SV on the effectiveness of a catalytic oxidation system is shown in Figs. 3 and 4, respectively. The data shown in these figures are for a fluidized-bed catalytic oxidation system. The waste gas treated by this unit contained 10–200 ppmv (parts per million by volume) of mixed VOCs, including aliphatic, aromatic, and halogenated compounds. It can be clearly seen from Figs. 3 and 4 that DE is a function of chemical composition of a stream under a given SV and temperature. As the incineration temperature is increased with fixed SV, DEs increased linearly for most mixtures. A decrease of DEs was observed as the SV was increased.

In designing a catalytic oxidation system, temperature and SV are not the only variables that must be considered. The emission stream composition and catalyst type must be evaluated simultaneously because the type of catalyst chosen for a system places practical limits on the types of compound that can be treated. For example, waste gases containing chlorine and sulfur can deactivate noble metal catalysts such

Table 1
Effluent Characteristics for emission stream 2[a]

Company	Glaze Chemical Company	Plant contact	Mr. John Leake
Location (Street)	87 Octane Drive	Telephone No.	(999)555-5024
(City)	Somewhere	Agency contact	Mr. Efrem Johnson
(State, Zip)		No. of Emission Streams Under Review	7

A. Emission Stream Number/Plant Identification	#2/#1 Oven Exhaust		
B. HAP Emission Source	(a) metal coating oven	(b)	(c)
C. Source Classification	(a) process point	(b)	(c)
D. Emission Stream HAPs	(a) toluene	(b)	(c)
E. HAP Class and Form	(a) organic vapor	(b)	(c)
F. HAP Content (1,2,3)[b]	(a) 550 ppmv	(b)	(c)
G. HAP Vapor Pressure (1,2)	(a) 28.4 mm Hg at 77°F	(b)	(c)
H. HAP Solubility (1,2)	(a) insoluble in water	(b)	(c)
I. HAP Adsorptive Prop. (1,2)	(a) provided	(b)	(c)
J. HAP Molecular Weight (1,2)	(a) 92 lb/lb-mole	(b)	(c)
K. Moisture Content (1,2,3)	2% volume	P. Organic Content (1)[c]	none
L. Temperature (1,2,3)	120°F	Q. Heat/O_2 Content (1)	2.1 Btu/scf/20.6 vol %
M. Flow Rate (1,2,3)	20,000 scfm (max)	R. Particulate Content (3)	
N. Pressure (1,2)	atmospheric	S. Particle Mean Diam. (3)	
O. Halogen/Metals (1,2)	none/none	T. Drift Velocity/SO_3 (3)	
U. Applicable Regulation(s)	Assume 95% removal		
V. Required Control Level			
W. Selected Control Methods	Thermal incineration, catalytic incineration		

[a]The data presented are for an emission stream (single or combined streams) prior to entry into the selected control method(s). Use extra forms, if additional space is necessary (e.g., more than three HAPs), and note this need.

[b]The numbers in parentheses denote what data should be supplied depending on the data on lines C and E:

1 = organic vapor process emission
2 = inorganic vapor process emission
3 = particulate process emission

[c]Organic emission stream combustibles less HAP combustibles shown on lines D and F.

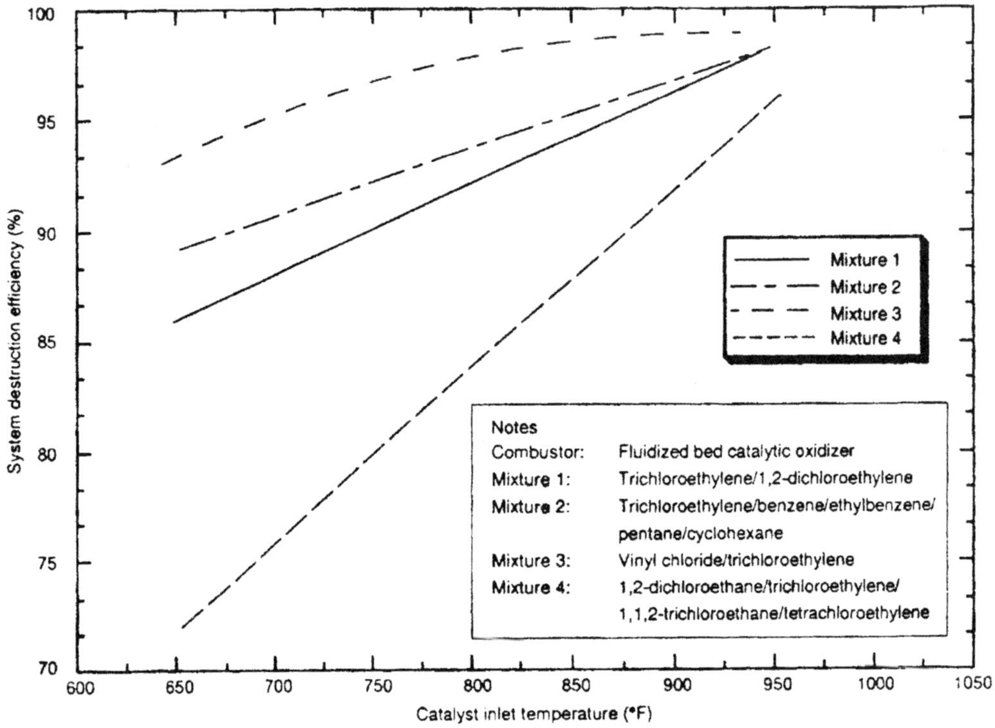

Fig. 3. Effect of temperature on destruction efficiency for catalytic oxidation at 10,500 h^{-1} space velocity. (From US EPA.)

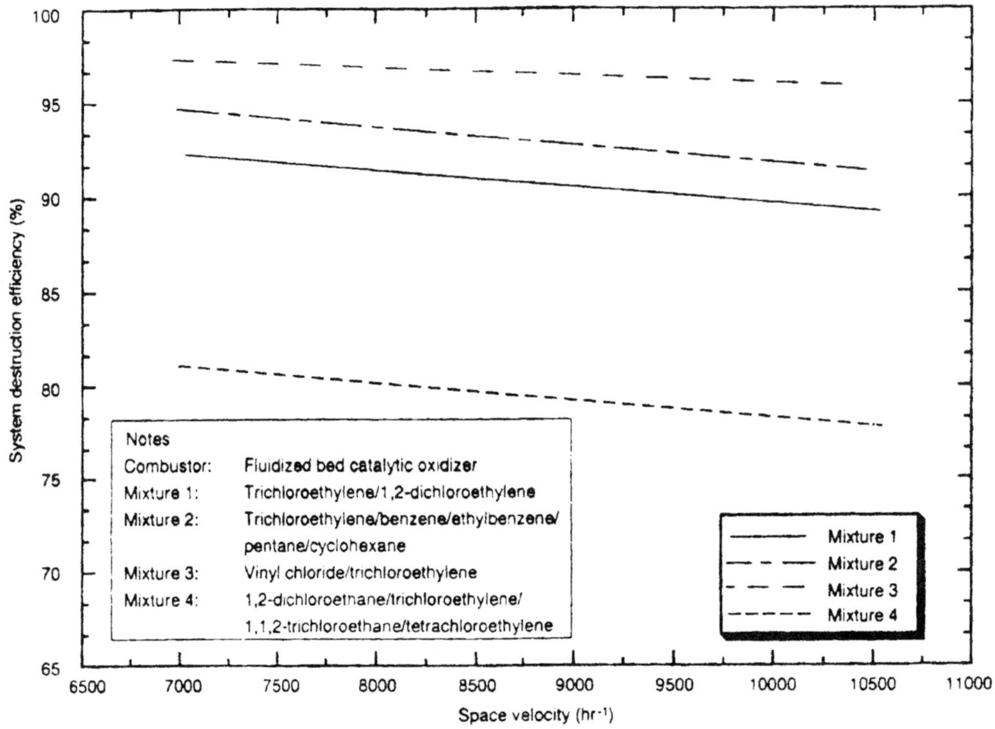

Fig. 4. Effect of space velocity on destruction efficiency for catalytic oxidation at 720°F.

Table 2
Destruction Efficiencies of Common VOC Contaminants in Fluidized-Bed Combustor

	Destruction efficiency at 650°F	Destruction efficiency at 950°F
	mean	mean
Cyclohexane	99	99+
Ethylbenzene	98	99+
Pentane	96	99+
Vinyl chloride	93	99
Dichloroethylene	85	98
Trichloroethylene	83	98
Dichloroethane	81	99
Trichloroethane	79	99
Tetrachloroethylene	52	92

Source: ref. 2.

as platinum. However, certain metal oxide catalysts can be used in the oxidation of chlorinated VOCs.

The control efficiencies of some common VOC contaminants are shown in Table 2 at two different operating temperatures for the fluidized-bed catalytic combustor discussed previously. As the data show, the destruction efficiency of a catalytic oxidation system can vary greatly for different contaminant types. The lowest DEs typically are seen for chlorinated compounds.

1.3. Applicability to Remediation Technologies

The applicability of catalytic oxidation depends primarily on emission stream composition. As described in Chapter 8 on thermal oxidation, waste gas composition will determine the auxiliary air and fuel requirements for combustion controls. These requirements in turn will have a strong influence on whether catalytic oxidation is an economical approach for controlling air emissions. The waste gas composition is also important in that for catalytic oxidation to be effective, the waste gas cannot contain catalyst poisons that would limit system performance.

Although catalytic oxidation has traditionally not been widely used to control halogenated hydrocarbons, improved catalysts make this application more feasible.

2. PRETREATMENT AND ENGINEERING CONSIDERATIONS

2.1. Air Dilution Requirements

In general, catalytic incineration (catalytic oxidation) is applied to dilute emission streams. If emission streams with high VOC concentrations are treated by catalytic incineration, they may generate enough heat upon combustion to deactivate the catalyst. Therefore, dilution of the emission stream with air is necessary to reduce the concentration of the VOCs. Dilution will be required if the heat content of an emission stream is greater than 10 Btu/scf for an air and VOC mixture and above 15 Btu/scf for inert and VOC mixture (2).

Typically, the concentration of flammable vapors in HAP emission streams containing air is limited to less than 25% of the lower explosive limit (LEL) (corresponding to a heat content of 176 Btu/lb or 13 Btu/scf) for safety requirements. To convert from Btu/lb to Btu/scf, multiply Btu/lb by the density of the emission stream at standard conditions (0.0739 lb/scf). Table 3 contains a list of LEL and upper explosive limits (UEL) for common organic compounds. In order to meet the safety requirement and to prevent damage to the catalyst bed, it is assumed in this handbook that catalytic incineration is directly applicable if the heat content of the emission stream (air and VOC) is less than or equal to 10 Btu/scf. For emission streams that are mixtures of inert gases and VOC (i.e., containing no oxygen), it is assumed that catalytic incineration is directly applicable if the heat content of the emission stream is less than or equal to 15 Btu/scf. Otherwise, dilution air will be required to reduce the heat content to levels below these cutoff values (i.e., 10 and 15 Btu/scf). For emission streams that cannot be characterized as air and VOC or inert gas and VOC mixtures, apply the more conservative 10 Btu/scf cutoff value for determining dilution air requirements. The dilution air requirements can be calculated from Eq. (1); note that the dilution air will change the emission stream parameters:

$$Q_d = [(h_e/h_d)-1]Q_e \tag{1}$$

where Q_d is the dilution air retirement (scfm), h_e is the heat content of the emission stream (Btu/scf), h_d is the desired heat content of the emission stream (Btu/scf), and Q_e is the emission stream flow rate (scfm).

2.2. *Design Variables*

Most catalytic incinerators currently sold are designed to achieve an efficiency of 95% (10). Table 4 presents suggested values and limits for the design variables of a fixed bed catalytic incinerator system to achieve 95% destruction efficiency. In selected instances, catalytic incinerators can achieve efficiencies on the order of 98–99%, but general guidelines for space velocities at these efficiencies could not be found. For specific applications, other temperatures and space velocities may be appropriate depending on the type of catalyst employed and the emission stream characteristics (i.e., composition and concentration). For example, the temperature of the flue gas leaving the catalyst bed may be lower than 1000°F for emission streams containing easily oxidized compounds and still achieve the desired destruction efficiency (4,5).

The destruction efficiency (DE) for a given compound may vary depending on whether the compound is the only VOC in the emission stream or part of a mixture of VOCs (5). The DE for a given compound in different VOC mixtures may also vary with mixture composition. Table 4 can be used to determine the ranges for temperature at the catalyst bed inlet (T_{ci}), temperature at the catalyst bed outlet (T_{co}), and SV for different catalysts based on the required DE.

When the catalyst bed inlet temperature is controlled at a proper level, the performance of a catalytic incinerator system depends greatly on both the temperature and pressure differential across the catalyst bed. The temperature differential or rise across the catalyst bed is the fundamental performance indicator for a catalytic incinerator system, as it indicates VOC oxidation efficiency. The pressure differential across the catalyst bed

Table 3
Flammability Characteristics of Combustible Organic Compounds in Air

Compounds	Molecular weight	LEL(% vol)	UEL(% vol)
Methane	16.04	5.0	15.0
Ethane	30.07	3.0	12.4
Propane	44.09	2.1	9.5
n-Butane	58.12	1.8	8.4
n-Pentane	72.15	1.4	7.8
n-Hexane	86.17	1.2	7.4
n-Heptane	100.20	1.05	6.7
n-Octane	114.28	0.95	3.2
n-Nonane	128.25	0.85	2.9
n-Decane	142.28	0.75	5.6
n-Undecane	156.30	0.68	
n-Dodecane	170.33	0.60	
n-Tridecane	184.36	0.55	
n-Tetradecane	208.38	0.50	
n-Pentadecane	212.41	0.46	
n-Hexadecane	226.44	0.43	
Ethylene	28.05	2.7	36.0
Propylene	42.08	2.4	11.0
Butene-1	56.10	1.7	9.7
cis-Butene-2	56.10	1.8	9.7
Isobutylene	56.10	1.8	9.6
3-Methyl-Butene-1	70.13	1.5	9.1
Propadlene	40.06	2.6	
1,3-Butadlene	54.09	2.0	12.0
Acetylene	2.5	100.0	
Methylacetylene	1.7		
Benzene	78.11	1.3	7.0
Toluene	92.13	1.2	7.1
Ethylbenzene	106.16	1.0	6.7
o-Xylene	106.16	1.1	6.4
m-Xylene	106.16	1.1	6.4
p-Xylene	106.16	1.1	6.6
Cumene	120.19	0.88	6.5
p-Cumene	134.21	0.85	6.5
Cyclopropane	42.08	2.4	10.4
Cyclobutane	56.10	1.8	
Cyclopentane	70.13	1.5	
Cyclohexane	84.16	1.3	7.8
Ethylcyclobutane	84.16	1.2	7.7
Cycloheptane	98.18	1.1	6.7
Methylcyclohexane	98.18	1.1	6.7
Ethylcyclopentane	98.18	1.1	6.7
Ethylcyclohexane	112.21	0.95	6.6

(continued)

Table 3 *(continued)*

Compounds	Molecular weight	LEL(% vol)	UEL(% vol)
Methyl alcohol	32.04	6.7	36.0
Ethyl alcohol	46.07	3.3	19.0
Ethyl alcohol	46.07	3.3	19.0
n-Propyl alcohol	60.09	2.2	14.0
n-Butyl alcohol	74.12	1.7	12.0
n-Amyl alcohol	88.15	1.2	10.0
n-Hexyl alcohol	102.17	1.2	7.9
Dimethyl ether	46.07	3.4	27.0
Diethyl ether	74.12	1.9	36.0
Ethyl propyl ether	88.15	1.7	9.0
Dilsopropyl ether	102.17	1.4	7.9
Acetaldehyde	44.05	4.0	36.0
Propionaldehyde	58.08	2.9	14.0
Acetone	58.08	2.6	13.0
Methyl ethyl ketone	72.10	1.9	10.0
Methyl propyl ketone	86.13	1.6	8.2
Diethyl ketone	86.13	1.6	
Methyl butyl ketone	100.16	1.4	8.0

Source: US EPA.

serves as an indication of the volume of catalyst present. The pressure drop decreases over time as bits of catalyst become entrained in the gas stream. To ensure proper performance of the system, it is recommended that both the temperature rise across the catalyst bed and the pressure drop across the catalyst bed be monitored continuously. Currently, most vendors routinely include continuous monitoring of these parameters as part of catalytic incinerator system package (8). However, some older units may not be so equipped; in this case, the reviewer should ensure that the incinerator is equipped with both continuous-monitoring systems.

In addition to catalyst loss, catalyst deactivation and blinding occur over time and limit performance. Catalyst deactivation is caused by the presence of materials that react with the catalyst bed. Blinding is caused by the accumulation of particulate matter on the catalyst bed surface decreasing the effective surface area of the catalyst (7,10). For these reasons, vendors recommend replacing the catalyst every 2–3 yr. Symptoms of catalyst loss include a decrease in pressure drop across the catalyst bed and a decrease in the temperature rise across the catalyst bed. Symptoms of deactivation and blinding include a decrease in the temperature rise across the catalyst bed. If a catalytic incinerator system exhibits these symptoms, the facility should take immediate action to correct these operational problems.

In a permit evaluation, determine if the reported values for T_{ci}, T_{co}, and SV are appropriate to achieve the DE by comparing applicant's values with the values in Table 4. However, it is important to keep in mind that the values given in Table 4 are approximate and a given permit may differ slightly from these values. The reported value for T_{ci} should equal or exceed 600°F in order to obtain an adequate initial reaction rate. To

Table 4
Catalytic Incinerator System Design Variables

Required destruction efficiency DE (%)	Temperature at the catalyst bed inlet[a] T_{ci}(°F)	Temperature at the catalyst bed outlet[b] T_{co}(°F)	Space velocity SV (h^{-1}) Base metal	Space velocity SV (h^{-1}) Precious metal
95	600	1,000–1,200	10,000–15,000[c]	30,000–40,000[c]
98–99	600	1,000–1,200	[d]	[d]

[a]Minimum temperature of combined gas stream (emission stream + supplementary fuel combustion products) entering the catalyst bed is designated as 600°F to ensure an adequate initial reaction rate.

[b]Minimum temperature of the flue gas leaving the catalyst bed is designated as 1000°F to ensure an adequate overall reaction rate to achieve the required destruction efficiency. Note that this is a conservative value; it is in general, a function of the HAP concentration (or heat content) and a temperature lower than 1000°F may be sufficient to achieve the required destruction level. The maximum temperature of flue gas leaving the catalyst bed is limited to 1200°F to prevent catalyst deactivation by overheating. However, base metal catalysts may degrade somewhat faster at these temperatures than precious metal catalysts.

[c]The space velocities given are designed to provide general guidance not definitive values. A given application may have space velocities that vary from these values. These values are quoted for monolithic catalysts. Pellet-type catalysts will typically have lower space velocities.

[d]In general, the design of catalytic incinerator systems in this efficiency range is done relative to specific process conditions.

Source: US EPA.

ensure that an adequate overall reaction rate can be achieved to give the desired DE without damaging the catalyst, check whether T_{co} falls in the interval 1000–1200°F. Then, check whether the reported value for SV is equal to or less than the value in Table 4. In some cases, it may be possible to achieve the desired DE at a lower temperature level. If a permit applicant uses numbers significantly different from those in Table 4, documentation indicating the rationale for this variance should accompany the application. In this case, the permit values should take precedence over those values given in Table 4. Otherwise, the applicant's design is considered unacceptable. In such a case, the reviewer may then wish to use the values in Table 4.

3. SUPPLEMENTARY FUEL REQUIREMENTS

Supplementary fuel is added to the catalytic incinerator system to provide the heat necessary to bring the emission stream up to the required catalytic oxidation temperature (T_{co}) for the desired level of DE. For a given T_{co}, the quantity of heat needed is provided by (1) the heat supplied from the combustion of supplementary fuel, (2) the sensible heat contained in the emission stream as it enters the catalytic incinerator system, and (3) the sensible heat gained by the emission stream through heat exchange with hot flue gases. If recuperative heat exchange is not practiced at a facility, then item (3) will be zero.

Because emission streams treated by catalytic incineration are dilute mixtures of VOC and air, they typically do not require additional combustion air. For purposes of this handbook, it is assumed that no additional combustion air is required if the emission stream oxygen content is greater than or equal to 20%.

Before calculating the supplementary heat requirements, the temperature of the flue gas leaving the catalyst bed (T_{co}) should be estimated to ensure that an adequate overall reaction rate can be achieved to give the desired DE without damaging the catalyst.

In other words, check whether T_{co} falls in the interval 1000–1200°F to ensure a high DE without catalyst damage. Equation (2) can be used to calculate T_{co}. This equation assumes a 50°F temperature increase for every 1 Btu/scf of heat content

$$T_{co} = T_{ci} + 50h_e \tag{2}$$

where h_e is the heat content of the emission stream (Btu/scf). In this expression, it is assumed that the heat content of the emission stream and the combined gas stream is the same. First, insert a value of 600°F for T_{ci} in Eq (2). Then, determine T_{co} if T_{co} is in the range of 1000–1200°F. If this is true, then the initial value of T_{ci} is satisfactory. If T_{co} is less than 1000°F, use Eq. (3) to determine an appropriate value for T_{ci} (above 600°F) and use this new value of T_{ci} in the calculation

$$T_{ci} = 1{,}000 - 50h_e \tag{3}$$

The value of T_{ci} obtained from Eq. (3) is then used in Eq. (4) for the determination of the auxiliary requirement.

Emission streams with high heat contents will be diluted based on the requirements discussed in Section 2.1. Therefore, values for T_{co} exceeding 1200°F should not occur.

For catalytic incinerators, a 50% efficient heat exchanger is assumed, whereas for thermal incineration, a 70% efficient exchanger is assumed. A 70% efficient heat exchanger for catalytic oxidation can result in excessive catalyst bed temperatures. Therefore, a 50% efficient heat exchanger is assumed for purposes of this discussion, although 70% efficient heat exchangers may be found on some streams.

To calculate supplementary heat requirements (based on natural gas as the fuel), the following simplified equation can be used for dilute emission streams that require no additional combustion air:

$$Q_f = \frac{D_e Q_e \left[Cp_{\text{air}}\left(1.1T_{\text{ci}} - T_{\text{he}} - 0.1T_r\right)\right]}{D_f\left[h_f - 1.1Cp_{\text{air}}\left(T_{\text{ci}} - T_r\right)\right]} \tag{4}$$

where Q_f is the fuel gas flow rate (scfm), D_e is the density of the emission stream (lb/ft^3[typically 0.0739 lb/ft^3]), D_f is the density of the fuel gas (0.0408 lb/ft^3 for methane at 77°F), Q_e is the emission stream flow rate (scfm), Cp_{air} is the average specific heat of air over a given temperature interval (Btu/lb-°F) (see Table 5), T_{ci} is the temperature of the combined gas stream entering the catalyst bed (°F), T_r is the reference temperature (77°F), T_{he} is the emission stream temperature after heat recovery (°F), and, h_f is the lower heating value of natural gas (21,600 Btu/lb).

Note that for the case of no heat recovery, $T_{he} = T_e$. The factor 1.1 is included in Eq. (4) to account for an estimated heat loss of 10% in the incinerator. The maximum emission flow rate should be used in Eq. (4) for determining supplementary heat requirements and, hence, will lead to a conservative design. In contrast to thermal incineration, there is no minimum supplementary heat requirement specified for catalytic incineration because no fuel is needed for flame stabilization. Depending on the HAP concentration,

Table 5
Average Specific Heats of Vapors

Temperature (°F)	Average Specific Heat, C_p (Btu/scf-°F)									
	Air	H_2O	O_2	N_2	CO	CO_2	H_2	CH_4	C_2H_4	C_2H_6
77	0.0180	0.0207	0.0181	0.0180	0.0180	0.0230	0.0178	0.0221	0.0270	0.0326
212	0.0180	0.0209	0.0183	0.0180	0.0180	0.0239	0.0179	0.0232	0.0293	0.0356
392	0.0181	0.0211	0.0186	0.0181	0.0181	0.0251	0.0180	0.0249	0.0324	0.0395
572	0.0183	0.0212	0.0188	0.0182	0.0183	0.0261	0.0180	0.0266	0.0353	0.0432
752	0.0185	0.0217	0.0191	0.0183	0.0184	0.0270	0.0180	0.0283	0.0379	0.0468
932	0.0187	0.0221	0.0194	0.0185	0.0186	0.0278	0.0181	0.0301	0.0403	0.0501
1112	0.0189	0.0224	0.0197	0.0187	0.0188	0.0286	0.0181	0.0317	0.0425	0.0532
1292	0.0191	0.0228	0.0199	0.0189	0.0190	0.0292	0.0182	0.0333	0.0445	0.0560
1472	0.0192	0.0232	0.0201	0.0190	0.0192	0.0298	0.0182	0.0348	0.0464	0.0587
1652	0.0194	0.0235	0.0203	0.0192	0.0194	0.0303	0.0183	0.0363	0.0481	0.0612
1832	0.0196	0.0239	0.0205	0.0194	0.0196	0.0308	0.0184	0.0376	0.0497	0.0635
2012	0.0198	0.0243	0.0207	0.0196	0.0198	0.0313	0.0185	0.0389	0.0512	0.0656
2192	0.0199	0.0246	0.0208	0.0197	0.0199	0.0317	0.0186	0.0400	0.0525	0.0676

Note: Average for the temperature interval 77°F and the specified temperature.
Based on 70°F and 1 atm
To convert to Btu/lb-°F basis, multiply by 392 and divide by the molecular weight of the compound. To convert to Btu/lb-mol°F, multiply by 392.
Source: ref. 4.

emission stream temperature, and level of heat recovery, supplementary heat requirements may be zero when heat recovery is employed.

Calculate T_{he} using the following expression if the value for T_{he} is not specified:

$$T_{he} = (HR/100)T_{co} + [1 - (HR/100)]T_e \quad (5)$$

where HR is the heat recovery in the heat exchanger (%) and, T_e is the emission stream temperature (°F).

4. ENGINEERING DESIGN AND OPERATION

4.1. *Flue Gas Flow Rates*

To calculate the quantity of catalyst required and cost of a catalytic incinerator, the flow rates of the combined gas stream entering and leaving the catalyst bed have to be determined. Equation (6) can be used to determine the inlet gas flow rate:

$$Q_{com} = Q_e + Q_f + Q_d \quad (6)$$

where Q_{com} is the flow rate of the combined gas stream entering the catalytic bed (scfm), Q_e is the flow rate of the emission stream (scfm), Q_f is the natural gas flow rate (scfm), and, Q_d is the dilution air requirement (scfm).

The flue gas volume change across the catalyst bed as a result of the catalytic oxidation of the HAP in the mixed gas stream is usually small, especially when dilute emission streams are treated. Therefore, the flow rate of the combined gas stream leaving the catalyst bed is approximately equal to the flow rate of the flue gas entering the catalyst bed at standard conditions:

$$Q_{fg} = Q_{com} \quad (7)$$

where Q_{fg} is the flow rate of the flue gas leaving the catalyst bed (scfm).

When calculating costs, a minimum Q_{fg} of 2000 scfm is typically used in catalytic incinerator design. Therefore, if Q_{fg} is less than 2000 scfm, assume Q_{fg} equals to 2000 scfm in cost calculations.

In some instances, operating costs are determined based on the actual flue gas flow rates. In these cases, the following equation can be used to convert standard condition flow rate (scfm) to actual flow rate (acfm):

$$Q_{fg,a} = Q_{fg}[(T_{co} + 460)/537] \quad (8)$$

where Q_{fga} is the flue gas flow rate at actual conditions (acfm).

4.2. *Catalyst Bed Requirement*

The total volume of catalyst required for a given DE is determined from the design space velocity as follows:

$$V_{bed} = \frac{60Q_{com}}{SV} \quad (9)$$

where V_{bed} is the volume of thecatalyst bed required (ft^3).

Table 6
Typical Pressure Drops for Catalytic Incinerators

Equipment type	Heat recovery HR(%)	Pressure drop *P* (in. H_2O)
Catalytic incinerator (fixed-bed)	0	6
Heat exchanger	35	4
Heat exchanger	50	8
Heat exchanger	70	15

Note: The pressure drop is calculated as the sum of the incinerator and heat-exchanger pressure drops.

Table 7
Comparison of Calculated Values and Values Supplied by the Permit Applicant for Catalytic Incinerator

	Calculated value (example case)[a]	Reported value
Continuous monitoring of temperature rise and pressure drop across catalyst bed	Yes	—
Supplementary fuel flow rate, Q_f	179 scfm	—
Dilution airflow rate, Q_d	0	—
Combined gas stream flow rate, Q_{com}	20,000 scfm	—
Catalyst bed volume, V_{bed}	40 ft^3	—

[a]Based on emission steam 2.

4.3. System Pressure Drop

The total pressure drop for a catalytic oxidizer depends on the type of equipment employed in the system as well as other design considerations. The total pressure drop required across a catalytic incineration system determines the waste gas fan size and horsepower requirements, which, in turn, determine the fan capital cost and electricity consumption.

An accurate estimate of system pressure drop would require complex calculations. A preliminary estimate can be made using the approximate values listed in Table 6. The system pressure drop is the sum of the pressure drops across the oxidizer and the heat exchanger.

The pressure drop can then be used to estimate the power requirement for the waste gas fan using the empirical relationship

$$\text{Power} = 1.17 \times 10^{-4} V \frac{\Delta P}{\varepsilon} \tag{10}$$

where Power is the fan power requirement (kWh), V is the waste gas flow rate (scfm), ΔP is the system pressure drop (inches of water column), and ε is the combined motor fan efficiency (dimensionless) (approx 60%).

5. MANAGEMENT

5.1. Evaluation of Permit Application

Table 7 can be used to compare the results from the calculations and the values reported by the permit applicant. The calculated values in Table 7 are based on the example case presented in Table 1. If the calculated values agree with the reported values, then the design and operation of the proposed catalytic incinerator system may be considered appropriate based on the assumptions used in this handbook.

In the case of a permit review for a catalytic incinerator, the following data at standard conditions (77°F, 1 atm) should be supplied by the applicant. The calculations in this chapter will then be used to check the applicant's values.

1. Reported destruction efficiency, $DE_{reported}$ (%)
2. Temperature of the emission stream entering the incinerator (oxidizer):
   If no heat recovery, T_e (°F)
   If emission stream preheated, T_{he} (°F)
3. Temperature of flue gas leaving the catalyst bed, T_{co} (°F)
4. Temperature of combined gas stream (emission stream plus supplementary fuel combustion products) entering the catalyst bed, T_{ci} (°F)
5. Space velocity through catalyst bed, SV (h^{-1})
6. Supplementary fuel gas flow rate, Q_f (scfm)
7. Flow rate of combined gas stream entering the catalyst bed, Q_{com} (scfm) (Note that if no supplementary fuel is used [i.e., $Q_f = 0$], the value of Q_{com} will equal the emission stream flow rate)
8. Dilution airflow rate, Q_d (scfm)
9. Catalyst bed requirement, V_{bed} (ft^3)
10. Fuel heating value, h_f (Btu/lb)

5.2. Operation and Manpower Requirements

The total annual cost (TAC) of a catalytic incinerator consists of direct and indirect annual costs. Direct annual costs include fuel, electricity, catalyst replacement operating and supervisory labor, and maintenance labor and materials.

Fuel usage is calculated in Section 3. Once the fuel gas flow rate is calculated, multiply it by 60 to covert flow rate from standard cubic foot per minute (scfm) to standard cubic foot per hour (scfh). The annual fuel usage can be calculated by multiplying the hourly fuel gas flow rate by the annual operating hours. Then, simply multiply the annual fuel usage by the cost of fuel to obtain this annual cost.

Electricity costs are primarily associated with the fan needed to move the gas through the incinerator. Use Eq. (11) to estimate the power requirements for a fan assuming a combined motor fan efficiency of 65% and a fluid specific gravity of 1.0:

$$F_p = 1.81\times10^{-4}(Q_{fg,a})(P)(\text{HRS}) \tag{11}$$

where F_p is the power needed for the fan (kWh/yr), $Q_{fg,a}$ is the total emission stream flow rate (acfm), P is the system pressure drop (in. H_2O) (from Table 6), and HRS is the operating hours per year (h/yr).

In general, catalyst replacement costs are highly variable and depend on the nature of the catalyst, the amount of poisons and particulates in the emission stream, the temperature history of the catalyst, and the design of the unit. Given that these parameters are so variable, it is not possible to accurately predict the catalyst replacement costs for a given application. For purposes of this handbook, it is assumed that the catalyst has a life-span of 2 yr. Based on this assumption, the catalyst replacement cost can be determined by multiplying the catalyst volume determined in Section 4.2 by the appropriate capital recovery factor (assuming a 2-yr life and 10% interest rate [i.e., CRF = 0.5762]) and the unit cost of catalyst replacement. The catalyst replacement cost can be estimated as $650/ft^3 for base metal oxide catalysts and $3000 for noble metal catalysts in 1990 (6).

The capital cost of a catalytic incinerator is estimated as the sum of the equipment cost (EC) and the installation cost. The equipment cost is primarily a function of the total emission stream flow rate and the heat-exchanger efficiency as well as the cost of auxiliary equipment.

After obtaining equipment costs, the next step in the cost calculation is to obtain the purchased equipment cost (PEC). The PEC is calculated as the sum of EC (incinerator and auxiliary equipment) and the cost of instrumentation, freight, and taxes. Appropriate factors can be applied to estimate these costs. After obtaining the PEC, the total capital cost (TCC) is estimated using the factors presented elsewhere (6).

Operating labor requirements are estimated as 0.5 h per 8-h shift. The operator labor wage rate is provided elsewhere (6). Supervisory cost is typically estimated as 15% of operator labor costs.

Maintenance labor requirements are estimated as 0.5 h per 8-h shift with a slightly higher labor rate to reflect increased skill levels. Maintenance materials are estimated as 100% of maintenance labor.

Indirect annual costs include the capital recovery cost, overhead, property taxes, insurance, and administrative charges. The capital recovery cost is based on an estimated 10-yr equipment life and subtracts out the initial catalyst cost, whereas overhead, property taxes, insurance, and administrative costs are percentages of the total capital cost.

5.3. *Decision for Rebuilding, Purchasing New or Used Incinerators*

Examples on how to make decisions on rebuilding, purchasing new incinerators or purchasing used incinerators have been presented by Gallo et al. (15), Moretti and Mukhopadhyay (16), Cooley (17), and Arrest and Satterfield (18). Many catalytic oxidizers are commercially available (19–21). Assistance can be a obtained from the US EPA (23, 24).

5.4. *Environmental Liabilities and Risk-Based Corrective Action*

Traditional approaches to environmental cleanups have been challenged by rising remediation and treatment costs, stricter regulatory compliance requirements, greater demands for protection of the public and the environment, and, of course, mounting business concerns, such as future legal and financial liabilities. Certain stakeholders seek “absolute” clean or zero concentrations of forcign chemicals in the environment. The goal of “clean” has become increasingly more elusive, as newer, improved instrumentation and analytical methods continually lower the detection limits.

Not only do the investigation and cleanup costs dramatically increase while the plant managers are chasing this ever diminishing target “clean” concentration, but the liability with respect to future "clean" standards is ever present, as the perception that the managers must attain zero concentration is still held by some stakeholders.

This is one of the areas where risk management and environmental compliance meet. A structured, planned approach to managing incidents is one of the most successful ways to control liabilities. The risk management process provides a framework for managing environmental liabilities. It is composed of the identification, quantification, and ultimate treatment of loss exposures. Two of the primary objectives of risk management are to control losses and to minimize the financial impacts resulting from the loss. There are several strategies that can be used singularly or in combination that enable the risk manager to accomplish these objectives. Indelicato has outlined these strategies as they pertain to the control of environmental liabilities or losses (22).

Many environmental concerns have been managed on a postloss basis using some form of health-based risk assessment or risk-based corrective action (RBCA). This approach provides strategies for managing environmental risk ranging from total cleanup to background level (or below a specified detection limit) to minor cleanup of "hot" spots with 30 yr of monitoring for any changes to the environment as a result of some low level of the contaminant being left in place.

The RBCA strategy provides for a cost-effective solution to minimizing the impacts to the public and the environment as a result of the contamination. Follow-up legal documentation from the state regulatory agency allows for a degree of certainty that the environmental liability is controlled and that business can proceed in a risk-managed manner.

6. DESIGN EXAMPLES

Example 1

Develop a calculation sheet for catalytic incineration that can be used for documentation of HAP emission stream characteristics and the catalytic incinerator system variables and for a permit view.

Solution

1. HAP emission stream characteristics: (Table 8) (see Note 1 below)
   1. Maximum flow rate, Q_e = ______________ scfm
   2. Temperature, T_e = ______________ °F
   3. Heat content, h_e = ______________ Btu/lb
   4. Oxygen content, O_2 = ______________ % (see Note 2 below)
   5. Required destruction efficiency, DE ______________ %
2. For a permit review, the following data should be supplied by the applicant: Catalytic incinerator system variables at standard conditions (77°F, 1 atm):
   1. Reported destruction efficiency, $DE_{reported}$ = ______________ %
   2. Temperature of emission stream entering the incinerator (oxidizer),
      T_e = ______________ °F (if no heat recovery)
      T_{he} = ______________ °F (if emission stream is preheated)
   3. Temperature of flue gas leaving the catalytic bed, T_{co} = ______________ °F
   4. Temperature of combined gas stream (emission stream plus supplementary fuel combustion products) entering the catalyst bed, T_{ci} = ______________ °F (see Note 3 below)
   5. Space velocity, SV = ______________ h^{-1}
   6. Supplementary fuel gas flow rate, Q_f = ______________ scfm

Table 8
HAP Emission stream Data Form[a]

Company ______	Plant contact ______
Location (Street) ______	Telephone No. ______
(City) ______	Agency contact ______
(State, Zip) ______	No. of Emission Streams Under Review ______

A.	Emission Stream Number/Plant Identification ______			
B.	HAP Emission Source	(a)______	(b)______	(c)______
C.	Source Classification	(a)______	(b)______	(c)______
D.	Emission Stream HAPs	(a ______	(b)______	(c)______
E.	HAP Class and Form	(a)______	(b)______	(c)______
F.	HAP Content (1,2,3)[b]	(a)______	(b)______	(c)______
G.	HAP Vapor Pressure (1,2)	(a)______	(b)______	(c)______
H.	HAP Solubility (1,2)	(a)______	(b)______	(c)______
I.	HAP Adsorptive Prop. (1,2)	(a)______	(b)______	(c)______
J.	HAP Molecular Weight (1,2)	(a)______	(b)______	(c)______
K.	Moisture Content (1,2,3)	______	P. Organic Content (1)[c]	______
L.	Temperature (1,2,3)	______	Q. Heat/O_2 Content (1)	______
M.	Flow Rate (1,2,3)	______	R. Particulate Content (3)	______
N.	Pressure (1,2)	______	S. Particle Mean Diam. (3)	______
O.	Halogen/Metals (1,2)	______	T. Drift Velocity/SO_3 (3)	______
U.	Applicable Regulation(s)	______		
V.	Required Control Level	______		
W.	Selected Control Methods	______		

[a] The data presented are for an emission stream (single or combined streams) prior to entry into the selected control method(s). Use extra forms, if additional space is necessary (e.g., more than three HAPs), and note this need.

[b] The numbers in parentheses denote what data should be supplied depending on the data on lines C and E:

1 = organic vapor process emission
2 = inorganic vapor process emission
3 = particulate process emission

[c] Organic emission stream combustibles less HAP combustibles shown on lines D and F.

7. Flow rate of combined gas stream entering the catalyst bed, Q_{com} = ______________ scfm
8. Dilution airflow rate, Q_d = ______________ scfm
9. Catalyst bed requirement, V_{bed} = ______________ ft^3
10. Fuel heating value, h_f = ______________ Btu/lb

The following should be noted:

1. If dilution air is added to the emission stream upon exit from the process, the HAP emission stream characteristics data required are the resulting characteristics after dilution.
2. The oxygen content listed above depends on the oxygen content of the organic compounds (fixed oxygen) and the free oxygen in the emission stream. Because emission streams treated by catalytic incineration are generally a dilute VOC and air mixture, the fixed oxygen in the organic compounds can be neglected.
3. If no supplementary fuel is used, the value for the temperature of combined gas stream entering the catalyst bed will be the same as that for the emission stream.

Example 2

Summarize the air dilution requirements for possible pretreatment of the emission stream to a catalytic incinerator (oxidizer).

Solution

For emission stream treatment by catalytic incineration, dilution air typically will not be required. However, if the emission stream heat content is greater than 135 Btu/lb or 10 Btu/scf for air plus VOC mixture or if the emission stream heat content is greater than 203 Btu/lb or 15 Btu/scf of inert gas plus VOC mixture, dilution air is necessary. For an emission stream that cannot be characterized as air plus VOC or inert gas plus VOC mixtures, assume that dilution air will be required if the heat content is greater than 12 Btu/scf. In such cases, refer to

$$Q_d = [(h_e/h_d) - 1]Q_e \tag{1}$$

Q_d = ______________ scfm

Example 3

A HAP emission stream documented in Table 1 is to be treated by a catalyst incinerator. A few important influent flue gas data are as follows:

Maximum flow rate, Q_e = 20,000 scfm
Temperature, T_e = 120°F
Heat content, h_e = 24.8 Btu/lb or 2.1 Btu/scf
Oxygen content, O_2 = 20.6%
Based on the control requirements for the emission stream, the required destruction efficiency DE = 95%.

Determine the air dilution requirement.

Solution

The air dilution requirements for possible pretreatment of the specific emission stream to a catalytic incinerator are presented in Example 2.
Because the heat content of the emission stream (h_e) is 2.1 Btu/scf (or 24.8 Btu/lb), no dilution is necessary.

Example 4

Outline the step-by-step procedure for the determination of catalytic incineration design variables, destruction efficiency, and related operational parameters for a permit review.

Solution

1. Based on the required DE, specify the appropriate ranges for T_{ci} and T_{co} and select the value for SV from Table 4.
   T_{ci} (minimum) = 600°F
   T_{co} (minimum) = 1000°F
   T_{co} (maximum) = 1200°F
   SV = ______________ h^{-1}

2. In a permit review, determine if the reported value for T_{ci}, T_{co}, and SV are appropriate to achieve the required DE. Compare the applicant's values with the values in Table 4 and check if the following hold:
   T_{ci} (applicant) ≥ 600°F
   1200°F ≥ T_{co} (applicant) ≥ 1000°F
   SV (applicant) ≤ SV (Table 4)

If the reported values are appropriate, proceed with the calculations. Otherwise, the applicant's design is considered unacceptable. The reviewer may then wish to use the values in Table 4.

Example 5

A catalytic incinerator (oxidizer) is to be used for treating a HAP emission stream documented in Table 1. Determine the following:

1. Destruction efficiency, DE
2. Temperature at the catalyst bed inlet, T_{ci}
3. Temperature at the catalyst bed outlet, T_{co}
4. Space velocity SV (assuming a precious metal catalyst is to be used)

Solution

The required destruction efficiency is 95%; therefore, the following hold:

T_{ci} (minimum) = 600°F
T_{co} (minimum) = 1000°F
T_{co} (maximum) = 1200°F
SV = 30000 h^{-1} (assume precious metal catalyst)

Example 6

Outline the step-by-step procedure recommended by the US EPA for the determination of the supplementary fuel requirements of a catalytic incinerator (oxidizer) treating dilute emission streams that require no additional combustion air.

Solution

1. Use the following equation to determine if T_{ci} = 600°F from Table 4 is sufficient to ensure an adequate overall reaction rate without damaging the catalyst (i.e., check if T_{co} falls in the interval 1000–1200°F):

$$T_{co} = 600 + 50h_e \tag{2}$$
$$T_{co} = \_\_\_\_\_\_\_\_\_\_\_ \ ^{\circ}F$$

If T_{co} falls in the interval of 1000 –1200°F, proceed with the calculation. If T_{co} is less than 1000°F, assume T_{co} is equal to 1000°F and use the following equation to determine an appropriate value for T_{ci}; and then proceed with the calculation:

$$T_{ci} = 1000 - 50h_e \tag{3}$$
$$T_{ci} = \_\_\_\_\_\_\_\_\_\_\_ \ ^{\circ}F$$

(Note: If T_{co} is greater than 1200°F, a decline in catalyst activity may occur as a result of exposure to high temperatures.)

2. Use the following equation to determine supplementary fuel requirements:

$$Q_f = \frac{D_e Q_e \left[Cp_{air} \left(1.1T_{ci} - T_{he} - 0.1T_r \right) \right]}{D_f \left[h_f - 1.1Cp_{air} \left(T_{ci} - T_r \right) \right]} \tag{4}$$

The values for the variables in this equation can be determined as follows:

Q_e	Input data
D_e	0.0739 lb/ft^3
D_f	0.0408 lb/ft^3
h_f	Assume a value of 21,600 Btu/lb (for natural gas) if no other information available
$C_{p_{air}}$	See Table 5 for values of $C_{p_{air}}$ at various temperatures.
T_{ci}	Obtain value from part "a" above or from permit applicant.
T_{he}	For the no heat recovery case, $T_{he} = T_e$. For the heat recovery case, use the following equation if the value for T_{he} is not specified:

$$T_{he} = (\text{HR}/10)T_{co} + [1 - (\text{HR}/100)]T_e \tag{5}$$

Assume a value of 50% for HR if no other information is available.

T_r 77°F

$Q_f = \_\_\_\_\_\_\_\_\_\_\_$ scfm

Example 7

Use the step-by-step procedure outlined in Example 6 to determine the flue gas flow rate Q_f in scfm when a catalytic incinerator is selected to treat the HAP emission stream documented in Table 1.

Solution

Because the emission stream is dilute (h_e = 2.1 Btu/scf) and has an oxygen concentration greater than 20%, Eqs. (2)–(4) are applicable.

1. Determine if T_{co} falls in the range 1000–1200°F
   T_{ci} = 600°F
   h_e = 2.1 Btu/scf (input data)
   $T_{co} = 600 + (50 \times 2.1) = 705$°F

   Because T_{co} is less than 1000°F, use Eq. (3) to calculate a required value for T_{ci}:
   $T_{ci} = 1000 - (50 \times 2.1) = 895$°F

Note that this inlet temperature results in $T_{co} = 1000°F$

2. Determine Q_f (assume recuperative heat recovery will be employed):

$Q_e = 20{,}000$ scfm
$T_r = 77°F$
$T_{he} = 560°F$ (based on HR of 50 percent)
$C_{p_{air}} = 0.253$ Btu/lb-°F
$D_e = 0.0739$ lb/scf
$h_f = 21{,}600$ Btu/lb

$$Q_f = \frac{0.0739(20,000)\left[0.253(984 - 560 - 7.7)\right]}{0.0408\left[21,600 - \left(1.1(0.253)(895 - 77)\right)\right]}$$

$Q_f = 179$ scfm

Example 8

It is assumed that dilute emission streams that require no additional combustion air will be treated by a catalyst incinerator. Outline a step-by-step procedure for determination of (1) the flow rate of combined gas stream entering the catalyst bed, (2) the flow rate of flue gas leaving the catalyst bed, and (3) the catalyst bed volume required for the treatment.

Solution

1. Determination of the flow rate of combined gas stream entering the catalyst bed. For dilute emission streams that require no additional combustion air, use

$$Q_{com} = Q_e + Q_f + Q_d \tag{6}$$
$$Q_{com} = \underline{\qquad\qquad\qquad} \text{ scfm}$$

2. Determination of the flow rate of flue gas leaving the catalyst bed
   a. Use the result from the previous calculaton:

$$Q_{fg} = Q_{com}$$
$$Q_{fg} = \underline{\qquad\qquad\qquad} \text{ scfm}$$

   If Q_{fg} is less than 2000 scfm, define Q_{fg} as 2000 scfm.

   b. Use Eq. (8) to calculate $Q_{fg,a}$.

$$Q_{fg,a} = Q_{fg}[(T_{co} + 460)/537] \tag{8}$$
$$Q_{fg,a} = \underline{\qquad\qquad\qquad} \text{ acfm}$$

3. Determination of the catalyst bed requirement. Use Eq. (9) to estimate the catalyst bed volume.

$$V_{bed} = 60\ Q_{com}/SV \tag{9}$$
$$V_{bed} = \underline{\qquad\qquad\qquad} \text{ ft}^3$$

Example 9

Use the step-by-step procedure outlined in Example 8 to determine the following when a catalytic incinerator treats the HAP emission stream documented in Table 1:

1. The flow rate of combined gas stream entering the catalyst bed, Q_{com}

2. The flow rate of flue gas leaving the catalyst bed, $Q_{fg,a}$
3. The catalyst bed volume, V_{bed}

Solution

1. Use Eq. (6) to determine Q_{com}
   $Q_e = 20{,}000$ scfm
   $Q_d = 0$ (because $h_e < 15$ Btu/scf)
   $Q_{com} = 20{,}000 + 179 + 0$
   $Q_{com} = 20{,}200$ scfm (rounded to three places)
2. Use Eq. (8) to determine $Q_{fg,a}$
   $Q_{fg} = Q_{com} = 20{,}200$ scfm
   $T_{co} = 1000°F$
   $Q_{fg,a} = 20{,}200\ [(1000 + 460)/537]$
   $Q_{fg,a} = 54{,}900$ acfm
3. Use Eq. (9) to determine V_{bed}
   $Q_{com} = 20{,}200$ scfm
   $SV = 30{,}000\ h^{-1}$ (Table 4)
   $V_{bed} = 60 \times 20{,}200/30{,}000$
   $V_{bed} = 40\ ft^3$

Example 10

Describe the methodology used for evaluation of a catalyst incinerator's permit application.

Solution

Compare the calculated values supplied by the applicant using the following table. If the calculated value for h_f, Q_e, Q_{com}, and V_{bed} differ from the applicant's values, the differences may be the result of the assumptions involved in the calculations. Discuss the details of the design and operation of the system with the applicant.

If the calculated and reported values are not different, then the design and operation of the system can be considered appropriate on the assumptions employed in this handbook.

Comparison of Calculated Values and Values Supplied by the Permit Applicant for Catalytic Incinerator

	Calculated value	Reporte value
Continuous monitoring of combustion temperature rise and pressure drop across catalyst bed	—	—
Supplementary fuel flow rate, Q_f	—	—
Dilution airflow rate, Q_d	—	—
Flue gas stream flow rate, Q_{com}	—	—
Catalyst bed volume, V_{bed}	—	—

NOMENCLATURE

ε	Combined motor fan efficiency (dimensionless) (approx 60%)
$C_{p_{air}}$	Mean specific heat of air (Btu/lb-°F)

D_e	Density of emission stream (lb/ft^3)
DE	destruction efficiency (%)
D_f	Density of fuel gas (lb/ft^3)
EC	Equipment cost ($)
F_p	Fan power requirement (kWh/yr)
h_d	Emission stream desired heat content (Btu/scf)
h_e	Emission stream heat content (Btu/lb or Btu/scf)
h_f	Supplementary fuel heating value (Btu/lb)
HR	Heat recovery in heat exchanger (%)
HRS	Operating hours per year (h/yr)
ΔP	Pressure drop (in. H_2O)
P	System pressure drop (in. H_2O)
Power	Fan power requirement (kW-h)
PEC	Purchased equipment cost ($)
Q_{com}	Flow rate of combined gas stream (scfm)
Q_d	Dilution air required (scfm)
Q_e	Emission stream flow rate (scfm)
Q_f	Supplementary fuel gas flow rate (scfm)
Q_{fg}	Flue gas flow rate (scfm)
$Q_{fg,a}$	Actual flue gas flow rate (acfm)
SV	Space velocity through catalyst bed (h^{-1})
TCC	Total capital cost ($)
T_{ci}	Temperature of gas stream entering catalyst bed (°F)
T_{co}	Temperature of flue gas leaving catalyst bed (°F)
T_e	Emission stream temperature (°F)
T_{he}	Temperature of emission stream exiting heat exchanger (°F)
t_r	Residence time (s)
T_r	Reference temperature (77°F)
V	Waste gas flow rate (scfm)
V_{bed}	Volume of required catalyst bed (ft^3)

REFERENCES

1. US EPA, *Survey of Control Technologies for Low Concentration Organic Vapor Gas Streams* EPA-456/R-95-003 US Environmental Protection Agency, Washington, DC, 1995.
2. US EPA, *Handbook: Control Techniques for Hazardous Air Pollutants*. EPA 625/9-86-014 (NTIS PB91-228809) US Environmental Protection Agency, Washington, DC, 1986.
3. US EPA, *Polymer Manufacturing Industry—Background Information for Proposed Standard.* EPA 450/3-85-019a (NITS PB88-114996). US Environmental Protection Agency, Washington, DC, 1985.
4. US EPA *Afterburner Systems Study*. EPA-R2-72-062 (NTIS PB212560). US Environmental Protection Agency, Washington, DC, 1972.
5. US EPA, *Parametric Evaluation of VOC/HAP Destruction Via Catalytic Incineration.* EPA-600/2-85-041 (NTIS PB85-191187). US Environmental Protection Agency, Washington, DC, 1985.

6. US EPA, *OAQPS Control Cost Manual*, 4th ed. EPA-450/3-90-006 (NTIS PB90-169954) US Environmental Protection Agency, Washington, DC, 1990.
7. US EPA, *VOC Control Effectiveness*. EPA Contract No. 68-02-4285, WA 1/022. US Environmental Protection Agency, Washington, DC, 1989.
8. US EPA, *Evaluation of Continuous Compliance Monitoring Requirements for VOC Add-on Control Equipment*. US EPA Contract No. 68-02-4464, WA 60, for Vishnu Katari, US EPA, SSCD, 1989.
9. US EPA, *Soil Vapor Extraction VOC Control Technology Assessment*. EPA-450/4-89-017 (NTIS PB90-216995). US Environmental Protection Agency, Washington, DC, 1989.
10. Telecon. Sink, Michael, PES, with Yarrington, Robert. Englehard Corp., Edison, NJ (1990) *Space Velocities for Catalysts, and Incinerator Efficiency*.
11. L. K. Wang, J. V. Krouzek, and U. Kounitson, *Case Studies of Cleaner Production and Site Remediation*. UNIDO Training Manual No. DTT-5-4-95. United Nations Industrial Development Organization, Vienna, 1995.
12. L. K. Wang, M. H. S. Wang, and P. Wang. *Management of Hazardous Substances at Industrial Sites*. UNIDO Registry No. DTT-4-4-95. United Nations Industrial Development Organization, Vienna, 1995.
13. US EPA, *Control of Air Emissions from Superfund Sites*. EPA/625/R-92/012. US Environmental Protection Agency, Washington, DC, 1992.
14. US EPA, *Organic Air Emissions from Waste Management Facilities*. EPA/625/R-92/003. US Environmental Protection Agency, Washington, DC, 1992.
15. J. Gallo, R. Schwartz, and J. Cash, *Environ. Protect.* 74–80 (2001).
16. E. C. Moretti, and N. Mukhopadhyay, *Chem. Eng. Prog.*, **89 (7)**, 20–25 (1993).
17. R. Cooley, *Environ. Protect.* **13**, 12–13 (2002).
18. H. C. Arrest and C. Satterfield. *Environ. Protect.*, **13**, 28–29 (2002).
19. Crawford Industrial Group, *Environ. Protect.*, **14 (1)**, 43 (2003).
20. Anon. *Chem. Eng.*, **107(9)**, 473–474 (2000).
21. Anon. *Environ. Protect.*, **14(2)**, 116–117 (2003).
22. G. Indelicato, *Environ. Protect.* 88–92 (2001).
23. US EPA. Air Pollution Control Technology Series Training Tool: Incineration. www.epa.gov. US Environmental Protection Agency, Washington, DC. 2004.
24. US EPA. Novel Nanoparticulate Catalysts for Improved VOC Treatment Devices. EPA Project 68D98152. US Enviornmental Protection Agency, Washington, DC. http://cfpub2.epa.gov. 2004.

10
Gas-Phase Activated Carbon Adsorption

Lawrence K. Wang, Jerry R. Taricska, Yung-Tse Hung, and Kathleen Hung Li

CONTENTS

1. INTRODUCTION AND DEFINITIONS

1.1. Adsorption

The phenomenon by which molecules of a fluid adhere to the surface of a solid is known as adsorption. Through this process, these solids or adsorbents can be selectively captured or removed from an airstream, gases, liquids, or solids, even at very small concentrations. The material being adsorbed is called the adsorbate and the adsorption system is called the adsorber (1–12).

A fluid's composition will change when it comes into contact with an adsorbent and when one or more components in the fluid are adsorbed by the adsorbent. The adsorption mechanism is complex. At all solid interfaces, adsorption can occur, but it is usually small unless the solid is highly porous and possesses fine capillaries. For a solid adsorbent to be effective, it should possess the following characteristics: large surface-to-volume ratio and a preferential affinity for the individual component of concern.

Adsorption can occur in a specific manner. It can be used effectively to separate gases from gases, solids from liquid, ions from liquid, and dissolved gases from liquid. For example, after a release of toxic gases such as sulfur dioxide and chlorine into a room at a wastewater-treatment plant, an adsorption unit can be used to remove the gases from air. Additionally, adsorption can be used to remove colloids or suspended solids from the liquids, as in decolorizing and clarifying a liquid. Adsorption is also used to improve the taste and odor of drinking water by removing dissolved gases from the water.

From: *Handbook of Environmental Engineering, Volume 1: Air Pollution Control Engineering*
Edited by: L. K. Wang, N. C. Pereira, and Y.-T. Hung © Humana Press Inc., Totowa, NJ

Three operations are commonly found in most processes involving adsorption: contact, separation, and regeneration. Initially, the adsorbent comes into contact with the fluid where the separation by adsorption results. Second, fluid that is not adsorbed is separated from the adsorbent. With a gas stream, this operation is completed as the gas stream passes through the adsorbent bed. Third, the adsorbent is regenerated, removing the adsorbate from the adsorbent.

1.2. Adsorbents

Commonly used adsorbents for selectively adsorbing certain gaseous constituents from gas streams include activated carbon, silica gel, alumina, and bauxite. The contaminated gaseous constituents should be adequately removed from airstreams for air pollution control (13–18). Commercially available adsorbents possessing adsorptive properties exist in great variety. Some of these adsorbents with their industrial uses are as follows:

- Activated carbon: solvent recovery, elimination of odors, purification of gases
- Alumina: drying of gases, air, and liquids
- Bauxite: treatment of petroleum fractions; drying of gases and liquids
- Bone char: decolorizing of sugar solutions
- Decolorizing adsorbents: decolorizing of oils, fats, and waxes; deodorizing of domestic water
- Fuller's earth: refining of lube oils and vegetable and animals oils, fats, and waxes
- Magnesia: treatment of gasoline and solvents; removal of metallic impurities from caustic solutions
- Silica gel: drying and purification of gases
- Strontium sulfate: removal of iron from caustic solutions

1.3. Carbon Adsorption and Desorption

In air pollution control, activated carbon is the most widely used adsorbent and is the focus of discussion in this chapter. Adsorbents such as silica gel or alumina are less likely to be used in adsorption systems for air pollution control; therefore, they are not discussed in this chapter.

Carbon adsorption is a process by which pollutants are selectively adsorbed on the surface of granular activated carbon beds. It has been shown that activated carbon is the most suitable adsorbent for the removal of organic vapors. Substantially all organic vapors in air at ambient temperature can be adsorbed by carbon, regardless of variation in concentration and humidity. Because the organic adsorbed has practically no vapor pressure at ambient temperature, the carbon system is particularly adapted to the efficient recovery of the adsorbed organic. As a result, the carbon systems can always be designed for operation without hazard, because the vapor concentration is always below the flammable range. At low concentration of organic solvent in the airstream, carbon systems can efficiently recover organic solvents.

Regeneration is a desorption process by which adsorbed volatile organic compounds (VOCs) are removed from the carbon beds either by heat desorption at a sufficiently high temperature (usually using steam) or by vacuum desorption at a sufficiently low vacuum pressure. Some of the adsorbed organic will remain activated after regeneration. It has been shown that during desorption for carbon regeneration, about 3–5% of organics desorbed on the virgin activated carbon is absorbed so strongly that it cannot be desorbed during regeneration (14–18).

Vapor adsorption by activated carbon occurs in two stages. These stages are described as the adsorption and saturation stages:

- Adsorption stage: During this initial stage, the carbon rapidly and completely adsorbs the vapor, but a stage is reached in which the carbon continues to adsorb but at decreasing rate.
- Saturation stage: During the process, a point is reached when vapor concentration leaving the carbon equals that of the inlet. This means that carbon is saturated and it has adsorbed the maximum amount of vapor at a given temperature and pressure. This saturation value is different for each vapor and carbon.

The carbon's saturation value must be determined experimentally. Dry air, which is saturated with a selected vapor or gas, is passed through the carbon at a known airflow rate and a known carbon weight. With constant pressure and temperature, the air is passed through the carbon. The weight of the carbon gradually increases as a result of the adsorption, and, finally, the carbon ceases to increase in weight. When this point is reached, the carbon is considered saturated with the adsorbate.

As previously discussed, activated carbon adsorbs all of the usual solvent vapors that have a low boiling temperature. As a result, it can be used to recover practically any single solvent or any combination of low-boiling solvents. Physical adsorption is limited to vapors that have a higher molecular weight than the normal components of air. Practically speaking, gases with molecular weight over 45 can be removed by physical adsorption. Most solvents used exceed this limit, except for methanol (10,12).

Retention capacity and breakpoint are important characteristics of activated carbon when considering activated carbon for air pollution control. The retention capacity of an activated carbon is expressed as the ratio of the weight of the adsorbate retained to the weight of the carbon. After initial saturation of an activated carbon with a selected absorbate, the retention capacity of activated carbon is the amount of this selected adsorbate that the carbon retains when pure air is passed through the carbon at a constant temperature and pressure. The retention capacity represents the weight of the particular gas or vapor that the carbon can completely retain.

The breakpoint of an activated carbon represents an adsorption stage when the retentive capacity of the carbon is reached. Adsorption is 100% initially when an air vapor mixture is passed over carbon, but as time passes, the retention capacity of the carbon is reached. As a result, traces of the vapor begin to appear, which is described as the breakpoint. Beyond the breakpoint, the removal efficiency of the carbon decreases rapidly. As the flow continues to pass over the carbon, an additional amount of vapor is adsorbed, but the amounts of vapor in the exit air increases and eventually equals that in the inlet, at which time the carbon is saturated at the particular operating conditions.

Carbon adsorption is an exothermic process. An exothermic process is a physicochemical process during which heat is liberated and the temperature of the adsorbent bed increases. As a result, it may be necessary to provide cooling for the carbon bed.

2. *Adsorption Theory*

A variety of theories have been set forth to explain the phenomenon of selective adsorption of certain vapors or gases, but the exact mechanism is still being disputed.

In 1916, Langmuir proposed that adsorption is the result of a chemical combination of the gas with the free valence of atoms on the surface of the solid in the monomolecular layer (11). Another theory proposed that the adsorbents exerted strong attractive forces, resulting in the formation of many adsorbed layers. Pressure is applied to the lower layers by the higher layers and the attractive force on the surface. Other investigators have shown that adsorption is the result of the liquefaction of the gas and its retention by capillary action in the exceedingly fine pores of the adsorbing solid. Mostly likely, adsorption is superimposed. For example, the adsorption power of activated charcoal is mainly the result of molecular capillary condensation, whereas the adsorption power of silica gel is mainly the result of capillary condensation. However, it must be noted that the method of preparing the solid adsorbent and the nature of the gas or vapor will affect the adsorption power of the solid adsorbent.

The adsorption isotherms for a carbon represent the equilibrium adsorption capacity of carbon. The isotherms relate the amount of VOC adsorbed (adsorbate) to the equilibrium pressure (or concentration) at constant temperature. Typically for activated carbon, as the molecular weight of the adsorbate increases, the adsorption capacity of the activated carbon increases. Additionally, the chemical characteristics of the compound can affect the adsorption. Unsaturated compounds and cyclic compounds are more completely adsorbed than either saturated compounds or linear compounds. The adsorption of capacity of a carbon virtually, for all adsorbates, is enhanced at lower operating temperatures and at higher VOC concentrations. The vapor pressure of the VOC also influences the adsorption capacity. VOCs with lower vapor pressures are more easily adsorbed than those with higher vapor pressures. For VOCs, the vapor pressure is inversely proportional to the molecular weight of the compound. Thus, the heavier VOCs will tend to be more easily adsorbed than the lighter VOCs. This characteristic is not true for very heavy volatile compounds; hence, carbon adsorption is not recommended for compounds with molecular weights above 130 lb/lb-mol.

At equilibrium, there are several factors that determine the quantity of hazardous air pollution (HAP) in a gas stream that is adsorbed on activated carbon. These factors include the adsorption temperature and pressure, the specific compound being adsorbed, and the carbon characteristics (e.g., pore size and structure). The equilibrium adsorptivity defines these relationships. For a given constant temperature, a relationship exists between the mass of adsorbate (i.e., HAP) per unit weight of adsorbent (i.e., carbon) and the partial pressure of HAP in the gas stream. Adsorption isotherms are developed and fitted to a power curve using

$$W_e = k(P_{\text{partial}})^m \tag{1}$$

where W_e is the equilibrium adsorptivity (lb adsorbate/ lb adsorbent), P_{partial} is the partial pressure of the HAP in the emission stream, and k and m are empirical parameters.

The partial pressure is calculated as

$$P_{\text{partial}} = (\text{HAP}_e)(14.696 \times 10^6) \text{ psia} \tag{2}$$

The Freundlich equation, for example, is this type of equation because it is only valid for a specified adsorbate partial pressure range and curves are fitted for the equation. In Eq. (1), the equilibrium adsorptivity, W_e, represents the maximum amount of adsorbate

Table 1
Parameters for Selected Adsorption Isotherms[a]

Adsorbate	Adsorption Temperature (°F)	Isotherm parameters k	m	Range of isotherm[b] (psia)
1. Benzene	77	0.597	0.176	0.0001–0.05
2. Chlorobenzene	77	1.05	0.188	0.0001–0.01
3. Cyclohexane	100	0.508	0.210	0.0001–0.05
4. Dichloroethane	77	0.976	0.281	0.0001–0.04
5. Phenol	104	0.855	0.153	0.0001–0.03
6. Trichloroethane	77	1.06	0.161	0.0001–0.04
7. Vinyl chloride	100	0.20	0.477	0.0001–0.05
8. *m*-Xylene	77	0.708	0.113	0.0001–0.001
	77	0.527	0.0703	0.001–0.05
9. Acrylonitrile	100	0.935	0.424	0.0001–0.015
10. Acetone	100	0.412	0.389	0.0001–0.05
11. Toluene	77	0.551	0.110	0.0001–0.05

Note: [a]Each isotherm is of the form: $W_e = kP^m$. (See text for definition of terms).
Data are for adsorption on Calgon-type "BPL" carbon (4 × 10 mesh).
[b]Equations should not be extrapolated outside of these ranges
Source: US EPA.

the carbon can retain at a given temperature and partial pressure. When designing a carbon bed system, the system must be such that equilibrium of the carbon with adsorbate is not reached, because this would result in excessive emissions and bed breakthrough. Usually, the carbon beds are operated to be taken off-line when the HAP concentration in the bed reaches about 50% of the equilibrium. As a result, the actual bed capacity is less than the equilibrium capacity. This actual capacity of carbon system is commonly referred to as the effective or working capacity (W_c). Generally, the working capacity is 50% less than the equilibrium capacity. Adsorption isotherm parameters for selected organic compounds are presented in Table 1. If no information is available on the working capacity (W_c), it is common practice to use 50% of the equilibrium adsorptivity (W_e) as the default value. If no information is available on W_e or W_c, the default value of 0.100 can be used for W_c.

3. CARBON ADSORPTION PRETREATMENT

Depending on the HAP influent characteristics, three possible pretreatments are cooling, dehumidification, and high VOC reduction, which may or may not be needed, prior to the gaseous-phase carbon adsorption.

3.1. Cooling

Lower temperatures provide for a more favorable condition for adsorption of VOCs. When emission stream temperatures are significantly higher than 130°F, a heat exchanger may be used to lower the temperature of the emission stream to 130°F or less.

3.2. Dehumidification

Emission streams can contain both water vapor and VOCs. In a carbon bed, water vapor competes with VOCs for adsorption sites on the carbon surface. When the humidity level exceeds 50% (relative humidity) in the emission stream, the efficiency of the adsorption may be limited for a dilute emission stream. Under conditions when the concentration of HAP, exceeds 1000 ppmv, relative humidity above 50% can be tolerated. Likewise, when the HAP concentration is less than 1000 ppmv, the relative humidity should be reduced to 50% or less (3).

Generally, dehumidification of an emission stream is accomplished by either cooling–condensing or by diluting the emission stream. The amount of water vapor in the emission stream can be lowered by cooling and condensing the water vapor in the emission stream. Typically, cooling and condensing of the emission stream can be accomplished by using a shell-and-tube-type heat exchanger. Dilution is another alternative available for dehumidification. This alternative can be used when the dilution air humidity is significantly less than the emission stream. The drawback to this alternative is that it increases the airstream flow, which, in turn, increases the size of the adsorber system. As a result, the dilution alternative may not be cost-effective. Another drawback is that the removal efficiency of the carbon adsorber, which is a constant outlet device, will be decreased.

3.3. High VOC Reduction

For safety reasons, the designer must consider the reduction of VOC in the air emission stream. This reduction should be considered when the flammable vapors are present in emission streams and the VOC and air mixture exceeds 25% of the lower explosive limit (LEL) for the VOC. This percentage may be raised to a range of 40–50% of the LEL, if proper monitoring and controls are used. Another reason for considering VOC reduction is because the heat released during adsorption of the VOC could increase the bed temperature. For the examples in this handbook, it will be assumed that the VOC and air mixture will be limited to less than 25% of the LEL. Some of the VOCs commonly found in emission stream are listed in Table 2.

4. DESIGN AND OPERATION

4.1. Design Data Gathering

As previously described, data are compiled on a HAP Emission Stream Data Form and the required HAP control is determined by the applicable regulations (7–13,19,20). The data provide the necessary information to perform the calculations for the required HAP control. For carbon adsorbers, the size (and purchase cost) of the system depends on the following parameters:

1. The mass loading of the VOC
2. The volumetric flow rate of the emission stream with VOC
3. The adsorption time
4. The working capacity of the carbon bed

The two most important parameters for sizing and determining the cost of the carbon adsorption system are mass loading and volumetric flow rate of the VOC. Using mass

Table 2
Flammability Characteristics of Combustible Organic Compounds in Air[a,b]

Compound	Mol. Wt.	LEL[a] (% vol)	UEL[a] (% vol)
Methane	16.04	5.0	15.0
Ethane	30.07	3.0	12.4
Propane	44.09	2.1	9.5
n-Butane	58.12	1.8	8.4
n-Pentane	72.15	1.4	7.8
n-Hexane	86.17	1.2	7.4
n-Heptane	100.20	1.05	6.7
n-Octane	114.28	0.95	3.2
n-Nonane	128.25	0.85	2.9
n-Decane	142.28	0.75	5.6
n-Undecane	156.30	0.68	
n-Dodecane	170.33	0.60	
n-Tridecane	184.36	0.55	
n-Tetradecane	208.38	0.50	
n-Pentadecane	212.41	0.46	
n-Hexadecane	226.44	0.43	
Ethylene	28.05	2.7	36.0
Propylene	42.08	2.4	11.0
Butene-1	56.10	1.7	9.7
cis-Butene-2	56.10	1.8	9.7
Isobutylene	56.10	1.8	9.6
3-Methyl-butene-1	70.13	1.5	9.1
Propadiene	40.06	2.6	
1,3-Butadiene	54.09	2.0	12.0
Acetylene	2.5	100.0	
Methyl acetylene	1.7		
Benzene	78.11	1.3	7.0
Toluene	92.13	1.2	7.1
Ethyl benzene	106.16	1.0	6.7
o-Xylene	106.16	1.1	6.4
m-Xylene	106.16	1.1	6.4
p-Xylene	106.16	1.1	6.6
Cumene	120.19	0.88	6.5
p-Cumene	134.21	0.85	6.5
Cyclopropane	42.08	2.4	10.4
Cyclobutane	56.10	1.8	
Cyclopentane	70.13	1.5	
Cyclohexane	84.16	1.3	7.8
Ethyl cyclobutane	84.16	1.2	7.7
Cycloheptane	98.18	1.1	6.7
Methyl cyclohexane	98.18	1.1	6.7
Ethyl cyclopentane	98.18	1.1	6.7
Ethyl cyclohexane	112.21	0.95	6.6
Methyl alcohol	32.04	6.7	36.0
Ethyl alcohol	46.07	3.3	19.0

continued

Table 2 *(Continued)*

Compound	Mol. Wt.	LEL[a] (% vol)	UEL[a] (% vol)
n-Propyl alcohol	60.09	2.2	14.0
n-Butyl alcohol	74.12	1.7	12.0
n-Amyl alcohol	88.15	1.2	10.0
n-Hexyl alcohol	102.17	1.2	7.9
Dimethyl ether	46.07	3.4	27.0
Diethyl ether	74.12	1.9	36.0
Ethyl propl ether	88.15	1.7	9.0
Diisopropyl ether	102.17	1.4	7.9
Acetaldehyde	44.05	4.0	36.0
Propionaldehyde	58.08	2.9	14.0
Acetone	58.08	2.6	13.0
Methyl ethyl ketone	72.10	1.9	10.0
Methyl propyl ketone	86.13	1.6	8.2
Diethyl ketone	86.13	1.6	
Methyl butyl ketone	100.16	1.4	8.0

[a]LEL: lower explosive limit; UEL: upper explosive limit.
Source: US EPA.

loading of the VOC, the designer determines the carbon requirement, whereas the volumetric flow rate of the VOC-laden emission stream allows the designer to determine the size of the vessels housing the carbon, the capacities of the fans and motors required, and the diameter of the internal ductwork. Although the other two parameters are less significant than mass loading and volumetric flow rate, they will influence the design and cost of a carbon adsorption system. Further discussions on these parameters are presented in detail in later sections.

4.2. *Type of Carbon Adsorption Systems*

A variety of industries utilize carbon adsorption systems for pollution control and/or solvent recovery. The operational mode is usually batch and can involve multiple beds. The five types of adsorption system are (1) fixed regenerative beds, (2) disposable/rechargeable canisters, (3) traveling bed adsorbers, (4) fluidized adsorbers, and (5) chromatographic baghouses. The first two types are the most common and are described in the following subsections.

4.3 *Design of Fixed Regenerative Bed Carbon Adsorption Systems*

Figure 1 illustrates a typical two-bed regenerative carbon adsorption system. A two-step process can be performed to design a fixed-bed carbon adsorption system. The following procedures assume a horizontal system. In the first step, the carbon requirement, C_{req}, is estimated based on expected inlet HAP loading, the adsorption time, the number of beds, and the working capacity of the carbon. Equation (3) shows this relationship:

$$C_{req} = (M_{HAP}\,\theta_{ad}\,[1 + (ND/NA)])/W_c \tag{3}$$

where C_{req} is the total amount of carbon required (lb), M_{HAP} is the HAP inlet loading (lb/h), θ_{ad} is the adsorption time (h), ND is the number of beds desorbing, NA is the

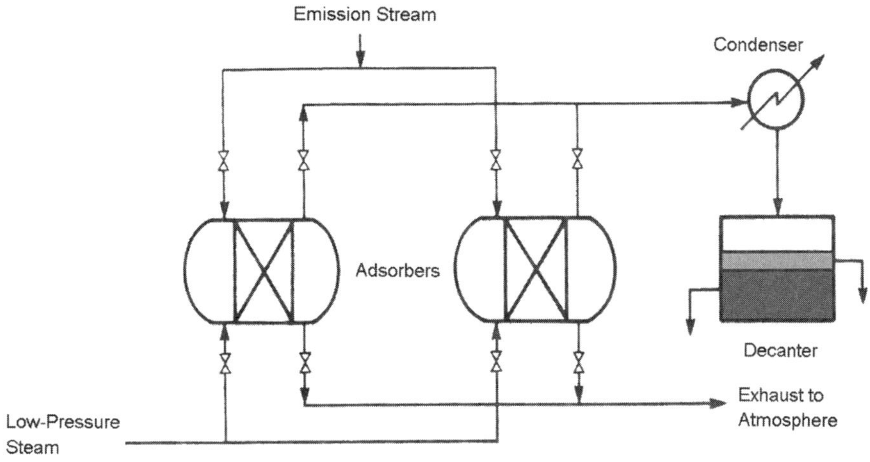

Fig. 1. Typical two-bed regenerative carbon adsorption system (*Source*: US EPA).

number of beds adsorbing, and W_c is the working capacity of carbon (lb HAP/lb of carbon). Equation (4) is used to determine M_{HAP}:

$$M_{\mathrm{HAP}} = 6.0 \times 10^{-5}\ (\mathrm{HAP}_e)(Q_e)(D_{\mathrm{HAP}}) \tag{4}$$

where M_{HAP} is the HAP inlet loading (lb HAP/h), $\mathrm{HAP_e}$ is the HAP emission stream concentration (ppmv), Q_e is the HAP emission stream flow rate (scfm), and D_{HAP} is the gas density of the HAP (lb/ft^3). The factor 6.0×10^{-5} is obtained by multiplying 60 min/h by 1 part/1,000,000. This factor is used to convert minutes to hours and ppmv to parts.

In the second step, the vessel size containing the carbon is determined. Equations (5)–(7) are used to obtain the necessary dimensions (D_v, L_v, and S). These equations do not provide the carbon bed dimensions, but the vessel dimensions assuming that the carbon occupies one-third of the vessel volume and horizontal orientation of the vessel.

The diameter and length of the adsorption vessel are determined using Eqs. (5) and (6), respectively. Under some applications, Eqs. (5) and (6) may yield unrealistic vessel dimensions, such as a vessel with a small diameter and long length. By adjusting the value for emission stream bed velocity, U_e, more practical dimensions can be obtained. For example, if the diameter is too small and the length too long, the value for U_e can be increased by enlarging the diameter and shortening the length of the vessel. In most cases, the value for U_e should not exceed 100 ft/min. If further adjustment is still needed, the vendor should be contacted to obtain more specific design information for a given application.

$$D_v = 0.127\ C'_{\mathrm{req}}\ U_e/Q'_{e,a} \tag{5}$$

$$L_v = 7.87\ (Q'_{e,a}/U_e)^{\,2}/C'_{\mathrm{req}} \tag{6}$$

where D_v is the diameter of the vessel (ft), C'_{req} is the carbon required per vessel (lb), U_e is the emission stream bed velocity (ft/min), with a default value of 85 ft/min, $Q'_{e,a}$ is the emission stream flow rate per adsorbing bed (acfm), and L_v is the vessel length (ft).

Equation (7) is used to obtain emission stream flow rate per total adsorbing system, ($Q_{e,a}$). The emission stream flow rate per adsorbing bed, $Q'_{e,a}$, is then obtain by dividing $Q_{e,a}$ by the number of beds adsorbing (NA):

$$Q_{e,a} = Q_e\,(T_e + 460)/537 \tag{7}$$

$$Q'_{e,a} = Q_{e,a}/\text{NA} \tag{7a}$$

where T_e is the emission stream temperature (°F). Because of the trucking restrictions, the diameter of the vessel (D_v) and the vessel length (L_v) are limited to 12 ft and 50 ft, respectively. After the diameter and length of the vessel are determined, the vessel surface area is calculated using Eq. (8):

$$S = \pi\, D_v\,(L_v + D_v/2) \tag{8}$$

In this handbook, the density of the HAP can be determined using

$$D_{\text{HAP}} = PM/RT \tag{9}$$

where P is the system pressure (atm) (usually 1.0), M is the HAP molecular weight (lb/lb-mol), R is the gas constant (0.7302 ft^3 atm/lb-mol °R), and T is the temperature (°R).

Regeneration of carbon beds is commonly accomplished using steam, followed by condensation. The quantity of steam required is dependent on the required removal efficiency (outlet concentration) and how much material (adsorbate) is to be removed (desorbed) from the carbon bed. The steam provides heat and carries media. The steam raises the bed temperature to its regeneration temperature and provides heat for the desorption process to occur. Approximately 60–70% of steam is required to carry the desorbed VOCs. Complete desorption of the carbon is not usually accomplished because it is not cost-effective; acceptable working capacities of adsorption can be achieved without utilizing large quantities of steam. A general rule of thumb for a solvent recovery system requires 0.25–0.35 lb of steam per pound of carbon. This steam usage ratio can be increased for applications where the outlet VOC concentrations need to be fairly low.

In this handbook, a HAP outlet concentration of 70 ppmv can be achieved after regeneration at the steam ratio of 0.30 lb of steam per pound of carbon. To achieve a HAP outlet concentration of 10–12 ppmv, a steam ratio of 1.0 lb of steam per pound of

Table 3
Carbon Adsorber System Efficiency Variables

	Steam requirement for regeneration, St (lb steam/lb carbon)	
	0.3	1.0
Outlet HAP concentration, HAP_o (ppmv)	70	10–12
Adsorption cycle time,[a] θ_{ad} (h)	2	2
Regeneration cycle time,[a] θ_{reg} (h)	2	2

[a]In some instances, cycle times may be considerably longer than the values given here. The values in this table are approximate, not definitive.

Source: ref. 7.

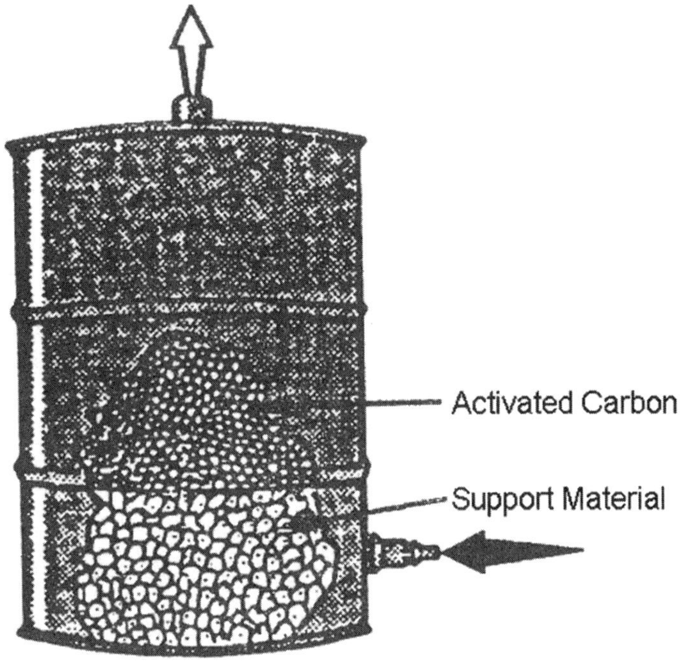

Fig. 2. Carbon canisters (*Source:* US EPA).

carbon will be required. Table 3 provides outlet concentrations, adsorption cycle times, and regeneration cycle times at these two ratios.

Equation (10) is used to determine the flow rate of steam required for regeneration:

$$Q_s = \text{NA}\ [\text{St}(C'_{\text{req}})/(\theta_{\text{req}} - \theta_{\text{dry-cool}})]/60 \quad (10)$$

where Q_s is the steam flow rate (lb/min), NA is the number beds adsorbing, St is the steam regeneration rate (lb steam/lb carbon), C'_{req} is the carbon requirement per adsorbing bed (lb), $\theta_{\text{dry-cool}}$ is the cycle time for drying and cooling the bed (h), and θ_{reg} is the regeneration cycle time (h).

The regeneration cycle time, θ_{reg}, is dependent on the time required to regenerate, dry, and cool the bed. Prior to placing a bed on-line, time must be allowed for drying and cooling the bed. This time can be as few as 15 min (0.25 h). To prevent the carbon from being fluidized in the bed, steam flow rates are limited to less than 4 lb of steam/min/-ft^2(Q_s/A_{bed}). In the case where Q_s/A_{bed} exceeds 4 lb/min-ft^2, the regeneration cycle time, θ_{reg}, or steam ratio, St, can be modified to prevent fluidization of the carbon. The cross-sectional area of the bed, A_{bed}, is obtained by dividing the emission stream flow rate per adsorbing bed ($Q_{e,a}$) by emission stream velocity (U_e):

$$A_{\text{bed}} = Q'_{e,a} / U_e = (Q_{e,a} / \text{NA}) / U_c \quad (10a)$$

4.4. *Design of Canister Carbon Adsorption Systems*

Figure 2 shows a canister carbon adsorption system. This system is normally used to control intermittent lower-volume airstreams. Additionally, carbon canister systems also are used when the expected volume of VOC recovery is fairly low, because these systems cannot be desorbed at the site and must be either land filled or shipped back to the vendor's desorption facility. As a result, canister systems do not receive any recovery credits. Another characteristic of the carbon canister system is that the effluent from the canister is usually not monitored continuously (via an FID, for example). Therefore, operators of canister systems do not have a clear indication of when a breakthrough occurs or when the system stops removing the VOC from the airstream.

Becuase carbon canister systems are not desorbed on site and are fairly self-contained units equipped with vessels, piping, flanges, and so forth, the fundamental variable to be determined in designing a canister system is the carbon requirement. Examining Eq. (3) with ND (number of beds desorbing) being zero, the total amount of carbon required is dependent on VOC inlet loading (M_{HAP}), total adsorption time (θ_{ad}), and working capacity of the carbon (W_c). Therefore, Eq. (3) becomes

$$C_{\mathrm{req}} = M_{\mathrm{HAP}}\, \theta_{\mathrm{ad}} / W_c \tag{11}$$

From Eq. (11), the total amount of carbon (C_{req}) required for a canister can be determined. The first step is to determine the HAP density (D_{HAP}) using Eq. (9). The second step is to calculate HAP inlet loading, M_{HAP}, using Eq. (4) and the D_{HAP}. These steps are shown as follows:

Step 1

$$D_{\mathrm{HAP}} = \mathrm{PM/RT} \tag{9}$$

Step 2

$$M_{\mathrm{HAP}} = 6.0 \times 10^{-5}\, (\mathrm{HAP}_e)(Q_e)(D_{\mathrm{HAP}}) \tag{4}$$

The value of M_{HAP} is substituted into Eq (11) and the total amount carbon is then calculated. It is assumed in this handbook that this amount of carbon will yield a removal efficiency of 90%. The required canister number (RCN) is determined by dividing the total amount of carbon required by the amount of carbon contained in each canister (typically 150 lb). To ensure sufficient carbon, the quotient is rounded up to the next whole canister. Once the number of canisters is determined, the design of the canister system is considered complete in this handbook and costing of the system can be performed.

4.5. *Calculation of Pressure Drops*

Figure 3 shows the relationship between the pressure drop and carbon bed depth at various air velocities. The relationship holds true for any type of carbon adsorption system.

4.6. *Summary of Application*

At a remediation site, a granular activated carbon (GAC) system is a likely candidate to be used in a control system because it is a point source having a low concentration of VOCs emitting into the atmosphere. This system is characterized by its relatively low capital cost, ease of installation, and ability to control a variety of VOCs. Usually, the outlet VOC concentrations are required to be less than 10–50 ppmv. Additionally, GAC system can be either regenerable or disposable types. Another benefit of this system is

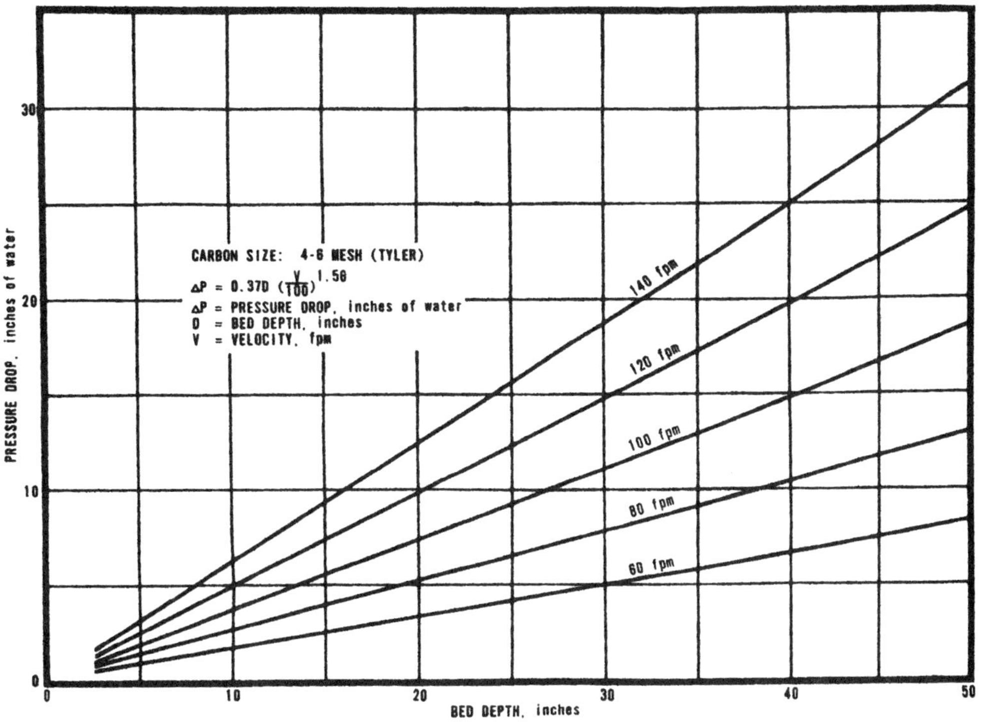

Fig. 3. Pressure drop versus carbon bed depth at various air velocities (Sources: Union Carbon Corporation and US EPA).

that it is one of the most widely used control technologies, and subsequently, much technical information is available from numerous vendors. Design considerations for GAC systems include characteristics of the emission stream, molecule weight of the adsorbate, and the pretreatment requirements.

An important limitation of the GAC system is that only compounds with molecular weights in the 50–200 g/g-mol range are effectively adsorbed by the system. Another characteristic of the GAC system is that the pollutants are not destroyed, only transferred from one medium to another, inevitably leaving solid or liquid waste after treatment with the GAC system. As a result of this transfer, GAC systems are often used in industry to capture and recycle valuable pure VOCs. At remediation projects using GAC systems, VOCs usually are not of sufficient purity or high value to warrant recycling, and as a result, the disposal of the adsorbed VOCs is almost always the final step.

Condensers, incinerator, or combustion engines become competitive in cost-effectiveness with GAC systems when VOC concentration in the air emission stream exceeds 1000 ppmv. Effectiveness of GAC systems depends on temperature, pressure, and moisture content of the emission stream. At high temperature and pressure, GAC systems are less effective. Additionally, GAC systems require low humidity in the emission stream because water binds to the active sites in the carbon, reducing the system's

Table 4
Applicability of CAS Selected Contaminants

Contaminant class	Examples	CAS typically effective	Comments
Aromatics	Benzene, toluene	Yes	Standard application of GAC
Aliphatics	Hexane, heptane	Yes	Standard application of GAC
Halogenated hydrocarbons	Chloroform	Yes	Standard application of GAC
Light hydrocarbons (MW < 50 or BP <20ºC)	Methane, Freon	No	Will not adsorb
Heavy hydrocarbons (MW > 200 or BP >200ºC)	Glycols, phenols	No[a]	Will not desorb or will not be adsorbed due to steric constraints
Oxygenated compounds	Ketones, aldehydes	No[b]	Fire hazard
Certain reactive organics	1,1,1-Trichloroethane, organic acid	No	Will react with and degrade GAC
Bacteria	Coliform	Yes	Requires silver-impregnated GAC
Radioisotopes	^{131}I	Yes	Requires coconut-shell carbon
Certain inorganics	Hydrogen sulfide, ammonia, hydrochloric acid	Yes	Requires impregnated GAC
Mercury	—	Yes[c]	Requires impregnated GAC

[a]Nonregenerable carbon systems may work.
[b]Not all oxygenated compounds are a problem.
[c]High levels of sulfur dioxide may blind the charcoal and reduce Hg removal efficiencies.
Source: US EPA.

capacity. Some problems the emissions stream may pose for the GAC system are plugging, fouling, and corroding the system. However, these problems can be overcome with pretreatment devices, which will increase the total system cost.

Some of carbon adsorption technology limitations can be alleviated with pretreatment. The emission stream prior to carbon adsorption system should have low solids and particulates to prevent fouling and plugging of the system. The emission stream should contain less than 1000 ppmv inorganics. Particulate filter can be used to lower levels in the emission stream prior to adsorption system. Relative humidity of emission stream prior to carbon adsorption system should be below 50%. As relative humidity in the emission stream rises above 50%, the efficiency of VOC adsorption decreases rapidly. By increasing the temperature of the emission stream, the relative humidity of the stream can be lowered, but this can affect the removal efficiency. At a moderate temperature range from 100ºF to 130ºF, VOC adsorption occurs readily on carbon.

Table 4 summarizes the effectiveness of the carbon adsorption system (CAS) for various classes of compound, and Table 5 provides the adsorption capacities of carbon for some HAPs. Also, adsorption capacities of carbon can be calculated as a function of temperature and inlet concentration. Additionally, it is known that CAS is very effective for radon gas reduction (15, 21), but at this time, GAC's adsorption capacity for radon gas reduction is unknown.

Table 5
Reported Operating Capacities for Selected Compounds

Compound	Average inlet concentration (ppmv)	Adsorption capacity[a] (lb VOC/100 lb carbon)
Acetone	1000	8
Benzene	10	6
n-Butyl acetate	150	8
n-Butyl alcohol	100	8
Carbon tetrachloride	10	10
Cyclohexane	300	6
Ethyl acetate	400	8
Ethyl alcohol	1000	8
Heptane	500	6
Hexane	500	6
Isobutyl alcohol	100	8
Isopropyl acetate	250	8
Isopropyl alcohol	400	8
Methyl acetate	200	7
Methyl alcohol	200	7
Methylene chloride	500	10
Methyl ethyl ketone	200	8
Methyl isobutyl ketone	100	7
Perchloroethylene	100	20
Toluene	200	7
Trichloroethylene	100	15
Trichlorotrifluoroethane	1000	8
Xylene	100	10

[a]Adsorption capacities are based on 200 scfm of solvent-laden air at 100°F (per hour).
Source: US EPA.

The VOC removal efficiency of gas phase carbon adsorption systems has been compared by Moretti and Mukhopadhyay (22) with that of catalytic oxidation, flaring, condensation, absorption, heaters, biofiltration, membrane separation, and ultraviolet (UV) oxidation.

4.7. Regeneration and Air Pollution Control of Carbon Adsorption System

A schematic of a standard fixed-bed CAS is shown in Fig. 4. Typically in a three-bed CAS, two beds are adsorbing, and the third is desorbing. Steam is typically used to regenerate the carbon. Most organic solvents are stripped from the carbon with high temperature and water vapor. The condensed water leaving the system carries the captured organics. Additional treatment is required to separate the captured organics from the water prior to its disposal. An alternative to using steam for regeneration is to use an inert gas to reactivate the carbon. After regeneration, the captured organics must be separated from the inert gas. The inert gas systems are initially more expensive than steam regeneration systems. The benefits of regenerating with inert gas are that it consumes less energy and recovers a purer solvent. These benefits may provide sufficient

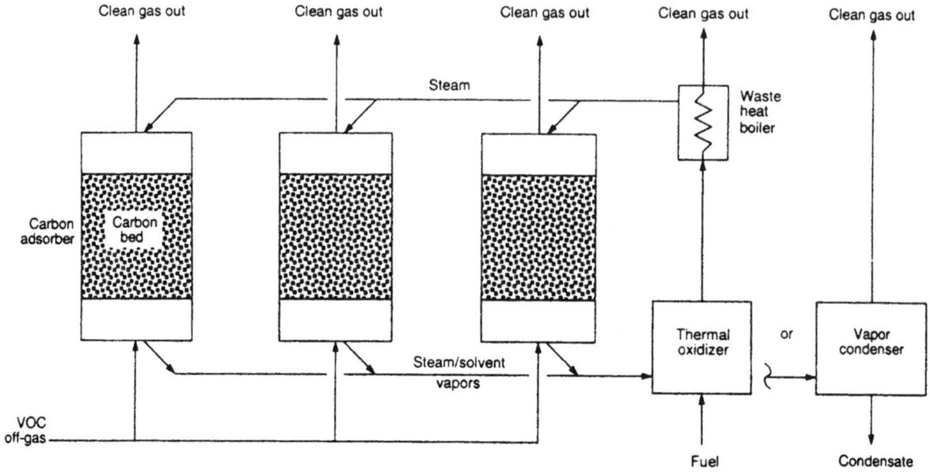

Fig. 4. Schematic diagram of carbon adsorption system with on-site batch regeneration (*Source*: US EPA).

cost savings to make an inert gas regeneration system economically feasible. As shown in Fig. 4, a combination of thermal oxidizer and waste heat boiler or a vapor condenser may be adopted for purifying the effluent (containing steam/solvent vapors) during regeneration. Recent developments in CAS operation and continuous on-site regeneration can be found in the literature (14–18).

A GAC control system typically consists of the following major components: pretreatment equipment (dehumidifier, absorbers, particulate filters, etc.), piping to carry the emission stream to the adsorbent, adsorption beds or canisters and piping to carry the discharge to other add-on controls or to stack and regeneration unit (if the unit is utilizing regeneration technology). The regeneration unit incorporates either multiple fixed beds or a moving bed. A multiple fixed-bed system has several parallel fixed beds operating while some of the beds are being regenerated. A moving-bed system, less commonly used, regenerates carbon at one point while adsorbing at another. The typical operational cycle for carbon bed is adsorption, heat regeneration, drying, and cooling. An important consideration for regeneration is that the amount of heat supplied for regeneration must exceed the heat released during adsorption.

4.8. *Granular Activated Carbon Versus Activated Carbon Fiber*

The activated carbon fiber (ACF) is relatively new in the US market (23,24), but it is widely used in Japan. The ACF offers a higher adsorption rate, longer life, a lower pressure drop, and smaller size in comparison with granular activated carbon (GAC).

The US Army Corps of Engineers Construction Research Laboratory reports a recent development of an ACF vapor-recovery system, which consists of a metal vessel containing an ACF cloth rolled up and inserted into cylinders (25). This newly developed recovery system could help generators of HAPs reduce pollution abatement costs up to 50%.

4.9. Carbon Suppliers, Equipment Suppliers, and Service Providers

The activated carbon suppliers, carbon adsorption system manufacturers, carbon desorption (regeneration) system manufacturers, regeneration service providers, and disposal service providers can be found in the literature (26,27). The new test methods for the activated carbon industry (28), training opportunities (29), and more GAC information (30) can be found in the literature.

5. DESIGN EXAMPLES

Example 1

Toluene is known to be a HAP in a contaminated gaseous emission stream, at 1000 ppmv 90°F stream temperature. The partial pressure of the HAP may be determined by using Eq. (2) and the working capacity of activated carbon by using Eq. (1).

Solution

The following contaminated air emission stream was obtained from an industrial plant:

HAP	=	toluene
Stream temperature, T_e	=	90°F
HAP emission stream concentration, HAP_e	=	1000 ppmv

Because the adsorption isotherms for toluene are presented in Table 1, the default value for a working capacity of 0.100 is not needed in this example. Table 1 presents the isotherms for toluene at an adsorption temperature of 77°F, which is lower than the stream temperature of 90°F in this example. Assume that this difference in temperature does not significantly affect the calculation results. The working capacity value, W_c, is usually 50% of the equilibrium capacity (W_e). Using Eq. (1) and values from Table 1, the W_e is calculated as follows:

$$W_e = k\,(P_{\text{partial}})^m \tag{1}$$

where

$$P_{\text{partial}} = (\text{HAP}_e)\,(14.696 \times 10^{-6}) \tag{2}$$

$$P_{\text{partial}} = (\text{HAP}_e)\,(14.696 \times 10^{-6}) = 1{,}000 \text{ ppmv } (14.696 \times 10^{-6}) = 0.0147 \text{ psia}$$

From Table 1, $k = 0.551$, and $m = 0.110$. Then, substituting into Eq. (1) yields

$$W_e = (0.551)\,(0.0147 \text{ psia})^{0.110}$$

$$W_e = 0.346 \text{ lb toluene/lb carbon}$$

Because the working capacity, W_c, is usually 50% of equilibrium capacity (W_e),

$$W_c = 0.50\,W_e = 0.50\,(0.346 \text{ lb toluene / lb carbon}) = 0.173 \text{ lb toluene/lb carbon}$$

Because the adsorption isotherm for toluene was available (Table 1), the default value for working capacity of 0.100 is not needed in this example.

Example 2

Outline the step-by-step procedures for design of a fixed-bed regenerative carbon adsorption system for control of an emission stream containing a hazardous air pollutant.

Solution

Step 1

Use Eq. (3) to calculate the amount of carbon required (C_{req}):

$$C_{req} = (M_{HAP}\, \theta_{ad}\, [1 + (ND/NA)])/W_c \quad (3)$$

$$C_{req} = \underline{\qquad\qquad\qquad} \text{ lb carbon}$$

The carbon required per vessel (C'_{req}) is determined by dividing carbon required (C_{req}) by the number of beds adsorbing (NA).

$$C'_{req} = C_{req}/NA$$

Step 2

If the HAP inlet loading (M_{HAP}) is not given, use Eq. (4) to calculate it.

$$M_{HAP} = 6.0 \times 10^{-5}\, (HAP_e)(Q_e)(D_{HAP}) \quad (4)$$

$$M_{HAP} = \underline{\qquad\qquad\qquad} \text{ lb/hr}$$

where

$$D_{HAP} = PM/RT \quad (9)$$

$$D_{HAP} = \underline{\qquad\qquad\qquad} \text{ lb/ft}^3$$

Step 3

The vessel diameter, D_v, vessel length, L_v, and the vessel size, S, are obtained using Eq. (5), (7), and (8), respectively.

$$D_v = 0.127\, C'_{req}\, U_e/Q'_{e,a} \quad (5)$$

where

$$Q_{e,a} = Q_e\, (T_e + 460)/537 \quad (7)$$

$$Q_{e,a} = \underline{\qquad\qquad\qquad} \text{ acfm}$$

The emission stream flow rate per adsorbing bed ($Q'_{e,a}$) is obtained by dividing the HAP emission stream by the number of beds adsorbing (NA)

$$Q'_{e,a} = Q_{e,a}/NA$$

$$Q'_{e,a} = \underline{\qquad\qquad\qquad} \text{ acfm/bed}$$

$$D_v = \underline{\qquad\qquad\qquad} \text{ ft}$$

$$L_v = 7.87\, (Q'_{e,a}/U_e)^2/C_{req} \quad (6)$$

$$L_v = \underline{\qquad\qquad\qquad} \text{ ft}$$

$$S = \pi\, D_v\, (L_v + D_v/2) \quad (8)$$

$$S = \underline{\qquad\qquad\qquad} \text{ ft}^2$$

Example 3

Design a two-bed regenerative carbon adsorption system for removal of HAP from a gaseous emission stream. Given data are as follows:

Maximum flow Rate, Q_e	=	15,000 scfm
Temperature, $T_e = 90°F$	=	(90 + 460) °R
System pressure, P	=	1 atm
HAP molecular weight, M	=	92 lb/lb-mol
HAP emission stream concentration, HAP_e	=	1000 ppmv
Gas constant, R	=	0.7302 ft³atm/lb-mol°R

Solution

This example is based on a two-bed system and the stream characteristics from Example 1. Because the M_{HAP} is not given and the HAP concentration is given, the first step is to calculate the D_{HAP} from

$$D_{HAP} = PM/RT \quad (9)$$

$$D_{HAP} = (1)(92 \text{ lb/lb-mole})/\ (0.7302 \text{ ft}^3 \text{ atm/lb-mole°R})(550 \text{ °R})$$

$$D_{HAP} = 0.23 \text{ lb/ft}^3$$

M_{HAP} is then calculated using

$$M_{HAP} = 6.0 \times 10^{-5}\ (HAP_e)(Q_e)(D_{HAP}) \quad (4)$$

$$M_{HAP} = 6.0 \times 10^{-5}(1{,}000 \text{ ppmv})(15{,}000 \text{ scfm})(\ 0.23 \text{ lb/ft}^3)$$

$$M_{HAP} = 207 \text{ lb/h}$$

The adsorption time (θ_{ad}) and regenerative time (θ_{reg}) are obtained from Table 3. Because $\theta_{ad} \geq \theta_{reg}$, a two-bed system may be utilized. Using Eq. (3) and the W_c from Example 1, the carbon requirement C_{req} can be estimated.

$$C_{req} = (M_{HAP}\ \theta_{ad}\ [1 + (ND/NA)])\ /\ W_c \quad (3)$$

$$C_{req} = ((207 \text{ lb/h})(2\text{h})[1 + 1/1])\ /\ (0.173 \text{ lb toluene / lb carbon})$$

$$C_{req} = 4{,}786 \text{ lb carbon}$$

The carbon requirement per bed is then obtained:

$$C'_{req} = 4{,}786 \text{ lb carbon}/2 = 2{,}393 \text{ lb}$$

The estimated carbon requirement is rounded to the nearest 10 lb:

$$C'_{req} = 2{,}390 \text{ lb}$$

The vessel diameter, D_v, vessel length, L_v, and the vessel size, S, are obtained using Eq. (5), (6), and (8), respectively.

$$D_v = 0.127\ C'_{req}\ U_e\ /\ Q'_{e,a} \quad (5)$$

where

$$Q_{e,a} = Q_e\ (T_e + 460)\ /\ 537 \quad (7)$$

$$Q_{e,a} = (15{,}000 \text{ scfm})\ (90 + 460)\ /\ 537$$

$$Q_{e,a} = 15{,}363 \text{ acfm}$$

The emission stream flow rate per adsorbing bed ($Q'_{e,a}$) is obtained by dividing the HAP emission stream by the number of beds adsorbing (NA):

$$Q'_{e,a} = Q_{e,a}/\mathrm{NA}$$

$$Q'_{e,a} = 15{,}363 \text{ acfm} /2 = 7{,}681.5 \text{ acfm}$$

To determine the diameter and length of vessel, assume an emission stream bed velocity (U_e) of 85 ft/min.

$$D_v = 0.127\ (2{,}390 \text{ lbs})\ (85 \text{ ft/min})/\ 7{,}681.5 \text{ acfm} = 3.4 \text{ ft}$$

$$L_v = 7.87\ (Q'_{e,a} / U_e)^2 / C'_{\text{req}} \qquad (6)$$

$$L_v = 7.87\ (7{,}681.5 \text{ acfm} /85 \text{ ft/min})^2/2{,}390 \text{ lb} = 27 \text{ ft}$$

$$S = \pi D_v\ (L_v + D_v /2) \qquad (8)$$

$$S = \pi\ (3.4 \text{ ft})\ (27 \text{ ft} + (3.4/2)) = 307 \text{ ft}^2$$

The vessel diameter and length are somewhat unrealistic. Increasing the bed velocity, U_e, to 100 ft/min will incrcase the diameter and shorten the length. It may be beneficial to contact a vendor to obtain assistance in selecting size of the vessel.

Example 4

Outline the methodology for the determination of the pretreatment requirements of a gas-phase carbon adsorption system.

Solution

The methodology for determining the carbon adsorption pretreatment requirements for an emission stream is outlined in the following three steps:

Step 1: Cooling Consideration

T_e = ___________°F

When the temperature of the emission stream is higher than 130°F, a heat exchanger is needed to lower the temperature to below 130°F or less. Refer to a suitable reference for the calculation procedures.

Step 2: Dehumidification Consideration

R_{hum} = ________%

When the relative humidity is above 50% and the HAP concentration is less than 1000 ppmv, a condenser may be used to cool and condense the water vapor in the emission stream, which will reduce the relative humidity of the emission stream.

Step 3: High VOC Concentration Consideration

HAP_e = ________ppmv

When the flammable vapors are present in the emission stream, they must be limited to below 25% of their LEL.

LEL = _______ppmv (*see* Table 2)

$$25\%\ \text{LEL} = 0.25\text{LEL} = \_\_\_\_\_\_\_\text{ppmv}$$

Carbon beds have a maximum practical inlet concentration for HAP of 10,000 ppmv. Greater inlet concentrations may not able to be treated by carbon.

Example 5

Outline the methodology for determining the pretreatment requirements of a gas-phase carbon adsorption system. Address the following questions:

1. Is cooling necessary if the temperature of the air emission stream is 90°F?
2. Is dehumidification necessary if the relative humidity of the air emission stream is less than 50%?
3. If the HAP concentration in the air emission stream is 1000 ppmv of toluene, is it considered to be a "high VOC concentrations," which must be reduced prior to carbon adsorption treatment?

Solution

Because the T_e = 90°F, R_{hum} = 50%, and HAP_e = 1000 ppmv of toluene, cooling and dehumidification are not necessary. The HAP_e is 1000 ppmv of toluene. The concentration is below the 25% of the LEL for toluene (12,000 ppmv). Table 2 indicates that LEL (% vol) = 1.2; therefore, ppmv = 1.2% (10,000 ppmv/%) = 12,000 ppmv.

Example 6

The emission stream inlet HAP concentration is 1000 ppmv with a control requirement of 95%. Determine the adsorption cycle time, regenerative cycle time, and the steam requirement for carbon regeneration.

Solution

The air emission stream inlet HAP concentration is 1000 ppmv and a control requirement is 95% or 50 ppmv outlet concentration. From Table 3, at an adsorption cycle time of 2 h, a regeneration cycle time of 2 h, and 0.3 lb of steam/lb of carbon, the regeneration rate will provide an outlet concentration of about 70 ppmv, or slightly less than 95% of the control efficiency. This assumes that the bed does not get close to breakthrough. For the examples in this handbook, these values will suffice.

Example 7

Outline the procedures for determining the carbon requirement of a canister system (Fig. 2).

Solution

The carbon requirement for a canister system can be estimated using

$$C_{\text{req}} = M_{\text{HAP}}\,\theta_{\text{ad}}/W_c \tag{11}$$

$$C_{\text{req}} = \_\_\_\_\_\_\_\_\_\text{lb of carbon}$$

If the HAP inlet loading (M_{HAP}) is not given, it can be determined

$$M_{\text{HAP}} = 6.0 \times 10^{-5}\,(\text{HAP}_e)(Q_e)(D_{\text{HAP}}) \tag{4}$$

$$M_{\text{HAP}} = \_\_\_\_\_\_\_\ \text{lb of HAP/h}$$

To determine the required canister number (RCN), the C_{req} is divided by the amount of carbon contained in a single canister and rounded up to the next whole number. Typically, each canister contains 150 lb of carbon.

$$RCN = C_{req}/(150 \text{ lb carbon/canister})$$

$$RCN = \_\_\_\_\_\_\_\_$$

Example 8

Outline the step-by-step procedures for calculating the steam requirements (lb steam/min-ft^2), of a fixed-bed carbon adsorption system.

Solution

The step-by-step procedures for calculating the steam requirements of a fixed-bed regenerative carbon system are summarized as follows:

Step 1: Carbon Adsorber Efficiency

For a given HAP outlet concentration (HAP_e), the adsorption time (θ_{ad}), the regeneration time (θ_{reg}),and the steam requirement (St) can be determined using Table 3.

$$\theta_{ad} = \_\_\_\_\_\_\_\_ \text{h}$$

$$\theta_{reg} = \_\_\_\_\_\_\_\_ \text{h}$$

$$St = \_\_\_\_\_\_\_\_ \text{lb steam/lb carbon}$$

Step 2: Steam Required for Regeneration

Calculate the steam requirement using

$$Q_s = NA\ [St(C'_{req})/(\theta_{req} - \theta_{dry\text{-}cool})]/60 \quad (10)$$

$$Q_s = \_\_\_\_\_\_\_ \text{ lb/h}$$

Calculate Q_s/A_{bed}:

$$A_{bed} = Q'_{e,a}/U_e \quad (10a)$$

$$A_{bed} = \_\_\_\_\_\_\_ \text{ft}^2$$

$$Q_s/A_{bed} = \_\_\_\_\_\_\_\_\_ \text{lb steam/min-ft}^2$$

Typically, the Q_s/A_{bed} is limited to 4 lb steam/min-ft^2 to prevent fluidization of the carbon bed. Assume $Q_{dry\text{-}cool}$ = 0.25 h, if no information is available.

Example 9

Determine the steam requirements of a fixed-bed regenerative carbon adsorption system for air emission control.

The given data are as follows:

$$RE = 95\%$$

$$HAP_o = 50 \text{ ppmv}$$

$$St = 0.3 \text{ lb steam/ lb carbon (\textit{see} Table 3)}$$

$$\theta_{reg} = 2 \text{ h } (\textit{see} \text{ Table 3})$$

$$A_{bed} = 181 \text{ ft}^2$$

Solution

Assume $\theta_{dry\text{-}cool} = 0.25$ h.

Use Eq. (10):

$$Q_s = \text{NA }[\text{St}(\text{C}'_{req})/(\theta_{req} - \theta_{dry\text{-}cool})]/60 \quad (10)$$

$$Q_s = 1[0.3(2{,}390)/(2 - 0.25)]/60$$

$$Q_s = 6.84 \text{ lb steam/min}$$

$$Q_s/A_{bed} = (6.84 \text{ lb steam/min})/181 \text{ ft}^2$$

$$Q_s/A_{bed} = 0.0378 \text{ lb steam/min ft}^2$$

Because Q_s/A_{bed} is less than 4 lb steam/min-ft^2, fluidization in the carbon bed is not expected.

Example 10

Canister carbon systems are typically used for emission stream flow rates less than 2000 scfm. The adsorption time in this example is based on the total volume of recovered solvent. The HAP pollutant is acetone and the given data are as follows:

Maximum flow rate, Q_e	=	2000 scfm
Temperature, T_e	=	90°F
Relative humidity, R_{hum}	=	40%
Required removal efficiency, RE	=	90%
HAP emission stream concentration, HAP_e	=	700 ppmv
Adsorption time, θ_{ad}	=	40 h

Determine the HAP density (D_{HAP}), the HAP inlet loading (M_{HAP}), the carbon requirement (C_{req}), and the required carbon canister number (RCN) for proper treatment of the air emission stream.

Solution

The HAP density, D_{HAP}, is calculated first using

$$D_{HAP} = \text{PM/RT} \quad (9$$

$$D_{HAP} = (1 \text{ atm})\ (58 \text{ lb/lb-mole})/(0.7302 \text{ ft}^3 \text{ atm/lb-mole°R})(460{+}90)(\text{°R})$$

$$D_{HAP} = 0.144 \text{ lb/ ft}^3$$

Using Eq. (4), the inlet HAP loading, M_{HAP}, is determined:

$$M_{HAP} = 6.0 \times 10^{-5}\ (\text{HAP}_e)(Q_e)(D_{HAP}) \quad (4)$$

$$M_{HAP} = 6.0 \times 10^{-5}\ (700 \text{ ppmv of acetone})(2{,}000 \text{ ft}^3\text{/min})\ (0.144 \text{ lb/ft}^3)$$

$$M_{HAP} = 12.10 \text{ lb acetone/h}$$

The carbon requirement, C_{req}, is determined using

$$C_{req} = M_{HAP}\ \theta_{ad}/W_c \quad (11)$$

The working capacity value, W_c, is usually 50% of the equilibrium capacity (W_e). Using Eq. (1) and values from Table 1, the W_e is calculated as follows:

$$W_e = k\,(P_{\text{partial}})^m \qquad (1)$$

where

$$P_{\text{partial}} = (\text{HAP}_e)\,(14.696 \times 10^{-6}) \qquad (2)$$

$$P_{\text{partial}} = (\text{HAP}_e)\,(14.696 \times 10^{-6}) = (700\ \text{ppmv})\,(14.696 \times 10^{-6}) = 0.01029\ \text{psia}$$

Then, from Eq. (1), the equilibrium capacity is obtained.

$$W_e = k\,(P_{\text{partial}})^m \qquad (1)$$

From Table 1, $k = 0.412$ and $m = 0.389$.

Substituting these values into Eq.(1) yields the equilibrium capacity:

$$W_e = (0.412)\,(0.01029\ \text{psia})^{0.389}$$

$$W_e = 0.06945\ \text{lb acetone/lb carbon}$$

Because the working capacity, W_c, is usually 50% of equilibrium capacity (W_e),

$$W_c = 0.50\,W_e = 0.50\,(0.069\ \text{lb acetone/lb carbon}) = 0.0345\ \text{lb acetone/lb carbon}$$

The carbon requirement is calculated using

$$C_{\text{req}} = M_{\text{HAP}}\,\theta_{\text{ad}}/W_c \qquad (11)$$

$$C_{\text{req}} = (12.10\ \text{lb of acetone/h})(40\ \text{h})/(0.0345\ \text{lb acetone/lb carbon})$$

$$C_{\text{req}} = 14{,}029\ \text{lb of carbon}$$

Typically, each canister contains 150 lb of carbon; therefore, the required canister number (RCN) is calculated as follows:

$$\text{RCN} = (14{,}029\ \text{lb carbon})/(150\ \text{lb carbon/canister})$$

$$\text{RCN} = 93.5\ \text{canisters, therefore use 94 canisters}$$

NOMENCLATURE

θ_{ad}	Adsorption cycle time (h)
$\theta_{\text{dry-cool}}$	Bed drying and cooling time (h)
θ_{req}	Regeneration cycle time (h)
A_{bed}	Bed area (ft^2)
C_{req}	Amount of carbon required (lb)
C'_{req}	Amount of carbon required per vessel (lb)
D_{HAP}	Density of HAP (lb/ ft^3)
D_v	Vessel diameter (ft)
HAP_e	HAP emission stream concentration (ppmv)
HAP_o	HAP outlet stream concentration (ppmv)
k	Isotherm empirical parameter
LEL	Lower explosive limit (%)

L_v	Vessel length (ft)
m	Empirical parameter or slope of equilibrium curve
M	Molecular weight (lb/lb-mol)
M_{HAP}	HAP inlet loading rate (lb/h)
NA	Number of beds adsorbing
ND	Number of beds desorbing
P	System pressure drop (atm)
P_{partial}	Partial pressure (psia)
Q_d	Dilution air required (scfm)
Q_e	Emission stream flow rate (scfm)
$Q_{e,a}$	Actual emission stream flow rate (acfm)
$Q'_{e,a}$	Actual emission stream flow rate per adsorbing bed (acfm)
Q_s	Steam flow rate, (lb/min)
R	Gas constant, (atm/lb-mole°R)
RCN	Required canister number
R_{hum}	Relative humidity (%)
S	Vessel surface area (ft^2)
St	Steam regeneration rate (lb steam/lb carbon)
T	Temperature (°R)
T_e	Emission stream temperature (°F)
UEL	Upper explosive limit (%)
U_e	Emission stream velocity (ft/sec)
W_c	Carbon bed working capacity (lb HAP/lb carbon)
W_e	Carbon bed equilibrium capacity (lb HAP/lb carbon)

REFERENCES

1. US EPA, *OAQPS Control Cost Manual*, 4th ed. EPA 450/3-90-006 (NTIS PB90-169954). US Environmental Protection Agency, Washington, DC, 1990.
2. Carlos Nunez of US EPA AEERL to Michael Sink, Pacific Environmental Services, Inc., *Memorandum with attachment.* US Environmental Protection Agency, Washington, DC, 1989.
3. Karen Catlett of US EPA OAQPS to Carlos Nunez of US EPA AEERL, *Memorandum with attachment.* US Environmental Protection Agency, Washington, DC, 1989.
4. Michael Sink of Pacific Environmental Services, Inc. to Al Roy of Calgon Corp., *Memorandum,* 1989.
5. Pacific Environmental Services, Inc. *Company data for the evaluation of continuous compliance monitors,* 1989.
6. US EPA, *Soil Vapor Extraction VOC Control Technology Assessment.* EPA-450/4-89-017 (NTIS PB90-216995). US Environmental Protection Agency, Washington, DC (1989).
7. US EPA, *Handbook: Control Technologies for Hazardous Air Pollutants.* EPA 625/6-91-014 (NTIS PB91-228809). US Environmental Protection Agency, Cincinnati, OH, 1991.
8. Michael Sink of Pacific Environmental Services, Inc to U. S. Gupta of Calgon Corp. Private Communication,1991.
9. US EPA, *Carbon Adsorption for Control of VOC Emissions: Theory and Full Scale System Performance.* EPA 450/3-88-012. US Environmental Protection Agency, Research Park, NC, 1988.
10. A. Turk, and K. A. Brownes, *Chem Eng.* **58**, 156–158 (1951).

11. S. Glasstone, *Physical Chemistry*, Princeton, NJ (1946).
12. US EPA, *Air Pollution Engineering Manual*, 2nd ed., EPA AP-40 987. US. Environmental Protection Agency, Washington, DC, 1973.
13. US EPA, *Handbook: Control Techniques for Fugitive Emissions from Chemical Process Facilities.* EPA 625/R-93/005. US Environmental Protection Agency, Cincinnati OH, 1994.
14. L. K. Wang., L. Kurylko and O. Hrycyk US patents. 5122165 and 5122166, 1992.
15. L. K. Wang, M.H.S. Wang and P. Wang, *Management of Hazardous Substances at Industrial Sites,* UNIDO Registry No. DTT-4-4-95. United Nations Industrial Development Organization, Vienna, Austria, 1995, p. 105.
16. L. K. Wang, J. V. Krouzek and U. Kounitson, *Case Studies of Cleaner Production and Site Remediation,* Training Manual No. DTT-5-4-95. United Nations Industrial Development Organization, Vienna, Austria, 1995, p. 136.
17. L. K. Wang, L. Kurylko and O. Hrycyk US patent 5399267, 1996.
18. L. K. Wang, L. Kurylko and O. Hrycyk, US patent 5552051, 1996.
19. US EPA, *Control of Air Emissions from Superfund Sites.* EPA 625/R-92/02. US Environmental Protection Agency, Washington, DC, 1992.
20. US EPA, *Bioremediation of Hazardous Waste Sites: Practical Approaches to Implementation.* EPA 600/K-93/002. US Environmental Protection Agency, Washington, DC, 1993.
21. J. Wilson, *Water Cond. and Purif.* 102–104 (2001).
22. E. C. Moretti, and N. Mukhopadlyay, *Chem. Eng. Prog.*, **89(7)**, 20–26 (1993).
23. T. Maeda, *Seminar Proceeding of Japanese Water Filtration Association* (1998).
24. N. Kikui, *Water Cond. and Purif.*, 104–107 (2001).
25. Anon., *Ind. Waste-water* 18–19 (2001).
26. Anon., *Water Eng. and Manage.*, **149(12)**, 22 (2002).
27. Anon., *Public Works* **133(5)**, 375 (2002).
28. M. Greenbank, H. Nowicki, and H. Yute, *Water Cond. and Purif.,* **45(2)**, 98–103 (2003).
29. US EPA, Air Pollution Control Technology Series Training Tool: Carbon Adsorption. www.epa.gov. US Environmental Protection Agency, Washington, DC. 2004.
30. L. K. Wang, Y. T. Hung and N. K. Shammas (eds.). *Physicochemical Treatment Process.* Humana Press, Totowa, NJ, 2005.

11
Gas-Phase Biofiltration

Gregory T. Kleinheinz and Phillip C. Wright

CONTENTS

1. INTRODUCTION

Biofiltration is the use of microorganisms, immobilized on a biologically active solid support, to treat chemicals in an airstream. Although the term implies a physical process, the process is biochemical and will not likely be changed in the near future. Biofilters have been used for volatile organic compound (VOC) abatement, mitigation of odor-causing compounds, and in conjunction with other treatment technologies (i.e., soil vapor extraction). With recent changes in US air regulations, increased pressure has been placed on industries that emit chemicals into the air. Biofilters have been an increasingly popular choice as a treatment option because of their low operating cost and relatively low capital costs compared to other technologies. Biofilters operate under the premise that contaminants in the airstream partition into an aqueous layer on the solid support, where it is bioavailable and then degraded by the microbial community present. Complete metabolism of an organic compound yields carbon dioxide and water, which is then moved out of the biofilter. In general, conventional biofilters have been the most successful in applications with low flow rates and relatively low concentration of contaminants. Table 1 lists some of the industries that have used biofilters.

Biological treatment methods have been widely used by industry to mitigate environmental contamination throughout the 20th century. However, only recently has biofiltration gained acceptance in the United States as a viable treatment alternative for

From: *Handbook of Environmental Engineering, Volume 1: Air Pollution Control Engineering*
Edited by: L. K. Wang, N. C. Pereira, and Y.-T. Hung © Humana Press Inc., Totowa, NJ

Table 1
Industries That Have Used Biofiltration for Air Pollution Control

Animal facilities, large-scale	Painting operations, large-scale
Automotive	Petrochemical manufacturing
Chemical manufacturing	Petroleum
Coatings	Plastics manufacturing
Composting	Printing
Ethanol production	Pulp and paper
Food processing	Rendering
Fragrance	Semiconductor
Iron foundries	Sewerage treatment
Landfill gas extraction	Wood products

air emissions. Some of the impetus for this adoption was the Clean Air Act Amendments (CAAA) that were put into place in 1990. This brought air emissions into the forefront of legislative and regulatory agencies throughout the United States. In addition to formal regulations, a lack of tolerance has been seen in recent years for unregulated odorous emissions. These types of odorous emission are typical of wastewater-treatment facilities and are largely unregulated. Prior to their adoption in the United States, biofilters had enjoyed much success in Europe, particularly The Netherlands, as a viable treatment alternative to a variety of air emission issues. In fact, the first biofilters are rumored to date back several hundred years to the mitigation of odors from outdoor privys; the first US patent was granted in 1957 to Pomeroy (1). However, early systems often used porous soil materials as a solid support, and primitive piping systems are used for airflow through the beds. These first attempts were moderately effective, but they were prone to channeling and poor air distribution.

Biofiltration has come a long way since 1957 and the market is expected to increase in the future. It has been estimated that the biofiltration industry would be over $100 million dollars in 2000 (2). To our knowledge, these numbers have not been verified, although biofiltration companies in the United States have seen unprecedented growth over the past 5 yr. Given the comparable capital costs of biofilters and low operating costs relative to competing technologies, it is likely that the market will continue to grow and evolve. For example, many wastewater-treatment facilities were constructed in the 1970s and were placed at the outskirts of their respective communities. As a result of urban growth and development, residential housing, offices, and businesses now surround these once semirural locations. With increased exposure to populations of people, wastewater-treatment facilities are under increasing pressure to mitigate odors on site. Because it is usually not practical to move the facility, odor-control technologies must be implemented on-site. It should also be noted that the wastewater industry is not the only industry being impacted by decreased tolerance for odorous air emissions.

2. TYPES OF BIOLOGICAL AIR TREATMENT SYSTEM

2.1. *General Descriptions*

In conventional packed-bed biofilters (Fig. 1), the vessel contains a layer, often 1–1.5 m thick, of some type of filter material such as compost or peat. The waste gas, which is

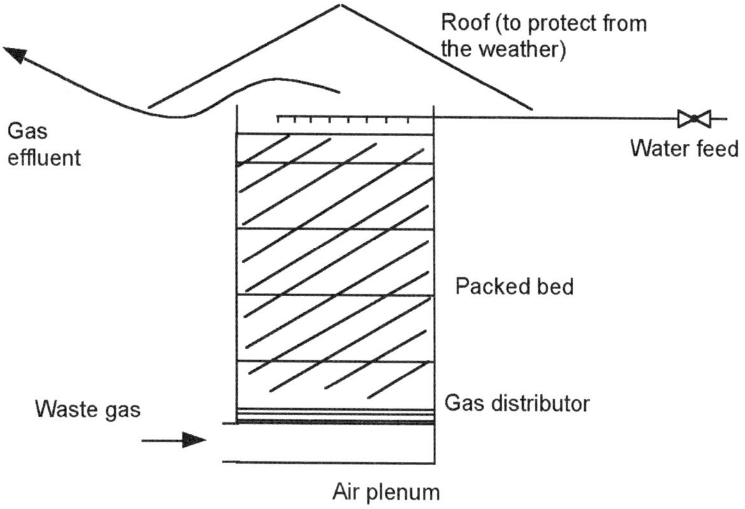

Fig. 1. Schematic of conventional packed-bed biofilter.

usually prehumidified to help prevent bed dryout, percolates up through this packed bed. Water sprays, or drip feeds (*see* Fig. 1), are positioned over/in the bed to add extra moisture to also prevent dryout, to provide a source for pH control, or to supply additional nutrients. The bed is run in a minimum liquid condition to reduce pressure drop, avoid wastage, and reduce entrainment of bacteria and production of anaerobic zones; that is, the interparticle space is largely air and the water phase is stationary on the surface of the solid support. Microorganisms are fixed within a biofilm on the solid support in this type of application. Airflow may be either upflow or downflow depending on the engineering at the site and results of pilot studies. Both airflow directions have demonstrated successes and failures, with other factors being more critical to the success of the system.

Trickle-bed reactors differ from conventional packed-bed biofilters in that the packing material is often synthetic packing *(see* Fig. 2), such as tellerettes or Pall rings, and the liquid feed into the column is much greater. The liquid phase, after trickling through the column, passes into another tank to allow settling of solids and additional biodegradation before being pumped back. The interparticle space is largely waterfilled, with the waterphase flowing through the media. Microorganisms are fixed within a biofilm on the solid support in this type of application. Airflow in this type of system is usually upflow, or countercurrent to the water flow.

The third treatment method, bioscrubbing, involves absorption of the target species into a liquid that is sprayed countercurrently to the gas flow in a tower contactor (*see* Fig. 3). The liquid phase containing the target species is then pumped around to an activated sludge tank (*see* Fig. 3) where the biodegradation occurs by using freely suspended microbes. The liquid phase is then pumped back to the absorber tower's spray feed system.

A large number of disadvantages (summarized in Table 2) prevent the widespread development of biological waste gas abatement, despite its advantages *(see* Table 6) (3–5,6).

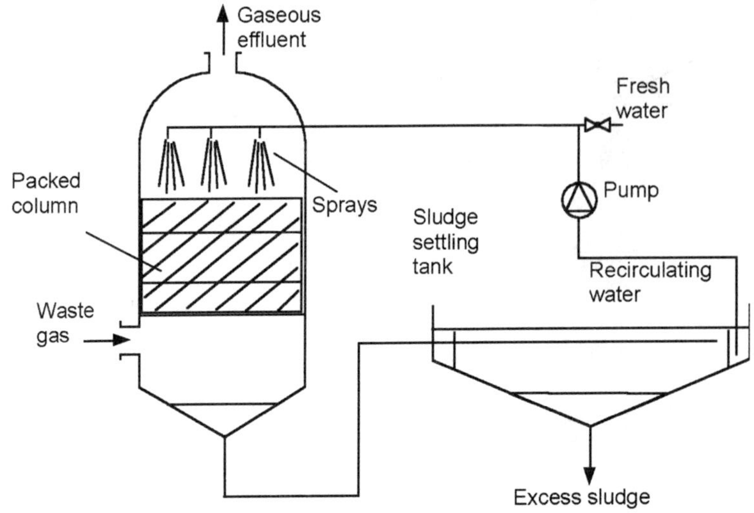

Fig. 2. Schematic of conventional trickle-bed biofilter.

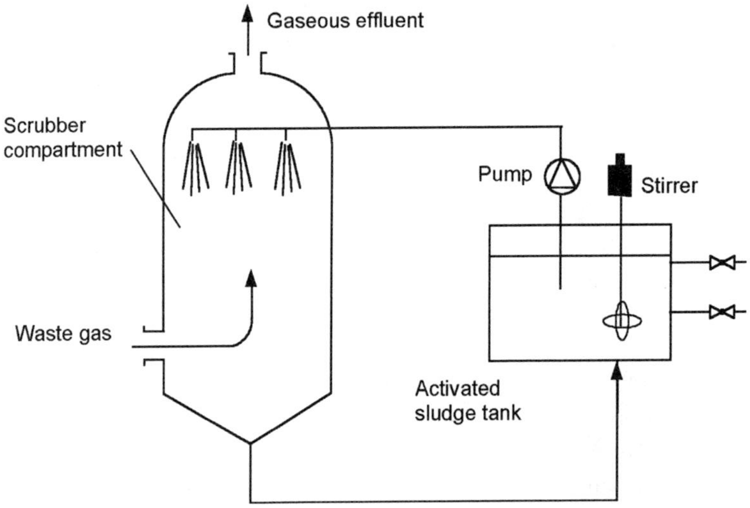

Fig. 3. Schematic of a conventional bioscrubber.

2.2. *Novel or Emerging Designs*

As discussed previously, biofilters have several limitations. Among these limitations are problems with high influent concentrations and toxicity, high flow rates and low retention times, and low solubility compounds with poor degradation. Generally, biofilters have been applied to airstreams containing high flows with low levels of contaminants. Also, biofilters have been traditionally applied to situations where the airstream contains relatively soluble compounds. However, because these limiting factors have been long realized, there has been substantial development of technologies that

Table 2
Disadvantages of Conventional Biological Odor Abatement Technologies

Conventional biofilters	Conventional tricklebed biofilters	Conventional bioscrubbers

• Packing is usually nonhomogeneous, often preventing a uniform gas distribution → shortcircuiting.
• Low specific gas flow (average for compost beds 150 m^3 gas/h/m^2 bed, max. up to 500 m^3 gas/h/m^2 bed).
• Aging phenomenon, resulting in:
  • Lumping.
  • Drying out
  • Developing of anaerobic zones because of moisture accumulation.
  • Development of shrink cracks
  • Bed compaction
  • Difficulty in maintaining an even bed pH
	• Biological overgrowth leading to increased Δp_{bed}.
	• Low specific area to reduce Δp_{bed}.
	• Drain water has to be continuously separated from excess biomass before being recycled.
	• Fresh water must be constantly fed to the system because of losses.
	• Nonhomogeneous temperature and concentration profiles.
		• More energy intensive than conventional packed-bed biofiltration.
		• Because of the large amount of liquid, there is a danger of active microorganisms being carried away.
		• More sensitive than packed-bed biofilters to feed fluctuations.
		• Operation takes place in more than one unit. The sludge tank often requires extra stirring and oxygenation.
		• Periodic removal of sludge.

address these limitations, so that biofilters may be used in a wider range of applications. If gas-phase biofiltration is going to receive increased takeup industrially, it is vital that the stability, efficiency, and range of operating conditions are improved. This section briefly addresses a number of the potential emerging technologies.

2.2.1. Pollutant Solubility in the Aqueous Phase

As described earlier, pollutant solubility may be an issue dictating method choice. Some researchers have attempted to address the limitations of water solubility by using surfactants in the biofiltration beds (7,8). The theory is that the increased solubility of the chemicals in the bed will increase partitioning into the liquid phase and thus make the chemicals more bioavailable. Lab studies have met with some success in applications dealing with chemicals produced by the forest products industry. In addition to attempts to increase solubility, changing the airflow rate has also been attempted. Because the process generally sets the airflow rates, the changes have to be made prior to the biofiltration system. These changes in flow rate are accomplished via adsorption/desorption systems; that is, high-flow, low concentrations are adsorbed on to a suitable substrate (i.e., activated carbon) and are then desorbed at a lower flow rate and possibly higher concentration (9). By lowering airflow rates, the biofilter has more contact time with the chemicals and high degradation rates. The total loading on the system can be precisely manipulated to achieve the highest degradation possible. Several companies are currently marketing systems that operate on this principle. In addition, the automotive paint industry has used this adsorption/desorption technology to trap airborne pollutants and send high concentrations to thermal oxidizers.

2.2.2. *Mobilized-Bed Biofilters*

As listed in Table 2, conventional technologies may be limited because of mass transfer or mixing limitations. Three-phase fluidized (or mobilized) beds may be an alternative to conventional packed-bed biofilter and absorber/scrubber/trickle-bed methods. They have a number of inherent advantages for multiphase contacting, such as good interphase mixing and heat and mass transfer performance. This contactor type also removes the disadvantages of poor moisture and temperature control inherent in other vapor-phase biofiltration systems.

There are some limited studies into this area (e.g., ref. 10); however, more work is required before their widespread use is acceptable, particularly in relation to process control and biological support matrices. Having said that, there are some industrial examples of mobile-bed types of biofilters/bioscrubbers, such as the SC Bioreactor ™ system in the United Kingdom (Waterlink/Sutcliffe Croftshaw Ltd, Lancashire, UK).

2.2.3. *Integrated/Train Processing*

Some preliminary lab studies have been conducted which combine biological treatment technologies into "treatment trains" for the treatment of complex waste streams containing chemicals with very different chemical properties (11,12). These systems combine the benefits of other reactor systems such as liquid reactors or chemical catalytic reactors (i.e., fast degradation rates or the ability to degrade more complex species) with biofilters for the removal of highly volatile compounds such as methanol and 2-propanol. By treating systems with "treatment trains," airstreams with over 10,000–15,000 ppmv of VOCs can be successfully treated at > 95% efficiency.

As an example, one of the possibilities is to use catalytic combustion to partially deconstruct the VOC molecules. Catalytic combustion is often not suitable alone, as the by products are often toxic in themselves. Therefore, suitable downstream treatment is important, and biofiltration offers a cost-effective route (12).

2.2.4. *Extremophilic Systems*

The operating window of many biofiltration systems is being widened by the application of so-called extremophiles, which thrive under conditions that normal microorganisms may find intolerable. For example, temperatures of over 60–80°C have been demonstrated, as have extremes of pH (both high and low), tolerance to high concentrations of pollutants, and extremely high salinity.

3. OPERATIONAL CONSIDERATIONS

3.1. General Operational Considerations

In order to understand biofilter operation, we must look at some important terminology related to the operation of biofilters. The term "empty bed residence time"(EBRT) refers to the amount of time some unit of influent air would take to pass through the empty biofilter bed space. In general, this is expressed as

$$\text{EBRT} = \left(\frac{V_b}{A_f} \right) \tag{1}$$

where V_b is the volume of the biofilter bed (m^3 or ft^3) and A_f is the airflow rate (m^3/h or cfm). The EBRT is always larger than the true residence time of the air passing through the biofiltration system. This is because the solid-support medium occupies a significant amount of the total area in the bed. The EBRT should not be used as a true measure of treatment time because of the highly variable nature of the solid-support material. The "true bed residence time" (TBRT) can be expressed as

$$\text{TBRT} = \left(\frac{V_b \times M_p}{A_f} \right) \tag{2}$$

where M_p is the medium porosity. Medium porosity can be anywhere from 20% to 80% depending on the intraparticle (space within individual particles) and interparticle (space between different particles) porosity. Porosity can be defined as

$$M_p = \left(\frac{V_s}{V_{\text{ss}}} \right) \tag{3}$$

where V_s is the volume of a given space and V_{ss} is the volume of solid-support material. The porosity of a biofiltration medium can be determined via a simple displacement experiment in a volumetric cylinder or via more sophisticated methods such as gas chromatography and the use of inert gas flow through experiments (13).

The EBRT or TBRT are usually analogous values and are directly related to the performance of the biofiltration unit. Industrial biofilters have TBRTs that can be as short as 15 s and as long as over 1 min (14). These times are usually a function of the design of the system relative to the concentration and formulation of the contaminants in the airstream. More recalcitrant, less water soluble, and so on, compounds require longer residence times. The longer the EBRT or TBRT, the better the removal of the biofilter. However, the airflow rate at most facilities is dictated by air-change rates in buildings or by the process from which the air is derived. Thus, as a designer of biofilter systems, one's only method of changing the EBRT or TBRT is to manipulate the size of the biofiltration unit. This may appear to be fairly simple; however, cost and space may not make this an easy proposition.

When evaluating the levels of contaminants to be treated, the most utilized measurement is volumetric mass loading (VL). Volumetric loading is defined as

$$\text{VL} = \left(\frac{Af \times C_I}{Vf} \right) \tag{4}$$

where C_I is the concentration of influent (g/m^3). Typically the range of VL is 10–160 g/m^3/h. Although loading is important in assessing a biofilter's needs in terms of size,

Table 3
Elimination Capacity Values for Several Biofilter Applications

Chemical	Maximum elimination capacity ($g^{-3}h^{-1}$)	Ref.
Acetone	280	12
BTEX	30	15
Hydrogen sulfide	130	16
JP-4, jet fuel	65	17
Methanol	300	18
MEK	120	19
α-Pinene	35	20
Styrene	100	21
Toluene	100	22

and so on, the term "removal efficiency" (RE) is used to express the percentage of the influent chemicals removed by the system. RE is defined as

$$RE = \left(\frac{C_I \times C_0}{V_f} \right) \times 100 \tag{5}$$

where C_0 is the concentration of the effluent (g/m^3). The term "elimination capacity" is utilized to express the overall effectiveness of the biofiltration unit and is generally expressed as

$$EC = \frac{(C_I \times C_0) \times A_f}{V_f} \tag{6}$$

or simply as

$$EC = (RE) \times (VL) \tag{7}$$

Elimination capacity is the best measure of overall biofilter performance, although, in some instances, effluent concentrations only are used for regulatory compliance. These are used for compliance purposes because many permits are based on the total mass that may be released regardless of effectiveness of the treatment system being used. A usual necessary (legislation dictated) RE will be in the range of > 95–99%, but at low influent loads, the REs will be approx 100%. However, as the loading increases, the RE will drop below 100%. This is called the "critical load" and is used in pilot systems to help size full-size units for optimal performance. Table 3 lists some ECs for a variety of chemicals being treated via different biofiltration systems.

Generally, commercial biofilters will remove anywhere from 10 to 280 $g^{-3}h^{-1}$. The higher removal is typical observed in highly water-soluble and easily degraded compounds such as acetone and methanol, whereas lower rates are observed with more complex and less water-soluble compounds such as α-pinene.

3.2. Biofilter Media

The choice of a solid-support medium for a biofiltration system could be the most critical decision in the design of these treatment systems. Solid-support media may be bioactive or inert in origin. As will be described below, the choice will also be dictated by the type/configuration of biofilter chosen.

All good biofilter support media share several common characteristics. These include the ability to support microbial growth on the surface of the particles. Materials that have rough surfaces, significant intraparticle porosity, and no inhibitory properties to bacteria are generally good at supporting a microbial population. The ideal situation is that they are resistant to breakdown and subsequent compaction. The breakdown and compaction of the media leads to numerous operational problems and requires that the media be replaced more often, thus adding cost. Often materials such as perlite are added as an aid to stop bed compaction. The medium should possess adequate water-holding capacity: usually between 40% and 70% for bioactive media and between 30% and 60% for some inerts. Unless the biofilter is of unique design, the media should possess a pH of between 6 and 8 and, ideally, would have some buffering capacity. The cost of the media relative to its lifetime should be acceptable to the operator. Each type of medium has a different cost and lifetime associated with it. It is critical that this be considered in the design, as media replacement can be a significant portion of the operating costs of a biofilter.

3.2.1. *Bioactive Media*

Some advantages of natural biofilter media are the relatively low cost and its ready availability. Natural materials such as compost and wood chips are readily available. However, they often vary significantly in their composition from one time/place to another. Bark chips can be an effective medium, but the choice of wood species is very important. For example, Douglas fir bark resists degradation more than pine bark and would save the operator the cost of media replacement and operations via lower energy costs. Although natural media have several advantages, they often encounter problems with breakdown and compaction that lead to channeling and large pressure drops across the systems. Once the media starts to break down, it can lead to significant increases in operating costs as a result of increases in energy costs. Most importantly, the degradation of natural media can lead to poor performance of the system in terms of removal efficiency. Natural solid support media can range in price from $\$10/ft^3$ to more than \$75 for such items as bagged bark.

3.2.2. *Inert Media*

Inert media has one obvious advantage: It does not break down, as natural material will. The life is often much longer and there is little, if any, degradation of the media because of microbial activity or chemical effects in the system. This allows for long-term operation and very consistent operational parameters (i.e., flow rates, pressure drops, etc.). Nevertheless, inert media may have several disadvantages. Inert material can be much more expensive than the more readily available natural material (although not always). Furthermore, many inert materials do not have much in the way of inherent nutritional value (N and P) for supporting microbial populations and thus rely more on the addition of these materials. Synthetic materials can range in cost from $\$40/ft^3$ for lava rock to $> \$100/ft^3$ for ceramic or plastic supports.

When a biofilter is being designed for a particular application, it is critical to evaluate the physical properties of the chemicals in that application to select the best solid support. It is not uncommon for a biofilter vendor to sell a "proprietary" media with

Table 4
Summary of Important Properties of Common Biofilter Materials

Property	Compost	Peat	Soil	Activated carbon, perlite, and other inert materials	Other inert materials
Natural micro-organisms' population density	High	Medium–low	High	None	None
Surface area	Medium	High	Low–medium	High	High
Air permeability	Medium	High	Low	Medium–high	Very high
Assimilable nutrient content	High	Medium–high	High	None	None
Pollutant sorption capacity	Medium	Medium	Medium	Low–high[a]	None to high[b], very high[a]
Lifetime	2–4 yr	2–4 yr	>30 yr[c]	>5 yr	>15 yr
Removal efficiency	Low	Low	Medium	N.A.[d]	N.A.
Maintenance requirements	High	High	Low	N.A.	N.A.
Space requirements	Medium	Medium	High	N.A.	N.A.
Substance adaptability	Low	Low	Medium	N.A.	N.A.
Cost	Low	Low	Very low	Medium–high	Very high

[a]Activated carbon;
[b]Synthetics coated with activated carbon;
[c]ref. 23;
[d]N.A. = not reported.
Source: Data from refs 23–25

some biofilter designs. Although these media may be appropriate, they are certainly not appropriate for all applications. In fact, proprietary media are often very costly and do not perform any better than other more readily available media. In one application, a proprietary media costing several hundred dollars a cubic foot failed in 17 d, resulting in significant downtime of the biofiltration system. The bottom line is to make sure you are aware of the needs of your exact application and pick a medium that addresses your application's needs and special circumstances (if any). Further discussion and summary of solid media choice is described in Table 4.

3.3. Microbiological Considerations

Although there are numerous engineering considerations to be aware of when designing a biofiltration system, one should always remember that these considerations would be meaningless without an active microbial population. The premise of conventional biofiltration is that a chemical passes through the biofilter bed and is transferred from the air phase to the liquid phase that surrounds the solid-support materials. This liquid phase is a biofilm where the microorganismss degrade the chemical of interest. Primarily, two forces affect the flux of chemicals from the air phase to the liquid phase. These are the aqueous solubility of the chemical and the rate of microbial metabolism in the biofilm.

Because the degradation of target compounds always occurs in the liquid phase, biofilters must maintain a hospitable environment for the microbes present in the biofilm. Generally, biofilters operate at a neutral pH of 6–8. However, some applications require low pH systems (pH of approx 2), such as the use of *Thiobacillus* species to oxidize hydrogen sulfide and other reduced-sulfur compounds. At neutral pHs, numerous genera have been identified in operational biofilters, including *Pseudomonas, Alcaligenes, Xanthomonas*, and several others. Although these organisms have been implicated in biofilter operation, there is likely to be a consortium active in a successful biofilter working together to degrade the chemicals of interest.

It is generally accepted that many types of microorganism contribute to the overall degradation of the chemicals in the system. This includes bacteria, protozoa, and fungi. Although microbial metabolism is required for destruction of the target chemical, too much metabolism can lead to biomass overgrowth and subsequent clogging because of the biofilter bed. To compound this issue, filamentous fungi can cause significant decreases in performance with only modest increases in growth because of their highly filamentous nature. Thus, when considering the growth of these systems, it is desirable to achieve a balance among chemical input, microbial growth, and microbial death. The sum of this would be a constant microbial population that could be maintained consistently over a relatively long period of time.

There has been some debate regarding the effectiveness of inoculating biofiltration units with microorganisms. It is safe to say that synthetic media require some sort of microbial inoculum. However, natural media may or may not require such inoculum. The capabilities of the indigenous microorganisms should be evaluated at bench/pilot scale to determine if they possess the required metabolic capabilities. Should the necessary organisms be present, classical microbial ecology theory suggests that the microbes most adapted (fastest degraders or most capably of surviving in the system) will outcompete those less adapted. Although inoculating may not harm a biofilter system, it may be a waste of time and resources. Conversely, inoculating synthetic media with specially selected microbes (from a laboratory enrichment for example) may significantly increase degradation rates. This inoculum may not grow in the system at a steady-state level and may lead to an overgrowth in the system and subsequent operational problems.

3.4. Chemical Considerations

It has been shown that malodorous gases often contain a rich "cocktail" of chemical species (5). Such typical compounds include hydrogen sulfide (H_2S), mercaptans, volatile organic and inorganic compounds (VOCs and VICs), volatile fatty acids, aromatic and aliphatic compounds, and chlorinated hydrocarbons. These gases can obviously pose an environmental threat in addition to their unpleasant odor. Therefore, the chemical nature of these compounds is important when choosing a biofiltration option, if possible. This section discusses the most important issues to take into account when examining the pollutant one is trying to abate.

Table 5
Comparison of Biodegradability of Various Chemicals

Rapidly degradable	Slowly degradable	Very slowly degradable
Alcohols	Hydrocarbons	Tricholorethylene
Aldehydes	Phenols	Trichlorethane
Ketones	Methylene chloride	Carbon tetrachloride
Esters	Mercaptans	Polyaromatic hydrocarbons (PAHs)
Ethers	Hydrogen sulfide	CS_2
Organic acids	Nitroaromatics	Monoterpenes
Amines		

Source: ref. 5.

3.4.1. *Biodegradability*

It has been reported that not all VOCs (5), and indeed other classes of compounds, are easily biodegradable. This results in incompatibility of the technology for all pollutant chemicals. As environmental legislation becomes tighter, more novel and efficient technologies for gas treatment will become necessary. The comparison of the relative ease of biodegradation of a number of typical pollutants is presented in Table 5.

A number of research challenges exist to ensure the total removal of pollutants. The "big picture" is how to modify existing bioreactors for the removal of major pollutants. The problem, notably with recalcitrant compounds such as trichloroethylene and polyaromatic hydrocarbons (PAHs), is that the size of the reactor to provide exit air of an approved standard is often enormous. In the rare areas of high land availability, this inefficient use of space is not a problem. Emerging technologies are being developed to solve this problem, as discussed later in this chapter.

3.4.2. *Solubility*

In developing design considerations for biofilters, or assessing if biofiltration is an appropriate treatment technology, there are numerous chemical considerations. One of the most important chemical parameters is the aqueous solubility of the compound(s) of interest. Because the biodegradation in biofiltration systems occurs in an aqueous biofilm, it is critical that the chemical be able to partition into this phase. Once the chemical is in the liquid phase, it is bioavailable, but not before. Chemical structure is also an important parameter to consider because some structures are more susceptible to biodegradation than others. Microbes can degrade chemicals at very different rates (*see* Table 5). For highly water-soluble compounds, the rate of biodegradation in the biofilm can be directly related to the rate of chemical movement from the air phase to the aqueous phase. For compounds that are not very water soluble, the rate of diffusion from the air phase to the liquid phase may limit biodegradation (26). It is desirable to have the rates of biodegradation, and so on, be correlated to the residence time of airflow through the biofilter; that is, generally the more water-soluble the compound, the more rapidly it is degraded in the biofilter and the shorter the residence time required. Conversely, the less water-soluble compounds require longer residence times because of the limiting effect of chemical diffusion. One additional consideration is the toxicity of the chemical on the microbial flora of the biofilter. Some highly water-soluble compounds, such as ethanol, may pose problems if introduced in too high a concentration; that is, the rate of solubility into the biofilm is greater than the rate of biodegradation, causing an accumulation in the biofilm and a toxic effect on the microbes (25). This toxic effect then causes a decrease in performance and a degradation of the microbial flora in the system. However, this can be addressed in some cases by using preacclimated highly tolerant microbial species.

Acidity may build up in the medium as a result of the oxidation of compounds containing sulfide, chloride, and so forth, which will yield an inorganic acid. These may be removed by water flushing at regular intervals or by using a buffering agent such as sodium hydroxide, magnesium hydroxide, calcium hydroxide, and so forth.

3.5. Comparison to Competing Technologies

As can be seen in Table 6, the odor-control techniques can be broken down into two broad categories: (1) physical/chemical: adsorption, absorption, and catalytic combustion; and (2) biochemical: biofiltration and bioscrubbing. When deciding on an odor-control strategy, a number of factors must be considered. These factors include flow rates, type and concentration of malodorous compounds, level of particulate matter, and stability of flows and concentrations. A decision also can be made based on comparing the lifetime costs of various treatment processes. As indicated earlier, biofiltration is an established technique offering the advantages of high efficiency with generally low operational and capital costs. The technology is based on utilization of immobilized bacteria or fungi in a conventional packed-bed reactor. The operation relies on absorption of the vapor-phase pollutant into a wet biofilm surrounding the solid media. Subsequently, biocatalytic oxidation takes place by means of the immobilized microbial species.

4. DESIGN CONSIDERATIONS/PARAMETERS

4.1. Predesign

It is important first to examine the pollutant gas to be treated. Important parameters that need to be assessed are the compounds that are present in the gas stream and their concentrations. Second the volumetric or mass flow rate and temperature of the gas stream to be treated is required. Ideally, it is of great use in the design process if one can obtain a history or a quantitative prediction of how these variables will vary, both temporally and particularly for the constituents, how much the relative concentrations will vary, and if any other compounds are likely to be present. If at all possible, it is ideal, if a bench-scale and/or a pilot-scale study could be undertaken, to obtain a relationship between the volumetric pollutant loading (usually expressed as $g/m^3_{gas}/h$) and the bed elimination capacity (EC, expressed in $g/m^3_{gas}/h$). A balance is required between the EC and the actual amount of pollutant removed. Often regulations state that a certain percentage of pollutant must be removed rather than an actual EC.

From the point of view of mineralization of the pollutant, the kinetics of such a process are likely to follow an inhibition-type model form. These types of model are unstructured kinetic model generally developed, or extended, from the Monod equation for substrate uptake (e.g., Haldane/Andrews, Levenspiel). The influence of the

Table 6
Summary of VOC Abatement Technologies

Technique	Advantage	Disadvantage
Reformation of the process	• Mostly removes the need to treat the VOC	• Nearly always impossible to remove ALL of the offending VOC
Absorption (scrubbing)	• Low capital cost • Reasonably high efficiency • Method is economic at high airflow rates • Good also for trapping particulates.	• High operating cost • Poor performance at unsteady state and relatively low pollutant concentrations
Adsorption	• Relatively high efficiency, especially for hydrocarbon-based systems • Compounds are recoverable	• High capital cost, especially if the unit is regenerable • Often large units required • Cost versus efficiency works best for narrow operating ranges • Prior removal of dusts and mists
Incineration (noncatalytic)	• Reliable • Good for varying concentrations and types of VOCs	• Very high capital cost • Unwanted byproducts (often toxic themselves)
Incineration (catalytic)	• Lower temperatures and higher efficiency than conventional incineration process	• Very high capital cost • Unwanted byproducts (often toxic themselves)
Masking agents	• Low capital and operating costs	• Do not remove VOC, simply "hide" it • Very specific • Unreliable (no adsorption)
Dilution and dispersion	• Inexpensive	• Nonpositive control • Not a removal technique.
Biological methods (e.g., biofiltration and bioscrubbing)	• Proven technology • Low operating cost • Good performance at low concentration of pollutant	• Variation in efficiency depending on pollutant • Not flexible to changes in gas stream concentration and loading • Poor performance at high loadings or with complex organic materials

Source: ref. 6.

inhibition term becomes more pronounced as the concentration of pollutant rises (*see also* Fig. 1).

Depending on the results of the study, it may be important to multistage the treatment process. This can be because, during the biodegradation process, some of the primary compounds or their degradation products may be recalcitrant. In this way, it may be possible to obtain, for example, a high EC for one compound, with 99+% removal, yet still be faced with approx 100% of another compound or metabolic intermediate. Intermediates can often be as environmentally dangerous as the primary compounds. Thus, to treat these other species, it may be economically (both from a capital cost and running cost point of view) or operationally attractive to have different stages, or even separate biofilters/bioscrubbers in the process.

4.2. Packing

Depending on the exact pollution application and bioreactor configuration (biofilter vs bioscrubber, vs biotrickling filter) a choice as to the appropriate packing material will need to be made (*see* Table 4). However, despite a number of the materials listed on Table 4 having a natural biological population, it still may be advisable in some cases for this population to be supplemented by "designed" or preacclimated microorganisms to result in less start-up time and potentially more stable long-term operational effectiveness.

5. CASE STUDIES

5.1. High-Concentration 2-Propanol and Acetone

It is often possible to continuously extend the range of biofiltration by use of high preacclimated microorganisms and extremophiles. For example, to treat 25,000 m^3 of high-concentration 2-propanol (IPA) and its intermediate acetone with 95% removal, the design of such a biofilter is as follows.

A bench-scale investigation reveals that it is possible to treat this stream with an inlet concentration of 15 g/m^3 of IPA (2-propanol) with a final maximum EC of 280 $g/m^3/h$. Thus, $C_I = 15\ g/m^3$ and so $C_0 = 0.05 \times 15 = 0.75\ g/m^3$

Now,

$$\mathrm{EC} = \frac{(C_I \times C_0)A_f}{V_f} = 280 \tag{8}$$

and so

$$V = \frac{(C_I \times C_0)A_f}{\mathrm{EC}} = \frac{(15 - 0.75)25000}{280} = 1272\ \mathrm{m}^3 \tag{9}$$

If we make the bed a typical depth of 1.5 m per stage and stack the bed two stages deep, then the cross-sectional area to treat this pollutant flow is

$$A_{\mathrm{bed}} = \frac{1272}{2 \times 1.5} = 424\ \mathrm{m}^2 \tag{10}$$

and, in a square configuration, this leads to an approx 21-m × 21-m square bed.

For this type of operation, at a high EC and pollutant load, a preacclimated microbial consortium would be needed (from the bench study), and an inert microbial support may, therefore, be an option. In this case, the amount of support medium can be calculated as follows, based on a 0.45 voidage:

Volume of packing = (1 – 0.45) × 1272 = 700 m^3 of packing.

Table 7
Microbial, pH, and Moisture Content Averages During the Pilot Study

	Microbial count	pH of solid support	Solid-support moisture content
July average	1.1 E8 CFU/g	6.2	35% (w/w)
August average	2.7 E5 CFU/g	6.6	30%
September average	2.9 E6 CFU/g	6.7	28%
Take down	3.4 E6 CFU/g	7.1	31%

Table 8
Airborne Chemicals Monitored During the Pilot Study

Parameter	Overall removal	Influent mean (ppm)	Effluent mean (ppm)
VOCs	93.7%	16.33 (±5.39)	0.94 (±5.72)
H_2S	100%[a]	0.06 (±0.22)	0.00 (±0.00)
Ammonia	81.6%	19.67 (±6.50)	3.61 (±5.14)

[a]Very low concentrations of H_2S.

Table 9
Estimated and Actual Costs of the Two Air Treatment Systems

	Chemical system	Biofilter
Estimated capital cost	$1,224,000	$941,000
Estimated annual O&M	$194,000	$45,000
Actual capital Cost	n/a	$1,120,000[a]
Actual O&M (First year)[b]	n/a	$45,000

[a]Includes all engineering, lava rock, and so forth.
[b]Unit has been operating for 1.5 yr.

A decision on mode of operating, such as upflow or downflow of the polluted air and method of delivery of liquid/nutrients, would subsequently need to be decided.

5.2. General Odor Control at a Municipal Wastewater-Treatment Facility

The following is a case study of a successful biofilter application for odor control of a low concentration but chemically diverse airstream. The case study describes the reasons for an air treatment system, the cost comparisons for a competing technology, a description of the decision-making process, and outcomes of the process. It should be noted that there are numerous ways to go about choosing your treatment system and this is one of many possible successful routes. However, this example does illustrate the great potential cost savings of biofiltration technology. Additional information is presented in Tables 7–10 and Fig. 4.

The Neenah–Menasha Sewerage Commission owns and operates a regional 13 mgd waste-water treatment facility serving a population of 55,000 in northeast Wisconsin. The plant serves the cities of Neenah and Menasha, Waverly Sanitary District, Town of

Table 10
Parameters Monitored During Full-Scale Operation

Parameter	Overall removal	Influent mean (ppm)	Effluent mean (ppm)	Overall average
VOCs	65%	32	11	n/a
H_2S	100%[a]	0.5	0	n/a
NH_3	100%	14	0	n/a
Solid-support microorganisms	n/a	n/a	n/a	3.1×10^6 CFU/g
Solid-support moisture content	n/a	n/a	n/a	29%[b]
pH of Solid support	n/a	n/a	n/a	6.9
Airflow rate	n/a	n/a	n/a	46,500 cfm
Influent air relative humidity	n/a	n/a	n/a	99+%
Influent air temperature	n/a	n/a	n/a	52–85°F[c]
TBRT	n/a	n/a	n/a	40 s

[a]Very low concentrations of H_2S.
[b]90% of water holding capacity of the solid support.
[c]Temperature range; largely dependent on season.

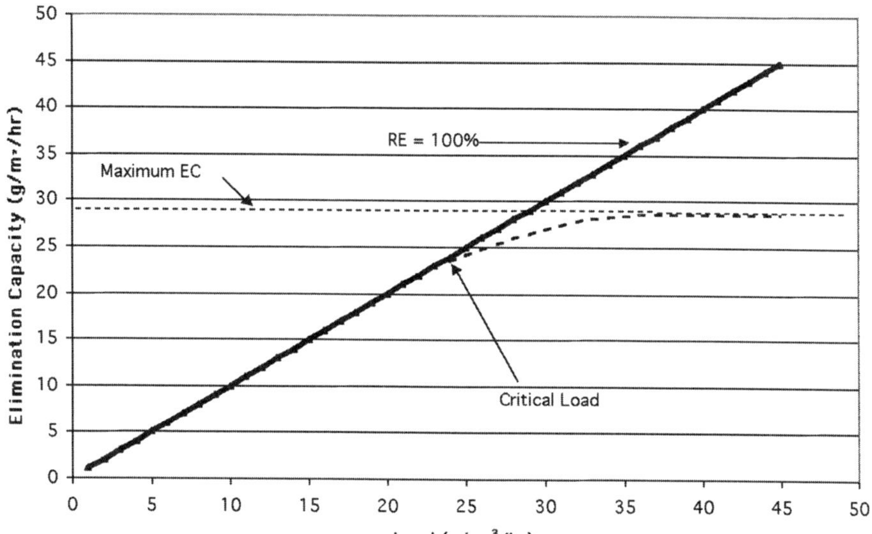

Fig. 4. Load versus elimination curve. The difference between the elimination capacity and the loading of the system is the RE of the system. (From G. Kleinheinz, unpublished.)

Menasha Utility District, and Town of Neenah Sanitary District. Major industrial contributors to the plant include U.S. Papers, Gilbert Papers, Galloway Dairy, and 20 pretreatment regulated industries. The treatment facility is located on the shore of Little Lake Butte des Morts and is surrounded on the remaining three sides by residential homes and a city park. Approximately 100 homes are within a 500-ft radius of the facility. The facility was originally constructed in the 1930s. Shortly after start-up of an expanded facility in 1986, residents began complaining about odors. In 1990, the commission authorized an odor survey that determined the main source of odors to be from the headworks and biosolids dewatering area.

Two types of vapor-phase odor-control technology were given serious consideration: wet chemical scrubbing and biofiltration. Wet chemical scrubbing is very effective in removing ammonia, hydrogen sulfide, and organic-related odors. However, the major challenge in the design of a wet chemical scrubber is minimization of chemical use and cost. Multistage systems accomplish this best, but these systems are still slaves to stoichiometry. Although effective, operation and maintenance (O&M) costs are high as well as capital costs. Biofiltration was also considered. Because our objective was a reliable low O&M cost system, biofiltration was a viable alternative.

Two vendor-offered biofilter systems, each with proprietary media and guaranteed performance, were considered. Vendor "A" offered a combination of a proprietary mixture of organic material (estimated 3- to 5-yr life) installed over ceramic balls. The estimated capital cost was $675,000 to $1,125,000 plus engineering, installation, and ducting. Media replacement cost was $75/yd^3. Vendor "B" offered a specially engineered compost media with a 5-yr guarantee and 12–20% the overall size as a typical biofilter. Their media replacement cost was $200/yd^3. Neither option was desirable to the client.

Rather than proceed with a "turnkey" vendor-supplied biofilter system, the client chose to characterize the odor constituents in the airstream and to pilot test a biofilter to demonstrate the system's performance. Lava rock was selected as the media because of its potential for long life, thus significantly reducing O&M expenditures.

After over 4 mo of operation, the 56-ft^3 pilot-scale biofilter showed excellent performance. There was no visible degradation of the solid support and biomass levels were consistent, which indicated that lava rock would likely be an effective long-term solid support. Although there were data collected on VOCs, H_2S, and NH_3 there was also a more subjective "smell test" performed by local residents, commissioners, and other interested parties. All of these tests demonstrated that the biofilter was effective in eliminating a significant portion of the objectionable odors in the airstream. Whereas the ammonia, hydrogen sulfide, and VOCs were chosen for monitoring, it was impossible to determine what portion of the total odor these chemicals actually contribute. Because extensive air analysis work indicated that there were hundreds of chemicals in the airstream, the client chose also to conduct subjective tests for odor removal. Because each chemical in the complex airstream has a different dispersion rate in air (odor threshold), it would be nearly impossible to characterize the removal of each of these chemicals from the pilot-scale system. Because each person defines "odor" differently, the client thought it was important to gain input on the pilot-scale system from local residents (who initially complained of the odors) and from the commissioners who will decide on funding for a full-scale system. All residents who smell-tested the system agreed that it significantly reduced the odors from the airstream.

As a result of the success of the pilot-scale biofilter, the cost of a 45,000-cfm chemical scrubber was compared to a lava-rock-based biofilter. The biofilter was to be constructed in two existing unused 100-ft-diameter steel tanks with an existing concrete floor/foundation and aluminum cover. The estimated and actual costs are compared next.

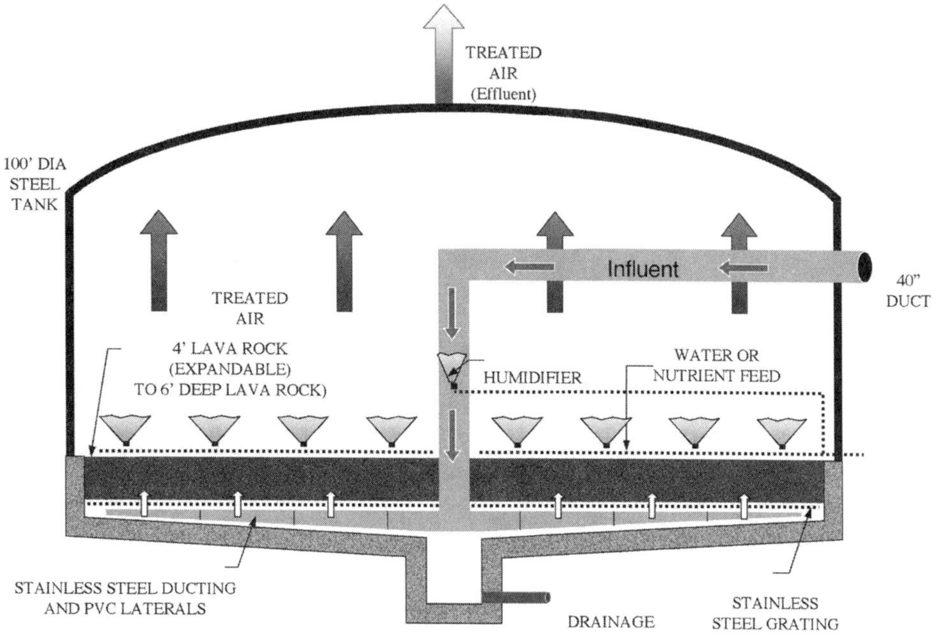

Fig. 5. Full-scale biofilter design from the above case study. (From G. Kleinheinz, unpublished.)

Based on these costs and the pilot-scale demonstration, the biofilter system was selected. Figure 5 shows a cutaway view of the basic design. Each existing steel tank was retrofitted to hold 4-ft-deep lava rock media. Stainless-steel grating was used to support the rock. PVC piping was used to distribute the foul air throughout the tank floor. A spray system using nonpotable water was used to keep the lava rock moist. Approximately 20 gpm/unit was provided to keep the lava rock moist. All drainage was collected and returned to the headworks for treatment. A chemical feed pump was provided to allow for the addition of nutrients if needed. The biofilter exhaust would exit the biofilter through the hatch openings on the aluminum covers. At a flow rate of approx 45,000 cfm, the units were sized to have an approximate empty bed residence time of 1.4 min. The lava rock has a porosity of approximately 50% for an actual residence time of approx 42 s. Based on pilot testing, this should allow for further air handling capacity in the future if needed.

The pilot test results demonstrated that biofiltration was a capital-cost-competitive, low-O&M-cost solution for effective odor control at this site. The biofilter provided the added benefit (over chemical scrubbing) of not requiring the on-site storage of large amounts of toxic chemicals. Test data allowed for a properly sized system specific to the odor constituents, rather than force-fitting a vendor system to the site. The pilot test allowed for local politicians and area residents to sample the air quality from the unit, which allowed their buy-in to the technology.

The unit has operated for over 2 yr with virtually no odor complaints from local residents. Given the relatively low cost of operation and the success in terms of public relations and odor mitigation, this application of biofiltration has been a success.

6. PROCESS CONTROL AND MONITORING

As these systems contain living entities, it is vital that proper process monitoring is carried out to ensure the long-term stability of the process. For example, if the bed dries out too much and/or high concentrations of pollutant or extremes of temperature or pH are experienced, then this may lead to a severe decrease in performance or, in the worst case, complete bed failure (i.e., pollutant breakthrough). The control and monitoring of biofiltration systems is highly variable, from little, if any, monitoring to complete monitoring of all operational and process parameters. Because the process at the facility usually dictates the airflow rate, it is often not considered a controllable variable. However, it is often important to monitor flow rates to verify that fans and the distribution system are operating properly. Generally, the more monitoring conducted, the more the operator understands about the treatment system. More importantly, the more monitoring that is conducted, the more likely that the operator will identify any upsets or changes in the system before they become operations problems that can lead to downtime. By identifying potential issues early, it is easier to correct them prior to serious damage to the microbes or equipment in the system. Although extensive monitoring is a "best case" scenario, it is often not practical or economical for some facilities. In these cases, the operator must make changes regarding which parameters to.

Often the cost-to-need ratio dictates the level of monitoring that is performed at a site; that is, if the biofiltration is for odor control only and the facility has a relatively small air treatment budget, it may choose to do minimal monitoring. Conversely, if a biofilter is being used to treat chemicals that are a regulated discharge, such as some VOCs, it may be more important for the facility to monitor the system more stringently. When a system used for odor control goes out of service, it often leads to some odor complaints for the operator, but few regulatory problems. When a system treating regulated chemicals goes out of service, it may mean that the facility will exceed its discharge permit and this could cause the facility to shut down or pay fines to exceed permit discharge levels.

Table 10 lists some parameters that are often monitored and the information that the monitoring provides the operator. If there is a need to compromise on some monitoring, the operator should use the information that is known about the process stream to help decide which parameters are most critical for that application. For example, if your system were treating a significant amount of reduced-sulfur compounds, the pH would be a critical factor to monitor because of the large amount of toxic products produced by the oxidation of reduced-sulfur compounds. In general, for biofilter systems, moisture and pH distributions are vital pieces of information.

7. LIMITATIONS OF THE TECHNOLOGY

As discussed previously, there are several "traditional" limitations to biofiltration technology, such as high concentrations of chemicals, size of some units, microbial capabilities, and process air temperatures. However, although these have been traditional limitations, recent work in both biofilter design and operation has helped overcome some of these problems of the past.

As mentioned earlier, operators are largely responsible (along with designers) for successful operation of biofilters. A biofilter operator needs to be aware of the operation parameters of the system to avoid such issues as drying out of the bed, compaction and overgrowth of microorganisms, and pH decrease, to name just a few.

As more is being understood about microbial population dynamics in these systems, the operational window for biofiltration systems is continually widening. It is imperative that when biofiltration is being considered as a treatment technology, all factors be considered prior to design and start-up, so that some potential limitations can be overcome. By using knowledgeable planning, the success stories of biofiltration will continue to expand in both total number and diversity of applications.

8. CONCLUSIONS

Biofiltration technologies are gaining wider acceptance as a viable air treatment technology. Biofilters are not applicable to all airstreams; however, recent development of biofiltration technology has seen an ever-increasing range of applications. Recent research and development of these systems has led to a better understanding of sizing, operational, and microbiological aspects of the treatment process. Biofilters are no longer the "black box" in which treatment takes place. We are now able to understand the complex chemical and biological interaction that takes place in these systems to better design them for a myriad of applications previously not considered appropriate applications of biofiltration.

It is imperative that biofilters be sized and properly fitted to their intended application. Too often, one biofilter design is adapted to many different applications with less than satisfactory results. Although the same design may be applicable for several applications, it is important that each application be evaluated on its needs and specific characteristics. These characteristics include airflow to be treated, concentration of chemicals in the airstream, temperature of the airstream, biodegradability of the contaminants, and so forth. Once these considerations, and possibly others, are evaluated, the choice to go with a biofilter can then be made. Once biofilters are decided upon as the treatment method, the designers can work on sizing, geometry, solid-support material, and so forth. depending on the characteristics of the airstream. It is imperative that the unit be properly installed and "fit" to the specific application. A bench- or pilot-scale trial is highly recommended in this context.

Once the biofilter is operational, a monitoring protocol must be implemented that allows for the evaluation of performance and for the notification of the operator of any upsets in the system. Because these are biological systems, it is imperative to find small problems before they become large problems that require downtime of the system.

In principle, biofilters are very simple methods of air treatment. However, increased understanding of the engineering and microbiology involved in the process has made them one of the more difficult treatments systems to operate effectively; that is, it takes a good understanding of engineering, the process stream being treated, and the microbiology in the system to allow for the long-term operation of these systems. If properly designed, operated, monitored, and maintained, a biofilter should allow for many years of cost-effective air treatment. This cost-effective operation will likely save the operator a significant amount (tens to hundreds of thousands of dollars, or more) in operational costs over its lifetime when compared to alternative treatment technologies.

Table 11
Parameters That Are Monitored during Various Biofiltration Applications

Parameter	Relative Importance[a]	Relative Cost[a]	Critical Information Provided
Concentrations/removal of target compounds in the airstream.	3	4 or 5	Critical information to assess the "performance" of the system. However, for complex odor applications complex monitoring may not be as valuable as the smell test at the site. This is more critical for applications where total emissions are part of a permitting or discharge process. Often the most costly and requires the most capital equipment of the monitoring parameters.
Microbial counts	2	3	Since the biofilter is a living system, this is often a cost-effective method to assess the overall health of the system. Large increases in numbers can be problematic as it may result in clogging of the system. Large decreases in counts may indicate an accumulation of toxic intermediates, changes in the airstream, or a lack of nutrients or moisture. Counts are generally greater than 1.0×10^6 Colony Forming Units (CFU) per gram of solid support.
Moisture content of the solid support	1	2	Inexpensive parameter to monitor and critical to good chemical partitioning and microbial growth. It is usually desirable to have the moisture content stable and as close to the moisture-holding capacity of the solid support as possible.
Nutrients (N and P)	1	3	Critical to the proper growth of microorganisms in the biofilter. Proper nutrient level have been shown to be a critical factor in efficient biofilter operation. Usually samples are collected and sent to a laboratory that does these analyses, thus making it a relatively easy parameter to monitor.
pH of solid support	1	2	pH of the solid support is very important to monitor due to potential acidic intermediates which are produced by biological oxidations. Most biofilters operate in a pH range of 6–8.
Porosity/integrity of the solid support	4 or 1[b]	2	The porosity of the solid support is important to determine to calculate the actual residence time in the system. It is critical to monitor this parameter for natural solid support materials as they will degrade over time. By catching the degradati on of this material early it may help the operator avoid system failures and unexpected down time.

Pressure drop across biofilter bed	2	1	One of the most critical factors to monitor and very inexpensive. Increases in pressure drop across a biofilter bed can indicate microbial overgrowth on the solid support. This overgrowth can lead to log-order inc-reases in microbial growth and increasing pressure drops. These large pressure drops can lead to large increases in electrical costs due to increased work by the motors to move air through the system. Since these electrical costs are often one of the largest operating expenses, large pressure drops can drastically increase operational costs. Large pressure drops can be a prelude to complete system failure.
Relative humidity (RH) of the influent air	2	1	This is an inexpensive parameter to monitor. Since the influent portion of many large-scale systems is difficult to access, this assures the operator the influent zone possesses adequate moisture. Influent air should be 99.9% RH for best operation.
Temperature	1	1	Easy and inexpensive parameter to monitor. Used to help assess if temperature changes can be a contributing factor to changes in biofilter performance. Generally, the closer to 70ºF the influentair is, the better performance your system will have.

[a]Relative scale is 1–5 with 5 being the most important or costly and 1 being the least important or costly.

[b]This is very important (4) if the solid support is a natural material like wood chips or bark. However, it is less important for materials that do not breakdown readily like many of the synthetic solid supports.

NOMENCLATURE

A	= Area (ft^2 or m^2)
C_I	= Influent concentration (g/m^3 or lb/ft^3)
C_0	= Effluent concentration (g/m^3 or lb/ft^3)
V_b	= Volume of biofilter bed (m^3 or ft^3)
A_r	= Airflow rate (m^3/minute or cfm)
EBRT	= Empty-bed residence time (s or min)
TBRT	= True bed residence time (s or min)
V_{ss}	= Volume of solid support (m^3 or ft^3)
M_p	= Media porosity (%)
V_s	= Volume of a given space (m^3 or ft^3)
VL	= Volumetric loading
RE	= Removal efficiency
EC	= Elimination capacity
Δp_{bed}	= Pressure drop across bed (kPa or psi)

REFERENCES

1. R. D. Pomeroy, US Patent 2,793,096, 1957.
2. J. M. Yudelson, *Remed. Manage.* **13**, 20–25 (1997).
3. S. P. P. Ottengraf, *Biological Systems for Waste Gas Elimination.* Elsevier Science, Amsterdam, 1987.
4. S. P. P. Ottengraf and R. M. M. Diks, an Int. Symp, 1991.
5. H. Bohn, *Chem. Eng. Prog.* 34 (1992).
6. P. C. Wright, Doctoral dissertation, University of New South Wales, Sydney, 1997.
7. A. J. Lewis, and G. T. Kleinheinz, *Proceedings of the 93rd Annual Meeting of the Air and Waste Management Association* (2000).
8. S. Dhamwichukorn, G. T. Kleinheinz, and S. T. Bagley, *J. Ind. Microbiol. Biotechnol.* **26**, 127–133 (2001).
9. G. T. Kleinheinz, B. A. Niemi, J. T. Hose et al., *Proceedings of the 92nd Annual Meeting of the Air and Waste Management Association* (1999).
10. P. C. Wright and J. A. Raper, *J. Chem. Technol. and Biotechnol.* **73**, 281–291 (1998).
11. M. R. Manninen, G. T. Kleinheinz, and S. T. Bagley, *Proceedings of the 92nd Annual Meeting of the Air and Waste Management Association* (1999).
12. M. Leethochawalit, M. T. Bustard, V. Meeyoo, et al., *Ind. Eng. Chem. Res.* **40**, 5334–5341 (2001).
13. G. T. Kleinheinz and S. T. Bagley, *J. Ind. Microbiol. Biotechnol.* **20**, 101–108 (1998).
14. G. Leson and A. M. Winer, *J. Air Waste Manage.* **41**, 1045–1054 (1991).
15. R. Kamarthi and R. T Willingham. *Proceedings of 87th Annual meeting of the Air and Waste Management Association, Air and Waste Management Association* (1994).
16. Y. Yang and E. R. Allen, *J. Air Waste Manage.* **44**, 83–868 (1994).
17. R. Ventera and M. Findlay, *Proceeding of the New England Environmental Exposition.* Longwood Environmental Management, Boston, MA (1991).
18. B. D. Lee, W. A. Apel, M. R. Walton, et al., *Proceedings of 89th Annual Meeting of the Air and Waste Management Association* (1996).
19. M. A. Deshussess, G. Hamer, and I. J. Dunn, *Environ. Sci. Technol.* **29**, 1059–1068 (1995).
20. M. Mohseni, D. G. Allen, and K. M. Nichols, *Tappi J.* **81**, 205–211 (1998).
21. A. P. Togna and B. R. Folsom, *Proceedings of 85th Annual Meeting of the Air and Waste Management Association, Air and Waste Management Association* (1992).
22. J. A. Don and L. Feestra, *Proceedings of Characterization and Control of Odouriferous Pollutants in Process Industries* (1984).
23. H. L. Bohn and R. K. Bohn, *Pollut. Eng.* **18**, 34 (1986).
24. A. H. Wani, R. M. R. Branion, and A. K. Lau, *J. Environ. Sci. Health. A.* **32**, 2027 (1997).
25. J. S. Devinny and D. S. Hodge *J. Air Waste Manage.* 45, 125–131 (1995).
26. G. T. Kleinheinz, Unpublished industrial data, 2002.
27. L. K. Wang, N. C. Pereira, and Y. T. Hung (eds.), *Biological Treatment Processes.* Humana Press, Totowa, NJ. 2005.
28. L. K. Wang, Y. T. Hung, and N. K. Shammas (eds.), *Advanced Biological Treatment Processes.* Humana Press, Totowa, NJ. 2005.
29. J. S. Devinny, M. A. Deshusses and T. S. Webster, *Biofiltration for Air Pollution Control.* Lewis Publishers, NY. 320 p., 1998.

12
Emerging Air Pollution Control Technologies

**Lawrence K. Wang, Jerry R. Taricska,
Yung-Tse Hung and Kathleen Hung Li**

CONTENTS

1. INTRODUCTION

The previous chapters discussed the principal technologies whereby the emissions of gaseous and particulate air pollutants can be controlled. Other control technologies also exist, and some of these are briefly presented in this chapter along with a discussion on vehicle air pollution and its control.

These additional control technologies may serve as the principal means of pollutant abatement or in a secondary role to augment the performance of other pollution control technologies. This secondary role is usually played either in a pretreatment step (i.e., prior to the main control process or technologies, as in the case of gravity separators being utilized to remove heavier particulates prior to a fabric filtration step) or in a post-treatment step (i.e., after the main control process or technologies, as in the case of entrainment separators being utilized to remove scrubbing liquid escaping with cleaned gases from gas scrubbers). Such supportive control systems as used for pre-treatment and post-treatment purposes should never be underestimated in their contribution to the

From: *Handbook of Environmental Engineering, Volume 1: Air Pollution Control Engineering*
Edited by: L. K. Wang, N. C. Pereira, and Y.-T. Hung © Humana Press Inc., Totowa, NJ

total pollution control system. Because the implementation of a single control technology may not in itself be adequate and may even directly cause other emission problems, it is often desirable to consider combinations of several different control techniques so that the final system selected is optimized from both economic and environmental considerations.

2. PROCESS MODIFICATION

Process modification should be utilized as the first and last steps when planning to control air pollution emissions. In most processes, there are many ways to obtain the desired end product. One or more of these alternatives may eliminate or, at least, reduce the emission of pollutants. Combustion operations are perhaps the best known in this regard. Boilers have been redesigned to reduce nitrogen oxide (NO_x) emissions by permitting the use of recirculated air, reducing hot zones, and eliminating flue gas quenching. Automobile engines are good examples in which redesign to eliminate cold spots and to recirculate vapors has reduced hydrocarbon (HC) emissions, and changes in timing, electronic charge distribution, and air-to-fuel ratios have improved fuel economy while reducing emissions. Furthermore, an automobile engine using the advanced fuel-cell technology is being developed in the United States for emission reduction.

High-efficiency control devices may need to be installed to meet regulations if process alterations do not result in an adequate reduction of pollution quantities. However, process improvements can change the character of the emissions to make their control easier. An ideal control device would close the process loop and return valuable product to the system. In these situations, it may be necessary to modify the process system so that it can successfully accept the returned material.

Combustion processes are examples of systems that can be modified to produce fewer pollutants (e.g., NO_x and HC), to accept return of recovered pollutants (e.g., HC), and to eliminate formation of pollutants. Elimination of fly ash and SO_2 can be accomplished, for example, by conversion from solid fuel (coal) to gaseous fuel (natural gas, which is methane, or compressed natural gas [CNG], which is propane). This, however, simply relocates the pollution control facilities because adequate gaseous fuels currently are not available, although they could be in the future if produced from coal, solid waste, or some other abundant raw material. Air pollution control would then be required at these conversion facilities. Liquefaction of these raw materials can also produce a low-pollution fuel, but the same constraint applies (i.e., air pollution control facilities will be required at the conversion site).

3. VEHICLE AIR POLLUTION AND ITS CONTROL

3.1. *Background*

Transportation vehicles in the United States have been the single largest source of air pollution emissions. In 1972, it was estimated that nearly 104 million tons of transportation source pollutants were released, which amounted to 48.7 weight percent (wt%) of total air pollution emissions. For the same year, the emission concentration of the various transportation-source pollutants by weight percent was 74.5% carbon monoxide (CO), 15.6% hydrocarbons (HCs), 8.4% nitrogen oxides (NO_x), 0.8% particulates, and 0.6% sulfur found in typical gasoline supplies in the United States. Heat emission rates of large automobiles traveling at high speeds are also enormous and reach values of 750,000 Btu/h per vehicle—over 500 times that required to maintain a comfortable temperature in a typical room or office.

In terms of fuel consumption, these transportation devices use about 140 billion gal of motor fuel in the United States each year. The only more abundantly used liquid is water. Seventy-four percent of motor fuel is consumed in highway use, of which cars account for 52%, trucks 21%, and buses and motorcycles 1%. Off-highway consumption breaks down as follows: aviation 13%, industry and construction 4%, and lawn and garden equipment 1.5%.

Motor fuels consist of gasoline, diesel, jet fuel, and liquefied petroleum gas (LPG), which is mainly butane. Cars use essentially only gasoline, whereas trucks use mostly gasoline (85% gasoline and 15% diesel on a consumption basis). Because gasoline accounts for about 75% of the total fuel consumption, most of this vehicle section discussion is related to gasoline engines. The current trend toward reducing the quantity of automotive pollutants is to encourage the use of smaller vehicles, which can give more miles per gallons, and to develop new nonpolluting vehicles using fuel-cell technology.

3.2. *Standards*

The August 8, 1977 Clean Air Amendments established the following emission standards, given in Table 1, to be met by automobiles in the United States. Depending on automobile size and type, these emissions in grams per miles may be equivalent to approximately

1.5 g/mile HC $\cong$ 120 ppm by volume
15 g/mile CO $\cong$ 6400 ppm by volume
2 g/mile NO_x $\cong$ 550 ppm by volume

No particulate emission limits have been set for automobiles, but opacity limits do exist for jet aircraft and diesel trucks. McKee (1) presents a general discussion on air quality and control, and Nevelle (2) presents some winning strategies for air pollution control using emerging technologies.

3.3. *Sources of Loss*

The maximum thermal efficiency of internal combustion engines (ICEs) is about 40%, making the overall actual automobile efficiency about 10%. In comparison, a large stationary boiler may have thermal efficiencies of over 70%, with an overall electrical generation efficiency of about 35%. From this, it is obvious that internal combustion engines could be replaced by more thermally efficient devices, but problems with mobility requirements would still exist.

In addition to the substantial thermal losses, the following list suggests other losses expressed as percent of overall efficiency. These values vary, depending on driving conditions, vehicle size and type, maintenance, and road and wind conditions:

1. Air filter element: Excess dirt can waste 20%.
2. Spark plugs misfiring can waste 12%.
3. Tires: Stiffer tires can save 5%.
4. Air conditioning can add 2–15% waste.

Table 1
Federal Certification Exhaust Emission Standards for Light-Duty Vehicle (Passenger Cars) and Light-Duty Trucks[a]

Vehicle type	Emission category	Vehicle useful life (5 yr/ 50,000 miles)						
		THC	NMHC	NMOG[b] (g/mile)	CO	NO_x	PM	HCHO[b]
LDV	Tier 0	0.41	0.34	0.165/0.1	3.4	1.0	0.20	0.018/0.018
	Tier 1	0.41	0.25	0.165/0.1	3.4	0.4	0.08	0.018/0.018
LDT1	Tier 0			0.165/0.1				0.018/0.018
	Tier 1		0.25	0.165/0.1	3.4	0.4	0.8	0.018/0.018
LDT2	Tier 0			0.165/0.1				0.018/0.018
	Tier 1		0.32	0.165/0.1	4.4	0.7	0.08	0.018/0.018
		Vehicle useful life (10 yr/100,000 miles)						
		THC	NMHC	NMOG (g/mile)	CO	NO_x	PM	HCHO
LDV	Tier 0			0.2/0.13				0.023/0.023
	Tier 1		0.31	0.2/0.13	4.2	0.6	0.10	0.023/0.023
LDT1	Tier 0	0.80	0.67	0.2/0.13	10	1.2	0.26	0.023/0.023
	Tier 1	0.80	0.31	0.2/0.13	4.2	0.6	0.10	0.023/0.023
LDT2	Tier 0	0.80	0.67	0.2/0.13	10	1.7	0.13	0.023/0.023
	Tier 1	0.80	0.40	0.2/0.13	5.5	0.97	0.10	0.023/0.023

[a]THC (total hydrocarbon), NMHC (nonmethane hydrocarbon), NMOG (nonmethane organic gases), HCHO (formaldehyde), CO (carbon monoxide), NOx (nitrogen oxides), PM (particulate matter), LDV (light-duty vehicle), LDT1 (light-duty truck 1), LDT2 (light-duty truck 2).

[b]Federal low emission standard/clean fueled vehicle standard.

Source: US EPA.

5. Automatic transmission can add 2–15% waste.
6. Power steering, brakes, and accessories can add 1% waste.
7. Emission control devices: 10% waste up to theoretical 3% savings

3.4. Control Technologies and Alternate Power Plants

The combustion of motor fuels in vehicles is accomplished by using essentially only internal combustion engines of either the spark or compression ignition design. These combustion operations can be modified to reduce emissions in the same way as the combustion examples noted in Section 2. Historically, these were the procedures first used and included recirculation of crank case vapors and air, eliminating cold spots, retarding timing, reducing the compression ratio, and the use of leaner fuel–air mixtures. These steps, however, carry the penalties of higher fuel consumption and poorer drivability.

Catalytic converters have been installed in many production model vehicles since 1975 to reduce pollution emissions. Thermal converters could be used for this, but have not because of their size and weight. The catalyst, to be most effective, should oxidize the incomplete products of combustion to produce carbon dioxide and water while reducing nitrogen oxides to elemental nitrogen and oxygen. Dual-acting catalyst mixtures and dual-catalyst beds are both used for this. The most common catalyst consists of the noble metals mixture of 70% platinum and 30% palladium in the shape of either pellets or a monolith (honeycomblike structure). Typical automobile converters are about 160 $in.^3$ in size.

The following subsections describe the various alternative engines under continuous evaluation by the automobile industry, one or more of which may provide the answer to reduced automotive air pollution and adequate transportation.

3.4.1. *Internal Combustion Engines*

Most of the current development centers on the reciprocating ICE. The catalytic converter is a piece of plumbing added after emissions have been produced. Catalysts in converters require that lead and other heavy metals be eliminated from the emissions to reduce poisoning of the noble metal catalysts. Accomplishing this, however, requires lowering the engine compression ratios so that the lower-octane fuel made without lead additives can be burned without causing knocking in the engine. Unfortunately, this also reduces the engine's efficiency.

New systems are continually being developed to improve combustion efficiency and lower emission. Most of these, in contrast to the catalytic or "after the emissions are produced" method, are accounted for by changes in the engine itself. Examples of these include the stratified charge systems, tapered cams, ultrasonic fuel atomizing, catalytic fuel cracking, and engines that operate on a variable number of cylinders.

The German Porsche's and the Japanese Honda's stratified charge systems both appear to be useful engine innovations for reducing emissions and improving fuel economy. A potential advantage of these systems is that they enable the engine to accept a wide range of fuels, which could make it possible to obtain better utilization of fuels from crude oil, coal, or solid-waste sources. The Porsche engine inlet as shown in Fig. 1 has two combustion chambers: main and auxiliary. A very weak mixture is supplied to the main chamber through the conventional inlet valve. A rich yet combustible mixture is injected into the auxiliary chamber where spark ignition occurs. The flame rapidly spreads from the calm auxiliary chamber into and throughout the turbulent weak mixture in the main chamber. Combustion occurs regularly and consistently over a wide range of speed and load conditions. This combustion procedure results in low emissions and high fuel economy.

The stoichiometric combustion of gasoline with air results in an air-to-fuel ratio of about 14.7 lb of air per pound of fuel. The Porsche engine runs well on air-to-fuel ratios ranging from 0.8 to 2.2 times the stoichiometric rates. Maximum power is actually attained at 0.9 times the stoichiometric ratio (3).

Honda's compound vortex controlled combustion (CVCC) stratified charge engine is similar to the Porsche system in that a precombustion chamber is used. The CVCC unit uses three valves (two standard valves plus one valve for the precombustion chamber) to cause the rich fuel–air mixture to ignite then swirl with minimum turbulence into the main chamber to complete the combustion.

The use of tapered cams can help to reduce emissions. Normal engine valves are opened and closed by cams that do not allow for any change in valve timing or lift with changes in engine speed. At higher speeds, it is desirable to open the valves earlier and keep them open longer to ensure the complete filling and emptying of the combustion

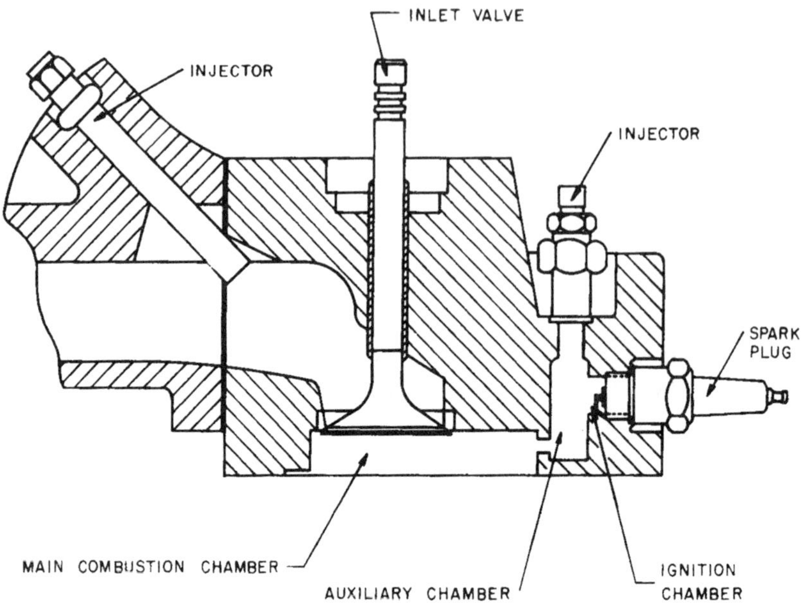

Fig. 1. Porsche combustion chamber inlet.

chamber. This, however, would cause increase overlap and rough operation at low speeds and idle. To overcome the problem, tapered cams operated by centrifugal governor control are being developed to provide variable valve timing and lift.

Carburetion has always been a problem in that all cylinders cannot be utilized with effectiveness by one atomization device per engine. As an alternative to having several carburetors per engine, systems that will improve atomization are being evaluated. Ultrasonic atomization is one system under study. Another is a catalytic carburetor, which cracks the gasoline to produce a more gaseous fuel for combustion instead of atomized liquid droplets.

One of the newest innovations is the development of a system to permit an eight-cylinder engine to operate on, for example, four, five, six, seven, or eight cylinders. This is accomplished by the use of an electronic control to close the valves of cylinders not being used. For example, during cruising periods, only four cylinders would normally be used, whereas when acceleration or load requirements increase, more cylinders would activate to provide the needed extra power.

3.4.2. Diesel Engines

Diesel engines appear to have the ability to meet the HC, CO, and NO_x standards of the 1980s. Because of their higher compression ratios, these reciprocating compression ignition engines are more efficient than gasoline engines and they can burn a cheaper grade of fuel owing to the positive-timed fuel-injection systems. These engines have not been widely accepted for automotive use because of their poorer performance and noise problems. It is the dominant engine for all land propulsion systems larger than the passenger automotive engine and in all ships except small pleasure craft. Diesel engines also serve a large portion of the emergency power generation systems.

At the higher compression ratios used in diesel engines, it can be seen from the following formula how greater brake horsepower (bhp) is developed:

$$\text{bhp} \propto (\text{mep})LAN \tag{1}$$

where mep is the mean effective pressure, L is the piston movement length, A is the cylinder cross-section area (displacement), and N is the number of power strokes per minute in all cylinders.

3.4.3. Steam Engines

The steam or Rankine external combustion engine (ECE) has a higher potential maximum efficiency than the ICE but has been plagued by problems with fluid selection and pumping. The liquid in a Rankine cycle is reversibly pumped into a boiler, where it is vaporized and then reversibly expanded to produce work. Rankine cycle efficiency η_R is the net reversible cycle work divided by boiler heat input:

$$\eta_R = W_{\text{Net}}/Q_{\text{Boiler}} \tag{2a}$$

In a rotating engine, work is being done on or by the power fluid every time it travels radially, as long as the fluid is constrained to rotate with the engine. Work done by the pump and added to the fluid per unit weight of fluid is

$$W_p = \Delta H = (\omega^2/2g_c\, R_C^{\,2} - R_B^{\,2}) \tag{3}$$

where ω is the angular velocity, g_c is the dimensional constant (32.174 ft-$\text{lb}_\text{m}/\text{lb}_\text{f}\,\text{s}^2$, or 1 k-m/N-$\text{s}^2$), and R_C and R_B are the radial distances to liquid levels of the condenser and boiler, respectively.

Referring to Fig. 2, the liquid pump work ΔH is $H_7 - H_6$ and vapor pump work ΔH is $H_4 - H_3$. The net reversible engine work is

$$W_{\text{Net}} = H_2 - H_4 \tag{4}$$

and the Rankine cycle efficiency can then be expressed as

$$\eta_R = (H_2 - H_4)/(H_1 - H_7) \tag{2b}$$

In this cycle using typical organic fluids as shown in Fig. 2, expansion after the boiler is into the superheated vapor region and the work is less than Carnot cycle work by the shaded areas. In Fig. 2, T_c is the fluid critical temperature. Regenerators can be used to improve the Rankine cycle efficiency, as shown in Fig. 3.

The steam produced in the engine is expanded in the turbine to drive the vehicle. Turbine stage efficiency η_t is the shaft power divided by energy available to the expander between the nozzle inlets H_2 and the diffuser inlet H_3, as expressed by

$$\eta_t = \text{Gross shaft power} / [(H_2 - H_3)\ (\text{Mass flow rate})] \tag{5}$$

The mass flow rate can be calculated using the critical fluid rate in a supersonic nozzle multiplied by the throat area of all nozzles (assume a discharge coefficient of 0.98).

Currently, steam engines can operate at 25–30 miles/gal on kerosene (depending on vehicle size and type), start up and run in less than 30 s, operate when the temperature is −20°C, and require relatively small boilers. A reciprocating steam engine rated at 100-horsepower (hp) output would represent a practically sized power plant for automobile utility.

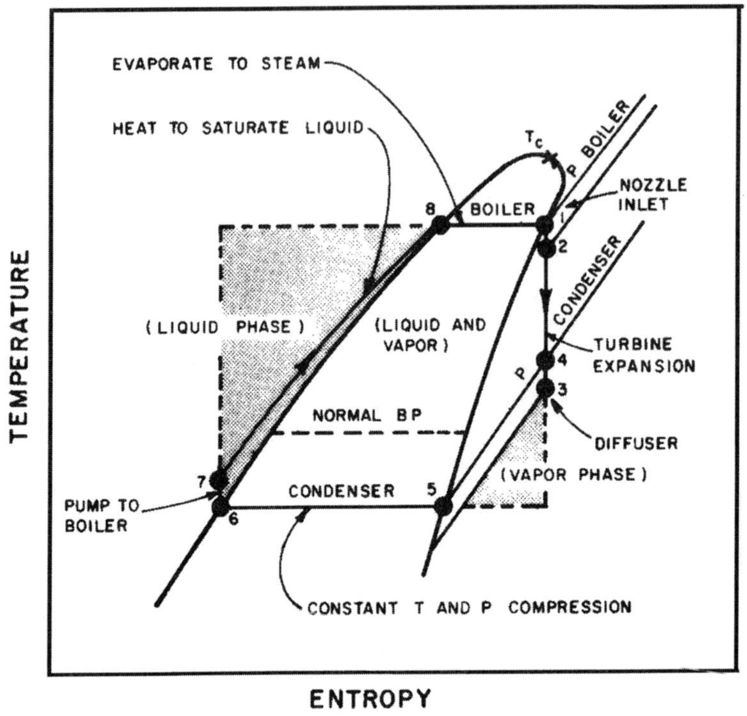

Fig. 2. Temperature–entropy diagram for Rankine cycle turbine engine with a rotating boiler.

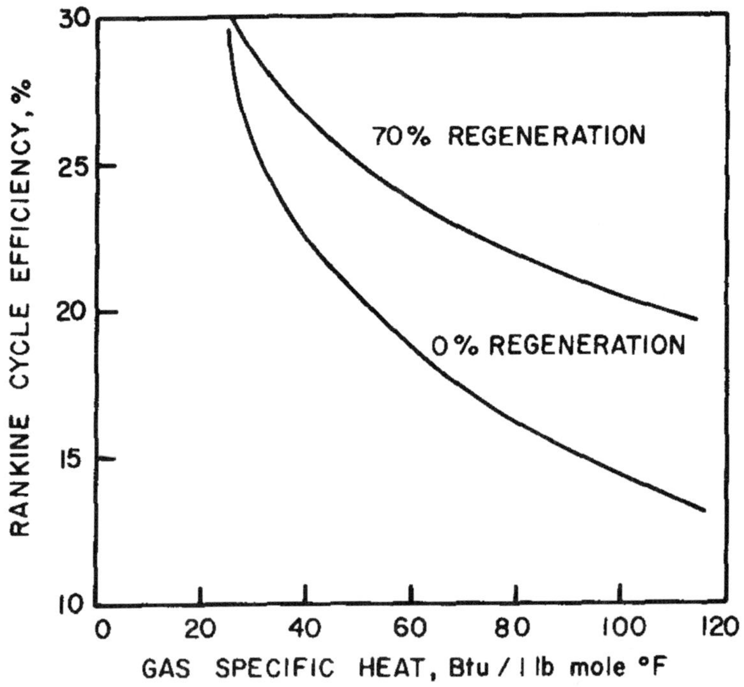

Fig. 3. Typical Rankine cycle efficiency as a function of working fluid vapor heat capacity.

3.4.4. Gas Engines

In the early 1950s, it was predicted that the gas turbine or Brayton cycle engine would replace the ICE within 10 yr. This has not occurred in the automobile field, but large power output units of this type have taken over in the propeller aircraft field. This system is inherently a high-power-output device that works well under constant load and constant speed. Its exhaust emission characteristics are superior to the spark-ignition reciprocating engine at a given energy output.

3.4.5. Rotary Engines

Rotary engines have had their ups and downs over the years but have still not become competitive with internal combustion engines. In 1974, rotary engines obtained less than 11 miles/gal, but by 1976, this had been improved by 40% after revising both the engine and afterburner. The configuration of the combustion chamber and intake port were modified, and the intake port timing was changed to allow use of leaner fuel–air mixtures. Exhaust gas temperatures are high so thermal reactors are used instead of catalytic converters, but these must be well insulated to keep the efficiency of operation high.

Interested readers may review refs. 4–6, which deal with various topics regarding automotive emissions and their control.

3.4.6. Fuel-Cell Engines

In 2002, US President George W. Bush announced his administration's support of fuel-cell development for automobiles. "Freedom Cooperative Automotive Research," or FreedomCAR (7–10), represents a major US energy policy direction change that is being strongly supported by the automobile industry. The ultimate success of FreedomCAR would create energy stability, energy security, and a lessened impact of transportation on our environment.

Fuel cells are actually a family of technologies. All types generate power by passing hydrogen over a catalyst to release electrons, which provide electrical power, and protons. The latter migrate through an electrolyte to a second catalyst, where they combine with oxygen to form water (9,10).

Automotive engineers have focused on proton-exchange membrane (PEM) fuel cells. They operate at relatively low temperatures, deliver high power/weight ratios, and could prove inexpensive to manufacture if researchers find a way to reduce the cost of the membrane.

More recent developments on fuel-cell-powered automobile engines are discussed in Section 10.

4. MECHANICAL PARTICULATE COLLECTORS

4.1. General

The term "mechanical collectors" is not a truly descriptive word that can be used to define a group of devices used to remove particulates from a gas stream. Devices such as cyclones and gravitational settling systems can be included in this category, yet, clearly, thesc units use centrifugal and gravitational forces, respectively, to perform the work, and mechanical assistance is only used to move the material in and out of the devices.

The title of this section, "Mechanical Particulate Collectors," is used mainly as a catchall to include a number of types of particulate collection equipment. Filters, cyclones, and scrubbers have been discussed in other chapters in this handbook. Cyclones operate on centrifugal and gravitational forces. Gravitational collectors themselves have not been covered and will be included in the following sections.

Particles are removed in scrubbers and filters by inertial impaction, interception, and phoretic forces. These forces, as related to various devices, will be discussed in Section 4.3. In addition to methods discussed in this section, the reader should note that particulate removal collectors have also been included in Section 5.

With respect to particles, the terms aerodynamic diameter and aerodynamic density are sometimes used to characterize particulate matter moving in a gas stream (more accurately) compared with the true properties of the material. Aerodynamic diameter, d_{pa}, is defined as

$$d_{\mathrm{pa}} = d_{\mathrm{p}} (\rho_p C/C_a)^{0.5} \tag{6}$$

where d_{p} is the particle diameter, ρ_p is the particle density, C is a correction factor [see Eq. (14)], and C_a is a correction factor applied to the aerodynamic diameter. Aerodynamic density, ρ_a, is the apparent density of suspended matter, which often ranges from 0.1 to 0.7 times the true density.

4.2. Gravitational Collectors

Much of the suspended and entrained particulate matter can be removed from the gas medium by providing a place where the particulates can settle out under gravitational force. In the absence of other forces, a force balance on a particle in still gas shows that, at steady state, the gravitational force F_G is essentially equal to the sum of the buoyant force F_B plus the drag force F_D:

$$F_G = F_B + F_D \tag{7}$$

For spherical particles, Eq. (7) can be rewritten using Newton's drag equation for F_D:

$$(1/6\pi)d^3 \rho_p g = (1/6\pi) d^3 \rho_g g + C_D \rho_g (V_p - V_g)^2 A/2 \tag{8}$$

where d is the diameter of the particle, ρ_p is the density of the particle, ρ_g is the density of the gas, g is gravitational acceleration, V_p is the velocity of the particle, V_g is the velocity of the gas, and A is the projected area of the sphere ($\pi d^2/4$). The dimensionless drag coefficient, C_D, is also known as the Fanning friction factor (f) and by other names. It is related to the drop Reynolds number, Re, by the approximations

$$C_D = 24 / \mathrm{Re} \qquad \text{for Re} < 0.1 \tag{9}$$

$$C_D = 18.5\ \mathrm{Re}^{-0.6} \qquad \text{for } 2 < \mathrm{Re} < 10^3 \tag{10}$$

$$C_D = 0.44 \qquad \text{for } 10^3 < \mathrm{Re} < 2 \times 10^5 \tag{11}$$

The drop Reynolds number is defined as

$$\mathrm{Re} = d(V_p - V_g)\, \rho_g / \mu_g \tag{12}$$

where μ_g is the viscosity of the gas.

For small spherical particles that are in the low-Reynolds-number region, Eq. (8) can be resolved to produce Stokes's equation for the terminal settling velocity V_s:

$$V_s = d^2 (\rho_p - \rho_g)\, g / (18\, \mu_g) \tag{13}$$

This yields good results for particles from about 3 to 30 μm in diameter. Particles with diameters less than 3 μm tend to slip through the gas molecules, and the terminal settling velocity must be corrected by multiplying Eq. (13) by the Cunningham slip correction factor C:

$$C = 1 + [(2T \times 10^{-4})/d]\ \{2.79 + 0.894 \exp - [(2.47 \times 10^3)(d) / T]\} \tag{14}$$

Note that in this equation, T is the absolute temperature in degrees Kelvin and D is the particle diameter in micrometers.

In settling devices, it is usually assumed that the particles fall in a quasistationary manner; that is, particles reach terminal-settling velocity instantaneously. However, it is necessary to consider the forward motion of the particles to make sure that the particles are not thrown out of the other side of the device. Entering particles are assumed to be moving at the same velocity as the entering gas. This makes it necessary to evaluate a non-steady-state force balance where the resultant force is essentially equal to the sum of gravitational force, and particle-stopping distance X_s is obtained by resolving this equation at low Reynolds number for spherical particles:

$$X_s = V_0\, d^2 \rho_p / (18 \mu_g) \tag{15}$$

where V_0 is the initial velocity. The distance for particles of less than 3 μm in diameter is obtained by correcting Eq. (15) by multiplying by C.

The size of particle that can be completely removed in a gravity separator can be found using

$$d_m = (36\, V_0\, h\, \mu_g / \rho_p\, g\, L)^{1/2} \tag{16}$$

where V_0, h, and L are shown in Fig. 4.

The settling chamber fractional efficiency η for specific size particles can be estimated using

$$\eta = 0.5\, V_s\, L / (V_0\, h) \tag{17}$$

4.3. Other Methods

Many forces, including gravity, which was just discussed, are available for use in particle collection devices. Systems that use centrifugal and electrostatic forces are covered in other chapters. This leaves devices that use forces such as inertial impaction phoretic forces, interception, and thermal, sound, and magnetic forces for operation. Additionally, there are hybrid systems utilizing various combinations of forces. Interception, which is the sticking of a particle that just grazed the collector as it passed, is often not distinguished from impaction. Magnetic forces are usually only used for very large material. Thermal precipitators and sonic agglomerates are specialty systems and have not been used for control of particulates on a large scale.

In Sections 4.3.1–4.3.3, inertial impaction, phoretic forces, and hybrid devices are briefly considered. In these subsections, particular emphasis will be placed on fine

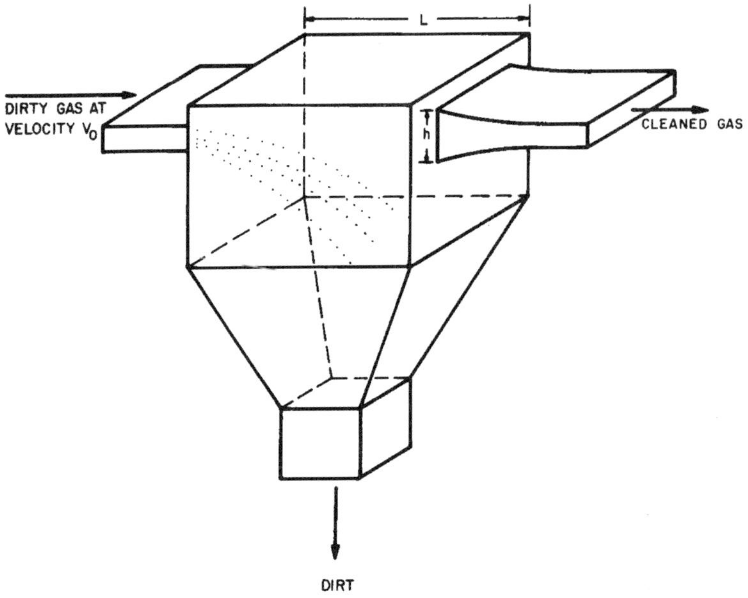

Fig. 4. Gravity settling chamber.

particle collection (< 3 μm), as practically all devices can remove larger particles. Particles from about 0.2 to 3 μm in diameter are the most difficult to remove, whereas particles with diameters smaller than 0.2 μm become easier to remove because of their diffusivity. Particle diffusivity can be estimated using the Stokes–Einstein equation:

$$D_{PM} = C\,K\,T\,/\,(3\,\pi\,\mu_g\,d) \tag{18}$$

where D_{PM} is the diffusivity of particle through continuous medium (cm^2/s), C is the Cunningham slip factor [Eq. (14), dimensionless], K is the Boltzmann's constant (1.38×10^{-16} g cm^2/s^2 K particle), T is the absolute temperature (K), and μ_g is the gas viscosity (g/cm s).

4.3.1. Inertia Impaction

Particles moving at high velocities toward a target often can be removed from the gas if the particles strike or impact on the target and do not become re-entrained into the gas stream. Wet scrubbers are good examples of devices utilizing inertial impaction and are covered in detail in another chapter. Collection by interception also takes place in this type of device. Diffusion is another important mechanism in particle removal by filtration.

The dimensionless impaction parameter ψ is an important indicator of impaction system effectiveness, and several forms of this parameter, which may be thought of as the ratio of particle stopping distance divided by the radius of the collector, are summarized here for the different systems.

For scrubbers operating in the Stokes region (0.04 < Re < 1.4) with atomized droplets as the collectors,

$$\psi = C \rho_p V d^2 / 9 \mu_g D_c \tag{19}$$

where V is the particle velocity relative to the target and D_c is the collector diameter (consider the larger droplet as the collector).

For impactor devices in which particles pass through a slit or opening and are captured by a plate-type collector,

$$\psi = (C \rho_p V d^2 / 18 \mu_g D_c)^{1/2} \tag{20}$$

where V is the particle velocity relative to the target and D_c is the slit width.

Efficiency of impactors can then be expressed for individual size particles as a function of the impaction parameter. Calvert (11) showed that the fractional collection efficiency η_p for spherical drops collected by droplets when $\psi > 0.2$ is

$$\eta_p \cong [\psi / (\psi + 0.7)]^2 \tag{21}$$

No impaction occurs on spheres when $\psi < 0.083$.

In contrast, particulate removal efficiencies can be predicted on an overall basis for impaction devices using empirical data. For example, Hesketh (12) showed that overall collection efficiency by weight percent E_o for a Venturi scrubber is

$$E_o = L + (1 - 3.47\ \Delta P^{-1.43})F \tag{22}$$

where L is the percentage of particles larger than 3 μm, F is the percentage of particles smaller than 3 μm, and ΔP is the Venturi scrubber pressure drop (inches of water). Furthermore, orifice scrubbers follow Eq. (22) if the orifice scrubber pressure force drop is divided by 2 to obtain the value of ΔP for use in the Venturi equation (22).

Equation (22) is applicable to open-throat Venturi systems in which the gases do not exceed 600°F and the particulates are somewhat wettable (i.e., they are not hydrophobic). The equation is also applicable for a wide range of materials because the data were obtained from flue gas, lime kiln, black liquor recovery, sinter furnaces, blast furnaces, foundry cupola, and terephthalic acid processing operations. The equation is based on the fact that all particles with diameter greater than 3 μm are captured according to Hesketh (13) with a penetration (one minus efficiency fraction) of

$$C_o / C_i = 3.47\ \Delta P^{-1.43} \tag{23}$$

where C_o / C_i is the ratio of concentration out to concentration in.

The pressure drop of the Venturi scrubbing system can be estimated using

$$\Delta P = V_t \rho_g A^{0.133} (L')^{0.78} / 1{,}270 \tag{24}$$

where ΔP is the Venturi pressure drop (inches of water), L' is the liquid-to-gas ratio (gal/1000 actual ft^3 wet gas leaving throat), ρ_g is the gas density downstream from the Venturi throat (lb/ft^3), and V_t is the throat gas velocity based on wet gas (in actual ft/min) downstream throat (ft/s), and A is the throat cross-section area (ft^2).

4.3.2. Phoretic Forces

Phoretic or radiometric forces include diffusiophoresis, Stephan flow, photoporesis, and thermophoresis, with diffusiophoresis being the most significant. These forces are exerted by a gas on particles in the gas because of nonuniformity of gas molecule energy and they are only effective on small, submicron-sized particles.

Diffusiophoresis is the net particle motion resulting from the diffusion of two or more types of gas molecule. For example, air molecules are heavier than water vapor molecules. If there is movement of water vapor into dry air (i.e., a net exchange of air–water molecules), particles would tend to move in the direction of the heavier air molecules.

Stephan flow is a form of diffusiophoresis that consists of the movement of particles caused by a net flow of gas molecules and is not usually considered separately. As an example, condensing vapor can cause the particles to move with the molecules to the water drop, where it is collected. Stephan flow is directed toward the liquid surface during condensation. Evaporating vapor causes the reverse and impedes collection.

Thermophoresis is the repulsion of particles by heated gas molecules on one side of the particles in the presence of a temperature gradient. *Photophoresis* is the motion of particles resulting from gas molecules rebounding from the illuminated (hotter) side of a particle. This side can be away from the light in the case of transparent particles or toward the light with opaque materials.

In order to optimize fine particle collection efficiency in any device, the phoretic forces must be considered and made to assist, not retard, collection. For the most part, precooling of extremely hot gases by saturating them to near the adiabatic saturation temperature is often the best rule of thumb to effectively take positive advantage of these forces. For the actual deposition rate ϕ of an aerosol on a spherical drop upon which vapor is condensing, Fuchs (14) presented the equation

$$\phi = [4\pi D_{GM} (C_\infty - C_s) N_0 R] / [C_0 \{\exp[D_{GM} (C_\infty - C_s) / D_{PM} C_0] - 1\}] \quad (25)$$

where D_{GM} is the gas diffusivity, D_{PM} is the particle diffusivity, C_∞ is the vapor concentration in the mass of gas, C_S is the vapor concentration on the drop surface, C_0 is the gas concentration, N_0 is the drop concentration, and R is the radius of the wet spherical droplet.

4.3.3. *Hybrid Collectors*

Numerous variations of the basic particulate collectors are made and sold. Most function like one of the basic systems, but there is one hybrid that deserves special mention. This is the combination of electrostatic charging with scrubbing to produce the charged wet scrubber (CWS). The removal of fine particulates (< 3 μm in diameter) is increased in this system by charging either the particulates and/or the collecting droplets.

Residual charges on particles leaving an electrostatic precipitator (ESP) improve scrubbing efficiency. Preliminary data (13) show that this improvement expressed as penetration, for a Venturi scrubber with charged particles, in contrast to Eq. (23) for uncharged particles is in the order of

$$C_o / C_i \cong 3.45 \times 10^{-7} V_t^{3.56} A^{0.145} (\rho_g \Delta P)^{1.78} \quad (26)$$

4.3.3.1. Charged Wet Scrubbers

These devices have been offered for commercial applications since 1975, and by early 1976, there were many commercial suppliers with others being developed. The charged wet scrubber (CWS) should be considered in applications in which wet scrubbers are desirable but may not give satisfactory collection efficiency in the submicron range. For example, it is noted in ref. 15 that under a similar diameter, dioctyl phthalate aerosol particles by 50 μm diameter water drops are computed to be approximately the following: inertial scrubbing, 25%; charged wet scrubbing (particle charged but not drops), 85–87%; charged wet scrubbing (particles and drops charged), 92–95%.

The capital investment and operating costs of CWSs are very similar to those of conventional wet scrubbers, yet they can yield collection efficiencies approaching those of electrostatic precipitators. Aerosol particles can be charged during generation by the subsequent contact and released from charged surfaces or by ion diffusion, for example, from a corona discharge. Normally, only small charges are present on particles, and in a homogeneous system, no net charge is apparent. However, some systems possess a natural charge that could be utilized. Examples of systems that could contain particulates with an initial inlet charge are scrubbers that follow electrostatic precipitators or scrubbers on high-temperature gases.

There are a number of complex factors that exist in charged wet scrubbers. Some of these are as follows:

1. The gas velocity and equipment physical geometrical size and arrangement establishes a residence time τ_{res} for the particle during which it is available to be removed from the gas.
2. Particles can be removed by impaction only if they are within a specific interception area and have an appropriate drag-to-viscous force ratio (impaction parameter); this results in a characteristic particle scrubbing cleaning time τ_{sc}.
3. Scrubbing drops are accelerated or decelerated to the gas velocity resulting in a specific scrubbing lifetime for the drops τ_{SR}.
4. Charged particles can interact with each other, and either neutralize or self-precipitate, resulting in a charged particle self-removal time τ_a.
5. Charged drops can also interact with each other or otherwise lose their charge, resulting in an effective charged drop lifetime τ_R.
6. As in an ESP, the particles must travel to the drops (or vice versa) or walls; this is dependent on the space charge and is expressed as a cleaning time τ_c.

It becomes apparent that it would be desirable to have long times for all the above with the exception of the cleaning times τ_{sc} and τ_c. τ_{res}, τ_{sc}, τ_{SR} are basic to all wet scrubbers. The other times become appropriate only when the particles and/or drops in the scrubber are charged.

For a wet scrubber to be effective, τ_{res} must be greater than τ_{sc}. Limiting the discussion to submicron-sized spherical particles in air near-normal conditions, it can be shown that particle scrubbing time τ_{sc} is a function of the relative velocity between the particle and drop *w* and particle radius *a*:

$$\tau_{sc} = 3\mu_c^{\,3} / w^3 N a^4 \rho_a^{\,2} \tag{27}$$

where μ_c is the gas viscosity divided by the Cunningham correction factor [Eq. (14)], *N* is the number of drops per unit volume, and ρ_a is the particle aerodynamic density. As *w* and *a* decrease, the time required for particle removal by scrubbing increase drastically; τ_{res} is simply the active volume divided by gas volumetric flow rate.

In the scrubber , τ_{SR} also should be greater than τ_{sc}. Assuming that the Stokes viscous drag theory is applicable and considering spherical particles in the absence of other forces, the inertial lifetime of the scrubbing drop τ_{SR} can be expressed as

$$\tau_{SR} = 2\,\rho_R R^2 / 9\mu \tag{28}$$

where R is the drop radius (cm), μ is the gas viscosity [g/(cm s)], and ρ_R is the drop density. The drop-scrubbing lifetime is often a scrubber particulate collection-limiting factor, but as the drop speed approaches the gas speed, the drop becomes a better gas absorber.

In CWSs, the remaining time constants are defined for submicron-sized particles:

$$\text{charged particle self-removal time} = \tau_a \equiv \varepsilon_0 / bqn \tag{29}$$

$$\text{charged drop lifetime} = \tau_R \equiv \varepsilon_0 / BQN \tag{30}$$

$$\text{charged drop particle cleaning time for oppositely charged particles} = \tau_c \equiv \varepsilon_0 / bQN \tag{31}$$

where ε_0 is the dielectric constant of free space [8.85×10^{-14} C/(cm V)], b is the particle mobility which from the Stokes model $\cong q/6\,\pi\,\mu_c a$, B is the drop mobility which from the Stokes model $= Q/6\,\pi\,\mu R$, q is the particle charge, Q is the drop charge, n is the particle number density, and N is the drop number density. These equations show that the characteristic times are related to mobilities and charge densities (charge density or space charge is charge times number concentration). The mobility of charged particles and drops are both charge and size dependent.

In a typical scrubbing situation, the concentration of particles in the inlet gas is relatively low compared to the concentration of collector drops in the scrubber. Therefore, the particulate space charge (qn) in a CWS would likely be less than the drop space charge (QN). Because of this, the factor limiting collection efficiency in a CWS is often the effective lifetime of the drop τ_R. A charged drop can be removed from the system by its own field long before its capacity to collect particles is significantly reduced. Note from the defining equations that increasing the charge on the drops decreases both cleaning time and drop lifetime. Assuming saturation charges on both drops and particles, the ratio of drop lifetime to cleaning time is

$$\tau_R / \tau_c = b / B \cong a / R \tag{32}$$

The effectiveness of a CWS versus an inertial wet scrubber can be compared by observing the ratio of their respective cleaning times. At conditions in which saturation charges resulting from impact charging in a field with intensity E_c are obtained,

$$q \cong 12\,\pi\,\varepsilon_0\,a^2\,E_c \tag{33}$$

$$Q \cong 12\,\pi\,\varepsilon_0\,R^2\,E_c \tag{34}$$

The cleaning time ratio becomes

$$\tau_c / \tau_{sc} = w^3\,\rho_a\,a^3 / 226\,\mu_c\,\varepsilon_0 R^2 E_c^{\,2} \tag{35}$$

For a CWS to perform better than an inertial wet scrubber, this cleaning time ratio must be greater than 1.0. Note the significance of this equation: As the particle radius a *decreases*, the charged scrubbing effectiveness *increases*.

The self-precipitation of particles increases the effectiveness of CWSs. As an example, both the calculated and the measured collection of positively charged aerosol particles upon negatively charged drops as a function of drop charge voltage are shown in Fig. 5. The dashed portion of the curve would be expected if no self-precipitation of particles occurred.

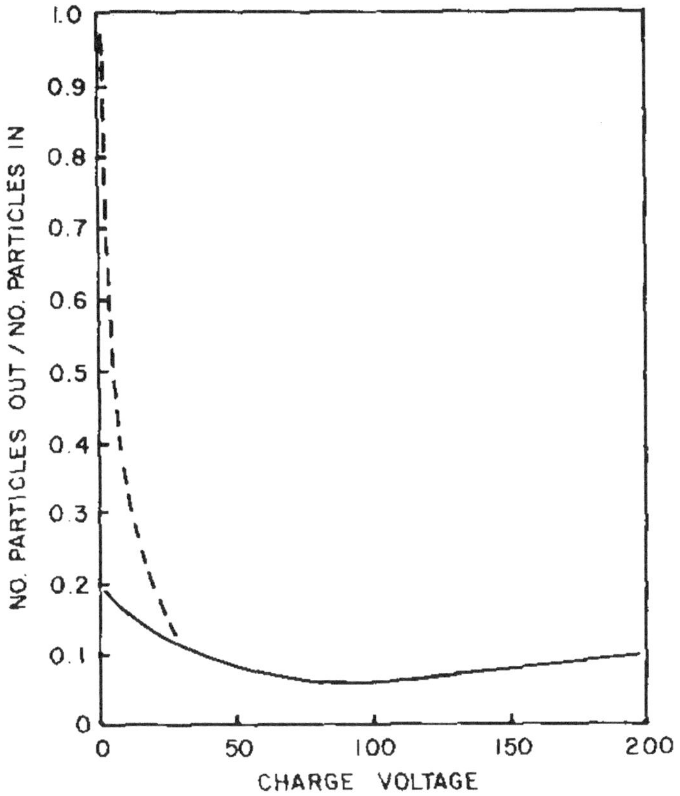

Fig. 5. Actual efficiency of charged wet scrubber as a result of self-precipitation of particles (solid line) with theoretical efficiency (dashed line).

Both the theoretical and experimental models show that for submicron-sized drops, the charging of particulates and drops greatly improves collection efficiency relative to an inertial wet scrubber. These CWSs approach the collection efficiency of a conventional high-efficiency ESP, which they are if one considers that some of the ESP electrodes are simply replaced by the collecting drops. The collection action is similar to ESP, and it appears that there is no significant difference whether drops and particles have the same charge, opposite charge, or are a mixture.

Although these systems yield ESP-like results, the capital investment and operating costs are comparable to those of wet scrubbers. Optimum procedures have not been established for the complete systems, but some details are known. For example, the optimum drop charge density, in coulombs per cubic centimeter, for the best collection efficiency (units in EQS system) appears to be about

$$Q_{opt} \cong [2.4\ \pi\ \mu R\ \varepsilon_0\ U\ N_0 / l\ ]^{1/2} \times 10^{-4} \quad (36)$$

where U is the mean gas velocity (cm/s), l is the length of the system (cm), and N_0 is the initial drop number density (number/cm^3). For example, in a typical scrubber with 7.6×10^6 drops/cm^3, 50 μm in radius, in air moving at 1000 cm/s in a 1000-cm-long system, the optimum charge density is about 2×10^{-10} C/cm^3, which is 1.3×10^9 electrons/cm^3.

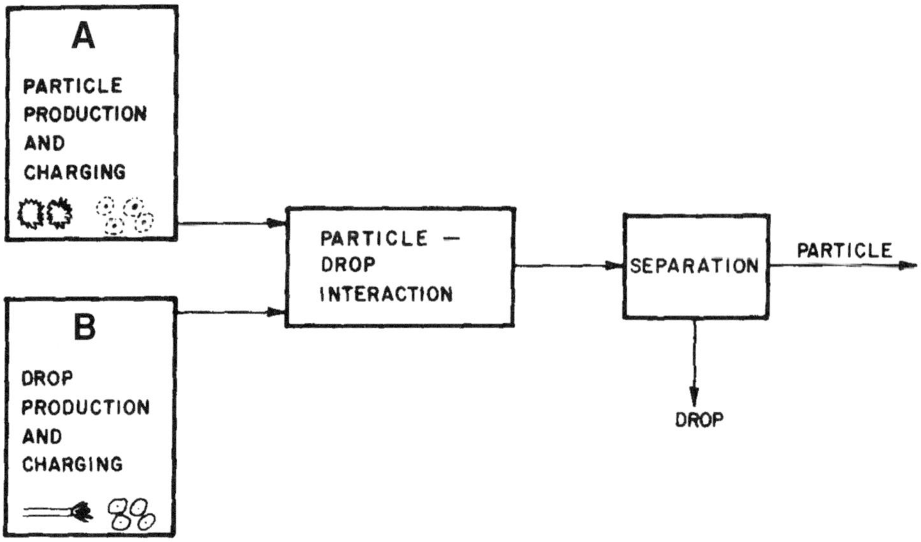

Fig. 6. Schematic of a charged-particle scrubbing system.

4.3.3.2. Types of Charged Wet Scrubber

There are numerous arrangements and configurations of CWSs that can be used for collecting submicron particles. In charged drop scrubber (CDS) units, all particles can have either a positive or negative charge, with all the water drops having the opposite charge. A variation of this is that the drops are bipolar (a mixture of both charges) to reduce the self-precipitation of drops. In this case, the particles can be either unipolar or bipolar.

Other arrangements include systems that charge particles only and not the drops. The collection in these devices has not been as effective as in the CDS. The space-charge precipitator is an example of this arrangement. In this unit, collection relies on precipitation on walls caused by self-fields induced by charged particles. Another example is the self-agglomerator, which increases the size of particles.

Systems that use solid collectors rather than liquid collectors include the electro-fluidized and electro-packed beds. These systems could be very useful and are considered wet scrubbing systems when a scrubbing liquid is used.

A charged-particle scrubbing system can be represented schematically (*see* Fig. 6). Physically, several of the boxes could be a single pieces of equipment. Conversely, a single box could represent several pieces of equipment. Systems noted as A, C, and D represent fairly conventional techniques. In system B, drops could be charged as produced. They also could be charged by ion impaction as in a conventional ESP or a system using similar precipitator technology. Interaction step C and separation of the drops with the collected particles from the gas step D could represent many of the standard wet scrubbing devices.

Production of charged drops consists of drop formation and drop charging. Currently, it is most common to generate the drops by mechanical atomization, by pneumatic atomization, by a combination of these, or by condensation. The main source for energy from drop formation would therefore be mechanical or thermodynamic. An

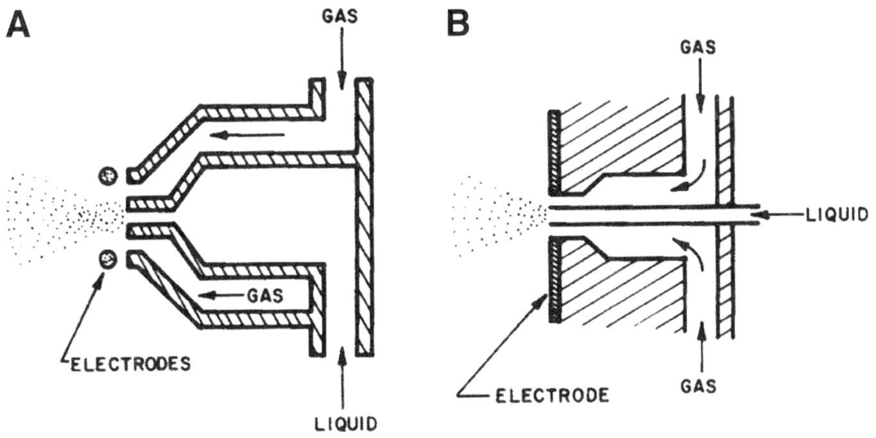

Fig. 7. Mechanical atomization with pneumatic assistance (used in charged wet scrubbers).

electrohydrodynamic spraying technique can be used to produce charged drops in a single operation and requires both mechanical and electrical energy. Other atomization procedures, such as acoustic atomization, can also be used if desired.

Pressure nozzles of various configurations can be used as mechanical atomizers to produce the drops. High-velocity gas striking streams, sheets, or sprays of liquid forms drops by pneumatic atomization. In either technique, charging can be accomplished by one of several procedures. High-potential electrodes can be placed in the area where the drops form to induce a charge on the drops. Another procedure is to produce a corona discharge so that the drops can be charged by ion impaction. The first procedure (induction charging) charges them after they are formed. The first procedure, to induce or influence charging, theoretically requires no electrical energy, and some actual values are noted later.

Figure 7 shows examples of induction charging using pneumatic atomization nozzles (mechanical atomization with pneumatic assistance). The first system (Fig. 7a) is a standard commercial nozzle with inducer electrodes added. Note that these electrodes can have the same charge or opposite charges and can also be at the same or different voltage potentials. Bipolar charged drops would be produced using equal voltages but different charges on the electrodes. A mixture of unipolar drops is generated using the same charge but different voltages on the electrodes. In one system, the electrodes were placed 1.6 mm from the nozzle body and spaced 4.7 mm (center to center) apart.

A multinozzle induction charging system such as shown in Fig. 7b has been constructed using a pneumatic atomizer with a nonconducting body. The gas orifice diameter was 7.5 mm.

In actual practice, current requirements of the inducer nozzles shown were about 1.2 μA per 100 V(root means square) for the single commercial test nozzle and about 0.2 μA per 100V for the multiple-nozzle system for inducer voltages up to about 350 V. Reports on electrohydrodynamic spraying nozzles indicate that a potential of 45 kV is used and requires about 0.25 kW energy per 1000 ft^3/min for charging. Corona drop charging units operate at about 27 kV.

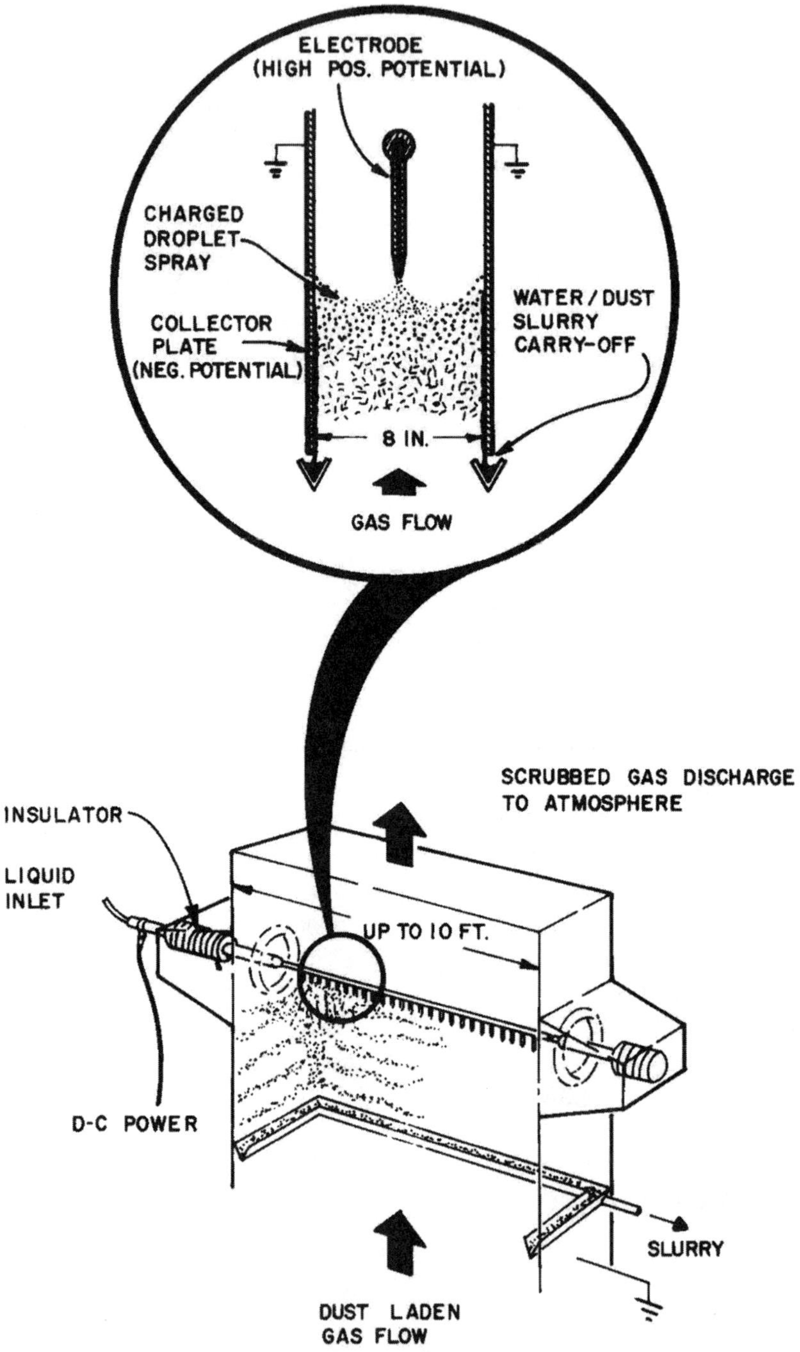

Fig. 8. TRW scrubbing liquid spray mechanism. (Courtesy of TRW Inc.)

The TRW system produces a charged spray as shown in Fig. 8. The dust is not charged in this process. The entire assembly is shown in place in the cutaway drawing in Fig. 9. Scrubbing liquid is introduced countercurrently to gas flow in a spray-tower-type

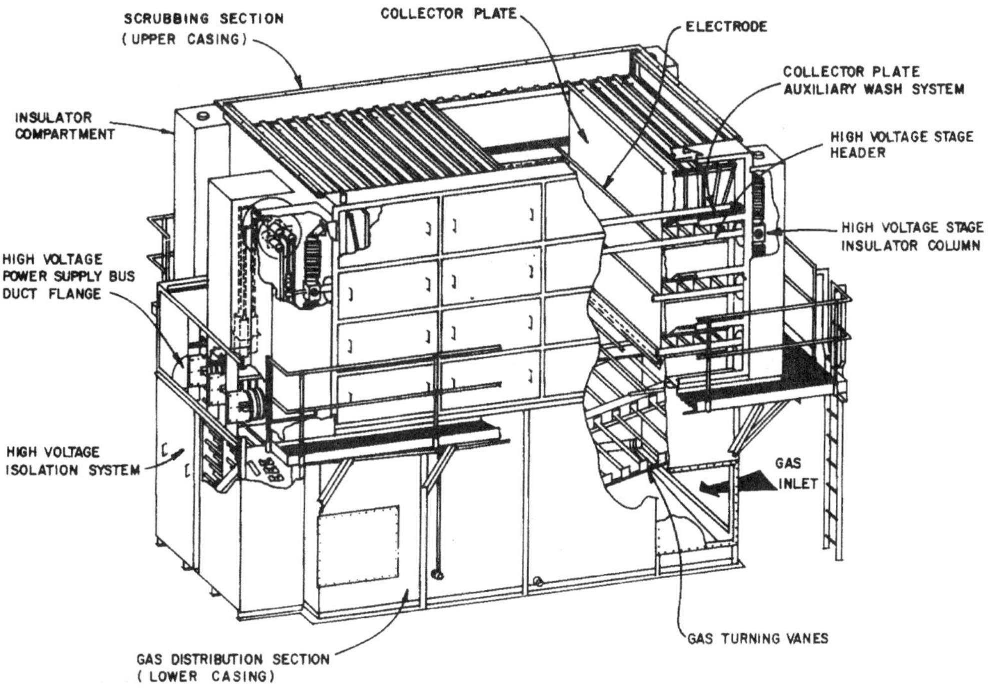

Fig. 9. TRW charged-droplet scrubber general arrangement. (Courtesy of TRW Inc.)

arrangement through the electrodes as shown. Dirty gas enters at the lower right and cleaned gas leaves at the top.

4.4. Use of Chemicals

It has been noted that many particle collection systems use liquids in the process to either serve directly as the collection medium or to assist indirectly by washing out the collected material. Water is the most common liquid used. The collected matter or addition of chemicals can change the collection efficiency of wet systems. For example, in Venturi scrubbers, wetting agents (surfactants) can improve collection efficiency by reducing the surface tension of the water. This changes the quality of atomization and enables hydrophobic material to become more easily wetted and collected. The collection efficiency of fly ash, which is slightly hydrophobic, can be improved compared with that predicted by Eq. (23) up to the amount predicted by (13)

$$C_o/C_i \cong 8.42 \times 10^{-8}\, V_t^{3.87}\, A^{0.157}\, (\rho_g / \Delta P)^{1.92} \tag{37}$$

4.5. Simultaneous Particle–Gas Removal Interactions

Wet scrubbing is useful for both particle removal by impaction and gas removal by absorption. If a chemical reaction occurs in addition to physical absorption, it is called chemical absorption. The absorption rate of a gas simultaneous with wet inertial impaction removal of particulates can be expressed using the Chilton and Colburn (16) concept of transfer units:

$$N_{OG} = \ln (Y_1 / Y_2) \quad (38)$$

where N_{OG} is the number of overall gas phase transfer units, Y_1 is the gas-phase concentration of solute in (mole fraction), and Y_2 is the gas phase concentration of solute out (mole fraction).

Considering specifically a Venturi-type inertia wet scrubber, simultaneous gas absorption expressions have been developed. Gleason (17) derived N_{OG} using the Nukiyama–Tanasawa atomization predictions for air–water systems and for gas-phase controlling:

$$N_{OG} = (244K' \, h \, L) / \{3{,}600 \, (16{,}050 + [1.4 \, L'^{1.5} \, V_t])\} \quad (39)$$

where K' is the overall mass transfer coefficient in velocity terms (ft/h), h is the active height of absorber involved, usually about 1 ft, L is the liquid-to-gas ratio (gal/1000 ft^3), and V_t is the throat velocity (ft/s).

Gleason (17) expresses the same factor for tower absorption units where the inertial scrubbing is a countercurrent operation and where Henry's law applies as

$$N_{OG} = \ln \{[(Y_1 - MX_2) / (Y_2 - MX_1)] + [(MG) / (F' - MG)]\} \quad (40)$$

where M is the slope of the equilibrium curve, X is the concentration of solute in liquid (mole fraction), F' is the liquid flow rate (mol/h-ft^2), and G is the gas flow rate (mol/h-ft^2).

The Venturi scrubber is a cocurrent collection device with low contact time and is essentially the most effective particle collection device (18), but only a fair absorber at best. If $N_{OG} = 1$ in a Venturi scrubber, the system is good. Absorption is both a function of contact area and contact time between liquid and gas phases and a function driving force. The driving force depends on the substances involved and the difference in concentration between the solute and each phase. To obtain good absorption in a Venturi scrubber, throat velocities as low as 50 ft/s (15 m/s) are used, but this shows that impaction effectiveness [i.e., impaction parameter Eq. (19)] must be sacrificed to improve absorption.

In practice, combinations of Venturi scubbers to remove particulates and countercurrent spray absorbers in series are used. Such systems using chemical scrubbants can reduce SO_2 concentration, for example, by over 90%. The number of overall transfer units for Venturi–spray-tower series combination is about 2.5, and this can be varied by operational changes.

Koehler (19) developed an expression for the case of simultaneous particle removal and SO_2 adsorption using an alkali hydroxide of MgO in an open throat Venturi scrubber in series with a countercurrent spray tower absorber. Using a similar system, Hesketh (20) was able to expand this expression to include scrubbant surface tension effects. The result shows that absorption efficiency drops as the contacting liquid surface tension is reduced from the normal value of 66 to 31 dyn/cm.

5. ENTRAINMENT SEPARATION

Devices that use liquids to remove gaseous and particulate pollutants require some method of keeping the scrubbing liquid from leaving as droplets with the cleaned gases. Slurry liquid droplets that leave in this manner usually evaporate in the exit ducts and stacks. The solids remaining, plus any unevaporated liquid, would be considered as particulate pollutants. The simultaneous removal of such secondary liquid pollutant and particulate matter can be accomplished by entrainment separation.

Entrainment separators often comprise two stages of liquid removal. The primary separation consists of removing the larger droplet by gravity, centrifugal, or other forces, as previously discussed. Meshes, packing, baffles (louvers), or some other method are used in the secondary separation to attempt to eliminate the remaining entrained liquid droplets from the exit gas stream. Sieve plates and tube banks have been considered for this purpose but are not commonly used.

Three basic kinds of knitted mesh of varying densities and voids (also known as mist eliminators or demisters) are used for entrainment separation. These include layers with crimp in the same direction, layers with crimp in alternate directions, and spirally wound layers. Typical design sizing consists of 10- to 15-cm-thick mesh with a density of 0.15 g/cm^3 operating at gas velocities of 0.3–5 m/s. These devices are operated so that gases flow upward or horizontally through the mesh and the removed liquid drains by gravity. A liquid release flow rate of about 2.5×10^{-3} g/s-cm^2 of mesh is a limiting factor. The pressure drop depends on flow rate and type of mesh but might be about 6 cm water for the type described under normal operation. A series of mesh collectors will remove drops as small as 1–5 μm by allowing them to coalesce in the first, then removing them in the second. Pressure drops in these systems reach 25 cm of water.

Baffles of the zigzag type can remove drops down to 5–8 μm in diameter at gas velocities of 2–3.5 m/s in high-efficiency units using staggered baffles. The pressure drops for these depend on the velocity and number of passes and average about 0.4 cm of water per pass. This is minimized by keeping a 1- to 3-cm spacing between passes. Common baffle arrangements have three or six passes.

A significant problem in entrainment separators is the re-entrainment of liquid. Re-entrainment occurs as a result of rupture of bubbles on top of the separator, creeping of liquid, and shattering of drops resulting from splashing. Calvert (21) reported that the minimum size drops resulting from re-entrainment are 40 μm in diameter, with the mass median drop diameters ranging from 80 to 750 μm. The smaller drops result from the shattering of drops. The transition between the two types occurs at about 250 μm.

Efficiencies of entrainment separators reach as high as 100% under ideal operating conditions. In a typical commercial operation, up to about 1% of the final-stage scrubbing slurry may be carried over to the mist eliminator, and the eliminators operate at up to 99.5% efficiency when re-entrainment is included.

6. INTERNAL COMBUSTION ENGINES

6.1. *Process Description*

Although the internal combustion engine (ICE) is only one part of a vehicle, its air emission must be properly treated by an air pollution control device. The ICE has now become process equipment for VOCs and controlled like any other point source to meet certain criteria (*see* Fig. 10).

In principle, the control device used on an ICE for a conventional automobile or truck engine is similar to that used on a thermal incinerator. The physical difference

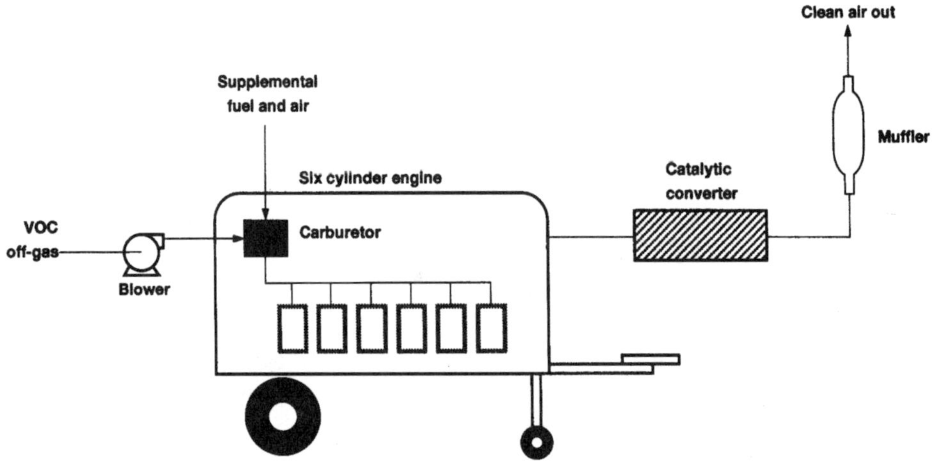

Fig. 10. Internal combustion engine-based VOC control system. (From US EPA.)

between the ICE and the incinerator is primarily in the geometry of the combustion chamber. Figure 10 shows a simplified diagram of a typical ICE-based system. The major components in an ICE-based system include the ICE (standard automobile or truck engine), supplemental fuel supply (usually propane or natural gas), carburetor, off-gas lines from remediation system, and additional air emission control devices (adsorbent bed, catalytic converter, etc.). Additionally, an ICE-based system requires a clean waste stream containing no acid and low levels of particulate matter; as a result, a pretreatment device may also be required.

Another requirement for an ICE-based system is supplemental fuel to support combustion when the VOCs in the airstream are insufficient. Supplemental fuel systems are required for start-ups, remediation projects with low VOC extraction rates, and sources such as soil vapor extraction (SVE) systems that produce changing VOC concentration over time. Combustion in an ICE system is possible when the concentration for VOCs in the air emission stream ranges from 60,000 to 100,000 ppmv at flow rates of 1.7–2.0 m^3/min or 60–70 acfm (22). Additionally, supplemental oxygen may be necessary to dilute the gas stream if the VOC level exceeds 25% of the lower explosive limit. The carburetor for automobile or truck engines must be modified to include two input valves: one for gaseous fuel (air emission stream containing VOCs) and the other for the supplemental liquid fuel.

The mobility of the ICE-based system is a major advantage of this technology. When the air emission stream provides sufficient energy to operate the system without the use of a supplemental energy source, then the mobility advantage of the ICE-based system is further enhanced. Another advantage of the ICE-based system is ready availability of parts for the automobile engine or truck engine and widespread knowledge of the operation and maintenance of the engine. Additionally, control of the emission stream from the ICE-based system can be accomplished with an off-the-shelf automobile catalytic converter, which is relatively inexpensive when compared to a custom-made unit.

6.2. Applications to Air Emission Control

Any point source of VOC can be controlled with an ICE-based system when the air emission stream meets certain criteria. For this alternative to be economically attractive, the air emission stream flow rate needs to be relatively small. The largest ICE-based system is capable of processing an emission stream with a flow rate up to a few hundred cubic feet per minute and a high concentration of VOCs. If the VOC concentration is less than 1000 ppmv, then supplemental fuel requirements become excessive and process becomes economically unfeasible. In California, ICE-based systems are commonly used for VOCs; as a result, the majority of manufacturers are located in that state (23). Relatively little information is available on the use of ICE technology on remediation sites, but it is feasible for these systems to be utilized at Superfund sites to control emission stream from small-scale SVE systems and from small-scale air strippers. Because their use is limited to small emission stream flows (several hundred cubic feet per minute) the available literature has focused on the use of ICE to control emission streams from SVE processes, capped-off landfills, and air stripping processes.

The ICE system becomes economically attractive when its use eliminates the need to run electrical power to the site because the engine may be used to run vacuum fans and other remediation equipment. Not only can these systems reduce utility costs, but they can achieve destruction removal efficiencies (DREs) of 99+% when a catalytic converter is incorporated into the system. Other advantages of the ICE system are their mobility and small size. Disadvantages of the ICE system are their limited capacity of less than several hundred cubic feet per minute, the noise levels emitted from the engine, and monitoring requirements for controlling air-to-fuel ratio so that the engine operates efficiently. The excessive noise levels can be controlled with sound-attenuating devices such as mufflers and enclosures. A computerized system can be used to control the air-to-fuel ratio so as to minimize monitoring costs.

Typically, ICE systems achieve removal efficiencies of 99% or greater. Pederson and Curtis (23) recently compiled results of several studies in which they listed removal efficiencies of different VOCs by ICEs from air emission streams from various SVE and air stripping systems. The results from this study are shown in Table 2. Information from case studies on ICE systems is summarized next.

6.2.1. ICE System Case Studies

VR Systems, an ICE vendor, supplies portable ICE systems that are designed to control air emission streams from SVE systems and tank-degassing systems. These ICE systems are designed to burn up to 100 kg/h (220 lb/h) of hydrocarbons. Additionally, these systems utilize liquid propane or natural gas as a supplementary fuel and have a computer system to control the air-to-fuel ratio to achieve higher DREs with fewer labor requirements.

Another vendor (Kerfoot Technologies, Mashpee, MA) provides the Soil-Scrub® process that is used with a heat-assisted SVE system. An ICE system was provided as primary control system for the emission from this SVE system. Additionally, the control system included a catalytic converter and GAC beds following the ICE system. The case studied examined the use of this system on gasoline-soaked soil. This soil was first encapsulated in plastic sheets and then was heated to 100°C. The final DRE achieved was

Table 2
Listing of Destruction Efficiencies of ICEs for SVE Systems

Volatile organic compound	Inlet concentration (ppm)	Discharge concentration after catalytic converter (ppm)	Removal efficiency (%)
Total hydrocarbon (THC)	26,000	140	99.46
	38,000	89	99.76
	68,000	160	99.72
	200,000	39	99.98
	318,832	16	99.99
Benzene	380	0.8	99.79
	470	1.8	99.66
	730	0.056	99.99
	785	0.63	99.92
	960	0.024	99.99
	995	ND[a] (<10 ppb)	99.99
	1,094	67	93.88
	1,400	0.13	99.99
Total xylenes	320	0.13	99.96
	360	0.080	99.98
Xylenes	114	0.7	99.39
	1,550	<11.5 ppb	99.99
Ethylbenzene	18	<0.5	—
	77	0.062	99.92
	91	ND (0.02)	100.00
Total petroleum hydrocarbon (TPH) non-methane	49,625	225	99.56
TPH	30,500	1.4	99.99
	34,042	14.5	99.95
	39,000	4.7	99.99
	65,450	30	99.95
Toluene	400	1.1	99.73
	720	0.024	99.99
	840	0.020	99.99
Methane HC	741	109	86.29

[a]Nondetectable.
Source: ref. 23.

99.9% over 36 h of operation. The remediation process achieved no detection for benzene, toluene, and xylenes and 82 ppm oil in the soil.

Robert Elbert & Associates used thermal vacuum spray aeration/compressive thermal oxidation system for remediating groundwater contaminated with gasoline. An ICE system was incorporated in the system to control the air emission. The system heated the water to 110 °F and applied to 12" of vacuum to preferentially evaporate the gasoline and extract the gasoline vapor, respectively. The extracted vapor was then sent to ICE system to control air emission stream. The remediation process could strip and oxidize 120 lb hydrocarbons per day and required approx 0.75 gallons of fuel per hr. The treated water had 32 ppm contaminants and the waste gas had 70 ppmv of contaminants. One limitation of the process was that an over-rich combustion condition could occur if the remediation process takes place in a well and the system is smothered with excessive water vapor. Another limitation of the process is that ICE system is sized based on the volumetric flow rate of the waste gas stream to be treated, which is limited to less than 1,000 cfm. Additional vendor information is summarized below (19, 20).

6.2.2. ICE Vendors

VR Systems manufactures the SVE–ICE system. The ICE unit in this system can range in size from 25 to 1000 scfm. The largest ICE unit consists of several engines in parallel that can destroy about 20 lb/h of hydrocarbon. Another SVE–ICE system, by RENMAR, can destroy 100–200 scfm of input gas per 300 cubic inches of engine capacity in their ICE unit (23). SVE or air stripping systems manufactured by RSI provides ICE that can handle VOC-laden air emission stream up to 80 scfm.

A comparison between the ICE process and other emerging and conventional processes for air pollution control is presented in Section 10. The innovative ICE process, in principle, is designed for removal of VOC only from any point source of air emission streams. It is not designed for removal of acids, particulate matter, or heavy metals.

7. MEMBRANE PROCESS

7.1. Process Description

Another emerging control process for VOC emission streams is membrane technology (24). The organic constituents are concentrated by the membrane module because the membranes that are selected are more permeable to organic constituents than to air. The driving force that causes the separation of the organic constituents from the air emission stream is the pressure difference across the membrane (25).

Figure 11 shows a schematic of a typical membrane separation process. The air emission stream is fed to the membrane module using either a blower or compressor. This process utilizes either a vacuum pump on the permeate side of the membrane module or a compressor before the membrane module as illustrated in Fig. 11a, b, respectively. The pump or compressor applies the pressure differential across the membrane module, which is the driving force for the separation of the feed gas into a concentrated stream (permeate) and depleted residue gas stream. The membrane allows most of the organic contaminants and some of the gas to permeate the membrane. The remaining gas (stripped off-gas) is either vented (if the organic concentration in the gas meets emission standards) or recycled to the VOC source (if the organic concentration in the gas does meet emission standards).

The permeate stream, which contains concentrated organics, must be treated further to either recover or dispose of the organic contaminants. Treatment of the permeate stream can accomplished in various ways. Figure 12 illustrates the process configuration for recovering contaminants. As shown in Fig. 12a, a carbon adsorption system is used to collect the solvents. Recovery of the solvent occurs during the carbon regeneration process in which steam is used to strip solvent from the carbon. The steam

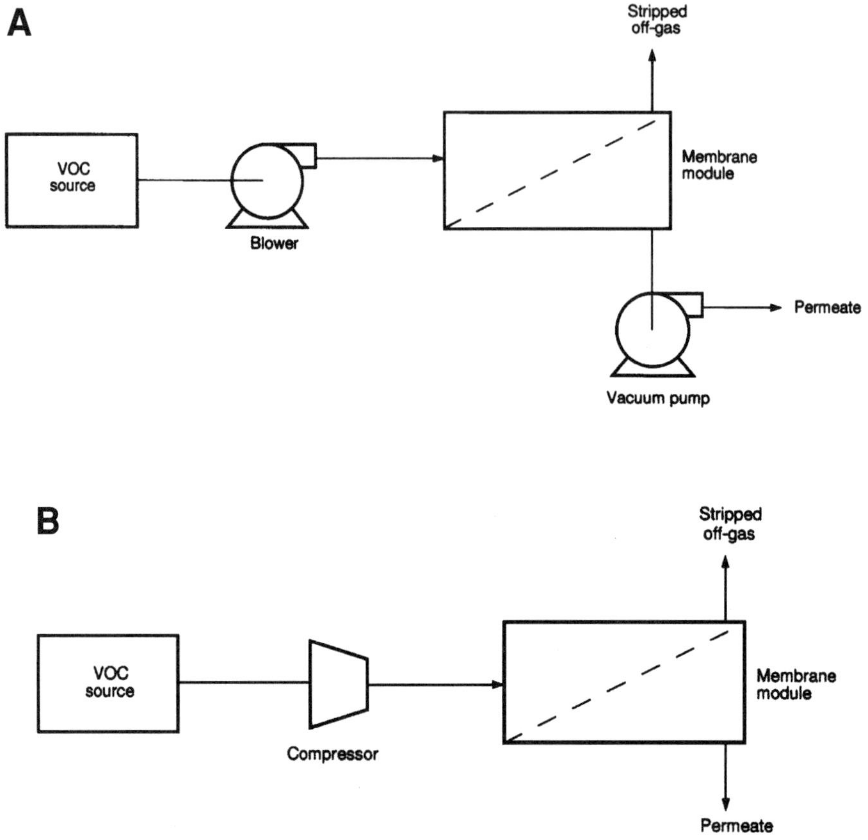

Fig. 11. (a) Membrane separation system with vacuum pump; (b), membrane separation system with compressor. (From US EPA.)

containing the solvent is condensed and the water/solvent (condensation) is then transferred to decanting unit. Recovery of the solvent occurs in the decanting unit where water is separated from the solvent.

Figure 12b shows a solvent recovery process that was developed and patented by Membrane Technology and Research, Inc. (MTR) to recover solvent from permeate gas generated from a membrane process. Another process is shown in Fig. 12c, in which a membrane system concentrates the solvent (permeate), directs it to condensation unit, and polishes the air-stripped gas with activated carbon to remove any residual VOCs. As an alternative to this process, permeate could be directed to an incinerator for destruction of the contamination.

The construction of the membrane consists of an ultrathin layer of a selective polymer supported on a porous sublayer. This selective polymer provides a barrier, whereas a microporous substructure provides mechanical strength for the module. Membrane material typically consists of rubber, Buna-n-nitrile, PVC, neoprene, silicone polycarbonate, and other polymer compounds. Some manufacturers produce a spiral-wound membrane module, in which the layers of the polymer are supported

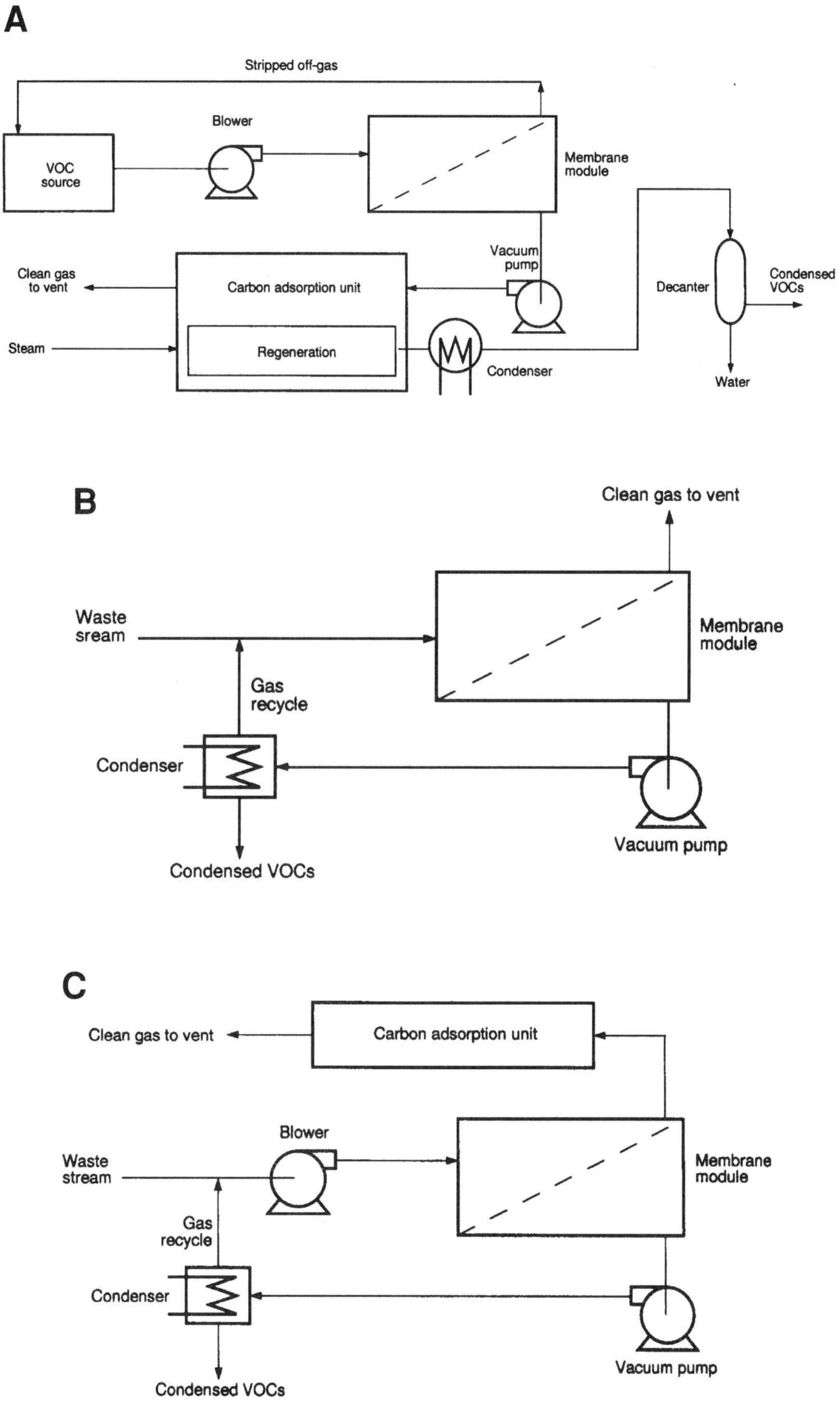

Fig. 12. (a) Membrane concentrator with carbon bed adsorption recovery system; (b) MTR single-stage membrane system; (c) single-stage membrane separation system with carbon bed adsorber polishing. (From US EPA.)

on a macroporous structure. Others produce hollow-tube membrane modules. Regardless of what type of membrane module is used, a compressor or vacuum pump is required to supply the pressure differential required for concentrating the organic contaminants.

7.2. *Application to Air Emission Control*

Membrane technology is an emerging control process for emission streams with volatile organic compounds. At this time, the theoretical aspect of the membrane technology is still being developed and there is only limited practical information available on the technology. Application of this technology would typically be used as a part of other control processes. Membrane technology would be used to concentrate VOC in the emission stream and thus reduce the flow. As a result, the downstream control equipment would be sized at a reduced flow capacity and the capital and operational costs for this downstream equipment would also be reduced.

A good application of membrane technology would be in an industry where recovery of high-quality product is required and where carbon adsorption process will not be applicable (e.g., recovery of aldehydes resulting from a fire hazard or 1,1,1-trichloroethane resulting from its reactivity with the carbon). Several options are available for treating a concentrated air emission stream from a membrane module. These options include condensation of the concentrate and recovery of the solvent. This option is especially practical when the solvent is expensive or recovery is a requirement. Incineration would be another option for treating the concentrated air emission stream for the membrane module. This option become feasible if the organic in emission stream has a high heating value and the solvent is inexpensive. Carbon adsorption may also be considered for treating the concentrated solvent stream from the membrane module, and solvent can be recovered using a steam regeneration system. The use of membrane technology can result in cost savings, reduction in the flow of the emission stream, and reduction of energy requirement for downstream incineration control process (26).

It has been reported that membrane technology is suitable for a low-volume off-gas stream with concentration of organics ranging from 0.05% to 20% (27). Additionally, membrane technology is very effective as a bulk concentrator. The VOC concentration in the permeate stream from the membrane module can be 10–50 times the VOC concentration of the inlet emission stream.

Two factors influence the control efficiency of membrane technology: permeability of the solvent in the emission stream and separation factor. The permeability of a solvent is further defined as the solvent flux across the membrane. For solvents, the permeability is related to its diffusivity and solubility. The permeability of organic vapors usually increases with an increase in concentration and at high pressures (28). Permeability data are reported by Baker et al. (28) and is a function of pressure and selectivity for various membrane materials and contaminants. Strathman et al. (29) and Peinemann et al. also reported membrane test data (30).

The separation factor describes the relative permeability of the solvent and gas. It is defined as the degree of concentration the membrane can achieve. The greater the separation factor, the more efficient the separation process. These two factors are dependent on the pressure ratio (permeate-side pressure to inlet pressure) and the membrane material. To achieve high removal efficiency, the membrane material should demonstrate high permeability and good selectivity for the solvents to be recovered. Additionally, the membrane should be durable and stable enough to withstand normal wear during operation.

One can optimize the membrane selectivity by choosing a balance between the capital cost of membrane area and the energy cost for pumping. Additionally, the optimum membrane selectivity is determined by choosing the lowest selectivity that will produce the desired permeate concentration. When the solvent flux is decreased (more membrane area per flow of solvent), the membrane selectivity is increased. By comparison, the energy requirement for a low-selectivity membrane is greater because more energy is required to pump a higher volume of gas to meet the permeate requirements (at fixed permeate pressure). Therefore, the selection of a membrane must be a balance between capital cost for the membrane (the greater the membrane area the higher the capital cost) and the operational cost (the greater the pumping rate, the higher the energy cost) (30).

Weller and Steiner (31) presented the fundamental mass and energy balance equations that govern the design and performance of a single-stage gas permeation system. Additionally, Pan and Habgood (32) performed analysis on a crossflow pattern that applies to the spiral-wound membrane. The assumption for these analyses may be simplified as follows: the permeability of both components constant, negligible pressure drop across flow paths, and negligible mass transfer resistances except for permeation through the membrane. In test studies, for most cases, the error introduced by assuming constant permeability was not found to be excessive (26).

A small membrane test unit can be utilized to generate representative samples of concentrated feed and filtrate that demonstrate pollutant separations. Such small membrane testing units are commercially available (33, 34).

The feasibility of using full-scale membrane separation process for control of has been studied and reported by Moretti and Mukhopadhyay (35).

Section 10 compares the membrane process with other emerging and conventional processes for VOC control. The innovative membrane process, at present, is only designed for removal of VOC from air emission streams. It is unknown whether or not the membrane process can remove any particulate matter (PM) or heavy metals from air emission streams, although it is a filtration process in principle.

8. ULTRAVIOLET PHOTOLYSIS

8.1. Process Description

Since 1988, ultraviolet (UV) light technology has been used for the destruction of toxic organics in aqueous solutions. UV light has also been used as a primary treatment process, but in some cases, it has been used in conjunction with ozone and hydrogen peroxide, which serve as oxidants (36).

Researchers have recently shown that UV photolysis of organics can be accomplished by utilizing a broad spectrum of high-intensity UV light. These experiments include treating water, air, and soil. Promoters of direct UV photolysis have claimed that it can be used to disintegrate toxic organic toxics into nontoxic byproducts. Ultraviolet

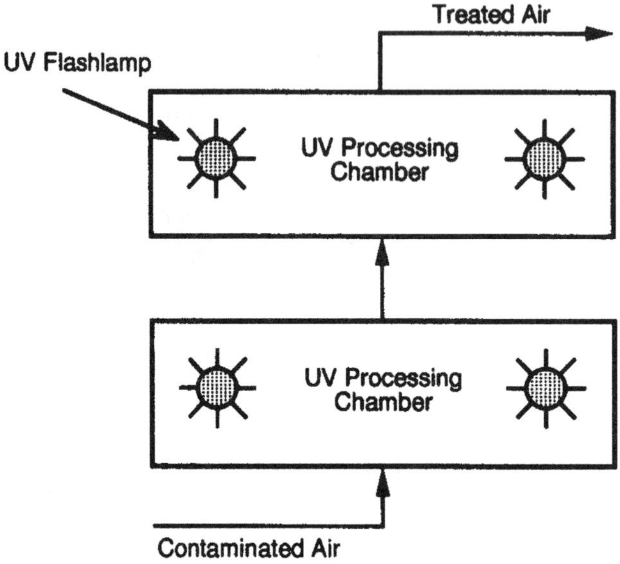

Fig. 13. Schematic of direct UV photolysis. (From US EPA.)

Energy Generators, Inc. (UVERG) has asserted that the Wekhof Direct UV Photolysis to be "both the most efficient and cleanest method of organics destruction in water, gas and in soil" (37).

Another manufacturer, Purus, Inc., has a direct UV photolysis process that can be used on-site for cleanup of organic contaminants. It also claims that the process converts organic contaminants into harmless byproduct. The Purus system utilizes xenon UV flashlamps. It also sells a commercial direct UV photolysis system that it claims treats contaminant air emissions.

Figure 13 shows a flow diagram for direct UV photolysis system. Air emission stream flows into one or more processing chambers that contain UV flashlamps, where the air emission stream is exposed to a broad spectrum of UV light. While in the chamber, the organic contaminants in the air emission absorb energy that causes the bonds of the organic molecules to break apart and release carbon atoms. It is proposed that, under ideal conditions, these released carbon atoms along with oxygen atoms present in the air emission stream can form carbon dioxide. When the analysis of the air emission stream in the chamber indicates that contaminant levels have been lowered to a sufficient level to meet emission standards, the air stream can then be released from the chamber.

8.2. Application to Air Emission Control

Volatile organic compounds contained in air emission may be destroyed using UV photolysis. Types of VOC could include volatile chlorinated organic compounds (e.g., trichloroethylene [TCE] and methylene chloride) and VOCs present in gasoline petroleum products (e.g., benzene and toluene). Literature reviews indicate that UV photolysis technology has not been used at Superfund sites. This technology may be appropriate in controlling air emissions containing toxic organic compounds released from wastewater and groundwater-treatment technologies. A possible application for the UV photolysis technology may be air emission streams from biological treatment process, air stripping process, and *in situ* remediation of soil by vacuum extraction process. Proponents of UV photolysis technology indicate the major advantage of this technology is that it destroys the toxic organic compounds rather than transferring the compound to another medium, such as activated carbon (38–42).

Studies on the UV photolysis technology have been limited. Because of the limited amount of information, it is unclear what range of conditions may be effective for this treatment technology. A study conducted on an air emission stream containing 300,000 ppb of TCE was lowered to 100 ppb by UV photolysis process that had an approx 3-s residence time (38). Until further studies establish the range of operating conditions for this technology, its application for Superfund sites cannot be predicted.

Currently, there are no specific sizing criteria presently available for treating air emissions by UV photolysis, but sizing is dependent on the following parameters: (1) the flow rate of the contaminated air emission stream, (2) the concentration of contamination in air emission stream, and (3) the refractoriness of the compounds. Currently, to size an UV photolysis system, pilot tests would be performed on a sample air stream to determine the appropriate size for a full-scale system.

The feasibility of using the UV oxidation system for VOC control has been studied by Moretti and Mukhopadhyay (35). The efficiency of UV oxidation has been compared with other air pollution control technologies, such as thermal oxidation, catalytic oxidation, flaring, condensation, adsorption, absorption, boilers and process heaters, biofiltration, and membrane separation.

Section 10 discusses a comparison between the UV process and other emerging and conventional processes for VOC control. The UV process is not technically feasible for removal of PM and heavy metals from air emission streams. It is very effective for VOC control (99.9% removal efficiency) in a very narrow influent VOC concentration range 200–300 ppmv in an air emission stream.

9. HIGH-EFFICIENCY PARTICULATE AIR FILTERS

9.1. Process Description

Medical, research, and manufacturing facilities use high-efficiency particulate air (HEPA) filters when they require 99.9% or greater particulate removal. Superfund site have not widely adopted the use of HEPA filters, which could be utilized as a PM polishing step on ventilation systems for either a building undergoing asbestos remediation or an enclosure for a solidification/stabilization mixing bin (39).

Typically, PM control system with HEPA filter (Figs. 14 and 15) has following the major components: (1) HEPA filter, (2) filter housing, (3) ductwork, and (4) fan. The filter housing unit requirements for a HEPA filter are dependent on the nature of PM being collected and on the number/arrangement of filters. To keep personnel who are removing the HEPA filters from having contact with the filters, a bag-out bag will be required in the filter housing unit. A filter housing unit with a bag-out bag is shown in Figs. 14 and 15.

Depending on the degree of control and allowable pressure drop across the filters, the arrangement of filters can be in parallel, series, or a combination of both. Parallel filters

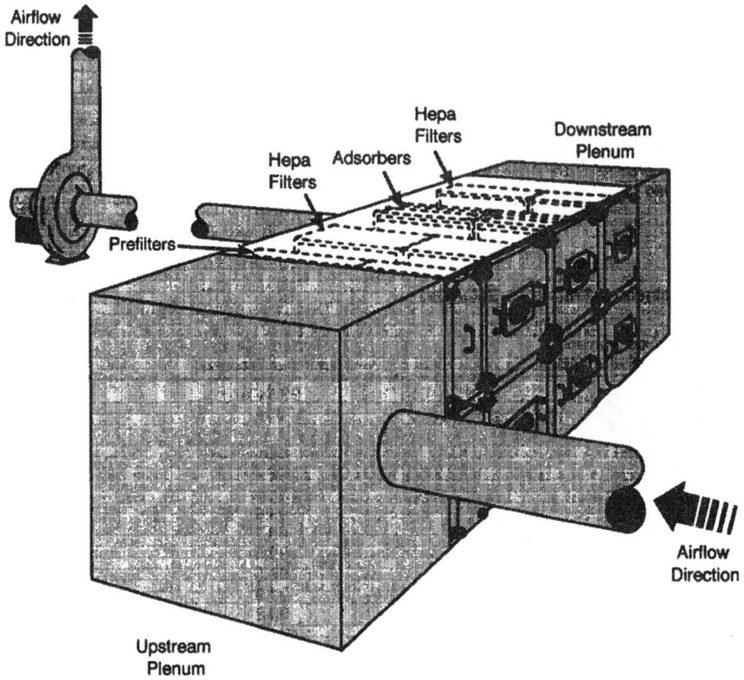

Fig. 14. Particulate matter control system employing HEPA filters. (From US EPA. and refs. 39, 43, and 44.)

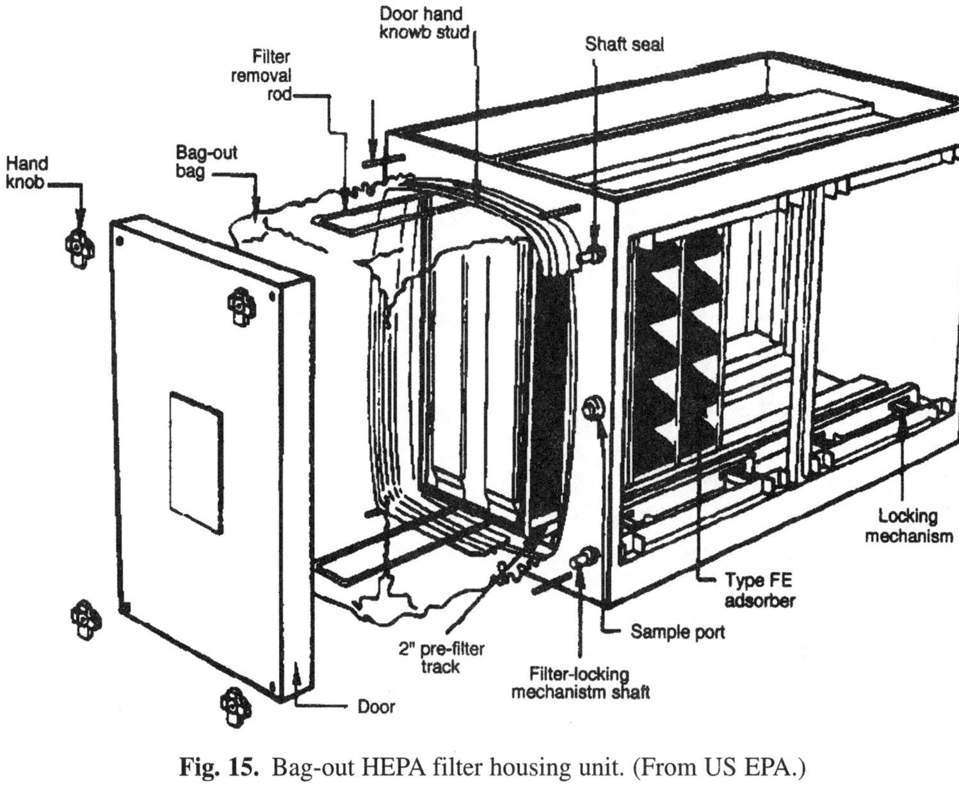

Fig. 15. Bag-out HEPA filter housing unit. (From US EPA.)

will have lower pressure drop than filters in series, whereas series filter will achieve a higher PM collection efficiency than filters in parallel. The housing unit for parallel filters will be larger than the housing unit for series filters.

9.2. Application to Air Emission Control

The use of HEPA filters for the control of PM provides advantages/disadvantages, which are listed in Table 3. A list of remediation technologies that are compatible with

Table 3
Advantages and Disadvantages of HEPA Filters

Advantages	Disadvantages
• Easy to operate. • Provide 99.9% or greater PM removal efficiencies.	• Many require prefilters for exhaust with high PM concentrations. • Required housing units are expensive and may be subject to corrosion. • Filters are subject to fouling by high-humidity exhaust gases. • Filters must be replaced periodically because of plugging caused by PM. • High power costs because of pressure drop across filter.

Table 4
Compatibilities of HEPA Filters with Remediation Technologies

Emission source	Qualifications
Asbestos removal from building	During the asbestos removal, HEPA filters must either be installed in the building ventilation system or in the negative air pressure system. A bag-out housing units will be required for HEPA filters and filter must be disposed of properly.
Enclosure ventilation system	Inlet air with high PM concentrations may require pre-filters. Bag-out housing units may be required for characteristics of PM (e.g, heavy metal or SVOC contamination). The lifetime of the HEPA filter will be shorten under high-humidity application.
Hoods or enclosures of solidification/ stabilization mixing bins	Lime used in the process may subject the HEPA housing to corrosion. Inlet air with high PM concentrations may require pre-filters. Bag-out housing units may be required for characteristics of PM (e.g. heavy metal or SVOC contamination). The lifetime of the HEPA filter will be shorten under high humidity application.

HEPA filters are provided in Table 4. It has been reported by vendors that HEPA filters can achieve efficiencies of 99.9% and up for particulates having a diameter of 0.3 μm. Table 5 provides parameters that will affect the efficiency and/or useful lifetime of HEPA filters. Using manufacturer pressure drop versus face velocity curves, each type of HEPA filter system can be sized. These curves can be used knowing the maximum allowable pressure drop across the filter and the airflow rate, then the type of filter and the filter arrangement can be determined. For example, HEPA filters are selected to control PM emissions in an exhaust gas flowing at 9000 acfm (2250 fpm for a 2-ft × 2-ft HEPA filter) and the maximum allowable pressure drop across the filters is 0.8 in. H_2O gage; then, 10 H2424B, 9 H2430B, and 18 H2323A HEPA filters must be used in parallel.

The cost of a HEPA filter is dependent on the following specific characteristics: (1) PM removal efficiency achievable and (2) allowable maximum face velocity across filter.

Additionally, the face velocity across the filter, PM loading rate, and the moisture loading rate onto the filters determine the useful life of the filters. The frequency of filter replacement can be determined for the HEPA filter system once the useful lives of the filters are estimated. The cost for HEPA filters generally ranges from \$20 to \$100/ft^2 of filter area. Additionally, the cost of housing unit can range from \$150 to \$5,000/ft^2 area. The cost of the housing unit is a function of the type of housing unit requirements (e.g., regular versus bag-out).

Section 10 summarizes the advantages, disadvantages, removal efficiencies, operating arranges, and cost-effectiveness of various emerging processes discussed in this chapter (*see* Table 6). A comparison between the emerging processes and conventional processes is also made. Briefly, the innovative HEPA filters are very efficient (99.9% removal efficiency) for removal of PM from air emission streams. Its removal efficiency for reducing VOC, acid gases, and heavy metals, however, has not been established.

10. TECHNICAL AND ECONOMICAL FEASIBILITY OF SELECTED EMERGING TECHNOLOGIES FOR AIR POLLUTION CONTROL

10.1. *General Discussion*

Today's emerging technologies may very well become tomorrow's conventional technologies, if research engineers and scientists continuously improve upon these technologies. This section will discuss the operating ranges, cost-effectiveness, and removal efficiencies of selected emerging technologies, namely ICEs, membrane process, UV process , high-efficiency air filters, and fuel-cell-powered engines. (Various mechanical particulate collectors (see Section 4) are also covered by the Wet and Dry Scrubbing chapter in this book and by a companion Humana book, *Advanced Air and Noise Pollution Control* [50].)

10.2. *Evaluation of ICEs, Membrane Process, UV Process, and High-Efficiency Particulate Air Filters*

10.2.1. *VOC Removal*

Table 6 compares selected emerging technologies (ICEs, membrane process, and UV process) with conventional technologies (carbon adsorption, thermal oxidation, catalytic oxidation, condensation, scrubbing, and biofiltration) in terms of their advantages and

Table 5
Parameters Affecting the HEPA Filter Efficiency and Lifetime

Parameter	Comments
Moisture	Moisture will blind the filter. The blinding will increase the pressure drop across the filter, which will eventually lead to filter failure because of excessive resistance.
PM	As the PM loading increases, the filter life decreases. Additionally, the change in pressure drop across the filters will be accelerated. As velocity increases across the filter, the PM control efficiency decreases. The higher the pressure drops across the filter, the shorter the filter life.

disadvantages when treating point sources of air pollution for VOC control. Figures 16 and 17 further illustrate the operating ranges (in terms of initial VOC concentrations of the influent air emission streams), the process removal efficiencies (RE), and relative cost-effectiveness of both emerging and conventional technologies for point-source VOC controls. From Tables 6 and 7 and Figs. 16 and 17, it appears that the two emerging technologies (ICEs and UV process) are excellent for point-source VOC controls. The RE of both ICEs and UV processes are higher than or equal to the RE of any conventional technologies, within their operating ranges (Fig. 16). The internal combustion engine technology is cost-effective only when the influent VOC concentration is extremely high (7000–100000+ ppmv), whereas the UV process is cost-effective only when the influent VOC concentration is in the moderate narrow range of 100–1000 ppmv (Fig. 17).

10.2.2. PM and Heavy Metal Removal

Table 7 compares a selected emerging technology (HEPA filters) with conventional technologies (fabric filtration, electrostatic precipitation, and scrubbing/absorption) in terms of their advantages and disadvantages when treating point sources of air pollution for PM control.

A further comparison between the same emerging technology (HEPA filters) and the same conventional technologies (fabric filtration, electrostatic precipitation, and scrubbing/absorption) in terms of their RE for point-source PM controls is presented in Table 8. The RE of HEPA filters for removal of PM (10 μm size) is greater than any other conventional technologies under evaluation. However, HEPA filters cannot remove acid from polluted gas streams (*see* Table 8).

The estimated RE for controlling toxic metals by various technologies is introduced in Table 9. HEPA filters are not recommended for toxic metals control because of lack of an established RE.

10.3. Evaluation of Fuel-Cell-Powered Vehicles for Air Emission Reduction

The US Department of Energy introduces how a fuel cell functions. In an electrochemical reaction with oxygen, hydrogen generates electricity and water inside a fuel cell. Electrical energy is generated by the fuel cell in three steps: (1) Hydrogen is fed

Table 6
Point-Source VOC Controls by Emerging and Conventional Technologies

Control	Applicable remediation technologies	Advantages	Disadvantages
Carbon adsorption	SVE, air stripping, thermal destruction, bioremediation, thermal desorption, solidification/stabilization, soil washing, etc.	Effective for gas streams with variable flow rates Effective for gas streams with variable VOC content Effective for gas streams with low VOC content	Bed fires may occur if oxygenated material is present and bed temperature rises because of heat of adsorption Spent carbon must be either regenerated or discarded Filters/mist eliminators may be needed for liquids or PM Not effective for low-molecular-weight compounds Less effective for high-humidity gas streams
Thermal oxidation (incineration)	All remediation technologies	Widely demonstrated technology Effective for a variety of VOCs No disposal concerns	Potential generation of PICs, acid gases Supplemental fuel required Only effective for combustibles
Catalytic oxidation (Incineration)	SVE, air stripping	Requires operating lower temperatures than thermal oxidizers No disposal concerns	Catalyst easily fouled or degraded High temperatures may cause burnout Low heat content of waste gas stream will require extra fuel Not effective for many chlorinated solvents
Condensers	SVE, air stripping, biotreatment	Effective for very high VOC concentrations Good for pretreatment of dilute streams prior to other controls	Performance somewhat sensitive to process conditions (flow rate, temperature) High utility costs
Internal combustion engines	SVE, air stripping	Compact units can provide usable power Well-developed technology	Limited capacity (<1000 scfm) Supplemental fuel may be expensive Easily fouled or corroded

Membranes	SVE, air stripping	Reduces waste stream volume Concentrates VOCs	Pretreatment only Limited data available
Operational controls	All remediation technologies	Improved removal efficiency for minimal cost	Requires knowledgeable operator
Soil filters/ biotreatment	SVE, air stripping	Low cost Simplicity May degrade semivolatile or nonvolatile organics May trap some metals Not susceptible to variations	Not effective on all VOCs May require large surface area, biologically sensitive to temperature and humidity High-pressure drop
Wet absorbers	Incineration, thermal destruction	Good for high-temperature gas streams Simple to operate Effective for wide variety of VOCs	Low RE for low VOC concentrations Effluent may pose disposal problems Susceptible to concentration, flowrate, and temperature changes, less efficient at low flow rates
UV	SVE, air stripping	High removal efficiencies No solvent or wastewater generated	Complex system Limited data available

Source: US EPA.

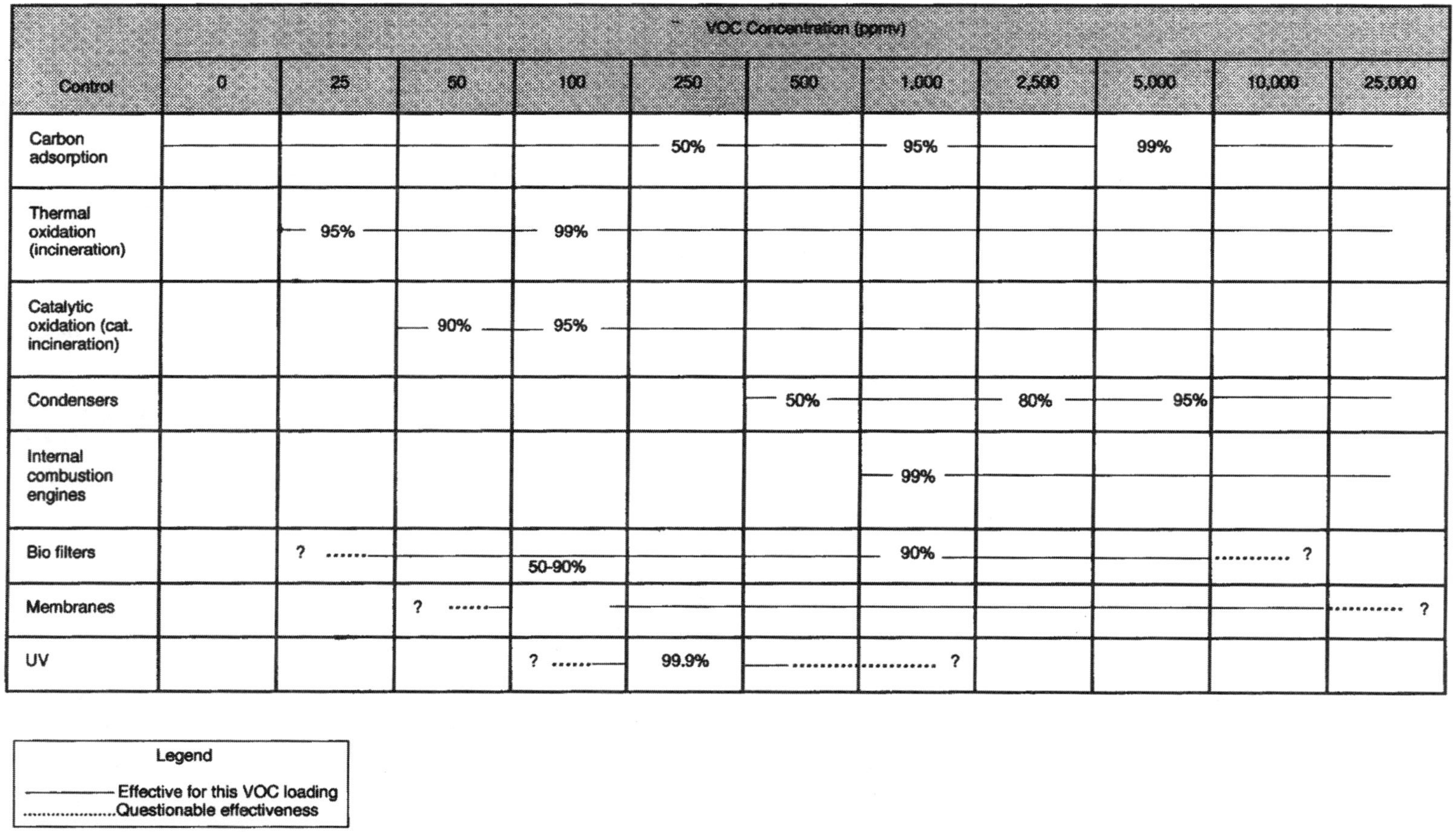
VOC Concentration (ppmv)
Control
0
25
50
100
250
500
1,000
2,500
5,000
10,000
25,000
Carbon adsorption
50%
95%
99%
Thermal oxidation (incineration)
95%
99%
Catalytic oxidation (cat. incineration)
90%
95%
Condensers
50%
80%
95%
Internal combustion engines
99%
Bio filters
?
50-90%
90%
?
Membranes
?
?
UV
?
99.9%
?
Legend
Effective for this VOC loading
Questionable effectiveness

Note: All values assume a well-designed, well-run system.

Fig. 16. Percent RE versus VOC loading of APCD for point source VOC controls. (From US EPA.)

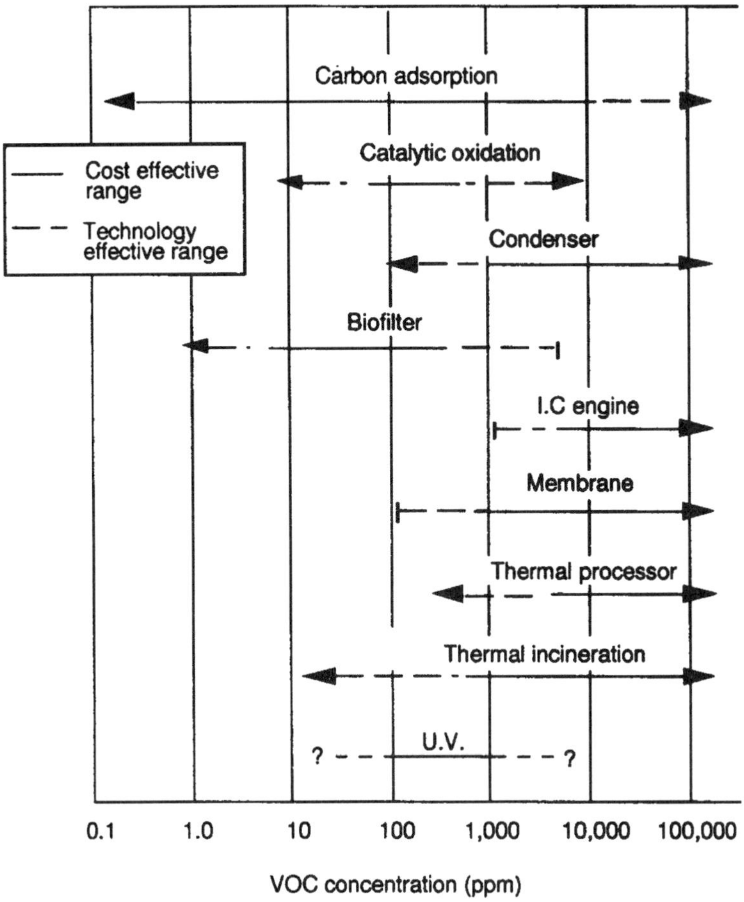

Fig. 17. Relative cost-effectiveness of APCD for point-source VOC controls. (From US EPA.)

into the anode, or electrically negative post of the fuel cell; (2) in the middle of the fuel cell, the "electrolyte" takes an electron from the hydrogen atom, using it to produce electricity; and (3) the cathode, the electrically positive post of the fuel cell, is where electrons recombine with the hydrogen and oxygen to yield water (9,10,44–46,48). With water as the end product, fuel-cell-powered automobiles will not pollute the air. The total amount of carbon dioxide gases (global warming gases) produced by the United States will be significantly reduced when these types of vehicle are commercially available at an affordable price.

The potential for development of fuel-cell-powered (or hydrogen-fueled) vehicles is very high. As seen next, three major automakers are developing their own hydrogen-fueled vehicles with promising results (45).

10.3.1. Toyota

Toyota Motor Corp. developed a fuel-cell hybrid sports utility vehicle that has an electric motor rated at 109 hp and 194 lb-ft of torque and is powered by

Table 7
Point-Source PM Controls by Emerging and Conventional Technologies

Control	Remediation	Advantages	Disadvantages
Baghouse (fabric filter)	Incineration Thermal Stabilization/ solidification Materials handling	Lower pressure drop than venturi for fine PM removal (i.e.,2–6 in. H_2O compared with > 40 in. H_2O for Venturi) Simple to operate Can collect electrically resistive PM	Cannot control a high-temperature (>550° F) stream without a precooler Cannot control highly humid stream Special fabric requirements Mechanical collectors required for large (>20 µm) particles
ESP	Incineration Thermal desorption	Low operating cost Low pressure drop (0.5 in. H_2O) Control very small (<0.1 µm) PM with high RE Can collect corrosive or sticky mists Wet ESPs can collect acid gases, PM	High initial capital nvestment Not readily adaptable to changing conditions, particle loading Space requirements greater than for baghouse or Venturi Conditioning agents may be needed to control resistive particles
Venturi scrubber	Incineration Thermal desorption Materials handling	Low initial investment Little space required Simple to operate, few moving parts Can control sticky, flammable, or corrosive matter with few problems Simultaneous collection of PM and gases RE is independent of resistivity	High operating cost because of pressure drop (>40 in. H_2O), especially for smaller PM Has wastewater and cleaning/disposal costs
Packed, spray, or tray tower absorbers	Incineration Thermal desorption Materials handling	Can be very effective, as part of control train Simple to operate, few moving parts Can control sticky, flammable, or corrosive matter with few problems Simultaneous collection of VOCs, larger PM, metals, and gases RE is independent of resistivity for Venturis, lower absorbers (wet ionizing)	Has wastewater and cleaning/disposal costs Not effective for streams < 1000 scfm

Dry scrubber: DSI, SDA	Incineration Thermal desorption Materials handling	Not wet slurry Effective for removal of PM, metals, and dioxins	Contaminated lime must be disposed
HEPA filter	Incineration Thermal desorption Materials handling	Extremely high PM removal rate Simple to operate, few moving parts Can control asbestos to very low levels	High-pressure drop Subject to fouling corrosion Used filters require proper disposal Requires prefilter for heavily loaded streams
Operational controls	All remediation technologies	Improved removal efficiency for minimal cost	Requires knowledgeable operator

Source: US EPA.

Table 8
Percent RE Range of APCD for Point-Source PM Controls

	Fly ash	Acid gases	<10 μm	>10 μm	Metals
Baghouses	—	—	99+[a]	99+[a]	90–95[b]
Wet scrubbers	—	95–99+	Low	—	40–50[b]
Venturi scrubbers	—	99	80–95	80–95	Variable
Dry scrubbers	—	95–99	99+	99+	95–99[b]
ESP	99+	—	99[c]	99[c]	85–99+[b]
Quench chambers	—	50%	—	—	—
HEPA filters	—	—	99.9+[d]	99.9+[d]	—

[a]Except for "sticky" particles.
[b]Lower removal efficiency for mercury.
[c]For resistive particles.
[d]With high-pressure drop.
Source: US EPA.

Table 9
Percent of RE /APCD for Toxic Metals Controls

	Pollutant				
Air pollution control device	Ba, Be	Ag	Cr	As, Sb, Cd, Pb, Ti	Hg*
WS[a]	50	50	50	40	30
VS-20[a]	90	90	90	20	20
VS-60[a]	98	98	98	40	40
ESP-1	95	95	95	80	0
ESP-2	97	97	97	85	0
ESP-4	99	99	99	90	0
WESP[a]	97	97	98	95	60
FF[a]: FF/WS[a]	95	95	95	90	50
PS[a]	95	95	95	95	80
SD/FF; SD/C/FF	99	99	99	95	90
DS/FF	98	98	98	98	60
ESP-1/WS; ESP-1/PS	96	96	96	90	80
ESP-4/WS; ESP-4/PS	99	99	99	95	85
VS-20/WS[a]	97	97	97	96	80
WS/IWS[a]	95	95	95	95	85
WESP/VS-20/WS	99	99	98	97	9
C/DS/ESP/FF; C/DS/C/ESP/FF	99	99	99	99	98
SD/C/ESP-1	99	99	98	95	85

**Note*: Flue gasses are assumed to have been precooled (usually in a quench).If gases are not cooled adequately, mercury recoveries will diminish, as will cadmium and arsenic recoveries to a lesser extent.

[a]APCD codes:
C = Cyclone
WS = Wet scrubber, including sleve tray tower, packed tower, bubbles cap tower
PS = Proprietary wet scrubber design (high-efficiency PM and gas collection)
VS-20 = Venturi scrubber,approx 2630 in W.G. p
VS-60 = Venturi scrubber, approx >60 in W.G. p
ESP-1 = Electrostatic precipitator: 1 stage
ESP-2 = Electrostatic precipitator: 2 stages
ESP-4 = Electrostatic precipitator: 4 stages
WESP = Wet electrostatic precipitator
IWS = Ionizing wet scrubber
DS = Dry scrubber
FF = Fabric filter (baghouse)
SD = Spray dryer (wet/dry scrubber)

nickel–metal hydride batteries and a hydrogen fuel cell, which also recharges the batteries. The FCHV-3, which is a modified Highlander SUV, accepts pure hydrogen as its fuel via special stations and has a top speed of 96 mph and a maximum range of 180 miles. The University of California at Irvine and the University of California at Davis each have one FCHV-3. Four more are expected to be delivered to the two schools in 2003 (45).

10.3.2. Honda

Honda Motor Co. has also developed a fuel-cell vehicle with an electric motor rated 80 hp and 201 lb-ft of torque powered by a fuel cell. The EV-Plus uses a supercapacitor instead of a larger, heavier battery to store some electricity for use during bursts of acceleration. Fueled by pure hydrogen via special stations, the vehicle has a top speed of 93 mph, and a maximum range of 170 miles. The City of Los Angeles owns one EV-Plus, and expects delivery of five more in 2003 (45).

10.3.3. Daimler-Chrysler

Daimler-Chrysler's fuel-cell car is the NECAR 5, which is based on the Mercedes-Benz A-Class. In 2002, the car made a much heralded 3262-mile cross-country trip from California to New York in 12 d. The NECAR 5 extracts hydrogen from methanol, a method which the company says takes up less space than pure hydrogen. The American automaker's vehicle has a 49-hp engine, a top speed of 100 mph, and a maximum range of 90 miles (45,48). The NECAR 5 will be distributed in the United States in late 2003.

NOMENCLATURE

a	Particle radius
A	Area
b	Particle mobility
B	Drop mobility
bhp	Brake horsepower
C	Cunningham slip correction factor
C_a	Cunningham correction factor for aerodynamic diameter
C_D	Coefficient of drag (dimensionless)
C_i	Mass concentration in
C_o	Mass concentration out
C_S	Vapor concentration on the drop surface
C_∞	Vapor concentration in mass of gas
CO	Carbon monoxide
D_m	Removable particle size, diameter
d	diameter
d_p	Particle diameter
d_{pa}	Aerodynamic particle diameter
D_{PM}	Diffusivity of particle in the medium
D_{GM}	Diffusivity of gas
E_c	Impact charging in a field with intensity
E_o	Overall collection efficiency

F	Percentage of fine particles < 3 μm
F_B	Buoyant force
F_D	Drag force
F_G	Gravitational force
F'	Liquid flow rate
f	Fanning friction factor
g	Gram, or gravitation acceleration
G	Gas flow rate
g_c	Gravitation acceleration constant
h	Height
H	Enthalpy
HC	Hydrocarbon
HCHO	Formaldehyde
K	Boltzmann's constant
K'	Overall mass transfer coefficient
k	Constant
L	Length, or percentage of layer particles over 3μm
LDV	Light-duty vehicle
LDT1	Light-duty truck 1
LDT2	Light-duty truck 2
l	length
L'	Liquid-to-gas ratio
M	Slope of equilibrium line
MgO	Magnesium oxide
N	Number; number of strokes per minutes, or number density
N_0	Drop concentration, or initial drop number density
n	Particle number density
N_{OG}	Number of overall gas transfer units
NMHC	Nonmethane hydrocarbon
NMOG	Nonmethane organic gases
NO	Nitrogen oxide
NO_x	Nitrogen oxides
p	Pressure
pH	Log hydrogen ion concentration
q	Charge on particle
Q	Heat energy, or charge on droplet
Q_{boiler}	Boiler heat input
R	Radial distance or radius of wet spherical droplet
R_B	Radial distance to liquid of boiler
R_C	Radial distance to liquid of condenser
Re	Drop Reynolds number
SO_2	Sulfur dioxide
T	Absolute temperature
THC	Total hydrocarbon
U	Mean gas velocity

V	Velocity of particle in scrubber or relative to the target
V_g	Velocity of gas
V_0	Initial horizontal velocity
V_p	Velocity of particle
V_s	Terminal settling velocity
V_t	Throat gas velocity
W	Work
W_{net}	Net reversible cycle work
W_p	Work by a pump
X	Concentration of solute in liquid
X_s	Stopping distance
x	Mole fraction in liquid
Y_1	Gas-phase concentration of solute in
Y_2	Gas phase concentration of solute out
Δ	Change
ε_0	Dielectric constant of a free space
μ	micron, 10^{-6}m
μ_c	Gas viscosity divided by Cunningham correction factor
μ_g	Gas viscosity
η	Fractional collection efficiency
η_p	Efficiency of impactor
η_R	Net reversible cycle work
η_t	Turbine stage efficiency
ρ_a	Apparent density of suspended matter
ρ_g	Gas density
ρ_p	Particle density
τ	Surface tension
τ_a	Charged particle self-removal time
τ_c	Cleaning time
τ_R	Charged particle lifetime
τ_{res}	Residence time
τ_{sc}	Scrubbing cleaning time
τ_{SR}	Particle scrubbing time
ϕ	Deposition rate
ψ	Inertia impaction parameter
ω	Angular velocity

REFERENCES

1. H. C. McKee, *Chem. Eng. Prog.* **97(10)**, 42–45 (2001).
2. A. Neville, *Environ. Protect.*, **13(10)**, 16–20 (2002).
3. T. K. Garret, *Environ. Sci. Technol.* **9(9)** (1975).
4. J. E. McEvoy, *Catalysis for the Control of Automotive Pollutants. Advances in Chemistry. Series.* No. 143. Chemical Society, Washington, DC, 1975.
5. G. S. Springer, and D. J Patterson, *Engine Emissions, Pollutant Formation and Measurement.* Plenum, New York, 1973.

6. Society Automotive Engineer, (1964) Part I, (1967) Part II, (1975) Part III, *Soc Auto Eng, Inc.*, NY.
7. Anon., *Chem. Eng. Prog.* **98(2)**, 21–22 (2002).
8. J. Hensel, *Environ. Protect.* **14(1)**, 12–13 (2003).
9. J. Winnick, *Chem. Eng. Prog.* Vol. **99(2)**, 9 (2003).
10. A. S. Brown, *Chem. Eng. Prog.* **99(2)**, 12–14 (2003).
11. S. Calvert, *Scrubber Handbook*, APT, Inc., 1972; US EPA Contract No. CPA 70–95. US Environmental Protection Agency, Washington, DC, 1972.
12. H. E. Hesketh, *P AICHE Symposium*, 1976.
13. H. E. Hesketh, *J. Air Pollut. Control Assoc.* 24(10), 939–942 (1974).
14. N. A. Fuchs, *The Mechanics of Aerosols*, Perrgamon, Oxford, 1964, Chap 2.
15. McIlvaine Co., *The Scrubber Manual*, McIlvaine Co., New York, 1976.
16. T. H. Chilton, and A. P. Colburn *Ind. Eng. Chem.* **26**, 1183 (1934).
17. R. J. Gleason, *Pilot Scale Investigation of Venturi-Type Contactor for Removal of SO_2 by the Limestone Wet-Scrubbing Process*. US EPA Contract No. EHSD-71–24. US Environmental Protection Agency, Washington, DC, 1971.
18. S. Calvert, *Chem. Eng.*, **84(17)**, 54 (1977).
19. G. R. Koehler, and E. J. Dober, *US EPA Flue Gas Desulfurization Symposium*, 1974.
20. H. E. Hesketh, *Pilot Plant SO_2 and Particulate Removal Study*. SIU-C Illinois Inst., Environ Inst., Quality Project No. 10.027, 1975.
21. S. Calvert, *Entrainment Separators for Scrubbers—Initial Report*. US EPA-650/2-74–119a US Environmental Protection Agency, Washington, DC, 1974.
22. Jim Sadler to Charles Albert Radian Corporation, *Personal communication*, 1991.
23. T. Pedersen and J. Curtis. *Soil Vapor Extraction Technology: Reference Handbook*. EPA/540/2–91/003 (NTIS PB91-168476). US Environmental Protection Agency, Washington, DC, 1991.
24. N. Wynn, *Chem. Eng. Progr.*, **97(10)**, 66–73, (2001).
25. L. K. Wang, J. V. Krouzek. and U. Kounitson. *Case Studies of Cleaner Production and Site Remediation*. Training Manual No. DTT-5-4-95. United Nations Industrial Development Organization, Vienna, 1995.
26. K. E. Hummel, and T. P. Nelson, *Test and Evaluation of a Polymer Membrane Preconcentrator*. EPA 600/2-90-016. US Environmental Protection Agency, Washington, DC, 1990.
27. McCoy and Associates, *The Air Pollution Consultant VOC Emission Control During Site Remediation*, McCoy and Associates, Lakewood, CO, 1991.
28. R. W. Baker, N. Yoshioka, J. Mohr et al., *J. Membr. Sci.* 31, 259–271 (1987).
29. H. Strathman, C. Bell, and K. Kimmerie, *Development of a Synthetic Membrane for Gas and Vapor Separation. Pure and Applied Chemistry* No. 50. Blackwell Scientific Publications, Oxford, 1986, p. 12.
30. K. V. Peinemann, J. Mohr. and R. W. Baker, *AICHE Symp. Ser.82,* American Institute of Chemical Engineers, New York, p. 250.
31. S. Weller, and W. A. Steiner, *Chem. Eng. Prog.*, **46**, 585–591 (1950).
32. C.Y. Pan and H. W. Habgood, *Ind. Eng. Chem.* **13**, 323–331 (1974).
33. Anon, *Chem. Eng. Prog.*, 98 (1), **29** (2002).
34. Anon, *Chem. Eng. Prog.*, 99(2), **74** (2003).
35. E. C. Moretti, and N. Mukhopadhyay, *Chem. Eng. Prog.* **89(7)**, 20–25 (1993).
36. K. A. Roy, *UV-oxidation Ttechnologies*, Tower-Borner, Glen Elly, IL, 1990.
37. A. Wekhof, *Environ. Prog.*, **10,** 4 (1991).
38. Purus, Inc., Product Brochure for UV Lamps, *On-site Organic Contaminant Destruction with Advanced Ultraviolet Flashlamps*, Purus, Inc., San Jose, CA, 1991.

39. US EPA, *Control of Air Emission from Superfund Sites.* EPA/625/R-92/012. US Environmental Protection Agency, Washington, DC, 1992.
40. H. E. Hesketh, *Air Pollution Control: Traditional and Hazardous Pollutants.* Technomic, Lancaster, PA, 1991.
41. B. Srikanth, *Water Qual. Prod.*, 24–25 (2001).
42. M. S. Timmons, *Water Conditi. Purifi.* 80–83 (2001).
43. US EPA, *Control Techniques for Fugitive VOC Emissions from Chemical Process Facilities.* EPA/625/R-93/005. US Environmental Protection Agency, Cincinnati, OH, 1994.
44. Flanders Equipment, *Flanders, Product Brochure.* Flanders Equipment, Washington, DC, 1984.
45. J. Mazurek, *USA Today*, pB1-B2, 30 January 2003.
46. Anon, *Chem. Eng. Prog.* **98(2)**, 21–22 (2002).
47. L. K. Wang, N. C. Pereira, and H. E. Hesketh (eds.), *Handbook of Environmental Engineering, Volume 1, Air and Noise Pollution Control.* Humana, Totowa, NJ, 1979, pp. 355–392.
48. US EPA. EPA/Daimler Chrysler/UPS Fuel Cell Delivery Vehicle Initiative. www.epa.gov. US Environmental Protection Agency, Washington, DC. (2004).
49. US EPA. Air Quality Planning and Standards. www.epa.gov. US Environmental Protection Agency, Washington, DC. (2004).
50. L. K. Wang, N. C. Pereira and Yung-Tse Hung (eds.). *Advanced Air and Noise Pollution Control.* Humana, Totowa, NJ. 2005.

Index

D